T0417599

NanoNutraceuticals

NanoNutraceuticals

Editor-in-Chief
Bhupinder Singh

Editors
Minna Hakkarainen
Kamalinder K. Singh

CRC Press is an imprint of the
Taylor & Francis Group, an **informa** business

CRC Press
Taylor & Francis Group
6000 Broken Sound Parkway NW, Suite 300
Boca Raton, FL 33487-2742

Printed on acid-free paper

International Standard Book Number-13: 978-0-8153-9992-6 (Hardback)

Visit the Taylor & Francis Web site at
http://www.taylorandfrancis.com

and the CRC Press Web site at
http://www.crcpress.com

Contents

Foreword

The global market of nutraceuticals has been growing quite steeply with over 7% rate of annual growth through the coming decade. The "Nutraceuticals Sector" was valued at $182.6 billion in 2015 and is projected to cross $300.0 billion by 2025, the figures almost competing with those of the "Pharmaceuticals Sector." This is happening due to continuous support to nutraceuticals received from conscious and literate consumers. Free access to the millions of websites about nutraceuticals provides enough information to make educated choice in using nutraceuticals.

Despite the tremendous success in the nutraceuticals market, the products face myriad challenges. The first and foremost is regulatory challenge, as today there are not adequate regulations that control the production and consumption, as well as quality and distribution, of such products worldwide. Another challenge is uniform nomenclature, as every country uses different names for these products, such as *nutraceuticals*, *dietary supplements*, *functional foods*, *health products*, *herbal supplements*, *nutritional supplements*, *natural products*, and so on, which sometimes confuses the consumers, regulators, producers, and also the physicians. Recognition is the third challenge, as despite such phenomenal popularity of these products, the Western medicine pundits are reluctant to recognize their usefulness. The academic community in the healthcare domain still has to accept and appreciate these products as mainstream healthcare providers, even though the success is more driven by the conscious consumers.

Some other challenges encompass their high dose levels, inadequate consistency and reproducibility, characterization and solubility problems, formulation challenges, stability issues at least over two years, and manufacturing issues on large scale for high-quality products. Further, much needs to be accomplished using nanonutraceuticals of biotechnologically-developed, genetically-modified, and tissue cultured products, and to find ways and manners to prove the efficiency of these products vis-à-vis those from naturally grown sources.

Application of nanotechnology will address several of such challenges currently being faced by nutraceuticals. It has immense promise to reduce their dose levels and to result in better and longer stability to the nutraceuticals. The formulations of the bioactive as nanostructured products will help in their superior characterization, improved patient acceptability, and, above all, high reproducibility of their therapeutic effectiveness.

On pharmacokinetic and pharmacodynamics fronts, nanotechnology can offer improved component solubility, enhancement of bioavailability, increasing permeation in biological milieu, and in turn better ADME parameters.

This volume on nanonutraceuticals will be a great resource for academicians, industrial researchers, students, and people on the road, who are driving this "nutraceuticals revolution." And, in this context, I am quite sure that it will also help to establish a holistic approach to healthcare, thus helping to resolve the challenges of chronic and lifestyle diseases that are attacking humanity like an epidemic.

Yashwant Pathak
Professor, College of Pharmacy
Dean, CoP Dean's Office
University of South Florida, USA

About the Series

Implying miniaturization, the word "nano" has become a household name today. Nanotechnology encompasses the study and application of functional nanomaterials in diverse fields even at their molecular levels. Virtually, nothing remains to be nanonized at the moment, whatever the domain may be. Traversing a voyage from the origin of its idea by Richard Feynman in 1959, nanotechnology has now entered into our day-to-day living, society and culture, shaping our future in a categorical manner. On one hand, nanotechnology is being explored to develop newer nanostructures, and on the other hand, the horizons of their promising applications are being widened manifold. Not merely as an "evolution", this has lately been turning out to as a "revolution" across different scientific disciplines and industrial sectors across the globe.

Notwithstanding the promising benefits of the cutting-edge technology, its applications in healthcare are considered as paramount. This perspective assumes immense significance, as the whole world is experiencing unprecedented biomedical challenges and changes, particularly in context with "3-***D***" succession of ***D***iseases, ***D***iagnosis and ***D***rugs. In line with futuristic predictions made by Robert Freitas in 1999 in his book on 'Nanomedicine', today it is ostensibly the fastest emerging offshoot of nanotechnology. Further integration of Nanomedicine with biological sciences, such as biotechnology, biochemistry, pharmaceutical sciences, biomedical engineering, bioimaging, nanomedical robotics and biomedical technology, has spurred the genesis of a whole contemporary field of "NanoBioMedicine". Employing various nanostructures and nanodevices, the science of nanobiomedicine primarily endeavors to improve upon the patient safety, efficacy, economy and compliance of varied therapeutic, preventive and diagnostic strategies. Not merely dealing with the nuances of interactions of nanostructures with biomolecular receptors, the modern discipline of nanobiomedicine also embraces the methods of design, engineering and technological transfer of nanoformulations of drugs and biomacromolecules, and fabrication of advanced biomedical appliances. Drawing apt inputs from almost all the frontiers of science and technology, nanobiomedicine today is heralding a new era in medicare with lots of hopes and promises for the suffering society.

Nanobiomedicine, of late, has been making rapid strides, eventually providing significant impetus towards scientific enthusiasm, research efforts, industrial anticipation and regulatory attention. Of late, several nanomedicinal products have been introduced in the commercial circulation, with a few thousands more in the pipeline, and yet many more in the early stages of developmental research. These have been documented to exhibit stellar benefits not only of enhanced surface-area-per-unit volume, but also of improved aqueous solubility, target-specificity, controlled release potential, biocompatibility, stealth characteristics, precise control of particle size and ability for drug combination therapy. Such nanoscaled products principally encompass different types of nanoparticles, nanocrystals, nanoemulsions, nano-conjugates, liposomes, dendrimers, nanogels, nanocosmeceuticals, nanocapsules, nanophytomedicines, lipid nanospheres and nanocomposites. A huge proportion of these innovative nanotech-enabled products have demonstrated enormous patient benefits and plausible solutions to the erstwhile insurmountable medical problems.

Alarmingly high incidences of inadequate efficacy, safety and patient compliance reported with the current crop of drugs, have been calling for developing such newer and therapeutically superior nanomedicinal formulations to address the present patient needs and market demands. Driven by the technological updates, increased gubernatorial support, recent notifications of newer regulatory guidance for drug products containing nanomaterials and enhanced preference for targeted therapeutics, the world has now been witnessing a rapid surge in such nanomedicines. Nanopharmaceuticals, therefore, are an outsized industry today, with its market crossing a few hundreds of billions of US dollars, over 500 companies and over 80 products worldwide. In the forthcoming decades too, these nanomedicines are anticipated to massively impact almost all the terrains of patient care, personalized medicine, clinical and veterinary practice and above all, the pharmaceutical market.

The ardent need in the current hour of shifting paradigms, accordingly, is to acquaint the professionals on advances in this vibrant science of nanobiomedicine, growing at a phenomenally striking pace.

A recent literature survey of the leading journals revealed a phenomenally high rate of publications in this multi- and interdisciplinary domain of nanobiomedicine. For a realm advancing as fast as this one, it is quite challenging to obtain a complete overview of the exciting developments, as most updated information on such nano-bio interfaces lies scattered in diverse journals. Owing to nonavailability of the standard treatise(s) on the subject, therefore, we have undertaken a grand task of bringing about a book series covering a gamut of subject topics spanning nanobiomedicine.

Accordingly, the current book series, "**Emerging Trends in NanoBioMedicine**", an anthology of four volumes, aspires to dexterously present the pertinent facts and figures on an array of novel subject topics. Each individual volume of the compendium is an assortment of choicest reviews and research articles, contributed by illustrious scientists and experts selectively drawn from across the world. The primary motivation behind this Herculean effort has been to bring forth high quality insight into latest research in the field as a multi-volume digest. It comprises of four volumes, *viz.*

Volume 1: NanoBioMaterials
Volume 2: NanoBioEngineering
Volume 3: NanoNutraceuticals
Volume 4: NanoAgroceuticals & NanoPhytochemicals

NanoBioMaterials, the first volume of the series, maps out an exhaustive overview of the contemporary research accomplishments of variegated nanomaterials exhibiting profound impact in improving treatment efficiency of various diseased states. This volume tacitly introduces and demonstrates close association of biology with material science, with intent for effective and safe use of nanomaterials for different biological applications. Being a promising, though burgeoning, realm of science, the treatise accentuates diverse applications of nanomaterials in the fields of drug delivery, bioimaging, tissue engineering and many more. It primarily focuses on innovations based on various nanobiomaterials, such as polymeric, metallic, lipidic, peptidal, carbon-based, and prodrug-conjugate-based, leading finally to produce next-gen biomedicinal products.

Highlighting challenges for fundamental research and opportunities for ground-breaking advances in nano-enabled bioengineering, the second volume of this series successfully unfolds the topic of **NanoBioEngineering**. Being a rational fusion of biology, engineering and nanotechnology, it presents breakthrough techniques and tools amid the existing armamentarium to utilize novel phenomena like microfluidics and nanotooling and nanomaterials like aptamers, nanocargos, nanofibers, nanocrystals, hyper-branched dendrimers, surface-engineered and superparamagnetic nanopaticles, for meeting the desired healthcare needs. The integrated volume, herein, provides a lucid account on far-reaching topics from diverse disciplines of life sciences and technology, including physics, chemistry, biology, engineering, biomedical technology and drug development.

The enormous potential of nutraceuticals in fostering human healthcare, the burning issues confronting their poor bioavailability and usage of nanotechnology principles for resolving these issues, have been adroitly deliberated upon in the third volume of the book series, **NanoNutraceuticals**. Provision of the requisite knowledge on a gamut of modern hi-tech approaches to improve their activity, especially using nanotechnology through a single textual source enshrined, this book volume can be instrumental in ameliorating the subdued popularity of nutraceuticals and lack of serious consideration by health regulatory agencies. Thus, the contents of this book volume have the capability to metamorphose the biological performance of these functional foods through their skilful delivery to the specific target using apt nanocarriers for safer and more efficacious management of the diseased states like cancer and neurodegenerative diseases, and for benefitting the human healthcare *per se*.

Volume four, **NanoAgroceuticals & NanoPhytochemicals**, is a testament to the sophisticated and unconventional extrapolation of nanotech applications in the field of agroceuticals and phytochemicals. Albeit still in its infancy, the subject matter encompassed in the book volume holds immense potential to bring forth tangible benefits to the farmers, food industry, patients and consumers. This book presents a cogent account on various nano-enabled delivery technologies of phytochemicals in agriculture to improve the storage shelf-life of crops, reduce the amount of sprayed chemicals through smart delivery of phytochemicals, sustain delivery of fertilizers, decrease the loss of nutrient(s) in fertilizers and augment percent harvesting yield through optimized water and nutrient management. The role of nanotechnology in the development of novel and innovative products out of agriculture products and herbal supplements, and their subsequent production, processing, storage and packaging, has also been explicitly highlighted in this book volume.

All these volumes put together highlight some of the most remarkable advances that have taken place recently in the vast field of nanobiomedicine, and reflect new challenges and issues that confront their further progress. The series seeks to offer the readers an apt tête-à-tête with fresh and original perspectives, epitomizing a vast multiplicity of experience, wisdom and vision, which is quite implausible to be made available collectively anywhere else.

Verily, in this context, I feel quite satiated and fascinated that the long-felt void of a repository on the much sought-after discipline of nanobiomedicine is being filled up by spearheading this current multi-volume book series. Although the field of nanobiomedicine has become too vast today, meticulous endeavor has been made to customize the contents of the book series and steer the novitiates as well as seniors on the recent and upcoming trends. Not only does this treatise embrace holistic applications of a plethora of promising nanoconstructs, especially in the fields of medical therapeutics and diagnostics, biotechnology, pharmacogenomics, human gene biology, clinical pharmacology, and so on, but it also touches upon their regulatory, safety and industrial scale-up issues.

At this juncture, I would like to take this opportunity to express my heart-felt gratitude to several people whose pooled efforts brought forth this dream of mega-series into real being. The list includes all the authors, editors, publishing house team, my research group, and research funding agencies, besides others.

First and foremost, I put across my heart-felt gratitude to the eminent and erudite authors for dexterously working on the chapters by bringing together the most comprehensive and updated information on wide-ranging subject(s). Not merely did these contributors meet with the stringent timelines, they also abided by a battery of editorial suggestions sportingly. They garnished and flavoured these chapters with apt perspectives, emanating from their own experiential wisdom. Earnest efforts have been made by them to encompass a diversity of graphic illustrations, presented immaculately as expository graphs, photographs, methodology flow charts, hierarchical tabulations, bar charts, and so on. Each chapter has been duly referenced to current literature, citing vital texts for further reading too.

As the designated Editor-in-Chief of the book series, I fall short of apposite words to express my deep appreciation to the distinguished editors of the four volumes for pouring in their invaluable and pertinent inputs from time to time. The sagacious, prompt and critical feedback provided by them as well as by the worthy reviewers of the individual chapters while spending their priceless time helped in ameliorating the quality standards of the manuscripts a great deal.

I put into record the highly painstaking, persistent and meticulous efforts of the management and staff members of the publisher, *viz.*, M/s CRC Press (Taylor & Francis Group), USA, in bringing out this mega compilation. Without their immense support and direction, it could not have been possible to bring forth this colossal series in a timely manner.

Our well-knit team of exceptionally diligent, efficient and perseverant researchers and post-docs deserve the special mention, as they were verily the spine of this massive and momentous project. Naming a few, while leaving many others, may demean their stellar contribution, for which I would always remain beholden.

I also seize this opportunity in acknowledging the University Grants Commission (UGC), New Delhi, India, for awarding this prestigious National UGC Centre of Excellence in "Applications of Nanomaterials, Nanoparticles & Nanocomposites" with focus area as "Biomedical Sciences" to the

Panjab University, Chandigarh, India. In fact, the drive to take up this marathon task for writing the explicit treatise was initiated after my assuming the charge of this Centre of Excellence, dedicated to explore and establish various nanobiomedicinal precepts. Thanks are also due to the tangible financial support provided by several other governmental agencies and pharma industrial houses to enable us to bring out tangible outcomes on diverse innovative nanostructured systems and in characterizing them. Several of such systematic works on nanopatterned structures, primarily employing Quality by Design (QbD)-based paradigms, have fetched exalted global recognitions by American Association of Pharmaceutical Sciences (USA), Select Biosciences (UK), StateEase Inc. (USA), Evonik AG (Germany), Minitab Ltd. (UK), CPhI-Asia, EBA-Oxford (UK), APTI (India) and many more.

With this, I invite and incite you to start this amusing expedition to the salubrious and picturesque world of nanobiomedicine with a missionary message, "*Think Tiny and Act Big.*" I earnestly wish, hope and pray that this four-volume book series on "**Emerging Trends in NanoBioMedicine**" will serve as a classic repertoire of essential knowledge and know-how in the domain to address the unmet needs of the scientists and practitioners from current and future generations.

Bhupinder Singh

Preface

Humans have always used natural resources for managing and treating ailments. The indigenous systems of medicines, employing herbal and mineral drugs, were in vogue well before the appearance of the modern system of medicine. Integrating the words, *nutrition* and *pharmaceuticals*, the term *nutraceuticals* describes the merchandise demonstrating promising therapeutic benefits and reduced side effects vis-à-vis those associated with corresponding pharmaceuticals. The list of such reported nutraceuticals, present and extractable from food substances, continues to grow. By virtue of their multiple beneficial effects in human health, these nutraceuticals have drawn the attention of medical, nutritional, and pharmaceutical practitioners not only for prevention and treatment of myriad diseases but also for improving the immunity of the subjects against an array of ailments. The scope of nutraceuticals is as broad as that of pharmaceuticals, yet it lacks serious consideration by critical regulatory agencies of the health industry.

Immense potential of nanotechnology has been successfully used by researchers for efficient targeted and/or temporal delivery of these nutraceuticals, enhancement of their systematic bioavailability, and amelioration of their biological activity. Economical encapsulation of nutraceuticals and sustainable development of their nanoformulations encompass some of the difficult concerns, which can be surmounted by applying nanotechnology in food and agriculture sectors. Nanonutraceuticals offer superior personalized health care for human well-being through efficient prevention and treatment of diverse diseased states. Nanotechnology introduces considerable opportunities in nutritional supplementation and is poised to invigorate the growth of the nutraceutical industry, already estimated at $285.0 billion in 2021.

The nutraceutical movement, nevertheless, has not been as vigorous as it deserves to be. One of the plausible reasons for its modest popularity is inaccessibility of vital information on a gamut of contemporary nanoconstructs through a single textual source. Hardly any coherent compendium is available on wide-ranging vistas and applications of nanonutraceuticals. The current book volume, accordingly, has been designed to comprise the rational principles of nanotechnology for skillful delivery of nutraceuticals with a definitive aim to transform their biological performance. It presents cutting-edge recent advances in this field, together with their vital futuristic outlooks. The nuances of several important nutraceuticals, augmented using nanotherapeutics strategies, are cogently presented including their role in reinforcing healthcare, current problems encountered in using these compounds, ostensible remedial measures, and the state-of-the-art technological know-how for developing their safe and effective nanostructured technologies.

Chapter 1 lucidly presents the concept of nutraceuticals and nonspecific biological therapies and the present challenges in the field. Use of nanotechnology has been presented as an invaluable tool to prepare formulations and to deliver and improve the bioavailability and uptake of these health-benefitting substances. Continuing on the same theme, Chapter 2 emphasizes how smart nanostructured materials and formulations can be designed and utilized to augment the bioavailability and activity of nutraceuticals, especially in light of two model phytochemicals, curcumin and sesamol. Chapter 3 further discusses various nanotailored delivery systems, including polymeric nanoparticles, solid lipid nanoparticles, polymeric micelles, nanoemulsions, multilayer emulsions, and inorganic nanoparticles. Here, the special focus has been on the application of nanotechnology to achieve effective delivery of chemo-preventive nutraceuticals for cancer therapy. Chapter 4 delineates the stellar benefits of combining nanotechnology and nutraceuticals for cancer therapeutics and intervention. It presents broad perspectives for utilizing different nanocarriers of anticancer natural bioactives, aiming at effective cancer management.

Chapter 5 highlights various strategies for counteracting oxidative stress by employing naturally occurring phytochemicals, rich in exogenous antioxidants. In addition, the chapter describes various novel nanoconstructs formulated and optimized for delivery of promising antioxidant bioactives, for instance, to increase the bioavailability and to attain targeted delivery. Chapter 6 describes the use of nutraceuticals, that is, "healing foods," along with nanotherapy, for holistic treatment of central nervous system disorders. Novel herbal formulations of different bioactive compounds and plant extracts are discussed, along with the developments and advances in their delivery using nanotechnology-enabled tools and techniques. Chapter 7 accentuates specifically the treatment of Alzheimer's disease using myriad nanonutraceuticals. The multifactorial components from neuro-inflammation, coupled with oxidative stress and enhanced acetylcholinesterase activity, has been touched upon with strategic integration of novel drug delivery, nutraceuticals, and nanotherapeutics, for attaining the final goal of not only slowing down the disease, but completely curing it.

Chapter 8 focuses on providing the knowledge as to how nanoemulsions can be utilized in food and nutraceutical industries to encapsulate, protect, and deliver nutrients, especially the lipophilic ones such as omega-3 fatty acids, vitamins, or probiotics. The current state of nanoemulsion-based formulations, ranging from preparation to biological fate, has been discussed. There is also a valuable discussion delineating their enormous potential in the futuristic research trends, such as developing "on-demand functional foods." Chapter 9 presents nanodelivery of probiotics to overcome their common pitfalls of low bioavailability, poor permeability, and inadequate delivery across the gastrointestinal tract. Development issues, safety, and regulatory aspects are discussed as well as the next-gen potential of nanoprobiotics outlined. Chapter 10 describes the therapeutic potential of resveratrol, a polyphenol stilbenoid present in grapes and red wines for cardio-protection and neuroprotection, as well as for its antioxidant, anti-inflammatory, anti-carcinogenic, and antiaging properties. Nano- and micro-formulations of resveratrol for its successful delivery have been elaborated and discussed.

Chapter 11 turns to vitamin D3, a well-known and vital nutraceutical. Low levels of this vitamin in the body is one of the most common deficiencies, as vitamin D3 has limited availability in our daily food, causing health issues, especially during the winters when the availability of sunlight, the important source of vitamin D3, is relatively limited. Nanodelivery of vitamin D3 could remove many of the hiccups in today's practices and provide better targeting, reduce dose, increase stability of the vitamin, decrease liver metabolism, and provide a tool to battle against diverse diseased states more effectively, including cancer, cardiovascular diseases, bone disorders, and malnutrition. Chapter 12 discusses curcumin, a versatile natural bioactive compound with multiple curative properties, including anticancer, antioxidant, and anti-inflammatory functions. Its plausible federal approval as a therapeutic agent is, however, prohibited by its low bioavailability. Nanotechnology is presented as a promising tool to circumvent the existing problems with delivery and utilization of curcumin as well as with other natural bioactives marked with low water solubility. Chapter 13 describes the multiple roles of iron in our body. The problems concerning its low bioavailability, and its delivery and uptake by our body as well as difficulties in finding effective ways to treat common problems such as iron-deficiency anemia, have been elaborated adroitly.

Chapter 14 unravels the remarkable potential of pomegranate as a source of prominent nutraceutical. Besides touching on numerous health benefits of its bioactive constituents, the formulation development and characterization methods of various nanoformulations of fruit components, especially those obtained from its seeds and peel, have also been discussed. This chapter also shares an updated overview of various publications and patents filed for its various nanostructured products.

Bhupinder Singh
Minna Hakkarainen
Kamalinder K. Singh

About the Editors

Professor Bhupinder Singh Bhoop, MPharm, PhD, DSt, FPAS, is globally acclaimed for his extensive research work on diverse nanostructured drug delivery systems, developed primarily using Quality by Design (QbD) paradigms. His work encompasses variegated nanoengineered systems, naive as well as functionalized, like SLNs, NLCs, SNELs, MWCNTs, polymeric nanoparticles, lipid-polymer hybrid nanoparticles, mixed nanomicelles, nanocapsules, nanosponges, nanobilosomes, liposomes and other vesicular systems, for a gamut of diseased states, particularly cancers, and cardiovascular, immunodeficiency and neurodegenerative disorders.

A prolific writer, he has over 280 original journal publications, 16 books, 1 directory, 52 book chapters, 5 patents and 3 technology transfers to his credit. He is on Editorial and Reviewers Board of scores of journals. A widely traveled scientist, he has delivered more than 275 invited talks across the globe, and has duly trained tens of hundreds of scientists from pharma industry on nanomedicines and QbD. He has completed 16 research projects, and guided 31 PhD's (previously and currently), 3 post-doctorates and 57 postgraduate scholars. Currently, he is serving as founder coordinator of UGC Centre for Excellence in Nano Biomedical Applications, and Coordinator, UGC Centre of Advanced Studies (Pharmaceutical Sciences), both at Panjab University, Chandigarh, India. He has been Chairman, University Institute of Pharmaceutical Sciences, and Dean, Faculty of Pharm Sciences, and a Fellow of Panjab University too. His work has fetched Professor Bhoop numerous awards and accolades at global and national levels.

Minna Hakkarainen, MSc, PhD, is a professor in polymer technology and head of the Division of Polymer Technology at KTH Royal Institute of Technology in Stockholm, Sweden. Her research interests are quite diverse, ranging from biomedicine and water purification to packaging using bio-based, biodegradable, and/or recyclable polymers and nanocomposites, graphene and carbon derivatives with targeted applications. She has led several national and international research projects and has published more than 135 highly cited scientific papers in international peer-reviewed journals as well as several book chapters.

Kamalinder K. Singh, MPharm, PhD, is a professor of pharmaceutical technology and drug delivery at School of Pharmacy and Biomedical Sciences, University of Central Lancashire (UCLan), Preston, UK. She leads the Nanomedicine and Innovative Technologies research group at UCLan within Faculty of Clinical and Biomedical Sciences. Her research interests include advanced drug delivery and drug targeting, particularly biocompatible nanoparticle-based platform technologies. She serves the international steering committee for *Handbook of Pharmaceutical Excipients*, wherein she has contributed 45 monographs. In addition, she has one book, 8 book chapters, more than 75 peer-reviewed high-impact research publications, 20 patents (five granted), and 250 invited lectures/presentations to her credit. She has handled 20 research and consultancy projects, and has guided 20 PhDs and one post-doc fellow.

Contributors

Suneera Adlakha
University Institute of Pharmaceutical Sciences
Panjab University
Chandigarh, India

Madhunika Agrawal
University Institute of Pharmaceutical Sciences
Panjab University
Chandigarh, India

Satyam Kumar Agrawal
Ayurgenomics Unit
CSIR–Institute of Genomics & Integrative Biology
New Delhi, India

Nida Akhtar
Department of Pharmaceutics
Rajiv Academy for Pharmacy
Mathura, India

Javed Ali
Department of Pharmaceutics
School of Pharmaceutical Education and Research
Jamia Hamdard
Hamdard University
New Delhi, India

Saket Asati
Department of Pharmaceutical Sciences
Dr. Harisingh Gour University
Sagar, India

Sanjula Baboota
Department of Pharmaceutics
School of Pharmaceutical Education and Research
Jamia Hamdard
New Delhi, India

Silki Chandel
Department of Pharmaceutical Sciences
Dr. Harisingh Gour University
Sagar, India

Candace Minhthu Day
Centre for Pharmaceutical Innovation and Development
School of Pharmacy and Medical Science
University of South Australia
Adelaide, Australia

Parneet Kaur Deol
G.H.G. Khalsa College of Pharmacy
Gurusar Sadhar, Ludhiana
Punjab, India

Padma V. Devarajan
Department of Pharmaceutical Sciences and Technology
Institute of Chemical Technology (Elite Status and Centre of Excellence)
Deemed University
Mumbai, India

Surbhi Dhawan
Department of Pharmaceutical Sciences
Maharshi Dayanand University
Rohtak, India

Anisha A. D'souza
Piramal Enterprises Ltd.
Mumbai, India

Bharti Gaba
Department of Pharmaceutics
School of Pharmaceutical Education and Research
Jamia Hamdard
Hamdard University
New Delhi, India

Sanjay Garg
Centre for Pharmaceutical Innovation and Development
School of Pharmacy and Medical Science
University of South Australia
Adelaide, Australia

Gargi Ghoshal
Dr. S.S. Bhantnagar University Institute of Chemical Engineering & Technology
Panjab University
Chandigarh, India

Chahinez Houacine
School of Pharmacy and Biomedical Sciences
University of Central Lancashire
Preston, United Kingdom

Vinod S. Ipar
Department of Pharmaceutical Sciences and Technology
Institute of Chemical Technology (Elite Status and Centre of Excellence)
Deemed University
Mumbai, India

Amit K. Jain
Bhagyoday Tirth Pharmacy College
Sagar, India

Ashay Jain
Dr. S.S. Bhantnagar University Institute of Chemical Engineering & Technology
Panjab University
Chandigarh, India

Priyanka Jain
Department of Pharmaceutical Sciences
Dr. Harisingh Gour University
Sagar, India

Shaila Jain
Bhagyoday Tirth Pharmacy College
Sagar, India

Vandita Kakkar
University Institute of Pharmaceutical Sciences
Panjab University
Chandigarh, India

Hemanth Kumar Kandikattu
Department of Medicine, Cardiology
University of Missouri
Columbia, Missouri, USA

O. P. Katare
University Institute of Pharmaceutical Sciences
Panjab University
Chandigarh, India

Indu Pal Kaur
University Institute of Pharmaceutical Sciences
Panjab University
Chandigarh, India

Farhath Khanum
Defence Food Research Laboratory
Mysore, India

Dinesh Kumar
IPR Division, Jubilant Generics
Noida, India

Shobhit Kumar
Department of Pharmaceutics
School of Pharmaceutical Education and Research
Jamia Hamdard
Hamdard University
New Delhi, India

Amit S. Lokhande
Department of Pharmaceutical Sciences and Technology
Institute of Chemical Technology (Elite Status and Centre of Excellence)
Deemed University
Mumbai, India

Krantisagar S. More
Department of Pharmaceutical Sciences and Technology
Institute of Chemical Technology (Elite Status and Centre of Excellence)
Deemed University
Mumbai, India

Sanju Nanda
Department of Pharmaceutical Sciences
Maharshi Dayanand University
Rohtak, India

Jasjeet K. Narang
Department of Pharmaceutics
Khalsa College of Pharmacy
Amritsar, India

Pradip Nirbhavane
University Institute of Pharmaceutical Sciences
Panjab University
Chandigarh, India

Ankit Parikh
Centre for Pharmaceutical Innovation and Development
School of Pharmacy and Medical Science
University of South Australia
Adelaide, Australia

Kamla Pathak
Department of Pharmaceutics
Pharmacy College Saifai
Uttar Pradesh University of Medical Sciences
Etawah, India

Nisha Rawat
University Institute of Pharmaceutical Sciences
Panjab University
Chandigarh, India

Komal Saini
University Institute of Pharmaceutical Sciences
Panjab University
Chandigarh, India

Simarjot Kaur Sandhu
University Institute of Pharmaceutical Sciences
Panjab University
Chandigarh, India

Amita Sarwal
University Institute of Pharmaceutical Sciences
Panjab University
Chandigarh, India

Rakesh Kumar Sharma
Defence Food Research Laboratory
Mysore, India

Sumit Sharma
University Institute of Pharmaceutical Sciences
Panjab University
Chandigarh, India

Teenu Sharma
University Institute of Pharmaceutical Sciences
UGC Centre of Advanced Studies
Panjab University
Chandigarh, India

Naveen Shivanna
Defence Food Research Laboratory
Mysore, India

Bhupinder Singh
University Institute of Pharmaceutical Sciences
UGC Centre of Excellence in Applications of Nanomaterials, Nanoparticles and Nanocomposites (Biomedical Sciences)
Panjab University
Chandigarh, India

Gurpreet Singh
University Institute of Pharmaceutical Sciences
Panjab University
Chandigarh, India

Kamalinder K. Singh
School of Pharmacy and Biomedical Sciences
University of Central Lancashire
Preston, United Kingdom

V. R. Sinha
University Institute of Pharmaceutical Sciences
Panjab University
Chandigarh, India

Yunmei Song
Centre for Pharmaceutical Innovation and Development
School of Pharmacy and Medical Science
University of South Australia
Adelaide, Australia

Vandana Soni
Department of Pharmaceutical Sciences
Dr. Harisingh Gour University
Sagar, India

Supriya Verma
University Institute of Pharmaceutical Sciences
Panjab University
Chandigarh, India

Awesh K. Yadav
Bhagyoday Tirth Pharmacy College
Sagar, India

Vaishali Yadav
Bhagyoday Tirth Pharmacy College
Sagar, India

1 Nutraceuticals: A Revolutionary Approach for Nano Drug Delivery

*Shaila Jain, Vaishali Yadav, Amit K. Jain, and Awesh K. Yadav**

CONTENTS

* Corresponding author.

1.1 INTRODUCTION

Nanotechnology is a multidisciplinary field of applied science that studies the design and application of nanometric (often 100 nm scale) functional structures. Nanotechnology has been gaining a noteworthy driving force in recent years, particularly for the treatment of cancer. Over the last five decades, researchers have shown considerable attention in mounting biocompatible, biodegradable non-immunogenic nanoparticles as safe, successful, and effective delivery of bioactives (Crommelin *et al.*, 2013). Nanotechnology has tremendous potential to remodel many allied disciplines such as food, pharmaceuticals, nutraceuticals, and cosmeceuticals. Increasing demand by informed consumers has led to innovative all-in-one delivery systems owing to prove a complementary tool in the treatment and prevention of diseases (Huang *et al.*, 2010). Nanotechnology holds considerable possibilities to transform the obtained therapeutic results.

Nutraceuticals are defined as nontoxic food components derived from foods that have systematically recognized credible haleness for the prevention and action of vulnerable diseases. Due to their antioxidant properties, nutraceuticals have shown great attention in the prevention and treatments of various ailments (Bennett *et al.*, 2012). A nutraceutical has surplus health advantages, along with essential nutritional value, and is used to prop up proper well-being, manage symptoms, and check the malignant processes by an established nonspecific physiological benefit pattern against chronic disease (Jain *et al.*, 2015). Nutraceuticals are widely used in various diseases due to possible health benefits (Jain *et al.*, 2017a). Phytochemicals such as lycopene, β-carotene, curcumin, ferulic acid, resveratrol, quercetin, and other carotenoids have fascinated the scientists owing to their intrinsic antioxidant property and advantages in enhancing the immunity against numerous diseases such as cancer, cardiovascular, neurodegenerative disorders like Alzheimer's disease, psoriasis, and inflammation (Thakur *et al.*, 2017). Moreover, the usage of carotenes, phospholipids, tocopherols, ascorbic acid and its esters are undoubtedly endorsed, as these are compounds of safe and reliable purity, relatively cheap, easily available, and are generally regarded as safe (GRAS) (Jain *et al.*, 2017b). However, inadequate bioavailability and poor solubility issues related with such phytochemicals, leading to their reduced biological activity, is considered a major impediment associated with the active molecules (Jain *et al.*, 2016).

There has been a persuasive growth in the reinforcement of nutraceuticals by way of nanotechnology. Effectual solubilization, encapsulation, and delivery, usually based on nanotechnology, bestow these biocompatible nutraceuticals with properties like significant absorption in lower doses, reduced frequency of administration, and improved therapeutic index. Nanotechnology often renders targeted delivery of the drugs and also demonstrates slow and sustained release from the formulation (Aswathy *et al.*, 2015; Aditya *et al.*, 2017).

1.2 HISTORICAL DEVELOPMENT

Nutraceutical, a blend of the words *nutrition* and *pharmaceutical*, was coined in 1989 by Stephen L. DeFelice. Hippocrates, the father of Western medicine, advised, "let food be thy medicine." Pliny the Elder also revealed the importance of daily nutrients for health benefits. The term *nutraceuticals* broadly includes a variety of products such as isolated nutrients, dietary supplements, specific diets, and processed foods such as cereals, soups, and beverages. In ancient time, our ancestors only relied on various plant materials for their health needs, continuously performed trial and error based research on various plant materials as food. A number of plants were revealed as wonderful drugs, while few proved to be poisonous for human health. The natural product and nutritional supplement industry experienced explosive growth during the last half of the 20th century, leaving it fragmented and in need of market leadership. In 1993, *nutraceutical* was coined to answer this call. Alchemists and Casimir Funk were among the major contributors who supported the modern concept of nutraceuticals by introducing the novel concepts of "elixier vitae" and "vitamins," respectively. Since ancient times, our ancestors have always been justifying the use of food according to disease state of body, advocating the treatment of various body ailments with the help of controlled food charts (Gupta *et al.*, 2010; Kannappan *et al.*, 2011).

1.3 BIOAVAILABILITY CONCERNS

While nutraceuticals are emerging as vital interest in disease prevention and treatment, the bioavailability of their key active ingredients has become a burning issue in the development of their oral formulations (Ting *et al.*, 2014). Nutraceuticals today are available abundantly in conventional dosage forms such as capsules, pills and powders for oral delivery, and tinctures for topical applications. In the conventional mode of delivery of nutraceuticals, a portion of the administered nutraceutical dose is absorbed, and a remaining portion may initiate nonspecific toxicity and adverse side effects, which in turn results in reduced bioavailability (Aditya *et al.*, 2017).

To triumph over these tribulations, the ideology of nanotechnology works to encapsulate the functional ingredients using food-grade or GRAS materials that can exhibit controlled-release behavior of nutraceuticals, which will eventually lead to their enhanced bioavailability (Huang *et al.*, 2010).

1.4 APPROACHES TO ENHANCE ORAL BIOAVAILABILITY OF NUTRACEUTICALS

The effectiveness of nutraceuticals to provide therapeutic or physiological benefits precisely depends on the bioavailability of active ingredients to the target site of action. Factors such as poor aqueous solubility, insufficient residence time in the gastrointestinal tract (GIT), instability to changing physiological environments, low permeability, and susceptibility to rapid metabolic transformation could significantly lower the efficacy of nutraceuticals for disease prevention. Many promising delivery systems have already been developed to enhance the therapeutic efficiency of nutraceuticals (Lee *et al.*, 2010; Li *et al.*, 2011). Nanocarriers for nutraceuticals and supplements exploit a variety of delivery systems including nanoparticles, solid lipid nanoparticles, niosomes, nanospheres, nanoliposomes, nanofibers, nanoemulsions, nanocapsules, and carbon nanotubes (Aklakur *et al.*, 2015).

To enhance the bioavailability of nutraceuticals, several delivery systems with functional properties have been designed to overcome aforementioned challenges. Some of the approaches that investigators have used when designing an optimum delivery system for nutraceuticals are as follows.

1.4.1 Protection of Labile Compounds

Upon oral administration, nutraceuticals are subjected to complex digestion processes that involve changes in physiological or physiochemical environments due to the dynamic environmental conditions of GIT, leading to instability of active ingredients. The factors, such as pH variations, ionic strength, enzymatic degradations, and mechanistic motilities, potentially contribute to the degradation of nutraceuticals. Therefore, delivery systems with protective mechanism could enhance the gastric permanence of labile biological components and, thus, improve upon oral dosing efficiency of nutraceuticals.

1.4.2 Extension of Gastric Retention Time

Inadequate gastric withholding time dominantly affects the partial absorption of nutrients, excessive compound excretion, and dwindles in the dose for supporting intestinal membrane fluidity (Hu *et al.*, 2012).

1.4.3 Intonation of Metabolic Activities

Amalgamation of physical and chemical enzyme inhibitors in the formulation seems to enhance the level of nutraceuticals in the systemic circulation, thereby preventing the changes produced as an upshot of metabolic modulation. However, the use of enzyme inhibitors possibly will have to be considered to steer clear of toxicity raised due to impaired detoxification activity.

1.5 DELIVERY OF NUTRACEUTICALS

1.5.1 Oral Delivery

Despite the low inherent oral bioavailability of many nutraceuticals, these still remain a strong preference as these follow the innate process of food and nutrient consumption in the body and are noninvasive. Innovative technologies and formulation strategies therefore play a vital role in the success of orally administered nutraceuticals by enhancing their solubility and permeation efficiency through GIT as well as defending them from rapid elimination. The path of nutraceuticals through intestine is represented in Figure 1.1. Curcumin, for instance, has been formulated as biodegradable nanoparticles with improved oral bioavailability (Shaikh *et al.*, 2009).

1.5.2 Dermal Delivery

Nutraceutical-derived antioxidant effects may be enhanced by the dermal application of bioactives, such as in the case with lutein (Palombo *et al.*, 2007). Nutraceuticals used topically with positive cosmetic benefits include co-enzyme Q10, genistein, curcumin, N-acetylcysteine, gluconolactone, and fucose-rich sulfated polysaccharides (Wijesinghe *et al.*, 2012).

1.5.3 Ophthalmic Delivery

A number of nutraceuticals, particularly co-enzyme Q10, vitamin E, and lutein, when administered ophthalmically have disease-modifying effects on pathologies of the eye owing to their various pharmacotherapeutic actions such as anti-cataract. Intraocular treatment success is, however, largely dependent on the residence time and permeability of the topically administered formulation. Simple inclusion of selected nutraceuticals dosed locally and concurrently with allopathic agents may extend the intraocular retention time, leading to synergistic clinical benefits, and prove to be safer alternatives for long-term ophthalmic therapy (Peng *et al.*, 2010).

1.6 CARRIERS FOR NUTRACEUTICAL DELIVERY

Nutraceuticals tend to cover a vast majority of the curative domain, particularly arthritis, cold and cough, sleeping disorders, digestion, depression, diabetes, etc. But these do not have much

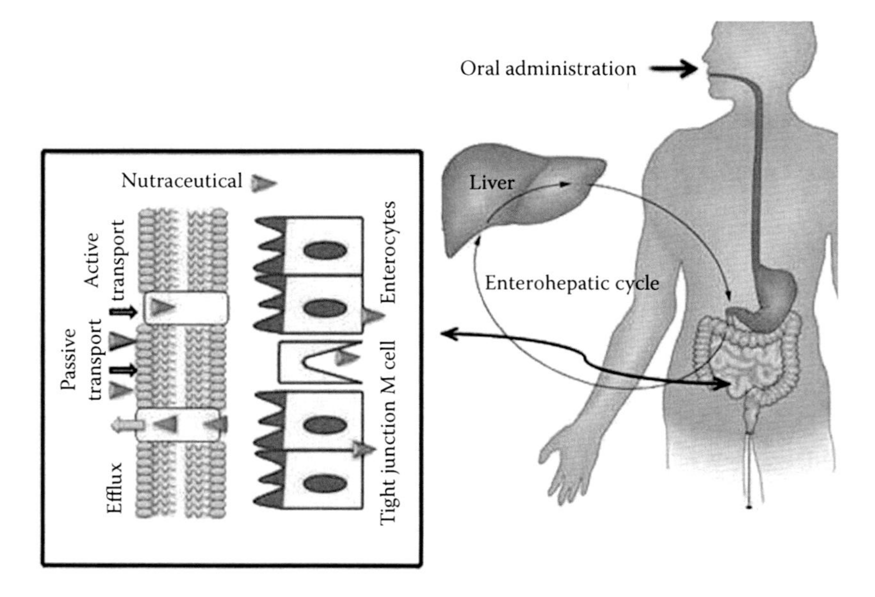

FIGURE 1.1 Passage of nutraceutical through intestine.

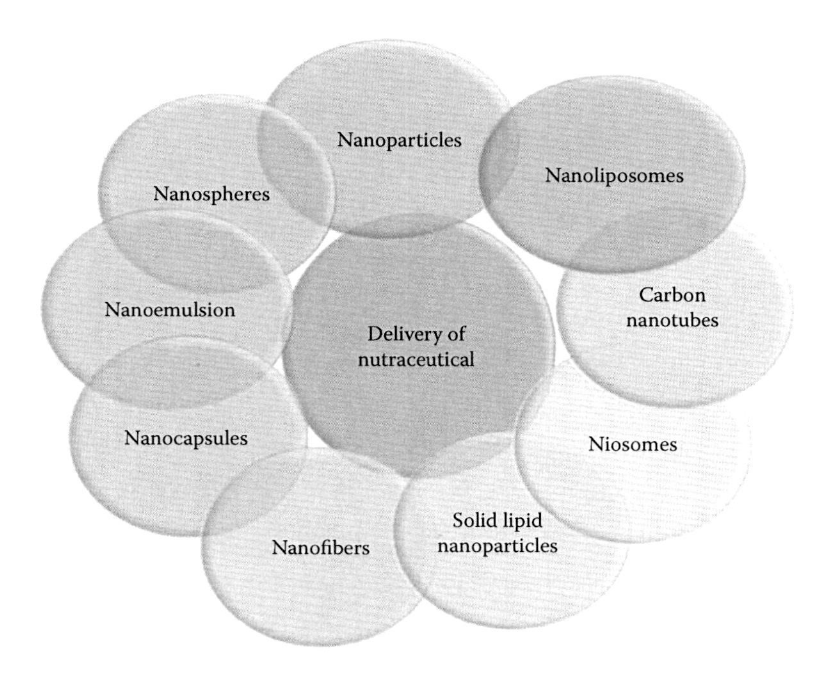

FIGURE 1.2 Nano drug delivery system employed for nutraceutical.

effectiveness due to above stated problems encountered with nutraceuticals as well as their formulation, calling thereby an effective carrier system for the delivery of nutraceuticals. A number of formulation approaches such as nanoparticles, solid-lipid nanoparticles, niosomes, nanospheres, nanoliposomes, nanofibers, nanoemulsions, nanocapsules, and carbon nanotubes encompass the resourceful deliverance of the encapsulated nutraceuticals (Figures 1.2 and 1.3). The developed formulations boost up pharmacological activity, stability, prevention from environmental degradation, protection from toxicity, and, hence, augmentation of therapeutic efficacy (Saraf *et al.*, 2010).

1.6.1 Nanoparticles

As the name suggests, these are the drug delivery systems with the particle size in nanometers. The main features for the synthesis of nanoparticles are the utilization of substances such as proteins, polysaccharides, and synthetic polymers (Britto *et al.*, 2012). The matrix substance should be selected depending on (a) dimension of particles, which is requisite; (b) intrinsic chattels of the drug like steadiness; (c) exterior distinctiveness as penetration; (d) extent of biodeterioration and biorapport; (e) drug liberation silhouette; and (f) antigenicity of formulation. The polymers employed for the preparation of nanoparticles can be natural or synthetic. The most frequently used natural polymers in preparation of polymeric nanoparticles are chitosan, gelatin, sodium alginate, and albumin, while the synthetic polymers are polylactides (PLA), polyglycolides (PGA), poly(lactide co-glycolides) (PLGA), polyanhydrides, polyorthoesters, polycyanoacrylates, polycaprolactone, polyglutamic acid, polymalic acid, etc. (Yadav *et al.*, 2011; Hu *et al.*, 2013; Aklakur *et al.*, 2015).

1.6.1.1 Advantages of Nanoparticles as Delivery Systems for Nutraceuticals

Nanoparticles provide broader applications for nutraceuticals owing to the following:

- Lesser dose of bioactive compound is needed.
- Superior bioavailability of active compound is obtained.
- Nanoparticles are mechanically and thermally more stable to changing pH.

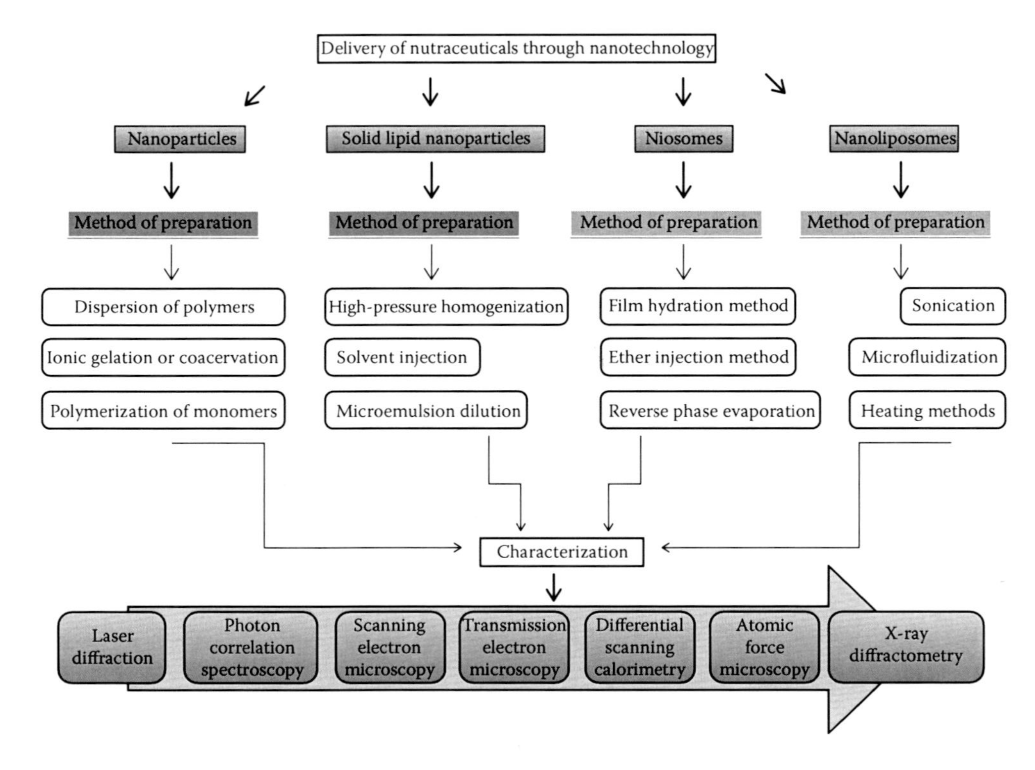

FIGURE 1.3 Method of preparation and characterization of nano drug delivery systems.

- Nanoparticles show better shelf-life.
- Matrix selection helps in formulating controlled release preparation.
- Attaching ligands to the surface of particle, site-specific targeting can be achieved.
- The formulation can be employed for oral, nasal, parenteral, intra-ocular administrations, etc. (Luo *et al.*, 2010).

As this delivery system offers diversity in their featured characteristics, many nutraceuticals have been formulated as nanoparticles. Upon extraction of *Curcuma longa L.*, a polyphenol compound is yielded as curcumin, which is not soluble in aqueous media. It has many pharmacotherapeutic actions including anticancer, anti-inflammatory, antioxidant, etc. Recent research has shown the action of curcumin against amyloid protein, which can be exploited in treatment of cognition-related disorders.

Conventional formulations of curcumin have shown hasty systemic purging, and to troubleshoot this effect, curcumin-loaded PLGA nanoparticles were formulated by high-pressure emulsification-solvent evaporation method. This formulation resulted in blood–brain barrier passage and increased retention time of curcumin (Tsai *et al.*, 2011). Curcumin is being widely exploited using various delivery systems resulting in improved properties of this bioactive compound (Figure 1.4). Similarly, vitamin D3 nanoparticles were prepared using oleoyl alginate ester. This formulation was able to cross cell membrane, showing superior action toward rickets disease, serving as an effective transporter. Thus, vitamin D loaded nanoparticles could be potentially used as a delivery system for nutraceuticals (Sun *et al.*, 2012). Furthermore, Table 1.1 epitomizes nanoparticles as carrier systems that are lucratively urbanized for delivering nutraceuticals.

1.6.1.2 Major Constraints in Nanoparticles Preparation

One crucial point while formulating nanoparticles is that their diminutive range and bulky facade area can pilot to particle gathering, which makes corporeal handling of nanoparticles intricate in

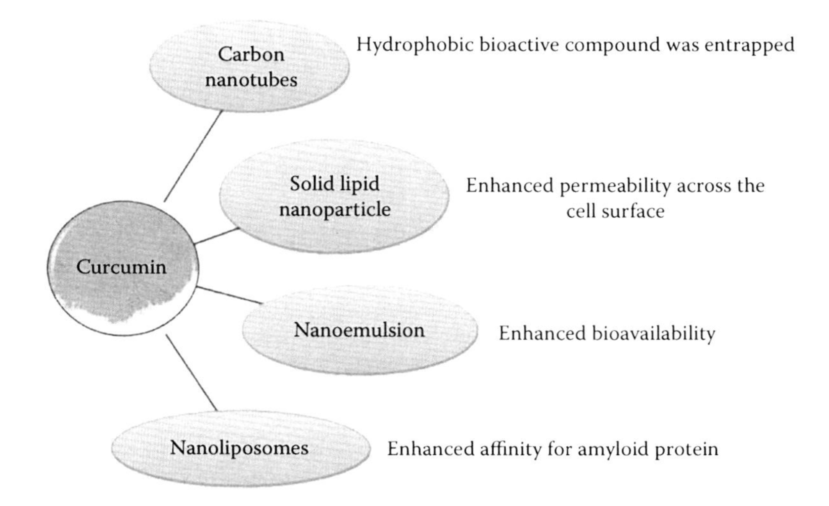

FIGURE 1.4 Curcumin used in various nano formulations and its advantages.

fluid and dehydrated forms. Moreover, partial drug encumber and burst liberation is another problem associated with small particle size and large surface area. Thus, to overcome such problems and surface amendment concerns, drug impediment approaches and release control parameters have to be considered, and should be surmounted before nanoparticles can be used clinically or made commercially available (Yadav *et al.*, 2012).

1.6.2 Solid Lipid Nanoparticles

These formulations were initiated in 1991 as exceptional delivery systems toward time-honored polymer and lipid based colloidal delivery services. The lipids, which are solid in nature, are added in these formulations and, if required, surfactants are also incorporated. The mean particle size of solid lipid nanoparticles (SLNs) ranges from 40–1000 nm. The configuration of SLNs might be contrasted with a symmetric brick wall. Lipids used in the preparation of SLNs are hard at room as well as body temperatures. Triglycerides, waxes, and hard fats are a few lipid materials, while lecithin, polysorbate 80, and polyvinyl alcohol are used as surfactants in SLNs preparation (Mitri *et al.*, 2011). The various methods employed for preparation of SLNs include ultrasonication, solvent injection, solvent diffusion, supercritical fluid extraction of emulsions, and spray-drying (Madureira *et al.*, 2015).

1.6.2.1 Advantages of SLN as Delivery Systems for Nutraceuticals

- Good biocompatibility
- Increased stability of drugs
- Protection of active ingredient against physical and chemical degradation
- Sustained release and drug targeting of active ingredients
- Possibility of oral, dermal, and parenteral drug administration
- Decreased potential for acute and chronic toxicity (Mitri *et al.*, 2011; Chen *et al.*, 2013)

Owing to the above-mentioned advantages, several nutraceuticals have been formulated as SLNs. One of the instances is quercetin, a hydrophobic molecule, bitter in taste and prone to degradation in alkaline condition; this limits its integration with food products. Thus, SLNs were produced using high

TABLE 1.1
Nanoparticles as Delivery System for Nutraceuticals

Nutraceuticals	Method of Preparation	Inference	Reference
α-lactalbumin	Desolvation process derived from Coacervation method	Controlled drug delivery was achieved	Mehravar *et al.*, 2009
Elsholtzia splendens	Ionic gelation	Improved antioxidant action	Lee *et al.*, 2010
Selenite	Chitosen	Low toxicity and improved antioxidant property	Luo *et al.*, 2010
Vitamine D3	Oleoyl alginate ester	Sustained rate in gastrointestinal fluid	Li *et al.*, 2011
Whey protein	Ethyl hexanoate and Propylene glycol	This formulation improved retention of bioactive compound	Giroux and Britten, 2011
Albizia chinensis	Solvent evaporation method	Improved solubility, permeability, and stability	Yadav *et al.*, 2012
Epigallocatechin-3-gallate	B-lactoglobulin	Antioxidant activity	Li *et al.*, 2012
Vitamin D	Zein	Hydrophobic nutrients can be delivered	Luo *et al.*, 2012
Vitamin C	Chitosan	This formulation can be successfully used as a stable medium to incorporate and transport vitamins	Britto *et al.*, 2012
Epigallocatechin-3-gallate	Caseinophosphopeptide and chitosan	Enhanced antioxidant activity	Hu *et al.*, 2013
Indole-3-carbinol and 3,30-diindolylmethane	Zein and zein/carboxymethyl chitosan	Improve their stability against harsh conditions and provide controlled release	Luo *et al.*, 2013
Soy protein	Ionic gelation method	Controlled release of hydrophobic nutraceuticals	Teng *et al.*, 2013
Soy protein	Folic acid	Enhanced delivery of anticancer drugs	Teng *et al.*, 2013
B-lactoglobulin	Carbodiimide-catalyzed approach	Desirable solubility, controlled release, and enhanced absorption to nutraceuticals was achieved	Teng *et al.*, 2013
A-lactalbumin	Disk-shaped nanoparticles were obtained upon partial hydrolysis of a-la by glu-c v8	Sensitive substances to the surroundings was reduced	Quintana *et al.*, 2013
Vitamin D	Soybean β-conglycinin	This formulation enhances hydrophobic drug delivery	Levison *et al.*, 2014
Fish oil	Zein via alcohol evaporation	Enhanced bio availability	Soltani and Madadlou, 2014
B-lactoglobulin	Low antisolvent content before mild evaporation	Controlled-release property was achieved	Teng *et al.*, 2014
Genistein	Nclusion-complexation	Enhanced bio availability	Semyonov, 2014
Caffeine	Desolvation with ethanol	Increased the stability of nanoparticles to digestion	Bagheri *et al.*, 2014
Rosmarinic acid	Hot melt ultrasonication method	Stability of bioactive compound was achieved	Campos *et al.*, 2014

(Continued)

TABLE 1.1 (CONTINUED)
Nanoparticles as Delivery System for Nutraceuticals

Nutraceuticals	Method of Preparation	Inference	Reference
Whey protein isolate, Beet pectin	Thermal treatment of protein-polysaccharide mixtures to induce the formation of protein nanoparticles	Protection against chemical degradation	Maya and McClements, 2014
Cocoa	Nanoprecipitation	Antiradical activity of the bioactive compound was preserved	Reyes *et al.*, 2014
Trans-resveratro	Cetyl palmitate and tween 60	Enhanced bio availability	Neves *et al.*, 2015
White tea extract	Nanoprecipitation method	Controlled release and antioxidant activity was preserved	Sanna *et al.*, 2015
Gammaoryzanol	Nanoprecipitation technique and solvent evaporation technique	Solvent evaporation technique representing better encapsulation	Ghaderi *et al.*, 2015
B-lactoglobulin	Internal gelation method	Higher cellular uptake and cytotoxicity	Ha *et al.*, 2015
Cruciferin	Cold gelation method	Increased stability of bioactive compound	Akbari and Wu, 2016

TABLE 1.2
Solid Lipid Nanoparticles as Delivery System for Nutraceuticals

Nutraceuticals	Lipid/Method Employed	Inference	Reference
Quercetin	Poly-d,l-lactide	Permeable to plasma membrane, highest loading efficiency and cost effectiveness, for the oral administration	Kumari *et al.*, 2011
Lutein	Cetyl palmitate, glyceryl tripalmitate, carnauba wax, caprylic/capric triglyceride, capryl glycoside, and tween 80	Poor solubility of across dermis was established	Mitri *et al.*, 2011
Caffeic acid	Chitosan, alginate, and pectin	Bioavailability was improved	Fathi *et al.*, 2013
Rosmarinic acid	Carnauba wax and tween 80	This formulation provided protection, improvement in stability, and increased bioavailability	Madureira *et al.*, 2015

pressure homogenization using lecithin and Span 20. This formulation increases the bio-accessibility and bioavailability of bioactive compound quercetin (Aditya *et al.*, 2014). Table 1.2 symbolizes SLNs that are usefully built-up for delivering nutraceuticals to enhance their various properties.

1.6.2.2 Major Constraints in the Preparation of SLNs

Upon formulation of SLNs, polymorphic transitions occur due to their instability, as they possess a tendency to form aggregates. These polymorphic transitions escort the transformation of lipid nanoparticles (from spheroid to disk-like), thus causing a palpable boost in particle face quarter. Hydrophobic pull occurs between bare nonpolar regions on particle surfaces contributing to aggregation and gelation, thus limiting their relevance. To triumph over this quandary, high levels of emulsifier are added that shape a shielding shell about the lipid nanoparticles (Jenning and Gohla, 2001).

1.6.3 Niosomes

Hydrophilic and lipotropic compounds that are excessively used in pharmaceuticals or nutraceuticals can be entrapped into a double layer structure of non-ionic surfactants in water, forming a self-agglomeration vesicle, called *niosomes* (Pando *et al.*, 2015). Unlike liposomes, which are prepared by phospholipid-based artificial vesicles, niosomes are prepared from nonionic surfactants; they are less expensive and are stable toward oxidative degradation. Polyoxyethylene alcohol, polyoxyethylene glycol alkyl ethers, alkyl ethoxylate, and alkyl phenol ethoxylate are a few nonionic surfactants employed in niosome preparation. The various methods that can be exercised to prepare niosomes include trans-membrane pH gradient, sonication, microfluidization, dried-reconstituted vesicles, and bubble method (Rentel *et al.*, 1999; Fraile *et al.*, 2015).

1.6.3.1 Advantages of Niosomes as Delivery Systems for Nutraceuticals

- Niosomes are osmotically active, chemically stable, and have long storage time.
- Niosomes possess hydrophilic heads; thus, their surface can be modified.
- Due to their nonionic nature they have high compatibility with natural organization and fewer side effects.
- They are biodegradable and non-immunogenic.
- These formulations can be loaded with lipotropic and hydrophilic compounds.
- They can advance the action of nutraceutical by shielding from biological environment, consequential superior accessibility, and put a ceiling on the bioactive moiety to the destined cell.
- These formulations are based on water media, accessibility for patients tends to enhance.
- In contrast to phospholipids, the management for surfactant do not necessitate any exceptional safety measures (Moghassemi and Hadjizadeh, 2014).

One of the natural polyphenol compounds is resveratrol, a nutraceutical with several benefits like cancer cure, reducing cardiovascular risk, and anti-inflammatory activity. One of the major limitations encountered with resveratrol is its photosensitive nature, as this compound occurs in cis and trans form, with only trans-resveratrol having the health benefits. Exposure of light causes inversion of trans form of resveratrol to cis form, which is inactive. Thus, niosomes containing resveratrol were prepared using thin film hydration method. Upon characterization, niosomal formulation exhibited high resveratrol entrapment efficiency and good stability (Pando *et al.*, 2013). Table 1.3 gives supplementary report of niosmes formulated with nutraceuticals to improve the properties of bioactive compound.

1.6.4 Nanospheres

One of the homogenous systems of nanoparticles, ranging between 10 and 200 nm in diameter, are nanospheres. These can be amorphous or crystalline in nature, where bioactive moiety is entrapped in the polymer matrix. Normally, methods are employed for the formation of nanospheres depending on the nature of polymer: emulsion polymerization, solvent evaporation, and nanoprecipitation

TABLE 1.3
Delivery of Nutraceuticals through Niosomes

Nutraceuticals	Lipid/Method Employed	Inference	References
Ovalbumin	Sucrose esters, cholesterol, and dicetyl phosphate	Oral vaccine delivery systems were prepared	Rentel *et al.*, 1999
Ellagic acid	Span 60, tween 60, polyethylene glycol 400, propylene glycol, and methanol	This formulation enhanced biopharmaceutical properties, high solubility, and high permeability of ellagic acid	Junyaprasert *et al.*, 2012
Resveratrol	Mechanical agitation and sonication method using span 80, span 60, and cholesterol	Protection of resveratrol has been achieved against photosensitive leading to an irreversible change from the trans to the inactive cis isomer	Pando *et al.*, 2013
Lactic acid	Sodium dodecyl sulfate	Lactic acid extraction agents in aqueous solutions was possible through this formulation	Fraile *et al.*, 2015
α- tocopherol	Thin film hydration method using Sorbitan monostearate and cholesterol	This formulation preserves adverse environmental and gastrointestinal conditions	Abaee and Madadlou, 2016
Chrysin and luteolin	Span 60	Enhanced bioavailability	Myung *et al.*, 2016

(Vilela *et al.*, 2012). The pharmaceutical industrial houses based on dietary supplements and nutraceuticals are now developing a new delivery system as nanospheres, which are nanosized particles from bioactive natural nonhazardous phospholipids and their constitution in the human body (Tan *et al.*, 2012).

1.6.4.1 Nanospheres Provide Nutraceutical Formulations with the Following

- Augmentation of biological stability, including solubility
- Upgradation of cellular transport
- Emancipation of the active component inside the cell
- Appreciably greater bioavailability and its biological action
- Site-specific targeting
- Improved half-life in the body
- Negligible side effects
- Superior therapeutic value

These meritorious features have made the utilization of nanospheres extensive across the globe. Enzyme immobilization is a classic technique for improving the stability of enzymes, but in the present scenario encapsulation is being used. Nanospheres are being employed for the encapsulation of enzymes via biomimetic silicification, engaging alpha chymotrypsin and a fungal protease from *Aspergilus oryzea*. Encapsulations appreciably augment the thermal and ultrasonic stabilities of enzymes, suggesting a range of assorted relevance for bioactive compounds (Madadlou *et al.*, 2010). Table 1.4 shows nanospheres as a recent and effective mode of delivering nutraceuticals.

TABLE 1.4
Delivery of Nutraceutical through Nanosphere

Nutraceutical	Lipid/Method Employed	Inference	References
Whey protein α-lactalbumin	V8 protease enzyme	Site specific delivery was obtained	Esmaeilzadeh *et al.*, 2011
Waxy maize starch	Nanoprecipitation using acetic anhydride, phthalic anhydride, fluorescein isothiocyanate, dibutyltin dilaurate, and N,N-dimethylaminopyridine	This formulation stabilized pickering emulsion via amphiphilic starch-based nanospheres	Tan *et al.*, 2012
Gallic, caffeic acids, catechin, and rutin	Cetyltrimethylammonium chloride and hydrogen tetrachloroaurate	Enhanced bioavailability	Vilela *et al.*, 2012

1.6.5 Nanoliposomes

Nanoscale lipid vesicles are referred to as *nanoliposomes* or merely nanometric adaptations of liposomes. The indispensable hydrophilic–hydrophobic interface, linking phospholipids and water molecules, is responsible for the formation of nanoliposomes (Aadinath *et al.*, 2016). Aggregation of vesicles upon storage is commonly observed when the nanometric size is formulated, which turn out to be microsized particles. However, these are supposed to have enough steadinesses to uphold their dimension and could be defined as bilayer lipid vesicles owning and preserving nanometric size ranges throughout the consumption of formulation. Passive or active targeting mechanisms are employed for targeted therapy via nanoliposomes. The various methods employed include sonication, extrusion, microfluidization, and heating methods (Xia *et al.*, 2014). Evaluation is decisive to ensure quality of product while using new techniques. The most significant consideration of their characterization embraces visual manifestation, size distribution, stability, zeta potential, lamellarity, and entrapment efficiency (Gulseren and Corredig, 2013; Shin *et al.*, 2013).

1.6.5.1 Advantages of Nanoliposomes as Delivery Systems for Nutraceuticals

- They increase bioavailability of the active ingredient.
- They improve the controlled release system of drug delivery.
- They enable precision targeting and demonstrate high therapeutic activity at the target site.
- They can be applied for pulmonary therapies, as gene delivery vectors for stabilization of sensitive biomaterials.
- They reduce toxicity and amplify resourceful allocation of bioactive resources.
- They exhibit longer transmission time in the bloodstream devoid of being renowned by macrophages.
- They prevent unwanted interactions with other molecules. (Mohammadi *et al.*, 2013; Lu *et al.*, 2014)

Astonishing benefits of nanoliposomes have made scientists across the globe further investigate the potential of this formulation toward bioactive compounds. Carotenoids have antioxidant properties and are therefore generally added in foods, but there seems to be a major problem of sensitivity of this molecule through radiance, heat, and oxygen. Thus, to overcome this problem, β-carotene-loaded nanoliposomes were prepared via ethanol injection method. This formulation has been successfully developed to deliver hydrophobic compounds (Zompero *et al.*, 2015). Table 1.5 further illustrates the carrier systems that are strategically urbanized built-up for delivering nutraceuticals.

TABLE 1.5
Delivery of Nutraceuticals through Nanoliposomes

Nutraceuticals	Method/Lipid Employed	Inference	Reference
Curcumin	Click chemistry was employed to produce curcumin-phospholipid conjugates	This formulation enhanced affinity for amyloid protein	Mourtas *et al.*, 2011
Crocin	Dehydration and rehydration method	This formulation improved cytotoxic effects	Mousavi *et al.*, 2011
Tea polyphenol	High-pressure homogenization	This formulation involved incorporation of tea catechins in the milk phospholipid bilayer more efficiently than in the case of a soy phospholipid bilayer	Gulseren and Corredig *et al.*, 2013
Vitamin D	Thin film hydration and sonication using lecithin and cholesterol	This formulation enhanced physical properties and storage stability	Mohammadi *et al.*, 2013
Curcumin	Ethanol injection method	This formulation enhanced bioavailability, with enhanced high mucoadhesive property, storage stability, and encapsulation efficiency	Shin *et al.*, 2013
Vitamin C	Double emulsion method with dynamic high pressure microfluidization	Provide long-term storage	Yang *et al.*, 2013
Allicin	Reverse-phase evaporation method	This formulation protected allicin from unfavorable conditions, such as light, heat and alkaline conditions, leading to an extension of the shelf life of this product, and can also weaken its offensive odor	Lu *et al.*, 2014
Beta-lactoglobulin	Reverse-phase evaporation method	This formulation provide gastrointestinal stability	Ma *et al.*, 2014
Vitamin E	Chitosan	This formulation protected against thermostability and provide feasibility for commercial usage	Xia *et al.*, 2014
Zataria multiflora	Thin-film evaporation, ethanol injection, and sonication methods	This formulation enhanced physical properties and storage stability	Khatibi *et al.*, 2015
Vitamin A palmitat	Thin-film hydration sonication method	Lecithin-cholesterol presented higher capabilities as a carrier for vitamin A palmitate encapsulation	Pezeshky *et al.*, 2016

1.6.6 Nanofibers

Nanofibers are the most exploited, novel nanocarrier systems for delivery as they can be formulated in a desirable shape using various synthetic and natural polymers. Nanofibers have specific physicochemical properties such as hefty surface-mass proportion, better-quality automatic performance, and high porosity. In the present scenario nanofibers have found wide application: drug delivery, tissue engineering, medical implants, environmental engineering and biotechnology (via membranes and filters), and energy (via solar cells and fuel cells). One use of nanofibers involves application as carrier systems for the delivery of antimicrobial agents, enzymes, drugs, colors, flavors, antioxidants, and other functional compounds (Ohgo *et al.*, 2003; Avci *et al.*, 2012). Nanofibers are employed in controlled release of bioactive moiety so that an unambiguous amalgam can be stabilized and concealed in these formulations for deliberate discharge

TABLE 1.6
Delivery of Nutraceuticals through Nanofibers

Nutraceuticals	Method of Preparation	Inference	References
Bombyx mori and *Samia cynthia* ricini	Electrospinning method using hexafluoroacetone	Nanoscale fibers from polymer solution were obtained	Ohgo *et al.*, 2003
Lactobacillus acidophilus	Electrospinning technology using poly vinyl alcohol	Thermal protection of probiotics in heat-processed foods	Fung *et al.*, 2011
Durum wheat straw	Electrospinning method using trifluoroacetic acid	Alternative raw material for the production of cellulose nanofibers as renewable source	Leyva *et al.*, 2011
Lawsonia inermis	Electrospinning technique using poly ethylene oxide and poly vinyl alcohol	Antibacterial action was investigated	Avci *et al.*, 2012
Aloe vera rind	Chemo mechanical method	Cellulose nanofibers were obtained	Cheng, 2014
Fish oil	Electrospinning technique using poly vinyl alcohol	Bioactivity was enhanced	Garcia-Moreno *et al.*, 2016

in suitable surroundings. The microporous structure of nanofibers favors cell adhesion, proliferation, relocation, and differentiation, all highly desired properties for tissue engineering applications. The following techniques are employed for their synthesis: electrospinning, self-assembly, and phase separation. Among these, electrospinning is mostly engaged as this process does not lead to deterioration of the active compound (Cheng, 2014; Garcia-Moreno *et al.*, 2016). Nanofibers are prepared from of the natural and synthetic polymers; some examples of natural polymer are collagen, hyaluronic acid, gelatin, chitosan, elastin, silk, and wheat protein. Examples of synthetic polymers include PLA, poly(ethylene terephthalate), PLGA, and PLGA-poly(ethylene glycol).

Electrospun nanofibers can be used for delivery of nutrients so as to guard bioactive moiety during storage and as a carrier for relocating the compounds to the objective situate in the system. The bioactive compounds are protected from pH, heat, and moisture so that they are released underneath appropriate circumstances on the objective situates (Fung *et al.*, 2011; Avci *et al.*, 2012; Rezaei *et al.*, 2015). Due to aforementioned reasons, various studies have been performed that use nanofibers as nutraceuticals delivery systems. Bioactive compounds can be easily incorporated in nanofibers utilizing the electrospinning method. One of the natural flavored extract from plant is eugenol, which has antibacterial action but seems to have limited use owing to its low stability and degradation via environmental exposure. Thus a nontoxic oligosaccharide, cyclodextrin, is used to encapsulate eugenol. Electrospun polyvinyl alcohol (PVA) nanofibers encapsulating eugenol and cyclodextrin were obtained via electrospinning to improve the thermal stability and slow release of eugenol in the fiber matrix (Kayaci *et al.*, 2013). Table 1.6 shows various bioactive compounds whose innate properties have improved on formulating with nanofibers.

1.6.6.1 Major Constraints in Nanofiber Preparation

The major limitation against the use of nanofiber is the sudden discharge of bioactive composite sooner than they accomplish their target. Also, nanofibers seem to interfere with the normal GIT function, as they are straight-away absorbed prior to attainment at the target site. Furthermore, elevated doses of synthetic surfactants lead to adverse effect too (Leyva *et al.*, 2011).

1.6.7 Nanoemulsion

Nanoemulsions (NE) are nano-sized emulsions; the droplet size lies between 20–200 nm (Donsi *et al.*, 2010). The minute particles of droplet lead to Brownian motion, which dominates gravitational force, giving stability of NE toward gravitational separation. Furthermore, small particle size decreases a variety of nonrepulsive forces, thus linking and clumping globules and contributing toward stability issues. The bioavailability concerning nutraceuticals seems to be enhanced when NE-based delivery system is used on account of the tiny element and high surface-to-volume ratio. NEs are considered as potential formulation for oral deliverance of lipotropic pharmaceutical and nutraceutical agents (Donsi *et al.*, 2010; Abbas *et al.*, 2015). The preparation of NE embraces two methods: high and low energy process. The former method controls the size distribution, while the latter depends on internal phase of the formulation. For the preparation of stable NE, solvent and emulsifiers play a decisive role; solvents that can be used in nutraceutical delivery system comprise hexadecane, sunflower oil, ethyl acetate, olive oil, and paraffin oil, while the commonly used emulsifiers take account of Tweens, Spans and sodium dodecyl sulfate (Silva *et al.*, 2012; McClements, 2013; Khalid *et al.*, 2017). These formulations are mainly prepared for enhancing oral fraction of lipophilic nutraceuticals within the GIT. As mentioned above, in the formulation of NE, small oil droplets are used to encapsulate the lipophilic components. In one study *in vivo* animal feeding model has been developed to observe the effect of lipophilic bioactive compound on emulsion composition and structure on the gastrointestinal fate. Heptadecanoic acid as a replica fatty acid was selected, as it is not the normal constitute of human body, and its increasing concentration in intestine can be measured easily as a function of absorption. The lipophilic compound used was Coenzyme Q10, which plays a significant role as biological antioxidant. The size and composition of droplets of nanoemulsion influence the amount of water insoluble nutraceuticals reaching systemic circulation (Cho *et al.*, 2014). Table 1.7 represents various formulations of nanoemulsion as delivery systems for nutraceuticals.

TABLE 1.7
Delivery of Nutraceuticals through Nanoemulsion and Carbon Nanotubes

Delivery System	Nutraceutical	Method/Emulsifier Employed	Inference	Reference
Nanoemulsion	Pea protein	Water-pea protein-sunflower oil mixture	Bioavailability was enhanced	Donsi *et al.*, 2010
	Capsaicin	Tween 80, Alginate sodium salt, Chitosan	Stability was improved	Choi *et al.*, 2011
	Curcumin	Purity Gum 136 Ultra was used	Bioavailability was enhanced	Abbas *et al.*, 2015
	Canthaxanthin ketocarotenoid	Sunflower oil, Sorbitan monododecanoate, polyoxyethylene, sorbitan monooleate and potassium sorbate	Very low surface tension for the whole system was achieved	Gharibzahedi *et al.*, 2015
Carbon nanotubes	Curcumin	Poly-D-lysine	Hydrophobic bioactive compound was entrapped	Sadeghi *et al.*, 2013
	Tobacco	Chemical vapor deposition	This formulation can regulate cell division	Khodakovskaya *et al.*, 2012
	Quercetin and Rutin	SWCNTs, SWCNT-OH, SWCNT-COOH	This formulation transports low soluble antioxidants to specific target with improved scavenging rate against the free radicals	Nichita and Stamatin, 2013

1.6.7.1 Advantages of Nanoemulsion as Delivery Systems for Nutraceuticals

- NE can be formulated with lipids and emulsifiers, which are of natural origin.
- Bioavailability of lipophilic compounds upon encapsulating with fats gets enhanced.
- Absorption is additionally accelerated, as it does not undergo intraluminal degradation.
- Formulation aspect can be expanded to creams, sprays, liquids, etc. (Choi *et al.*, 2011)

1.6.8 Nanocapsules

Nanocapsules are nanoparticles that have globular void shape, with dimension of about 0.2 μm, entrapping a bioactive compound. These delivery systems are also referred as hollow polymer nanostructures. The distinctiveness of this delivery system laid prominence on specific core and shell (Esmaeili and Ebrahimzadeh, 2015). Interfacial polymerization of monomers and preformed polymer methods are used for formulating nanocapsules.

Polymerization of monomers technique is basically employed for proteins and peptides, but not for bioactive compounds due to the presence of residual solvents that are proven to be toxic. The problem of organic solvent associated with the above technique is surmounted in preformed polymerization, which involves the usage of natural and synthetic polymers. The polymers used for preparation of nanocapsules are selected to improve curative profit and curtail side effects, with some of the important polymers including poly-Ɛ-caprolactone and polysaccharide gums (Santos *et al.*, 2016). Nanocapsules can load drugs, vaccines, diagnostics, and nutraceuticals, which can be administered by topical, oral, pulmonary, or parenteral routes. As bioactive compound is embedded in lipid core, it offers several advantages for delivering nutraceuticals: (a) tissue irritation is reduced at the deposition site; (b) bioactive compound is protected from external medium; and (c) drug encapsulation efficiency is enhanced. In one of the studies, curcumin and olive-oil were encumbered in nanocapsules using human serum albumin shell. These nanocapsules appreciably improved the entrapment efficiency as observed using a photodegradation study (Molina-Bolıivar *et al.*, 2015). Similar reports are enumerated in Table 1.8, wherein nanocapsules have successfully delivered the nutraceuticals with additive advantages.

TABLE 1.8
Delivery of Nutraceuticals through Nanocapsule

Nutraceuticals	Method of Preparation	Inference	Reference
Vitamin E	Nanoprecipitation method using polycaprolactone	This formulation showed good physical and chemical stability	Khayata *et al.*, 2012
Crataegus azarolus	Polymer deposition solvent evaporation method	Enhanced therapeutic action	Esmaeili *et al.*, 2013
Casein	Rapeseed and soy bean oils	Hydrophobic interaction were stabilized	Ghasemi and Abbasi, 2014
Egg albumin and soy protein	Emulsion preparation	Controlled release delivery system was obtained	Gupta and Ghosh *et al.*, 2014
Lycopene	Dextran and chitosan were used as matrix materials	Protection against moisture and thermal degradation	Perez-Masia *et al.*, 2014
Aloe vera	Emulsion and solvent penetration using polyethylene glycol–polybutylene adipate–polyethylene glycol polymer and olive oil	Enhanced medicinal effect was obatained	Esmaeili and Ebrahimzadeh, 2015
α-tocopherol	Emulsification-diffusion method	Enhanced therapeutic action	Perez *et al.*, 2015
Eugenol	Solvent displacement method using polycaprolactone	For enhanced therapeutic activity of eugenol in the treatment of periodontal infections	Pramod *et al.*, 2016

1.6.9 Carbon Nanotubes

This delivery system is comprised entirely of carbon particles, set through a sequence of simplest aromatic structure hooked on a tubiform configuration. Single-walled nanotubes (SWNTs) and multi-walled nanotubes (MWNTs) are the two categories of carbon nanotubes. SWNTs are made of a single sheet of barrel shaped grapheme, whereas MWNTs consist of numerous coordinated graphene sheets. The dimensions of SWNTs are 0.02 × 0.002 μm, while the second category has the dimension of around 1 × 0.05 μm. Carbon nanotubes (CNTs) possess featured properties that include extreme low mass, sequenced arrangement with immense automatic, voltaic, thermal, and facade area. These characteristics of the delivery system have been lately utilized in biomedical fields, too (Khodakovskaya *et al.*, 2012). CNTs have been used for their exceptional better sorption capacity and high stability in extraction procedure. Moreover, CNTs have been found to attain effective extraction scheme, as they have cave-like structure and can interact with a wide range of foreign molecules. CNTs have been utilized as solid-phase adsorbents for mining polyphenols from honey samples. This system demonstrates promising convenience for the scrutiny of honey and was used for improving the strength of polyphenols in honey extracts. Table 1.7 further represents CNTs as the carrier systems, which are quite advantageously developed for delivering nutraceuticals (Smith *et al.*, 2014).

1.7 CONCLUSION

Rather than relying on synthetic moieties for the treatment of catastrophic ailments, the world is revolving around the natural bioactive components providing high nutritional efficacy with negligible toxicity. To prevent spread of diseases and to be healthy, the new food products that have their penetration in body cells are now being continuously explored to improve quality of life. With the advancement of science and passage of time, nanonutraceuticals seem to be a contemporary and futuristic therapy of choice, specifically in field of medicine as well as food supplements. Today nanotechnology is considered a powerful tool in human care in terms of diagnosis, prevention, and treatment of complex diseased states such as cancer, neurodegenerative disorders, and cardiovascular diseases. Nanotechnology-mediated delivery of natural products and food supplements has proven extremely effective because nanoparticles rarely pose any toxicity to normal cells. Moreover, being biodegradable, these carrier systems offer full biocompatibility with the human body and provide a spectacular route for nutraceutical delivery. Nanotechnology has the potential to improve quality of foods by making them tastier, healthier, and more nutritious. Many nutraceuticals developments are currently in an elementary stage, yet with time and scientific know-how, these systems will unravel their wonderful promise for generating high value nutraceuticals.

1.8 FUTURE PROSPECTS

Nutraceuticals are vital and an essential requirement of human body, and in absence or undernourished supply, the body may suffer from various types of illness. Nutraceuticals play an important role in nourishing human health and improving quality of life. Advancements in utilization of nanotechnology have enormous potential to achieve higher concentrations of nutraceuticals necessary for efficacy against various diseases. Although South Asian nations currently hold only a burgeoning space among the market segment for nutraceuticals and food supplements, India is anticipated to be one of the biggest emerging markets of nutraceuticals in the future across the globe. Globalization of the nutraceutical and functional food industries presents significant challenges to stakeholders and researchers, one of which is the regulatory variance between countries in the marketplace. Notwithstanding the current trends and future challenges with nutraceuticals, these are slated to provide an equivalent alternative to medicine with phenomenal impact on improving the quality of life.

REFERENCES

Aadinath, W., Bhushani, A., and Anandharamakrishnan, C. Synergistic radical scavenging potency of curcumin-in-β-cyclodextrinin-nanomagnetoliposomes. *Mater Sci Eng C.* 2016; 64: 293–302.

Abaee, A., and Madadlou, A. Niosome-loaded cold-set whey protein hydrogels. *Food Chem.* 2016; 196: 106–113.

Abbas, S., Karangwa, E., Bashari, M., Hayat, K., Hong, X., Sharif, H. R. *et al.* Fabrication of polymeric nanocapsules from curcumin loaded nanoemulsion templates by self-assembly. *Ultrason Sonochem.* 2015; 23: 81–92.

Aditya, N. P., Espinosa, Y. G., and Norton, I. T. Encapsulation systems for the delivery of hydrophilic nutraceuticals: Food application. *Biotechnol Adv.* 2017; 1–22.

Aditya, N. P., Macedo, A. S., Doktorovova, S., Souto, E. B., Kim, S. H., Chang, P. S. *et al.* Development and evaluation of lipid nanocarriers for quercetin delivery: A comparative study of solid lipid nanoparticles (SLN), nanostructured lipid carriers (NLC), and lipid nanoemulsions (LNE). *Food Sci Technol.* 2014; 59: 115–121.

Akbari, A., and Wu, J. Cruciferin nanoparticles: Preparation, characterization and their potential application in delivery of bioactive compounds. *Food Hydrocoll.* 2016; 54: 107–118.

Aklakur, M., Rather, A., and Kumar, N. Nano delivery: An emerging avenue for nutraceuticals and drug delivery. *Crit Rev Food Sci.* 2016; 56: 2352–2361.

Aswathy, R. G., Sivakumar, B., Maekawa, T., and Kumar, D. S. Prospects of nano-nutraceuticals for better and healthier future. *J Nut Food Sci.* 2015; 5: 38.

Avci, H., Monticello, R., and Kotek, R. Preparation of antibacterial PVA and PEO nanofibers containing *Lawsonia inermis* (henna) leaf extracts. *J Biomater Sci Polym.* 2012; 24: 1815–1830.

Bagheri, L., Madadlou, A., Yarmand, M., and Mousavi, M. E. Spray-dried alginate microparticles carrying caffeine-loaded and potentially bioactive nanoparticles. *Food Res Int.* 2014; 62: 1113–1119.

Bennett, L. L., Rojas, S., and Seefeldt, T. Role of antioxidants in the prevention of cancer. *J Exp Clin Med.* 2012; 4: 215–222.

Britto, D., Moura, M. R., Aouada, F. A., and Mattoso, L. H. C. N,N,N-trimethyl chitosan nanoparticles as a vitamin carrier system. *Food Hydrocoll.* 2012; 27: 487–493.

Campos, D. A., Madureira, A. R., Gomes, A. M., Sarmento, B., and Pintado, M. M. Optimization of the production of solid Witepsol nanoparticles loaded with rosmarinic acid. *Colloids Surf B Biointerfaces.* 2014; 115: 109–117.

Chen, J., Dai, W. T., He, Z. M., Gao, L., Huang, X., Gong, J. M. *et al.* Fabrication and evaluation of curcumin-loaded nanoparticles based on solid lipid as a new type of colloidal drug delivery system. *Indian J Pharm Sci.* 2013; 75: 178–184.

Cheng, S., Panthapulakkal, S., Sain, M., and Asiri, A. Aloe vera rind cellulose nanofibers-reinforced films. *J Appl Poly Sci.* 2014; 40592: 1–9.

Cho, H. T., Trujillo, S. L., Kim, J., Park, Y., Xiao, H., McClements, D. J. Droplet size and composition of nutraceutical nanoemulsions influences bioavailability of long chain fatty acids and Coenzyme Q10. *Food Chem.* 2014; 156: 117–122.

Choi, A. J., Kim, C. J., Cho, Y. J., Hwang, J. K., and Kim, C. T. Characterization of capsaicin-loaded nanoemulsions stabilized with alginate and chitosan by self-assembly. *Food Bioprocess Technol.* 2011; 4: 1119–1126.

Crommelin, D. J. A., and Florence, A. T. Towards more effective advanced drug delivery systems. *Int J Pharm.* 2013; 454: 496–511.

Donsi, F., Senatore, B., Huang, Q., and Ferrari, G. Development of novel pea protein-based nanoemulsions for delivery of nutraceuticals. *J Agric Food Chem.* 2010; 58: 10653–10660.

Esmaeili, A., and Ebrahimzadeh, M. Polymer-based of extract-loaded nanocapsules Aloe vera L. selivery. *Synth React Inorg M.* 2015; 45: 40–47.

Esmaeili, A., Rahnamoun, S., and Sharifnia, F. Effect of O/W process parameters on Crataegus azarolus L nanocapsule properties. *J Nanobiotechnol.* 2013; 11: 16–25.

Esmaeilzadeh, P., Fakhroueian, Z., and Beigi, A. A. M. Synthesis of biopolymeric α-lactalbumin protein nanoparticles and nanospheres as green nanofluids using in drug delivery and food technology. *J Nano Res.* 2011; 16: 89–96.

Fathi, M., Mirlohi, M., Varshosaz, J., and Madani, G. Novel caffeic acid nanocarrier: Production, characterization, and release modeling. *J Nanomater.* 2013; 1–9.

Fraile, R., Geanta, R. M., Escudero, I., Benito, J. M., and Ruiz, M. O. Formulation of Span 80 niosomes modified with SDS for lactic acid entrapment. *Desalin Water Treat.* 2015; 56: 3463–3475.

Fung, W. Y., Yuen, K. H., and Liong, M. T. Agrowaste based nanofibers as a probiotic encapsulant: Fabrication and characterization. *J Agric Food Chem.* 2011; 59: 8140–8147.

Garcia-Moreno, P. J., Stephansen, K., Kruijs, J. V., Guadix, A., Guadix, E. M., Chronakis, I. S. *et al.* Encapsulation of fish oil in nanofibers by emulsion electrospinning: Physical characterization and oxidative stability. *J Food Eng.* 2016; 183: 39–49.

Ghaderi, S., Ghanbarzadeh, S., and Hamishehkar, H. Evaluation of different methods for preparing nanoparticle containing gammaoryzanol for potential use in food fortification. *Pharm Sci.* 2015; 20: 130–134.

Gharibzahedi, S. M. T., Razavi, S. H., and Mousavi, M. Optimal development of a new stable nutraceutical nanoemulsion based on the inclusion complex of 2-hydroxypropyl-β-cyclodextrin with canthaxanthin accumulated by dietzia natronolimnaea hs-1 using ultrasound-assisted emulsification. *J Dispersion Sci Technol.* 2015; 36: 614–625.

Ghasemi, S., and Abbasi, S. Formation of natural casein micelle nanocapsule by means of pH changes and ultrasound. *Food Hydrocoll.* 2014; 42: 42–77.

Giroux, H. J., and Britten, M. Encapsulation of hydrophobic aroma in whey protein nanoparticles. *J Microencapsul.* 2011; 28: 337–343.

Gulseren, I., and Corredig, M. Storage stability and physical characteristics of tea-polyphenol-bearing nanoliposomes prepared with milk fat globule membrane phospholipids. *J Agric Food Chem.* 2013; 61: 3242–3251.

Gupta, S. C., Kim, J. H., Prasad, S., and Aggarwal, B. B. Regulation of survival, proliferation, invasion, angiogenesis, and metastasis of tumor cells through modulation of inflammatory pathways by nutraceuticals. *Cancer Metastasis Rev.* 2010; 29: 405–434.

Gupta, S. S., and Ghosh, M. Preparation and characterisation of protein based nanocapsules of bioactive lipids. *J Food Eng.* 2014; 121: 64–72.

Ha, H. K., Kim, J. W., Lee, M. R., Jun, W., and Lee, W. J. Cellular uptake and cytotoxicity of β-lactoglobulin nanoparticles: The effects of particle size and surface charge. *J Anim Sci.* 2015; 28: 420–427.

Hu, B., Ting, Y., Yang, X., Tang, W., Zeng, X., and Huang, Q. Nanochemoprevention by encapsulation of epigallocatechin-3-gallate with bioactive peptides/chitosan nanoparticles for enhancement of its bioavailability. *Chem Commun.* 2012; 48: 2421–2423.

Hu, B., Ting, Y., Zeng, X., and Huang, Q. Bioactive peptides/chitosan nanoparticles enhance cellular antioxidant activity of (–)-epigallocatechin-3-gallate. *J Agric Food Chem.* 2013; 61: 875–881.

Huang, Q., Yu, H., and Ru, Q. Bioavailability and delivery of nutraceuticals using nanotechnology. *J Food Sci.* 2010; 75: R50–57.

Jain, A., Kesharwani, P., Garg, N. K., Jain, A., Nirbhavane, P., Ghanghoria, R., Banerjee, S., Iyer, A. K., and Mohd Amin, M. C. I. Nano-constructed carriers loaded with antioxidant: Boon for cardiovascular system. *Curr Pharm Des.* 2015; 21: 4456–4464.

Jain, A., Sharma, G., Kushwah, V., Garg, N. K., Kesharwani, P., Ghoshal, G., Singh, B., Shivhare, U. S., Jain, S., and Katare, O. P. Methotrexate and beta-carotene loaded-lipid polymer hybrid nanoparticles: A preclinical study for breast cancer. *Nanomedicine.* 2017b; 12: 1851–1872.

Jain, A., Sharma, G., Kushwah, V., Thakur, K., Ghoshal, G., Singh, B., Jain, S., Shivhare, U. S., and Katare, O. P. Fabrication and functional attributes of lipidic nanoconstructs of lycopene: An innovative endeavour for enhanced cytotoxicity in MCF-7 breast cancer cells. *Colloids Surf B Biointerfaces.* 2017a; 152: 482–491.

Jain, A., Thakur, D., Ghoshal, G., Katare, O. P., Singh, B., and Shivhare, U. S. Formation and functional attributes of electrostatic complexes involving casein and anionic polysaccharides: An approach to enhance oral absorption of lycopene in rats *in vivo*. *Int J Biol Macromol.* 2016; 93: 746–756.

Jenning, V., and Gohla, S. H. Encapsulation of retinoids in solid lipid nanoparticles (SLN). *J Microencapsul.* 2001; 18: 149–158.

Junyaprasert, V. B., Singhsa, P., Suksiriworapong, J., and Chantasart, D. Physicochemical properties and skin permeation of Span 60/Tween 60 niosomes of ellagic acid. *Int J Pharm.* 2012; 423: 303–311.

Kannappan, R., Gupta, S. C., Kim, J. H., Reuter, S., and Aggarwal, B. B. Neuroprotection by spice-derived nutraceuticals: You are what you eat. *Mol Neurobiol.* 2011; 44: 142–159.

Kayaci, F., Aytac, Z., and Uyar, T. Surface modification of electrospun polyester nanofibers with cyclodextrin polymer for the removal of phenanthrene from aqueous solution. *J Harz Mat.* 2013; 261: 289–294.

Khalid, N., Shu, G., Kobayashi, I., Nakajima, M., and Barrow, J. C. Formulation and characterization of monodisperse O/W emulsions encapsulating astaxanthin extracts using microchannel emulsification: Insights of formulation and stability evaluation. *Colloids Surf B Biointerfaces.* 2017; 355–365.

Khatibi, A. S., Misaghi, A., Moosavy, M. H., Amoabediny, G., and Basti, A. A. Effect of preparation methods on the properties of Zataria multiflora boiss. essential oil loaded nanoliposomes: Characterization of size, encapsulation efficiency and stability. *Pharm Sci.* 2015; 20: 141–148.

Khayata, N., Abdelwahed, W., Chehna, M. F., Charcosset, C., and Fessi, H. Stability study and lyophilization of vitamin E-loaded nanocapsules prepared by membrane contactor. *Int J Pharm.* 2012; 439: 254–259.
Khodakovskaya, M. V., Silva, K., Biris, A. S., Dervishi, E., and Villagarcia, H. Carbon nanotubes induce growth enhancement of tobacco cells. *J Am Chem Soc.* 2012; 6: 2128–2135.
Kumari, A., Yadav, S. K., Pakade, Y. B., Kumar, V., Singh, B., Chaudhary, A. *et al.* Development of biodegradable nanoparticles for delivery of quercetin. *Colloids Surf B Biointerfaces.* 2010; 80: 184–192.
Kumari, A., Yadav, S. K., Pakade, Y. B., Kumar, V., Singh, B., Chaudhary, A. *et al.* Nanoencapsulation and characterization of Albizia chinensis isolated antioxidant quercitrin on PLA nanoparticles. *Colloids Surf B Biointerfaces.* 2011; 82: 224–232.
Lee, J. S., Kim, G. H., and Lee, H. G. Characteristics and antioxidant activity of Elsholtzia splendens extract-loaded nanoparticles. *J Agric Food Chem.* 2010; 58: 3316–3321.
Leyva, B. M., Felix, F. R., Chavez, P. T., Wong, B. R., Cervantes, L. J., and Machado, D. S. Preparation and characterization of Durum Wheat (Triticum durum) straw cellulose nanofibers by electrospinning. *J Agric Food Chem.* 2011; 59: 870–875.
Li, Q., Liu, C. G., Huang, Z. H., and Xue, F. F. Preparation and characterization of nanoparticles based on hydrophobic alginate derivative as carriers for sustained release of vitamin D3. *J Agric Food Chem.* 2011; 59: 1962–1967.
Li, Y., Zheng, J., Xiao, H., and McClements, D. J. Nanoemulsion-based delivery systems for poorly water-soluble bioactive compounds: Influence of formulation parameters on polymethoxyflavone crystallization. *Food Hydrocoll.* 2012; 27: 517–528.
Lu, Q., Lu, P. M., Piao, J. H., Xu, X. L., Chen, J., Zhu, L. *et al.* Preparation and physicochemical characteristics of an allicin nanoliposome and its release behavior. *LWT - Food Sci Technol.* 2014; 57: 686–695.
Luo, Y., Teng, Z., and Wang, Q. Development of zein nanoparticles coated with carboxymethyl chitosan for encapsulation and controlled release of vitamin D3. *J Agric Food Chem.* 2012; 60: 836–843.
Luo, Y., Wang, T. T. Y., Teng, Z., Chen, P., Sun, J., and Wang, Q. Encapsulation of indole-3-carbinol and 3, 3′-diindolylmethane in zein/carboxymethyl chitosan nanoparticles with controlled release property and improved stability. *Food Chem.* 2013; 139: 224–230.
Luo, Y., Zhang, B., Cheng, W. H., and Wang, Q. Preparation, characterization and evaluation of selenite-loaded chitosan/TPP nanoparticles with or without zein coating. *Carbohydr Polym.* 2010; 82: 942–951.
Ma, J., Guan, R., Chen, X., Wang, Y., Hao, Y., Ye, X., and Liu, M. Response surface methodology for the optimization of beta-lactoglobulin nano-liposomes. *Food Funct.* 2014; 5: 748–754.
Madadlou, A., Iacopino, D., Sheehan, D., Djomeh, Z. E., and Mousavi, M. E. Enhanced thermal and ultrasonic stability of a fungal protease encapsulated within biomimetically generated silicate nanospheres. *Biochimica et Biophysica Acta.* 2010; 1800: 459–465.
Madureira, A. R., Campos, D. A., Fonte, P., Nunes, S., Reis, F., Gomes, A. M., Sarmento, B., and Pintado, M. M. Characterization of solid lipid nanoparticles produced with carnauba wax for rosmarinic acid oral delivery. *RSC Adv.* 2015; 5: 22665–22673.
Maya, A. I. J., and McClements, D. J. Biopolymer nanoparticles as potential delivery systems for anthocyanins: Fabrication and properties. *Food Res Int.* 2015; 69: 1–8.
McClements, D. H. Nanoemulsion-based oral delivery systems for lipophilic bioactive components: Nutraceuticals and pharmaceuticals. *Ther Deliv.* 2013; 4: 841–857.
Mehravar, R., Jahanshahi, M., and Saghatoleslami, N. Production of biological nanoparticles from lactalbumin for drug delivery and food science application. *Afr J Biotechnol.* 2009; 8(24): 6822–6827.
Mitri, K., Shegokar, R., Gohla, S., Anselmi, C., and Muller, R. H. Lipid nanocarriers for dermal delivery of lutein: Preparation, characterization, stability and performance. *Int J Pharm.* 2011; 414: 267–275.
Mohammadi, M., Ghanbarzadeh, B., Mokarram, R. R., Hoseini, M. Y., and Hamishehkar, H. Study of stability, zeta-potential, and steady rheological properties of nanoliposomes containing Vitamin D3. *J Med Counc I R Iran.* 2013; 36: 102–111.
Molina-Bolıivar, J. A., and Galisteo-Gonzaalez, F. Olive-oil nanocapsules stabilized by HSA: Influence of processing variables on particle properties. *J Nanopart Res.* 2015; 17: 391–414.
Mourtas, S., Canovi, M., Zona, C., Aurilia, D., Niarakis, A., Ferla, B. L. *et al.* Curcumin-decorated nanoliposomes with very high affinity for amyloid-b1-42 peptide. *Biomaterials.* 2011; 32: 1635–1645.
Mousavi, S. H., Moallem, S. A., Mehri, S., Shahsavand, S., Nassirli, H., and Nikouei, B. M. Improvement of cytotoxic and apoptogenic properties of crocin in cancer cell lines by its nanoliposomal form. *Pharm Biol.* 2011; 49: 1039–1045.
Myung, Y., Yeom, S., and Han, S. A niosomal bilayer of sorbitan monostearate in complex with flavones: A molecular dynamics simulation study. *J Liposome Res.* 2016; 26: 336–344.

Neves, A. R., Reis, S., and Segundo, M. A. Development and validation of a HPLC method using a monolithic column for quantification of trans-resveratrol in lipid nanoparticles for intestinal permeability studies. *J Agric Food Chem*. 2015; 63: 3114–3120.

Nichita, C., and Stamatin, I. The antioxidant activity of the biohybrides based on carboxylated/hydroxylated carbon nanotubes-flavonoid compounds. *Dig J Nanomater Biostruct*. 2013; 8: 445–455.

Ohgo, K., Zhao, C., Kobayashi, M., and Asakura, T. Preparation of non-woven nanofibers of Bombyx mori silk, Samia Cynthia ricini silk and recombinant hybrid silk with electrospinning method. *Polymer*. 2003; 44: 841–846.

Palombo, P., Fabrizi, G., Ruocco, V., Ruocco, E., Fluh, J., Roberts, R., and Morganti, P. Beneficial long-term effects of combined oral/topical antioxidant treatment with the carotenoids lutein and zeaxanthin on human skin: A double-blind, placebo controlled study. *Skin Pharmacol Physiol*. 2007; 20: 199–210.

Pando, D., Beltran, M., Gerone, I., Matos, M., and Pazos, C. Resveratrol entrapped niosomes as yoghurt additive. *Food Chem*. 2015; 170: 281–287.

Pando, D., Gutierrez, G., Coca, J., and Pazos, C. Preparation and characterization of niosomes containing resveratrol. *J Food Eng*. 2013; 117: 227–234.

Peng, C., Kim, J., and Chauhan, A. Extended delivery of hydrophilic drugs from silicone-hydrogel contact lenses containing Vitamin E diffusion barriers. *Biomaterials*. 2010; 31: 4032–4047.

Perez, M. J. G., Guerrero, Q. D., Silva, M. E., Sandoval, R. S., and Zambrano, A. Z. M. L. The effects of tocopherol nanocapsules/xanthan gum coatings on the preservation of fresh-cut apples: Evaluation of phenol metabolism. *Food Bioprocess Technol*. 2015; 8: 1791–1799.

Perez-Masia, R., Lagaron, J. M., and Lopez-Rubio, A. Morphology and stability of edible lycopene-containing microand nanocapsules produced through electrospraying and spray drying. *Food Bioprocess Technol*. 2014; 8: 459–470.

Pezeshky, A., Ghanbarzadeh, B., Hamishehkar, H., Moghadam, M., and Babazadeh, A. Vitamin A palmitate-bearing nanoliposomes: Preparation and characterization. *Food Biosci*. 2016; 13: 49–55.

Pramod, K., Alex, M. R. A., Singh, M., Dang, S., Ansari, S. H., and Ali, J. Eugenol nanocapsule for enhanced therapeutic activity against periodontal infections. *J Drug Target*. 2016; 24: 24–33.

Quintana, R. R. B., Covarrubias, M. A. V., Wilson, A. A. M., and Mundo, R. R. S. Alpha-lactalbumin hydrolysate spontaneously produces disk-shaped nanoparticles. *Int Dairy J*. 2013; 32: 133–135.

Rentel, C. O., Bouwstra, J. A., Naisbett, B., and Junginger, H. E. Niosomes as a novel peroral vaccine delivery system. *Int J Pharm*. 1999; 186: 161–167.

Reyes, C. N. Q., Jesus, E. R., Caballero, N. E. D., and Mendez, M. A. A. Development and characterization of gelatin nanoparticles loaded with a cocoa-derived polyphenolic extract. *Fruits*. 2014; 69: 481–489.

Rezaei, A., Nasirpour, A., and Fathi, M. Application of cellulosic nanofibers in food science using electrospinning and its potential risk. *Compr Rev Food Sci Food Saf*. 2015; 14: 269–284.

Sadeghi, R., Kalbasi, A., Emam-jomeh, Z., Razavi, S. H., Kokini, J., and Moosavi-Movahedi, A. A. Biocompatible nanotubes as potential carrier for curcumin as a model bioactive compound. *J Nanopart Res*. 2013; 15: 1931.

Sanna, V., Lubinu, G., Madau, P., Pala, N., Nurra, S., Mariani, A., and Sechi, M. Polymeric nanoparticles encapsulating white tea extract for nutraceutical application. *J Agric Food Chem*. 2015; 63: 2026–2032.

Santos, P. P., Flores, S. H., de Oliveira Rios, A., and Chiste, R. C. Biodegradable polymers as wall materials to the synthesis of bioactive compound nanocapsules. *Trends Food Sci Tech*. 2016; 53: 23–33.

Saraf, A. S. Applications of novel drug delivery system for herbal formulations. *Fitoterapia*. 2010; 81: 680–689.

Semyonov, D., Ramon, O., Shoham, Y., and Shimoni, E. Enzymatically synthesized dextran nanoparticles and their use as carriers for nutraceuticals. *Food Funct*. 2014; 5: 2463–2474.

Shaikh, J., Ankola, D. D., Beniwal, V., Singh, D., and Kumar, R. Nanoparticle encapsulation improves oral bioavailability of curcumin by at least 9-fold when compared to curcumin administered with piperine as absorption enhancer. *Eur J Pharm Sci*. 2009: 223–230.

Shin, G. H., Chung, S. K., Kim, T. J., Joung, J. H., and Park, J. H. Preparation of chitosan-coated nanoliposomes for improving the mucoadhesive property of curcumin using the ethanol injection method. *J Agric Food Chem*. 2013; 61: 11119–11126.

Silva, H. D., Cerqueira, M. A., and Vicente, A. A. Nanoemulsions for food applications: Development and characterization. *Food Bioprocess Technol*. 2012; 5: 854–867.

Smith, S. C., Ahmed, F., Gutierrez, K. M., and Rodrigues, D. F. A comparative study of lysozyme adsorption with graphene, graphene oxide, and single-walled carbon nanotubes: Potential environmental applications. *Chem Eng J*. 2014; 240: 147–154.

Soltani, S., and Madadlou, A. Gelation characteristics of the sugar beet pectin solution charged with fish oil-loaded zein nanoparticles. *Food Hydrocoll*. 2014; 43: 664–669.

Sun, F., Ju, C., Chen, J., Liu, S., Liu, N., Wang, K., and Liu, C. Nanoparticles based on hydrophobic alginate derivative as nutraceutical delivery vehicle: Vitamin D3 loading. *Artif Cells Blood Substit Immobil Biotechnol.* 2012; 40: 113–119.

Tan, Y., Xu, K., Liu, C., Li, Y., Lu, C., and Wang, P. Fabrication of starch-based nanospheres to stabilize pickering emulsion. *Carbohydr Polym.* 2012; 88: 1358–1363.

Teng, Z., Li, Y., Luo, Y., Zhang, B., and Wang, Q. Cationic β-Lactoglobulin nanoparticles as a bioavailability enhancer: Protein characterization and particle formation. *Biomacromol.* 2013; 14: 2848–2856.

Teng, Z., Li, Y., and Wang, Q. Insight into curcumin-loaded -Lactoglobulin nanoparticles: Incorporation, particle disintegration, and releasing profiles. *J Agric Food Chem.* 2013; 62(35): 8837–8847.

Teng, Z., Luo, Y., and Wang, Q. Carboxymethyl chitosan–soy protein complex nanoparticles for the encapsulation and controlled release of vitamin D3. *Food Chem.* 2013; 141: 524–532.

Teng, Z., Luo, Z., Wanh, T., Zhang, B., and Wang, Q. Development and application of nanoparticles synthesized with folic acid-conjugated soy protein. *J Agric Food Chem.* 2013; 61: 2556–2564.

Thakur, D., Jain, A., Ghoshal, G., Shivhare, U. S., and Katare, O. P. Microencapsulation of β-carotene based on casein/guar gum blend using zeta potential-yield stress phenomenon: An approach to enhance photostability and retention of functionality. *AAPS PharmSciTech.* 2017; 18: 1447–1459.

Ting, Y., Jiang, Y., Ho, C. T., and Huang, Q. Common delivery systems for enhancing *in vivo* bioavailability and biological efficacy of Nutraceuticals. *J Funct Foods.* 2014; 7: 112–128.

Tsai, Y. M., Chien, C. F., Lin, L. C., and Tsai, T. H. Curcumin and its nano-formulation: The kinetics of tissue distribution and blood–brain barrier penetration. *Int J Pharma.* 2011; 416: 331–338.

Vilela, D., Gonzalez, M. C., and Escarpa, A. Gold-nanosphere formation using food sample endogenous polyphenols for *in-vitro* assessment of antioxidant capacity. *Anal Bioanal Chem.* 2012; 404: 341–349.

Wijesinghe, W. A. J. P., and Jeon, Y. Biological activities and potential industrial applications of fucose rich sulfated polysaccharides and fucoidans from brown seaweeds: A review. *Carbohyd Poly.* 2012; 88: 13–20.

Xia, S., Tan, C., Xue, J., Lou, X., Zhang, X., and Feng, B. Chitosan/tripolyphosphate-nanoliposomes core-shell nanocomplexes as vitamin E carriers: Shelf-life and thermal properties. *Int J Food Sci Technol.* 2014; 49: 1367–1374.

Yadav, H. K. S., Nagavarma, B. V. N., Ayaz, A., Vasudha, L. S., and Shivakumar, H. G. Different techniques for preparation of polymeric nanoparticles—A review. *Asian J Pharm Clin Res.* 2012; 5: 16–23.

Yang, S., Liu, C., Liu, W., Yu, H., Zheng, H., Zhou, W., and Hu, Y. Preparation and characterization of nanoliposomes entrapping medium-chain fatty acids and vitamin C by lyophilization. *Int J Mol Sci.* 2013; 14: 19763–19773.

Zompero, R. H. F., Lopez-Rubio, A., Pinho, S. C., Lagaron, J. M., and Torre, L. G. Hybrid encapsulation structures based on rmbeta-carotene-loaded nanoliposomes within electrospun fibers. *Colloids Surf B Biointerfaces.* 2015; 134: 475–482.

2 Assigning Nanocouture to Phytochemical Nutraceuticals for Improved Biopharmaceutical Performance

Indu Pal Kaur, Vandita Kakkar, Parneet Kaur Deol, and Simarjot Kaur Sandhu*

CONTENTS

2.1 INTRODUCTION

Nutraceutical, a term coined in 1989 by Stephen L. DeFelice, founder and chair of the Foundation of Innovation Medicine, is a portmanteau of the words *nutrition* and *pharmaceutical* (Kalra, 2003). It is applied to the products ranging from isolated nutrients and dietary fibers to antioxidants and phytochemicals and herbal products (Das *et al.*, 2012). The global nutraceuticals market is expected to reach $278.96 billion by the end of 2021 from a valuation of $165.62 billion in 2014. It has been estimated that about two-third of the drugs presently in the market have their origin in natural molecules, with more than 50% share being held by nutraceuticals alone. In the period of 2005 to 2010, a total of 19 natural product-based drugs were approved for marketing worldwide, which included 7 natural products, 10 semi-synthetic natural products, and 2 natural product-derived drugs. Veregen® (Polyphenon E ointment), a defined mixture of catechins obtained from green tea, was the first herbal medicine to receive FDA approval in 2006 for topical use against genital warts (Farnsworth *et al.*, 1985; Misra *et al.*, 2011).

Phytochemicals, the chemical compounds that occur naturally in plants (phyto means "plant" in Greek), at times also find their use as nutraceuticals. Phytochemicals (e.g., carotenoids or flavonoids) may have biological significance but are not established as essential nutrients. They elicit

* Corresponding author.

either defensive or disease-protective properties, with the latter phenomenon specifically referred to as "preconditioning" or "hormesis" (Son *et al.*, 2008), which is a biphasic dose response of an agent to induce biologically opposite effects at different doses (Mattson, 2008), that is, stimulatory or beneficial effect at low doses and inhibitory or toxic effect at higher doses (Calabrese *et al.*, 2012). It is indicated that plants produce phytochemicals to dissuade insects and pests from eating them (Isman, 2006), and thus protect themselves. These "biopesticides," when ingested by humans in similar subtoxic doses, as present in the plants, induce mild cellular stress response, thereby promoting health and supporting conventional therapy for disease (Calabrese *et al.*, 2012). The majority of foods, such as whole grains, beans, fruits, vegetables, and herbs, contain phytochemicals or phytonutrients. These phytochemicals, either alone and/or in combination, have tremendous therapeutic potential in curing various ailments such as cancers and cardiovascular disorders or neurodegenerative diseases. Another interesting aspect of phytochemical action involves pluri-pharmacology, a multi-targeting ability of these molecules. The current paradigm for treatment of most of the aforementioned pathologies is to modulate the multiple targets that are implicated in these diseases. Such multi-target therapies were earlier considered undesirable as they are not specific and so were referred to as "dirty drugs" (Li *et al.*, 2014; Zheng *et al.*, 2014). Clinical experience with mono-targeted therapies for the control of complex disease states such as cancers and neurodegeneration indicates them to be costly and lacking the desired safety and efficacy (Csermely *et al.*, 2005; Boran *et al.*, 2010;). Many plant-based products, apart from accomplishing multi-targeting naturally, are inexpensive and possess a broader safety margin in comparison to their synthetic counterparts. The recent success of a number of such phytoconstituents has led many researchers to investigate the rationale and potential of these unspecific agents (Geldenhuys *et al.*, 2013).

Though the preventive role of phytochemicals was realized several decades earlier, exploration of their full potential and assignment of a "therapeutic status" to these phytochemicals is still awaited. This discrepancy is mainly attributed to the poor physicochemical profile of these agents. Most phytochemicals display poor solubility and/or permeability, rapid metabolism, poor gastrointestinal stability, instability at physiological pH, and active efflux in the gastrointestinal tract, resulting in a compromised oral bioavailability with erratic absorption profiles. High doses of phytochemicals, which are unsuitable for human consumption, are thus invariably required for elicitation of physiological effects. All these factors compound the problem of clinical translation of phytochemicals (Liu *et al.*, 2015).

Advent of nanotechnology has, however, opened new avenues for improving stability, solubility, and/or permeability of these challenging phytochemicals. "Nano" means dwarf in Greek, and nanotechnology is the science of creation and utilization of materials, devices, and systems through the control of matter on the nanometer length scale. Though nanotechnology was sought to be the province of basic and engineering sciences, it is now capturing the world of nutraceuticals including phytochemicals. Without changing the chemical structure of the nutrient, nanoprocessing successfully improves the way it acts.

Since the first featured article in *Food Nanotechnology* published in 2003, interest in this exciting research area has been growing tremendously. More featured articles have now been published to address the issues of food safety/security, functional foods for health promotion applications, nanoscale properties of food materials (Huang *et al.*, 2010), and structural design principles for the delivery of bioactive components (McClements *et al.*, 2009).

Several forms of nano-based drug delivery systems have been explored to enhance absorption of phytochemicals, delay their metabolic degradation, and prolong their circulation time. These refer to either the nanosizing of bioactives, per se, for improving their systemic availability, solubility, and their ability to be transferred across intestinal membrane into the blood (Shegokar *et al.*, 2010; Weiss *et al.*, 2006), or includes encapsulation of phytochemicals within nanocarriers to improve their bioprofile in general (Chen *et al.*, 2012). Polymeric nanoparticles, solid lipid nanoparticles, magnetic nanoparticles, metal and inorganic nanoparticles, quantum dots, polymeric micelles, phospholipid micelles, colloidal nano-liposomes, dendrimers, inorganic, and hybrid nanocarriers prepared using techniques such as high pressure homogenization, complex coacervation, co-precipitation, salting out, solvent displacement, solvent emulsification diffusion, supercritical fluid method, and self-assembly method (Gunasekaran *et al.*, 2014; Liu *et al.*, 2015) have been extensively studied and reported as successful systems for improving the phytochemical bio-performance. Nanotechnology has also been used to improve the stability of phytochemicals during processing, storage, and distribution.

Commercial success in this area can be clearly witnessed by the success story of Abraxane®, an approved FDA drug, and a solvent-free nano-version of the natural alkaloid, Taxol. It is more effective as well as less toxic and has been successful in addressing the solubility problem associated with Paclitaxel (Taxol). Herbasec® is a patented technology comprising liposomes-encapsulated standardized botanical extracts used in cosmetics for their antioxidant effects to prevent aging.

The existing list of patents on nano-tailored (nano-couture) phytochemical-based formulations also sets a stage for re-exploring and investigating therapeutic potential of phytochemicals for disease alleviation. This chapter endeavors to signify the curative potential of nano-coutured nutraceuticals supplemented with case studies on two phytochemicals: curcumin and sesamol (Table 2.1).

Curcumin is a widely researched molecule with thousands of research articles being published every year claiming its beneficial effects, though it is yet to gain the status of a "drug." This limitation is attributed to its highly compromised bioavailability. Sesamol, on the other hand, is a small and unique (with both aqueous and lipid solubility) antioxidant molecule minimally explored for its biological effects.

The basis for presenting these two molecules is our extensive expertise in assigning them with a pharmaceutical couture to result in nanoformulations manifesting significantly better biopharmaceutical profile, including therapeutic effects. We have been working with both of these agents for the last 15 years and have successfully assigned them a wide variety of curative activities, subsequent to their incorporation into solid lipid nanoparticulate (SLNs) systems. SLNs are submicron colloidal carrier systems composed of a physiological high melting point lipid as a solid core coated by and dispersed in an aqueous surfactant solution. They offer unique properties such as small size, large surface area, and high drug loading, all of which sum up to improve performance of encapsulated materials. Stability and safety are the important attributes that make SLNs a promising alternative to liposomes and polymeric nanoparticles, respectively.

Role of dietary antioxidants, in preclinical studies, is usually indicated as a pretreatment (protective) therapy. The latter does not have any clinical significance in the sense that administration of the agent is indicated prior to the disease occurrence, which is impractical and difficult to impress upon the general public. This chapter is intended to describe the efforts by researchers (including our own lab contribution) to "dress"/"couture" these phytochemicals into effective therapies using suitable nano-carriers. Figure 2.1 outlines an overview of the delivery systems attempted thus far with these two agents; only a few studies have been undertaken to couture sesamol except those from our lab.

TABLE 2.1
Source, Chemistry, and Physiochemical Properties of Curcumin and Sesamol

Source	Composition/Chemical Name	Structure and Molecular Weight	Log P	Solubility	Bioavailability
Rhizome of the herb *Curcuma longa*, family Zingiberaceae	Mixture of curcuminoids, containing approximately 94% diferuloylmethane, 6% demethoxycurcumin, 0.3% bisdemethoxycurcumin	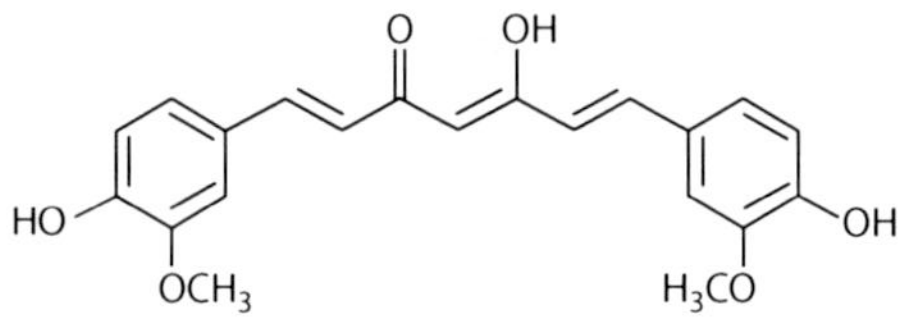 **Curcumin** Molecular Weight: 368.38 g/mole	~3.0	Insoluble in water at acidic and neutral pH; soluble in ethanol and acetic acid	≤1% (Anand *et al.*, 2007)
Roasted seeds of *Sesamum indicum*, family Pedaliaceae	3,4-(Methylenedioxy)phenol –	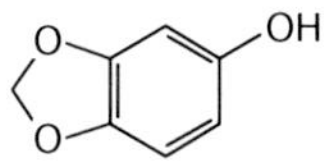 **Sesamol** Molecular Weight: 138.34 g/mole	1.29	38.8 ± 1.2 mg/mL at 25°C in water	35.5 ± 8.5% (Geetha *et al.*, 2015a)

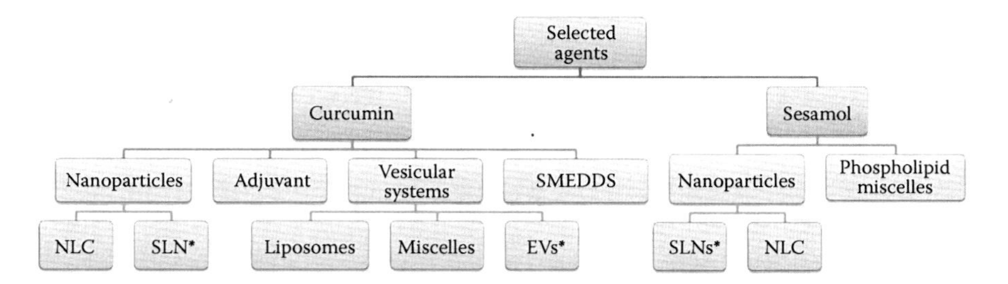

FIGURE 2.1 Delivery systems tried with the selected agents till date. NLC: Nanolipidic carriers, SLNs: Solid lipid nanoparticles, EVs: Elastic vesicles, SMEDDs: Self microemulsified drug delivery systems. * Systems reported by our lab.

2.2 CURCUMIN

Curcumin, referred to as "yellow gold," has gathered much attention due to its pleiotropic effects and more than 90 clinical studies (ongoing and recruiting) investigating its effects in a wide variety of human disorders such as Alzheimer's disease, asthma, irritable bowel syndrome, cancers, osteoarthritis, rheumatoid arthritis, depression, ulcerative colitis, osteosarcoma, proteinuric chronic kidney disease, and cystic fibrosis. Curcumin is the principal curcuminoid of the popular Indian spice turmeric and comprises of curcumin (94%), demethoxycurcumin (6%), and bisdemethoxycurcumin (0.3%) (Figure 2.2).

Curcumin exhibits potent antioxidant, anti inflammatory (Yu *et al.*, 2016), antimicrobial (Liu *et al.*, 2016), hepatoprotective (Singh *et al.*, 2014a), thrombosuppressive (Srivastava *et al.*, 1985), cardiovascular (Xiao *et al.*, 2016), hypoglycemic (Arun *et al.*, 2002), wound healing (Lina *et al.*, 2016), anti-fibrotic (Rivera-Espinoza *et al.*, 2009), anti-arthritic (Deodhar *et al.*, 1980), anti-tumoral (Aggarwal *et al.*, 2004), apoptosis-inducing (Kakkar *et al.*, 2012a), and anti-angiogenic properties (Shunsuke *et al.*, 2016). Its antioxidant activity is comparable to that of vitamin C and is 10 times higher than that of vitamin E (Motterlini *et al.*, 2000). The antioxidant properties of curcumin reside in its chemical structure, which comprises a phenolic group and a diketonic moiety in the same molecule (Dulbecco *et al.*, 2013). The hydroxyphenyl unit in curcumin has been reported to assign it an anti-inflammatory activity (Anand *et al.*, 2008). The anti-inflammatory effect of curcumin is partly mediated through inhibition of Iκβ kinase activity leading to suppression of NF-κβ activation. It is also reported to inhibit the expression of several NF-κβ dependent inflammatory chemokines and cytokines, such as IL-6, IL-2, and inflammation-promoting enzymes such as COX-2 and iNOS in Kupffer cells. Its anti-fibrotic activity due to the inhibitory effect on TGF-β is also known. This profibrinogenic cytokine plays a key role in promoting the activation of stellate cells to myofibroblasts through the production of extracellular matrix (Yao *et al.*, 2012). TGF-β is one of the main targets of curcumin, and its inhibition is likely to occur via suppression of NF-κβ (Maheswari *et al.*, 2006). A salient feature of turmeric/curcumin is that despite being consumed daily for centuries in Asian countries, no toxicity has been related to its consumption (Ammon *et al.*, 1991).

However, inspite of the promising multivariate activities of curcumin, its poor aqueous solubility, poor stability at physiological pH and alkaline pH, photolability, as well as its rapid metabolism and systemic elimination, limit its clinical application. Low oral bioavailability of curcumin is clear from the fact that a maximum serum concentration of only 0.36 ± 0.05 μg/mL was achieved in rats after an intravenous injection of 10 mg/kg. Similarly, 500 mg/kg oral dose in rats showed a maximum plasma concentration of 0.06 ± 0.01 μg/mL (Yang *et al.*, 2007) and 2 g/kg dose showed a maximum serum concentration of 1.35 ± 0.23 μg/mL at 1 h. Healthy human volunteers (weighing 50–75 kg) receiving a single dose of 2 g/kg curcumin (4 capsules of 500 mg each), on the other hand, showed an even lower serum concentration of 0.006 ± 0.005 μg/mL at 1 h (Shoba *et al.*, 1998). Garcea *et al.* (2004) reported that patients taking 3.6 g of curcumin a day (as a standard powder extract capsule supplied by Sabinsa Corporation) achieved negligible blood and liver levels.

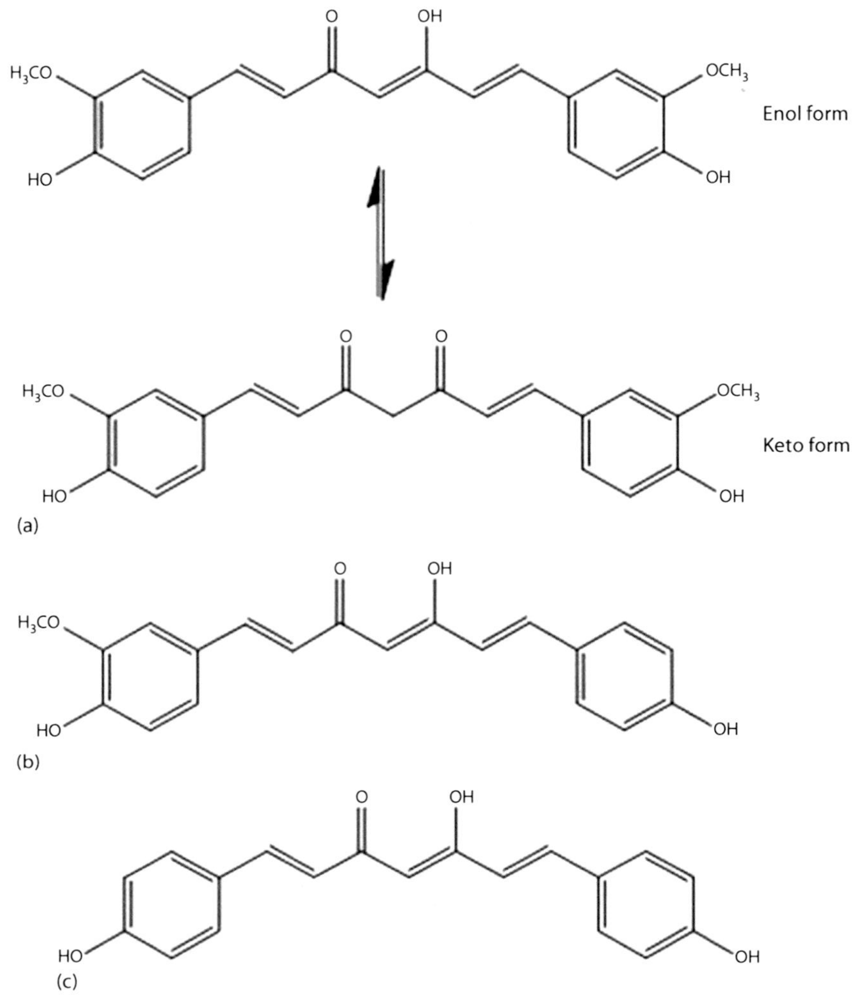

FIGURE 2.2 Chemical structure of (a) curcumin, (b) demethoxycurcumin, (c) bisdemethoxycurcumin.

Based on these observations, it is suggested that a person should consume large doses (about 12–20 g/day) of curcumin to achieve its therapeutic effects on the human body. That means one has to swallow 24 to 40 curcumin capsules (500 mg each) per day. These doses are high and practically impossible to be incorporated in clinical trials. Unbearable aftertaste to the palate, possibility of nauseatic feeling, GIT distress and perceived toxicity issues are the other reasons for not using such large doses (Garcea *et al.*, 2004).

As a result, numerous efforts have been made to improve bioavailability of curcumin. Use of adjuvants that can block the metabolic pathway of curcumin is the most common strategy for increasing the bioavailability of curcumin. The effect of combining piperine, a known inhibitor of hepatic and intestinal glucuronidation, was evaluated on the bioavailability of curcumin in healthy human volunteers. In humans receiving a dose of 2 g of curcumin alone, serum levels of curcumin were either undetectable or low. Concomitant administration of 20 mg of piperine with curcumin, however, produced much higher concentrations within 30 min to 1 h after drug treatment, as piperine increased the bioavailability of curcumin by 2,000% (Shoba *et al.*, 1998). However, inhibition of the glucuronidation pathway can have serious implications and health risks. Glucuronidation is a protective mechanism involved in the metabolism of commonly used drugs that helps to wash out several toxins from the body. Most elderly patients are on multiple drugs, at levels likely to be unsafely altered by inhibition of glucuronidation.

Attempts have also been made to improve solubility of curcumin by its chemical derivatisation (Hergenhahn *et al.*, 2003), complexation or interaction with macromolecules, for example, gelatin (Schranz, 1981), polysaccharides and proteins (Todd, 1991), and cyclodextrins (Tønnesen *et al.*, 2002). However, the slow process of complexation, high molecular weight of cyclodextrins, and pH of the processing medium may limit their practical utility. Paradkar *et al.* (2004) have reported curcumin–PVP solid dispersion but a problem with reproducibility of its physicochemical properties, its formulation into dosage forms, and the scale-up of manufacturing processes limits commercial applicability of the system (Paradkar *et al.*, 2004).

Another approach to improve the poor biopharmaceutical properties of curcumin is using nanocarriers. Nanocarriers have a small size (typically 10 to 100 nm), which besides enhancing solubilization, are useful for achieving targeted delivery too. Nanocarriers can improve the circulation time of the loaded therapeutic agent and may improve its accumulation at the pathological site exploiting the so-called enhanced permeation and retention (EPR) effect (Maeda *et al.*, 2006). During the last few decades, various types of nanocarriers have been investigated for improving the bioavailability of curcumin, and some have reached the stage of clinical evaluation and application. In line with this, Table 2.2 enlists details of various systems that were successful in increasing the bioavailability and improving targeted delivery of curcumin to tumors and other sites of disease, while Table 2.3 enlists ongoing/completed clinical trials on curcumin based nanosystems.

2.2.1 Studies on Curcumin from Our Laboratory

2.2.1.1 Solid Lipid Nanoparticles (SLN)

We developed curcumin-loaded SLNs (C-SLNs) with particle size of 134.6 nm and entrapment efficiency of 81.92 ± 2.91%. *In vivo* single dose oral pharmacokinetics performed with the developed C-SLNs (50, 25, 12.5, and 1 mg/Kg dose) using validated LC-MS/MS method in rat plasma revealed significant ($p < 0.05$) improvement in bioavailability (39, 32, 59, 155 times at 50, 25, 12.5, and 1 mg/Kg dose, respectively) (Kakkar *et al.*, 2011a). C-SLNs (Kakkar *et al.*, 2011b) were developed for their role in alleviating behavioral, biochemical, and histochemical changes associated with aluminium chloride ($AlCl_3$) (100 mg/Kg) induced toxicity symptoms similar in aetiology to Alzheimer's disease in male LACA mice. Treatment with free curcumin showed >15% recovery in membrane lipids (LPO) and 22% recovery in acetylcholinesterase (AChE) with respect to the $AlCl_3$ treated group. C-SLNs showed significantly better results (97.46% and 73% recovery in LPO and AChE) at a dose of 50 mg/kg, and the results were comparable ($p < 0.001$) to those achieved with rivastigmine. In another study, efficacy of C-SLNs (25–50 mg/Kg) was evaluated in experimental paradigm of cerebral ischemia (BCCAO model) in rats. Cerebral ischemic group (I/R) animals underwent weight loss and hyperthermia. C-SLNs treatment significantly increased the body weight and restored body temperature to normal, while free curcumin did not show any effect. I/R injury resulted in a significant increase in neurological deficit and a decrease in both ambulatory and rearing activity coupled with decreased muscle grip strength in comparison with sham control. Treatment with C-SLNs resulted in an improvement of 90% in cognition and inhibition of 52% of acetylcholinesterase in comparison to I/R group. Neurological scoring was also improved by 79%. Levels of superoxide dismutase, catalase, glutathione, and mitochondrial complex enzyme activities were significantly increased, while lipid peroxidation, nitrite, and acetylcholinesterase levels decreased ($p < 0.05$) after C-SLN administration. It is noteworthy to report the restoration of SOD, GSH, catalase, and mitochondrial complex enzyme levels equivalent to sham control values. Gamma-scintigraphic studies show 16.4 and 30 times improvement in brain bioavailability (AUC) upon oral and IV administration of C-SLNs versus solubilized curcumin (C-S) (Kakkar *et al.*, 2013a; Kakkar *et al.*, 2013b).

TABLE 2.2
Nanocarriers Used to Improve the Bioprofile of Curcumin

S. No.	Effect Achieved	Route/ Study Type	References
1.	**Solid Lipid Nanoparticles**		
	Slow drug release. Improvement in stability.	–	(Noack *et al.*, 2012)
	Sustained release for over 96 h. Increased apoptosis of 61.3 and 60.37% in MIA Paca-2 and Panc-1 cell lines in *in vitro* studies.	–	(Sutaria *et al.*, 2012)
	Prolonged inhibitory activity in cancer cells, as well as time-dependent increase in intracellular uptake in *in vitro* cell line studies.	–	(Sun *et al.*, 2013)
	Enhanced targeting of curcumin to lung and tumor, resulting in increased inhibition efficiency of curcumin from 19.5% to 69.3%.	i.p.	(Wang *et al.*, 2013)
	Significant decrease in tumor size and malignancy. Prolonged animal survival in comparison to free drug.	i.p.	(Zanotto-Filho *et al.*, 2013)
	Improved cytotoxicity at low dose in cancer cell lines in comparison to free curcumin.	–	(Rahman *et al.*, 2014)
	Significant improvement in neuromotor coordination when compared with 3-nitropropionic acid (3-NP)-induced Huntington rats. Significant reduction in mitochondrial swelling, lipid peroxidation, protein carbonyls, and reactive oxygen species.	Oral	(Sandhir *et al.*, 2014)
	Improved stability.	–	(Chow *et al.*, 2015)
	Upto 68.12% increase in BA.	Oral	(Gaur *et al.*, 2015)
	Prolonged stability at room and refrigerated conditions. Controlled drug release in simulated intestinal fluid. Significantly higher oral BA, and brain distribution of curcumin than free curcumin.	Oral	(Ramalingam *et al.*, 2015)
	12 folds increase in oral BA compared with marketed formulation of free Curcumin (Adcumin®).	Oral	(Shelat *et al.*, 2015)
	Significant reduction in levels of serum pro-inflammatory cytokines, including IL-6, TNF-α, and IL-1β as compared with free curcumin in sepsis induced mice.	i.p.	(Wang *et al.*, 2015)
	Significantly better healing of precancerous oral lesion in comparison to free curcumin mucoadhesive gel.	Topical	(Hazzah *et al.*, 2016)
	2.27 folds greater AUC_{0-t} than curcumin suspension with relative BA of 942.53%. High T_{max} and $t_{1/2}$. Significant improvement in effective permeability.	Oral	(Ji *et al.*, 2016)
	Sustained drug release. Improved permeability and antibacterial activity in *in vitro* studies.	–	(Jourghanian *et al.*, 2016)
	Improved BA, stability, and cytotoxicity against malignant glioma cells.	Oral	(Kumar *et al.*, 2016)
	Higher drug concentration in brain.	Intranasal	(Madane *et al.*, 2016)
	Improved biopharmaceutical properties.	Oral	(Pedro *et al.*, 2016)
	Improvement in permeability.	–	(Righeschi *et al.*, 2016)
	Prolonged release of curcumin.	–	(Behbahani *et al.*, 2017)
	Increase in intracellular uptake of curcumin in comparison to free curcumin. Increase in the expression of caspase-3, caspase-9 proteins, and the ratio of Bax/Bcl-2 resulting in increase in phototoxic effects of curcumin.	–	(Jiang *et al.*, 2017)

(*Continued*)

TABLE 2.2 (CONTINUED)
Nanocarriers Used to Improve the Bioprofile of Curcumin

S. No.	Effect Achieved	Route	References
2.	**Nanostructured Lipid Carriers**		
	Marked hypoacetylation of histone 4 (H4) at lysine 12 (K12) in the spinal cord while free curcumin did not change the H4K12 acetylation levels.	i.p.	(Puglia *et al.*, 2012)
	Significantly decreased neutrophil infiltration and TNF-α secretion resulting in reduced colonic inflammation.	Oral	(Beloqui *et al.*, 2016)
3.	**Polymeric Nanoparticles**		
	Increase in AUC and BA, and sustained release.	Oral	(Tsai *et al.*, 2012; Cheng *et al.*, 2013; Khalil *et al.*, 2013)
	Improvement in solubility.	–	(Maulick *et al.*, 2016)
	Effective inhibition of cell growth, induction of apoptosis, and arrest of the cell cycle in cervical cancer cell lines. Reduction in the tumor burden in a preclinical orthotopic mouse model of cervical cancer by decrease in oncogenic miRNA-21, suppression of nuclear β-catenin, and abrogating expression of E6/E7 HPV oncoproteins including smoking compound benzo[a]pyrene (BaP) induced E6/E7 and IL-6 expression.	Topical	(Zaman *et al.*, 2016)
	Sustained release of curcumin from N-carboxymethyl chitosan (NCC) coated curcumin-loaded SLN (NCC-SLN) in both acidic and intestinal fluid. Increased cytotoxicity and cellular uptake on MCF-7 cells. 6.3-fold and 9.5-fold higher lymphatic uptake and oral BA than that of curcumin solution, respectively.	Oral	(Baek *et al.*, 2017)
4.	**Vesicular Systems**		
	Improvement in BA.	Oral	(Chen *et al.*, 2012)
	Improvement in BA by 7.76 times.	Topical	(Li *et al.*, 2012)
	Improvement in BA.	Oral	(Zhang *et al.*, 2012)
	Improved percutaneous permeability.	Topical	(Jin-guang *et al.*, 2013; Madhavi *et al.*, 2013)
	10,000 fold increase in BA. High uptake by HepG2 human liver carcinoma cell line.	Oral	(Ucisik *et al.*, 2013)
	13×10^5 fold increase in curcumin solubility.	–	
	Significant down-regulation of NF-κB machinery. Induction of apoptosis of human pancreatic cells *in vitro* as compared to free curcumin.	–	(Hasan *et al.*, 2014)
	Enhanced cytotoxic activity and apoptotic rate against ovarian cancer A2780 cells compared with freely dispersed curcumin.	–	(Xu *et al.*, 2016)
5.	**Nanoemulsion**		
	Improved solubility.	–	(Kumar *et al.*, 2012)
	10 times higher AUC and 40 times higher C_{max} in comparison to free drug.	Oral	(Zhongfa *et al.*, 2012)

Note: BA: Bioavailability; AUC: Area under curve; C_{max}: Maximum plasma concentration; *MCF7*: Human breast adenocarcinoma cell line, T_{max}: Time for attaining maximum plasma concentration, $t_{1/2}$: Elimination half life, i.v.: intra venous, i.p.: intra peritoneal, TNF-α: Tumor necrosis factor-α, IL-6: Interleukin-6, IL-1β: Interleukin-1β, NF-κB: Nuclear Factor- κB.

TABLE 2.3
Details of Clinical Trials on Curcumin-Loaded Nanosystems

S. No.	Title	Sponsor	Intervention	Conditions	Phase	Reference
1	Safety, tolerability and pharmacokinetics of liposomal curcumin in healthy volunteers- a phase I dose escalation study.	Sigma Path Pharma Inc.	Liposomal curcumin	Drug safety	I	https://clinicaltrials.gov/ct2/show/NCT01403545
2	A phase IB dose escalation study on the safety, tolerability and activity of liposomal curcumin in patients with locally advanced or metastatic cancer.	Sigma Path Pharma Inc.	Liposomal curcumin	Patients with advanced cancer who have failed standard of care therapy	I II	https://clinicaltrials.gov/ct2/show/NCT02138955?term=curcumin+liposome&rank=2
3	Curcumin as a novel treatment to improve cognitive dysfunction in schizophrenia.	VA Greater Los Angeles Healthcare system	Curcumin nanoparticles	Schizophrenia, cognition, psychosis	I II	https://clinicaltrials.gov/ct2/show/NCT02104752?term=curcumin+nanoparticles&rank=1

Therapeutic potential of C-SLNs was also evaluated for hepatic disorders, post-induction in a model of carbon tetrachloride induced chronic hepatic injury in rats (Singh *et al.*, 2014a). C-SLNs (12.5 mg/Kg) significantly ($p < 0.001$) attenuated histological changes and oxidative stress and decreased the induction of alanine aminotransferase (ALT), aspartate transaminase (AST), and TNF-α. Effects were significantly better than free curcumin ($p < 0.001$), which was employed at an eight times higher dose of 100 mg/Kg, and silymarin, a well-established hepato-protectant (25 mg/Kg; $p < 0.001$).

In an *in vitro* study, C-SLNs were investigated for their anticancer effects (Kakkar *et al.*, 2012a). A 54–87% reduction in IC_{50} values was observed with developed C-SLNs, in comparison to free curcumin, in a panel of human cancer cell lines (HL-60, A549, PC 3) (Kakkar *et al.*, 2012a). Results demonstrated mechanisms similar to those claimed for free curcumin, including induction of cellular apoptosis by activation of caspases, release of cytochrome c, loss of membrane potential, blockage of nuclear factor kappa B (NF-κB) activation, and upregulation of TNF-α for C-SLNs (Kakkar *et al.*, 2012a). The extent of cell death achieved with C-SLNs in all these tests was significantly higher ($p < 0.001$) and was attributed to presentation of curcumin in dispersible/soluble form for enhanced permeability across cell surface. Further to this, the C-SLNs, with improved biopharmaceutical profile as compared to free drug, were reported to show protective effect in complete Freund's adjuvant (CFA)-induced arthritis in rats through attenuation of oxido-inflammatory and immunomodulatory cascade (Arora *et al.*, 2015).

2.2.1.2 Elastic Vesicles (EVs)

Free curcumin shows poor skin permeability on topical application. This was significantly improved in our lab by encapsulating curcumin in EVs using phospholipid and a suitable surfactant, and a conventional rotary evaporation sonication method (Agrawal *et al.*, 2015). Particle size and entrapment efficiency of the prepared EVs was 11.39 μm and 76.55%. The EV encapsulated curcumin (at a dose of 1, 3, 5, and 10 μmol) was investigated for its *in vivo* antiaging activity in UV-induced aging in mice. Microscopic and histological evaluation of skin, pinch test, and redox homeostasis of skin homogenates demonstrated almost 100% attenuation in photoaging at 10 μmol dose. Results were significantly ($p < 0.05$) better when compared with free curcumin and a marketed preparation (Agrawal *et al.*, 2010; Kaur *et al.*, 2010). This study was the first preclinical evidence for the use of topically delivered curcumin to attenuate photoaging. The developed system was also evaluated for its anti-inflammatory potential in animal models of both acute and chronic inflammation (Agrawal *et al.*, 2015). Results indicated similar (acute study) or significant effects when compared with a marketed diclofenac gel.

2.3 SESAMOL

Sesamol (5-hydroxy-1, 3-benzodioxole or 3,4-methylenedioxyphenol) is a phenolic component of sesame oil (Elleuch *et al.*, 2007), formed by hydrolysis of sesamolin during thermal oxidation (Mohamed *et al.*, 1998), and is a well-established antioxidant molecule (Chennuru *et al.*, 2013; Hemalatha *et al.*, 2013). The benzodioxole derivatives are known to possess anti-inflammatory, antitumor, antioxidant, and many other biological activities (Bhardwaj *et al.*, 2017; Yashaswini *et al.*, 2017).

It is reported to exhibit antimutagenic (Kaur *et al.*, 2000), anticancer (Kapadia *et al.*, 2002), antiplatelet (Changa *et al.*, 2010), estrogenic (Kaur *et al.*, 2012), hepatoprotective (Ohta *et al.*, 1994), photoprotective (Prasad *et al.*, 2005), radioprotective (Parihar *et al.*, 2006), neuroprotective (Angeline *et al.*, 2013), and antiaging activity (Sharma *et al.*, 2006). It has also been reported to be useful in atherosclerosis (Ying *et al.*, 2011), lipopolysaccharide (LPS) induced inflammatory response (Chu *et al.*, 2010), ortheoarthritis (Lu *et al.*, 2011), diabetes (Chopra *et al.*, 2010), Huntington's disease (Kumar *et al.*, 2009), cognition (Kuhad *et al.*, 2008), and wound healing (Shenoy *et al.*, 2011). Latter effects are attributed to the presence of the benzodioxole group in its ring (Figure 2.3),

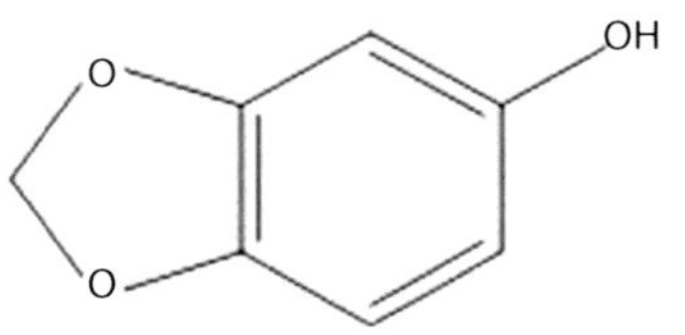

FIGURE 2.3 Chemical structure of sesamol.

which scavenges hydroxyl radicals to produce another antioxidant molecule, that is, 1,2-dihydroxy benzene (Kumagai *et al.*, 1991).

Although solubility and bioavailability of sesamol are significant and it matches the requirements of Lipinski's rule of five to be classified as a drug, its fast absorption and subsequent fast metabolism and elimination (Jan *et al.*, 2008) indicate a need to package it into a sustained/prolonged release carrier system to enhance its biopharmaceutical performance (Geetha *et al.*, 2015a). Further to this, Carcinogenic Potency Database indicates that sesamol has potential to cause forestomach cancers in rodents with TD_{50} values of 1.35 and 4.5 g/kg/day dose in mice and rats, respectively (Aggarwal *et al.*, 2004). This effect is attributed to its irritating effect on mucosa. Latter will be manifested both following oral administration and topical application, again reinforcing the need for incorporation into a controlled release system.

Puglia *et al.* (2017) reported that the sesamol loaded nanostructured lipid carrier dispersions were able to control the rate of sesamol diffusion through the skin, with respect to the free sesamol. Furthermore, encapsulated sesamol showed better and prolonged antioxidant activity. Another study (Yashaswini *et al.*, 2017) reported that encapsulation of sesamol within phospholipid micelles (PCS) resulted in better bioaccessibility (8.58%), transport across a monolayer of cells (1.5 fold), and cellular uptake (1.2 fold) in comparison to free sesamol (FS). Further, they also observed down-regulation of iNOS protein expression (27%), NO production (20%), ROS (32%), and lipoxygenase inhibition (IC_{50} = 31.24 μM) by PCS as compared to FS.

We have also tried to address issues of irritancy and fast elimination of sesamol. We have ascribed pharmaceutical couture to sesamol by encapsulating it within SLNs and floating beads and evaluated the developed systems against conditions such as menopause related emotional and cognitive CNS derangements (Kakkar *et al.*, 2011c), skin cancer (Geetha *et al.*, 2015b), gastric cancers, and sub-chronic liver damage (Singh *et al.*, 2014b).

2.3.1 Studies on Sesamol from Our Laboratory

2.3.1.1 Solid Lipid Nanoparticles (SLNs)

Unique properties of sesamol with aqueous solubility of 38.8 ± 1.2 mg/mL and log $P_{octanol/water}$ of 1.29 ± 0.01 result in a large tissue distribution (and volume of distribution) of this molecule (Geetha *et al.*, 2015a). Sesamol is well absorbed throughout the GIT and showed an oral bioavailability of 95.61%. However, it was observed to be widely distributed in rat tissue with the highest concentration in the kidneys followed by lungs, brain, and liver (Geetha *et al.*, 2015a). In spite of a favorable bioprofile, the wide distribution, small $t_{1/2}$, and fast clearance of sesamol indicate a need for packaging it into a suitable delivery system.

Sesamol-loaded SLNs developed in our lab (particle size of 122 nm and entrapment efficiency of 75.9 ± 2.91%) were investigated for brain targeting in menopause related emotional and cognitive central nervous system derangements (Kakkar *et al.*, 2011c). SLNs are passively uptaken via pinocytosis and endocytosis and actively by the endothelial cells into the brain tissue. Confocal images of rat brain sections (Figure 2.4) and gamma scintigraphic studies (Figure 2.5), in mice, administered with fluorescein labelled S-SLNs perorally, gave a direct evidence of their presence in the brain. Potential of S-SLNs to relieve cognition and anxiety in the surgical menopause model

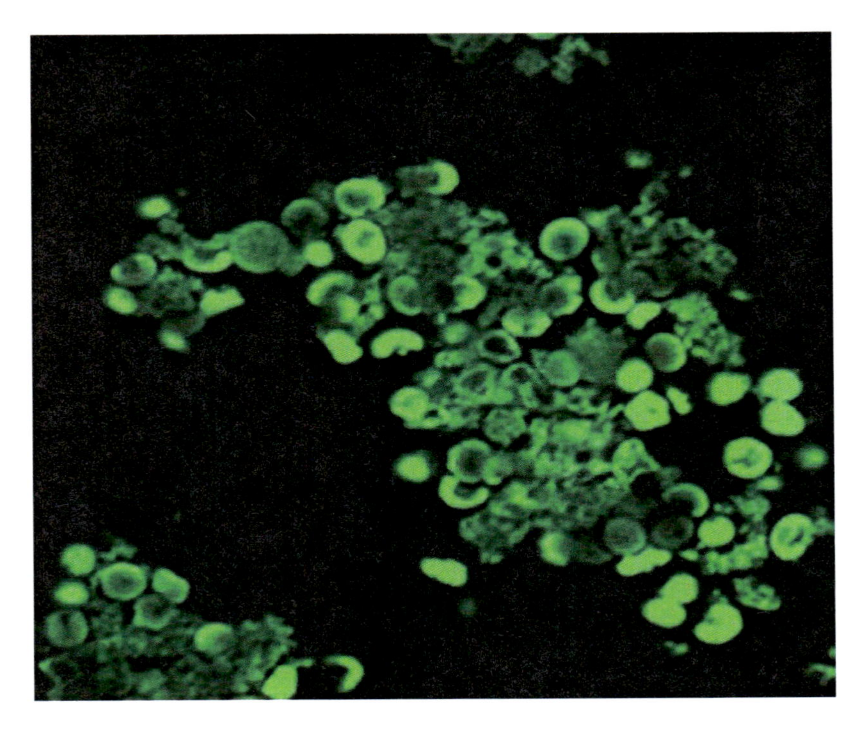

FIGURE 2.4 Confocal laser scanning microscopy (three dimensional view) of cryosection of rat brain 1 h after per oral administration of fluorescent marker loaded solid lipid nanoparticles. (Reprinted from Kakkar, V. *et al.*, *Rejuvenation Res.*, 14: 597–604, 2011c. With permission.)

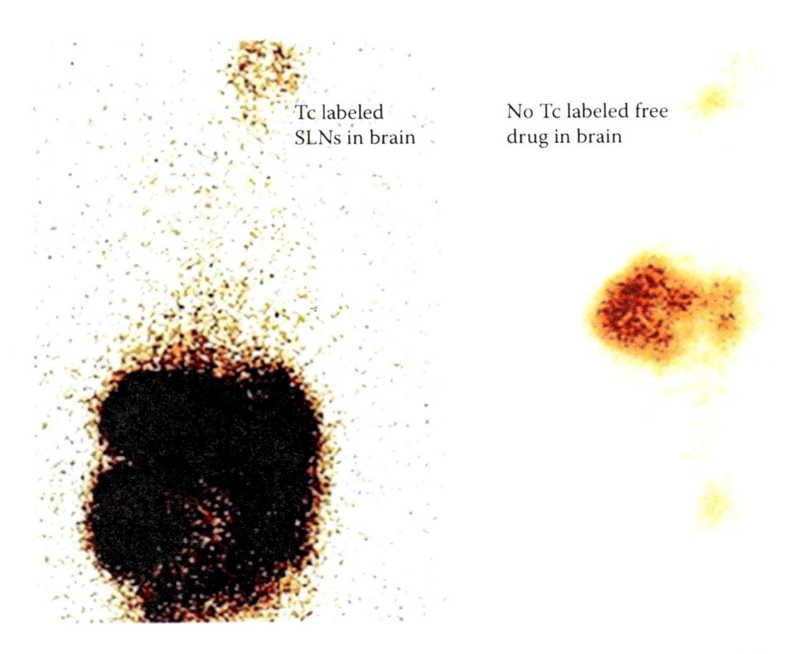

FIGURE 2.5 Gamma scintigraph of New Zealand rabbits 4 h after per oral administration of labelled solid lipid nanoparticles (SLNs) (left) and free drug (right). (Reprinted from Kakkar, V. *et al.*, *Rejuvenation Res.*, 14: 597–604, 2011c. With permission.)

of female wistar rats, as depicted by various animal behavioral tests, also confirmed effective delivery of sesamol to the brain when incorporated into SLNs (Kakkar *et al.*, 2011c).

Apart from being irritant in nature, high lipid and water solubility of sesamol limits its usefulness in topical delivery. When free sesamol was applied topically on isolated mice skin, it was found to enter systemic circulation immediately (no lag time) with significantly high flux (38.92 ± 0.62 μg/sqcm/h) across the skin and minimal retention in the epicutaneous tissue. To maintain high local concentration of sesamol in skin and target skin cancers, it thus needs to be encapsulated into a suitable delivery system. Sesamol SLNs were prepared with intent to reduce the irritant potential of free sesamol and to maximize its retention in the epidermal and dermal skin layers to exert a more pronounced and sustained local action. Incorporation into SLNs not only controls its release (lag phase of 1.5 h) but also provides a more significant local effect (3 times higher skin retention vis-à-vis free sesamol).

Single dose plasma pharmacokinetics studies in rats, using validated HPLC method, established that the amount of sesamol reaching plasma (C_{max}: 0.22 μg/mL at 4 h) was significantly less when S-SLN were applied topically in comparison to when the same dose of free sesamol was administered orally (0.44 μg/mL at 2 h), indirectly confirming an improved local bioavailability of sesamol from S-SLN (unpublished work).

Chemopreventive potential of developed S-SLNs was also evaluated in 7,2-dimethylbenz(a) anthracene (Bhardwaj *et al.*, 2016) induced mice model of skin cancer. Biochemical, histological, and molecular studies in term of fluorescent microscopy and western blotting confirmed its suitability in treatment of skin cancers. Sesamol administration (both in free and encapsulated form) was found to significantly decrease the tumor burden and lipid peroxidation level, and increase antioxidant levels, thereby hampering the development and promotion of skin tumors. Downregulation of bcl-2 and stimulation of bax protein expression on treatment with both free and encapsulated sesamol was responsible for the induction of apoptosis in tumor cells (Bhardwaj *et al.*, 2016). The developed SLNs were also evaluated for their usefulness in TPA-induced and benzo(a)pyrene initiated tumor, post induction in mice. At the end of S-SLN treatment, mouse epidermis showed normalization (in histology studies) of skin cancers. The biochemical studies indicated restoration of enzymatic (SOD, CAT) and non-enzymatic (GSH) antioxidant status of cancerous skin and a significant reduction in lipid peroxidation (Geetha *et al.*, 2015b).

An orally administered formulation of S-SLNs showed significant amelioration of CCl_4 induced sub-chronic hepatotoxicity in comparison to free sesamol. Improvement of histo-architecture, amelioration of serum liver injury markers (ALT, AST, and LDH), attenuation of oxidative stress (MDA, SOD, and GSH), and suppression of TNF-α confirmed the efficacy of S-SLNs both as a therapeutic (posttreatment) and a protective (co-treatment) for liver diseases. S-SLNs treatment was significantly more effective than sesamol (8 mg/kg). Effects were better or comparable to a well-established hepatoprotective and antioxidant agent, silymarin. It may further be noted that the employed and recommended dose of silymarin (25 mg/kg) is >3 times of that employed for S-SLNs (8 mg/kg) (Singh *et al.*, 2014b).

2.4 CONCLUSION

New drug discovery is a long, tedious, time-consuming, and costly (both in terms of money and work) process. Hence, revisiting the claims of traditional medicine, supplemented with proven *in vitro* and *in vivo* studies, and establishing a suitable mechanism of their action is a wise and rational alternative. However, various issues, for example, poor biopharmaceutical profile, high recommended dose, adverse effects, and erratic stability, hinder the translation of these agents to therapeutics, thus calling for the need to couture these phytochemicals using nanotechnology strategies. Presented studies may thus be considered as prototypes, which establish the importance of amalgamating rational formulation design with observational data from the folklore, complemented with scientific and precise *in vitro* and *in vivo* findings of the molecular biologists, pharmacologists, and biochemists to bring a new class of safe, cheap, and effective curatives to the market. The authors passionately and strongly feel the need to elevate the status of natural extracts and molecules from preventive dietary supplements to curative-therapeutics by an intelligent use of newer and existing drug delivery and dosage form design options. However, the potential for bioaccumulation and potential human health risks associated with the use of such nanostructured carrier systems is also a concern. While the proposed applications of nanotechnology are wide and varied, the purported claims need critical validation in terms of risks and benefits.

REFERENCES

Aggarwal, S., Takada, Y., Singh, S., Myers, J. N., and Aggarwal, B. B. Inhibition of growth and survival of human head and neck squamous cell carcinoma cells by curcumin via modulation of nuclear factor-kappaB signaling. *Int J Cancer.* 2004; 111: 679–692.

Agrawal, R., and Kaur, I. P. Inhibitory effect of encapsulated curcumin on UV-induced photoageing in mice. *Rejuvenation Res.* 2010; 13: 397–410.

Agrawal, R., Sandhu, S. K., Sharma, I., and Kaur, I. P. Development and evaluation of curcumin-loaded elastic vesicles as an effective topical anti-inflammatory formulation. *AAPS PharmSciTech.* 2015; 16: 364–374.

Ammon, H. P., Wahl, M. A. Pharmacology of *Curcuma longa. Planta Medica.* 1991; 57: 1–7.

Anand, P., Kunnumakkara, A. B., Harikumar, K. B., Ahn, K. S., Badmaev, V., amd Aggarwal, B. B. Modification of cysteine residue in p65 subunit of nuclear factor-kappa B (NF-kappaB) by picroliv suppresses NF-kappa B-regulated gene products and potentiates apoptosis. *Cancer Res.* 2008; 68: 8861–8870.

Anand, P., Kunnumakkara, A. B., Newman, R. A., and Aggarwal, B. B. Bioavailability of curcumin: Problems and promises. *Mol Pharm.* 2007; 4: 807–818.

Angeline, M. S., Sarkar, A., Anand, K., Ambasta, R. K., and Kumar, P. Sesamol and naringenin reverse the effect of rotenone-induced PD rat model. *Neuroscience.* 2013; 254: 379–394.

Arora, R., Kuhad, A., Kaur, I. P., and Chopra, K. Curcumin loaded solid lipid nanoparticles ameliorate adjuvant-induced arthritis in rats. *Eur J Pain.* 2015; 19: 940–952.

Arun, N., and Nalini, N. Efficacy of turmeric on blood sugar and polyol pathway in diabetic albino rats. *Plant Foods Hum Nutr.* 2002; 57: 41–52.

Baek, J. S., and Cho, C. W. Surface modification of solid lipid nanoparticles for oral delivery of curcumin: Improvement of bioavailability through enhanced cellular uptake, and lymphatic uptake. *Eur J Pharm Biopharm.* 2017; 117: 132–140.

Behbahani, E. S., Ghaedi, M., Abbaspour, M., and Rostamizadeh, K. Optimization and characterization of ultrasound assisted preparation of curcumin-loaded solid lipid nanoparticles: Application of central composite design, thermal analysis and X-ray diffraction techniques. *Ultrason Sonochem.* 2017; 38: 271–280.

Beloqui, A., Memvanga, P. B., Coco, R., Reimondez-Troitino, S., Alhouayek, M., Muccioli, G. G., Alonso, M. J., Csaba, N., and de la Fuente, M. A comparative study of curcumin-loaded lipid-based nanocarriers in the treatment of inflammatory bowel disease. *Colloids Surf B Biointerfaces.* 2016; 143: 327–335.

Bhardwaj, R., Kaur, I. P., Kaur, T., Deol, P. K., Kakkar, V., Vaiphei, K., and Sanyal, S. N. Sesamol induces apoptosis by altering expression of bcl-2 and bax proteins and modifies skin tumor development in Balb/c mice. *Anticancer Agents Med Chem.* 2017; 17: 726–733.

Bhardwaj, R., Kaur, I. P., Kaur, T., Deol, P. K., Kakkar, V., Vaiphei, K., and Sanyal, S. N. Sesamol induces apoptosis by altering expression of bcl-2 and bax proteins and modifies skin tumor development in Balb/c mice. *Anticancer Agents Med Chem.* 2016; doi:10.2174/1871520616666160819103249.

Boran, A. D., and Iyengar, R. Systems approaches to polypharmacology and drug discovery. *Curr Opin Drug Discov Devel.* 2010; 13: 297–309.

Calabrese, V., Cornelius, C., Dinkova-Kostova, A. T., Lavicoli, I., Paola, R. D., Koverech, A., Cuzzocrea, S., Rizzarelli, E., and Calabrese, E. J. Cellular stress responses, hormetic phytochemicals and vitagenes in aging and longevity. *Biochimica et Biophysica Acta.* 2012; 1822: 753–783.

Changa, C. C., Lu, W. J., Chiang, C. W., Jayakumar, T., Ong, E. T., Hsiao, G., Fong, T. H., Chou, D. S., and Sheu, J. R. Potent antiplatelet activity of sesamol in an *in vitro* and *in vivo* model: Pivotal roles of cyclic AMP and p38 mitogen-activated protein kinase. *J Nutr Biochem.* 2010; 21: 1214–1221.

Chen, F., and Coggins, P. C. *Handbook of meat, poultry and seafood quality.* Blackwell Publishing Ltd.; 2012.

Chen, H., Wu, J., Sun, M., Guo, C., Yu, A., Cao, F., Zhao, L., Tan, Q., and Zhai, G. N-trimethyl chitosan chloride-coated liposomes for the oral delivery of curcumin. *J Liposome Res.* 2012; 22: 100–109.

Cheng, K. K., Yeung, C. F., Ho, S. W., Chow, S. F., Chow, A. H., and Baum, L. Highly stabilized curcumin nanoparticles tested in an *in vitro* blood-brain barrier model and in Alzheimer's disease Tg2576 mice. *AAPS J* 2013; 15: 324–336.

Chennuru, A., and Saleem, M. T. S. Antioxidant, lipid lowering, and membrane stabilization effect of sesamol against doxorubicin-induced cardiomyopathy in experimental rats. *Biomed Res Int.* 2013; 934239.

Chopra, K., Tiwari, V., Arora, V., and Kuhad, A. Sesamol suppresses neuro-inflammatory cascade in experimental model of diabetic neuropathy. *J Pain.* 2010; 11: 950–957.

Chow, S. F., Wan, K. Y., Cheng, K. K., Wong, K. W., Sun, C. C., Baum, L., and Chow A. H. L. C. Development of highly stabilized curcumin nanoparticles by flash nanoprecipitation and lyophilization. *Eur J Pharm Biopharm.* 2015; 94: 436–449.

Chu, P. Y., Hsu, D. Z., Hsu, P. Y., and Liu, M. Y. Sesamol down-regulates the lipopolysaccharide-induced inflammatory response by inhibiting nuclear factor-kappa B activation. *Innate Immun.* 2010; 16: 333–339.

Csermely, P., Agoston, V., and Pongor, S. The efficiency of multi-target drugs: The network approach might help drug design. *Trends Pharmacol Sci.* 2005; 26: 178–182.

Das, L., Bhaumik, E., Raychaudhuri, U., and Chakraborty, R. Role of nutraceuticals in human health. *J Food Sci Technol.* 2012; 49: 173–183.

Deodhar, S. D., Sethi, R., and Srimal, R. C. Preliminary study on antirheumatic activity of curcumin (diferuloyl methane). *Indian J Med Res.* 1980; 71: 632–634.

Dulbecco, P., and Savarino, V. Therapeutic potential of curcumin in digestive diseases. *World J Gastroenterol.* 2013; 19: 9256–9270.

Elleuch, M., Besbes, S., Roiseux, O., Blecker, C., and Attia, H. Quality characteristics of sesame seeds and by-products. *Food Chem Toxicol.* 2007; 103: 641–650.

Farnsworth, N. R., Akerele, R. O., Bingel, A. S., Soejarto, D. D., and Guo, Z. Medicinal plants in therapy. *Bull World Health Organ.* 1985; 63: 965–981.

Garcea, G., Jones, D. J., Singh, R., Dennison, A. R., Farmer, P. B., Sharma, R. A., Steward, W. P., Gescher, A. J., and Berry, D. P. Detection of curcumin and its metabolites in hepatic tissue and portal blood of patients following oral administration. *Br J Cancer.* 2004; 8: 1011–1015.

Gaur, P. K., Mishra, S., Verma, A., and Verma, N. Ceramide–palmitic acid complex based curcumin solid lipid nanoparticles for transdermal delivery: Pharmacokinetic and pharmacodynamic study. *J Exp Nanosci.* 2015; 11: 1–16.

Geetha, T., Kapila, M., Prakash, O., Deol, P. K., Kakkar, V., and Kaur, I. P. Sesamol-loaded solid lipid nanoparticles for treatment of skin cancer. *J Drug Target.* 2015b; 30: 1–11.

Geetha, T., Singh, N., Deol, P. K., and Kaur, I. P. Biopharmaceutical profiling of sesamol: Physiochemical characterization, gastrointestinal permeability and pharmacokinetic evaluation. *RSC Adv.* 2015a; 15: 4083–4091.

Geldenhuys, W. J., and Schyf, C. J. V. Designing drugs with multi-target activity: The next step in the treatment of neurodegenerative disorders. *Expert Opin Drug Discov.* 2013; 8: 115–129.

Gunasekaran, T., Haile, T., and Nigusse, T. Nanotechnology: An effective tool for enhancing bioavailability and bioactivity of phytomedicine. *Asian Pac J Trop Biomed.* 2014; 4: S1–S7.

Hasan, M., Belhaja, N., Benachour, H., Barberi-Heyob, M., Kahn, C. J. F., Jabbari, E., Linder, M., and Arab-Tehrany, E. Liposome encapsulation of curcumin: Physico-chemical characterizations and effects on MCF7 cancer cell proliferation. *Int J Pharm.* 2014; 461: 519–528.

Hazzah, H. A., Farid, R. M., Maha, M. A., Nasra, M. M. A., Zakaria, M., Gawish, Y., El-Massik, M. A., and Abdallah, O. Y. A new approach for treatment of precancerous lesions with curcumin solid–lipid nanoparticle-loaded gels: *In vitro* and clinical evaluation. *Drug Deliv.* 2016; 23: 1409–1419.

Hemalatha, G., Pugalendi, K. V., and Saravanan, R. Modulatory effect of sesamol on DOCA-salt-induced oxidative stress in uninephrectomized hypertensive rats. *Mol Cell Biochem.* 2013; 379: 255–265.

Hergenhahn, M., Bertram, B., Wiessler, M., and Sorg, B. Curcumin derivatives with improved water solubility compared to curcumin and medicaments containing the same US 20030153512 A1. 2003.

Huang, Q., and Yu, H. Bioavailability and delivery of nutraceuticals using nanotechnology. *J Food Sci.* 2010; 75: 50–57.

Isman, M. The role of botanical insecticides, deterrents and repellents in modern agriculture and an increasingly regulated world. *Annu Rev Entomol.* 2006; 51: 45–66.

Jan, K. C., Ho, C. T., and Hwang, L. S. Bioavailability and tissue distribution of sesamol in rat. *J Agric Food Chem.* 2008; 56: 7032–7037.

Ji, H., Tang, J., Li, M., Ren, J., Zheng, N., and Wu, L. Curcumin-loaded solid lipid nanoparticles with Brij78 and TPGS improved *in vivo* oral bioavailability and *in situ* intestinal absorption of curcumin. *Drug Deliv.* 2016; 23: 459–470.

Jiang, S., Zhu, R., He, X., Wang, J., Wang, M., Qian, Y., and Wang, S. Enhanced photocytotoxicity of curcumin delivered by solid lipid nanoparticles. *Int J Nanomedicine*. 2017; 12: 167–178.

Jin-guang, C., Wei, L., and Yu, J. Preparation of curcumin ethosomes. *Afr J Pharm Pharacol*. 2013; 7: 2246–2251.

Jourghanian, P., Ghaffari, S., Ardjmand, M., Haghighat, S., and Mohammadnejad M. Sustained release curcumin loaded solid lipid nanoparticles. *Adv Pharm Bull*. 2016; 6: 17–21.

Kakkar, V., Bhushan, S., Kumar, G. S., and Kaur, I. P. Enhanced apoptotic effect of curcumin loaded solid lipid nanoparticles. *Mol Pharm*. 2012a; 9: 3411–3421.

Kakkar, V., and Kaur, I. P. Evaluating potential of curcumin loaded solid lipid nanoparticles in aluminium induced behavioural, biochemical and histopathological alterations in mice brain. *Food Chem Toxicol*. 2011b; 49: 2906–2913.

Kakkar, V., Mishra, A. K., Chuttani, K., and Kaur, I. P. Proof of concept studies to confirm the delivery of curcumin loaded solid lipid nanoparticles (C-SLNs) to brain. *Int J Pharm*. 2013a; 448: 354–359.

Kakkar, V., Mishra, M., Chuttani, K., and Chopra, K. Delivery of sesamol loaded solid lipid nanoparticles to the brain for menopause related emotional and cognitive central nervous system derangements. *Rejuvenation Res*. 2011c; 14: 597–604.

Kakkar, V., Muppu, S. K., Chopra, K., and Kaur, I. P. Curcumin loaded solid lipid nanoparticles: An efficient formulation approach for cerebral ischemic reperfusion injury in rats. *Eur J Pharm Biopharm*. 2013b; 85: 339–345.

Kakkar, V., Singh, S., Singla, D., and Kaur, I. P. Exploring solid lipid nanoparticles to enhance the oral bioavailability of curcumin. *Mol Nutr Food Res*. 2011a; 55: 495–503.

Kalra, E. K. Nutraceutical—Definition and introduction. *AAPS PharmSci*. 2003; 5: E25.

Kapadia, G. J., Azuine, M.A., Tokuda, H., Takasaki, M., Mukainaka, T., Konoshima, T., and Nishino, H. Chemopreventive effect of resveratrol, sesamol, sesame oil and sunflower oil in the Epstein-Barr virus early antigen activation assay and the mouse skin two-stage carcinogenesis. *Pharmacol Res*. 2002; 45: 499–505.

Kaur, A., Jindal, S., Kaur, I. P., and Chopra, K. Effect of sesamol on the pathophysiological changes induced by surgical menopause in rodents. *Climacteric*. 2012; 15: 1–12.

Kaur, I. P., and Agrawal, R. Topical curcumin formulation. 2335/DEL/2010.

Kaur, I. P., and Saini, A. Sesamol exhibits antimutagenic activity against oxygen species induced mutagenicity. *Mutat Res*. 2000; 470: 71–76.

Khalil, N. M., Do Nascimento, T. C., Casa, D. M., Dalmolin, L. F., De Mattos, A. C., Hoss, I., Romano, M. A., and Mainardes, R. M. Pharmacokinetics of curcumin-loaded PLGA and PLGA-PEG blend nanoparticles after oral administration in rats. *Colloids Surf B Biointerfaces*. 2013; 101: 353–360.

Kuhad, A., and Chopra, K. Effect of sesamol on diabetes-associated cognitive decline in rats. *Exp Brain Res*. 2008; 185: 411–420.

Kumagai, Y., Lin, L. Y., Schmitz, D. A., and Cho, A. K. Hydroxyl radical mediated demethylenation of (methylenedioxy)phenyl compounds. *Chem Res Toxicol*. 1991; 4: 330–334.

Kumar, A., Ahuja, A., Ali, J., and Baboota S. Curcumin-loaded lipid nanocarrier for improving bioavailability, stability and cytotoxicity against malignant glioma cells. *Drug Deliv*. 2016; 23: 214–229.

Kumar, A., Ahuja, A., Ali, J., and Baboota, S. Curcumin loaded nano globules for solubility enhancement: Preparation, characterization and *ex vivo* release study. *J Nanosci Nanotechnol*. 2012; 12: 8293–8302.

Kumar, P., Kalonia, H., and Kumar, A. Sesamol attenuate 3-nitropropionic acid-induced Huntington-like behavioral, biochemical, and cellular alterations in rats. *J Asian Nat Prod Res*. 2009; 11: 439–450.

Li, C., Zhang, Y., Su, T., Feng, L., Long, Y., and Chen, Z. Silica-coated flexible liposomes as a nanohybrid delivery system for enhanced oral bioavailability of curcumin. *Int J Nanomedicine*. 2012; 7: 5995–6002.

Li, F., Zhao, C., and Wang, L. Molecular-targeted agents combination therapy for cancer: Developments and potentials. *Int J Cancer*. 2014; 134: 1257–1269.

Lina, D., Xue, F., Xiaoqin, X., and Yiguang, J. Wound healing effect of an *in situ* forming hydrogel loading curcumin-phospholipid complex. *Curr Drug Deliv*. 2016; 13: 76–82.

Liu, Y., Cai, Y., Jiang, X., Wu, J., and Le, X. Molecular interactions, characterization and antimicrobial activity of curcumin–chitosan blend films. *Food Hydrocoll*. 2016; 52: 564–572.

Liu, Y., and Feng, N. Nanocarriers for the delivery of active ingredients and fractions extracted from natural products used in traditional Chinese medicine (TCM). *Adv Colloid Interface Sci*. 2015; 221: 60–76.

Lu, Y. C., Jayakumar, T., Duann, Y. F., Chou, Y. C., Hsieh, C. Y., Yu, S. Y., Sheu, J. R., and Hsiao, G. Chondroprotective role of sesamol by inhibiting MMPs expression via retaining NF-κB signaling in activated SW1353 cells. *J Agric Food Chem*. 2011; 59: 4969–4978.

Madane, R. G., and Mahajan, H. S. Curcumin-loaded nanostructured lipid carriers (NLCs) for nasal administration: Design, characterization, and *in vivo* study. *Drug Deliv.* 2016; 23: 1326–1334.
Madhavi, B. B., Vennela, K. S., Masana, P., and Madipoju, B. Enhanced transdermal drug penetration of curcumin via ethosomes. *Malaysian Journal of Pharmaceutical Sciences.* 2013; 11: 49–58.
Maeda, H., Wu, J., Sawa, T., Matsumura, Y., and Hori, K. Tumor vascular permeability and the EPR effect in macromolecular therapeutics: A review. *J Control Release.* 2006; 65: 271–284.
Maheswari, R. K., Singh, A. K., Gaddipati, J., and Srimal, R. C. Multiple biological activities of curcumin: A short review. *Life Sci.* 2006; 78: 2081–2087.
Mattson, M. Hormesis defined. *Ageing Res Rev.* 2008; 7: 1–7.
Maulick, C., Reena, J., Kumar, A., Sunil, V., and Sonal, M. Design of curcumin loaded polymeric nanoparticles-optimization, formulation and characterization. *J Nanosci Nanotechnol.* 2016; 16: 9432–9442.
McClements, D. J., Decker, E. A., Park, Y., and Weiss, J. Structural design principles for delivery of bioactive components in nutraceuticals and functional foods. *Crit Rev Food Sci Nutr.* 2009; 49: 577–606.
Misra, S., Tiwari, V., Kuhad, A., and Chopra, K. Modulation of nitrergic pathway by sesamol prevents cognitive deficits and associated biochemical alterations in intracerebroventricular streptozotocin administered rats. *Eur J Pharmacol.* 2011; 659: 177–186.
Mohamed, H. M. A., and Awatif, I. I. The use of sesame oil unsaponifiable matter as a natural antioxidant. *Food Chem.* 1998; 62: 269–276.
Motterlini, R., Foresti, R., Bassi, R., and Green, C. J. Curcumin, an antioxidant and anti-inflammatory agent, induces heme oxygenase-1 and protects endothelial cells against oxidative stress. *Free Radic Biol Med.* 2000; 28: 1303–1312.
Noack, A., Hause, G., and Mäder, K. Physicochemical characterization of curcuminoid-loaded solid lipid nanoparticles. *Int J Pharm.* 2012; 423: 440–451.
Ohta, S., Suzuki, M., Sato, N., Kamogawa, A., and Shinoda, M. Protective effects of sesamol and its related compounds on carbon tetrachloride induced liver injury in rats. *Yakugaku Zasshi.* 1994; 114: 901–910.
Paradkar, A., Ambike, A. A., Jadhav, B. K., and Mahadik, K. R. Characterization of curcumin-PVP solid dispersion obtained by spray drying. *Int J Pharm.* 2004; 271: 281–286.
Parihar, V. K., Prabhakar, K. R., Veerapur, V. P., Kumar, M. S., Reddy, Y. R., Joshi, R., Unnikrishnan, M. K., and Rao, C. M. Effect of sesamol on radiation-induced cytotoxicity in Swiss albino mice. *Mutat Res.* 2006; 611: 9–16.
Pedro, A. S., Villa, S. D., Caliceti, P., de Melo, S. A. B. V., Albuquerque, E. C., Bertucco, A., and Salmaso, S. Curcumin-loaded solid lipid particles by PGSS technology. *J Supercrit Fluids.* 2016; 107: 534–541.
Prasad, N. R., Menon, V. P., Vasudev, V., and Pugalendi, K. V. Radioprotective effect of sesamol on gamma-radiation induced DNA damage, lipid peroxidation and antioxidants levels in cultured human lymphocytes. *Toxicology.* 2005; 209: 225–235.
Puglia, C., Frasca, G., Musumeci, T., Rizza, L., Puglisi, G., Bonina, F., and Chiechio, S. Curcumin loaded NLC induces histone hypoacetylation in the CNS after intraperitoneal administration in mice. *Eur J Pharm Biopharm.* 2012; 81: 288–293.
Puglia, C., Lauro, M. R., Offerta, A., Crascì, L., Micicchè, L., Panico, A. M., Bonina, F., and Puglisi, G. Nanostructured lipid carriers (NLC) as vehicles for topical administration of sesamol: *In vitro* percutaneous absorption study and evaluation of antioxidant activity. *Planta Med.* 2017; 83: 398–404.
Rahman, M. H., Ramanathan, M., and Sankar, V. Preparation, characterization and *in vitro* cytotoxicity assay of curcumin loaded solid lipid nanoparticle in IMR32 neuroblastoma cell line. *Pak J Pharm Sci.* 2014; 27: 1281–1285.
Ramalingam, P., and Ko, Y. T. Enhanced oral delivery of curcumin from N-trimethyl chitosan surface-modified solid lipid nanoparticles: Pharmacokinetic and brain distribution evaluations. *Pharm Res.* 2015; 32: 389–402.
Righeschi, C., Bergonzi, M. C., Isacchi, B., Bazzicalupi, C., Gratteri, P., and Bilia, A. R. Enhanced curcumin permeability by SLN formulation: The PAMPA approach. *LWT-Food Sci Technol.* 2016; 66 475–483.
Rivera-Espinoza, Y., and Muriel, P. Pharmacological actions of curcumin in liver diseases or damage. *Liver Int.* 2009; 29: 1457–1466.
Sandhir, R., Yadav, A., Mehrotra, A., Sunkaria, A., Singh, A., and Sharma, S. Curcumin nanoparticles attenuate neurochemical and neurobehavioral deficits in experimental model of huntington's disease. *Neuromol Med.* 2014; 16: 106–118.
Schranz, J. L. Water-soluble curcumin complex US 4368208 A. 1981.
Sharma, S., Kaur, I. P. Development and evaluation of sesamol as an antiaging agent. *Int J Dermatol.* 2006; 45: 200–208.

Shegokar, R., Muller, R. H. Nanocrystals: Industrially feasible multifunctional formulation technology for poorly soluble actives. *Int J Pharm.* 2010; 399: 129–139.

Shelat, P., Mandowara, V. K., Gupta, D., and Patel, S. Formulation of curcuminoid loaded solid lipid nanoparticles in order to improve oral bioavailability. *Int J Pharm Pharm Sci.* 2015; 7: 278–282.

Shenoy, R. R., Sudheendra, A. T., Nayak, P. G., Paul, P., Kutty, N. G., and Rao, C. M. Normal and delayed wound healing is improved by sesamol, an active constituent of *Sesamum indicum* (L.) in albino rats. *J Ethnopharmacol.* 2011; 133: 608–612.

Shoba, G., Joy, D., Joseph, T., Majeed, M., Rajendran, R., and Srinivas, P. Influence of piperine on the pharmacokinetics of curcumin in animals and human volunteers. *Planta Med.* 1998; 64: 353–356.

Shunsuke, S., Yuki, Y., Sei, K., Masahiro, I., Keigo, K., Kazunori, O., Aki, K., Hiroyuki, Y., Chikashi, I., Masamitsu, T., I. Y., and Hiroyuki, S. A curcumin analog, GO-Y078, effectively inhibits angiogenesis through actin disorganization. *Anticancer Agents Med Chem.* 2016; 16: 633–647.

Singh, N., Khullar, N., Kakkar, V., and Kaur, I. P. Attenuation of carbon tetrachloride induced hepatic injury with curcumin loaded solid lipid nanoparticles. *BioDrugs.* 2014a; 28: 297–312.

Singh, N., Khullar, N., Kakkar, V., and Kaur, I. P. Hepatoprotective effects of sesamol loaded solid lipid nanoparticles in carbon tetrachloride induced sub-chronic hepatotoxicity in rats. *Environ Toxicol.* 2014b; 31: 520–532.

Son, T., Camandola, S., and Mattson, M. P. Hormetic dietary phytochemicals. *Neuromol Med.* 2008; 10: 236–246.

Srivastava, R., Dikshit, M., and Srimal, R. C. Anti-thrombotic effect of curcumin. *Thromb Res.* 1985; 40: 413–417.

Sun, J., Bi, C., Chan, H. M., Sun, S., Zhang, Q., and Zheng, Y. Curcumin-loaded solid lipid nanoparticles have prolonged *in vitro* antitumour activity, cellular uptake and improved *in vivo* bioavailability. *Colloids Surf B Biointerfaces.* 2013; 111: 367–375.

Sutaria, D., Grandhi, B. K., Thakkar, A., Wang, J., and Prabhu, S. Chemoprevention of pancreatic cancer using solid-lipid nanoparticulate delivery of a novel aspirin, curcumin and sulforaphane drug combination regimen. *Int J Oncol.* 2012; 41: 2260–2268.

Todd, P. H. Curcumin complexed on water-dispersible substrates US 4999205 A. 1991.

Tønnesen, H. H., Már Másson, M., and Loftsson, T. Studies of curcumin and curcuminoids. XXVII. Cyclodextrin complexation: Solubility, chemical and photochemical stability. *Int J Pharm.* 2002; 244: 127–135.

Tsai, Y. M., Chang-Liao, W. L., Chien, C. F., Lin, L. C., and Tsai, T. H. Effects of polymer molecular weight on relative oral bioavailability of curcumin. *Int J Nanomedicine.* 2012; 7: 2957–2966.

Ucisik, M. H., Küpcü, S., Schuster, B., and Sleytr, U. B. Characterization of CurcuEmulsomes: Nanoformulation for enhanced solubility and delivery of curcumin. *J Nanobiotechnology.* 2013; 11: 1–13.

Wang, J., Wang, H., Zhu, R., Liu, Q., Fei, J., and Wang, S. Anti-inflammatory activity of curcumin-loaded solid lipid nanoparticles in IL-1b transgenic mice subjected to the lipopolysaccharide-induced sepsis. *Biomaterials.* 2015; 53: 475–483.

Wang, P., Zhang, L., Peng, H., Li, Y., Xiong, J., and Xu, Z. The formulation and delivery of curcumin with solid lipid nanoparticles for the treatment of on non-small cell lung cancer both *in vitro* and *in vivo.* *Mater Sci Eng C.* 2013; 33: 4802–4808.

Weiss, J., Takhistov, P., and McClements, D. J. Functional materials in food nanotechnology. *J Food Sci.* 2006; 71: 107–116.

Xiao, J., Sheng, X., Zhang, X., Guo, M., and Ji X. Curcumin protects against myocardial infarction-induced cardiac fibrosis via SIRT1 activation *in vivo* and *in vitro. Drug Des Devel Ther.* 2016; 10: 1267–1277.

Xu, Y. Q., Chen, W. R., Tsosie, J. K., Xie, X., Li, P., Wan, J. B., He, C. W., and Chen, M. W. Niosome encapsulation of curcumin: Characterization and cytotoxic effect on ovarian cancer cells. *J Nanomater.* 2016; 2016: 1–9.

Yang, K., Lin, L., Tseng, T., Wang, S., and Tsai, T. Oral bioavailability of curcumin in rat and the herbal analysis from *Curcuma longa* by LC-MS/MS. *J Chromatogr B.* 2007; 853: 183e9.

Yao, Q. Y., Xu, B. L., Wang, J. Y., Liu, H. C., Zhang, S. C., Tu, C. T. Inhibition by curcumin of multiple sites of the transforming growth factor-beta1 signalling pathway ameliorates the progression of liver fibrosis induced by carbon tetrachloride in rats. *BMC Complement Altern Med.* 2012; 12: 115–126.

Yashaswini, P. S., Kurrey, N. K., and Singh, S. A. Encapsulation of sesamol in phosphatidyl choline micelles: Enhanced bioavailability and anti-inflammatory activity. *Food Chem.* 2017; 228: 330–337.

Yashaswini, P. S., Rao, A. G. A., and Singh, S. A. Inhibition of lipoxygenase by sesamol corroborates its potential anti-inflammatory activity. *Int J Biol Macromolec.* 2017; 94: 781–787.

Ying, Z., Kherada, N., Kampfrath, T., Mihai, G., Simonetti, O., Desikan, R., Selvendiran, K., Sun, Q., Ziouzenkova, O., Parthasarathy, S., and Rajagopalan, S. A modified sesamol derivative inhibits progression of atherosclerosis. *Arterioscler Thromb Vasc Biol.* 2011; 31: 536–542.

Yu, S., Wang, X., He, X., Wang, Y., Gao, S., Ren, L., and Shi, Y. Curcumin exerts anti-inflammatory and antioxidative properties in 1-methyl-4-phenylpyridinium ion (MPP+)-stimulated mesencephalic astrocytes by interference with TLR4 and downstream signaling pathway. *Cell Stress Chaperones.* 2016; 21: 697–705.

Zaman, M. S., Chauhan, N., Yallapu, M. M., Gara, R. K., Maher, D. M., Kumari, S., Sikander, M., Khan, S., Zafar, N., Jaggi, M., and Chauhana, S. C. Curcumin nanoformulation for cervical cancer treatment. *Sci Rep.* 2016; 6: 1–14.

Zanotto-Filho, A., Coradini, K., Braganhol, E., Schröder, R., de Oliveira, C. M., Simões-Pires, A., Battastini, A. M., Pohlmann, A. R., Guterres, S. S., Forcelini, C. M., Beck, R. C., and Moreira, J. C. Curcumin-loaded lipid-core nanocapsules as a strategy to improve pharmacological efficacy of curcumin in glioma treatment. *Eur J Pharm Biopharm.* 2013; 83: 156–167.

Zhang, L., Lu, C. T., Li, W. F., Cheng, J. G., Tian, X. Q., Zhao, Y. Z., Li, X., Lv, H. F., and Li, X. K. Physical characterization and cellular uptake of propylene glycol liposomes *in vitro*. *Drug Dev Ind Pharm.* 2012; 38: 365–371.

Zheng, H., Fridkin, M., and Youdim, M. From single target to multitarget/network therapeutics in Alzheimer's therapy *Pharmaceuticals.* 2014; 7: 113–135.

Zhongfa, L., Chiu, M., Wang, J., Chen, W., Yen, W., Fan-Havard, P., Yee, L. D., Chan, K. K. Enhancement of curcumin oral absorption and pharmacokinetics of curcuminoids and curcumin metabolites in mice. *Cancer Chemother Pharmacol.* 2012; 69: 679–689.

3 Emerging Field of Nanocarriers for Efficient Delivery of Chemopreventive Nutraceuticals

Madhunika Agrawal and Satyam Kumar Agrawal*

CONTENTS

3.1 INTRODUCTION

Nutraceuticals, such as curcumin, resveratrol, eugenol, piperene, propolis, and many others derived from nutritional or medicinal plants, may have chemopreventive activities, as already documented by studying in various *in vitro* and *in vivo* models (Salami *et al.*, 2013). Although nutritional chemoprevention has surfaced as a cost-effective effort in preclinical studies to manage most common chronic diseases including cancer (Bhuvaneswari and Nagini, 2005), the transformation of numerous phytochemicals from laboratory to clinics is not cheering. Transportation of therapeutically active compound to the anticipated site of action is the chief concern in the therapy of several diseases (Bhatia, 2016). Many reasons are credited to this restricted success, including incompetent systemic delivery and availability inside animal systems (Srivastava *et al.*, 2013).

* Corresponding author.

Advent of nanotechnology has excited science due to the extraordinary and unique characteristics of nanoparticles, which can be organized to exclusive conformation and functions (Chapman *et al.*, 2012). Various factors govern the functioning of nanoparticles in the body, such as morphological aspects regarding shape and size, that have a crucial role in circulation of nanoparticles in the system (Caldorera-Moore *et al.*, 2010). Nanoparticles are required to be in the body until they reach the specified target. The body's immune system may act as the responsible factor for removing nanoparticles from the system, if the latter were recognized as foreign bodies by the reticulo-endothelial system (RES). Moreover, liver, kidneys, or spleen can also contribute toward the elimination of nanoparticles (Caldorera-Moore *et al.*, 2010). Several reports have documented that particles greater in size than 7 μm are removed from the body by filtration through lungs capillaries (Illum *et al.*, 1982; Ikada, 1990). Further, the particles of the size range between 0.1–7 μm are distinguished by RES in the liver or spleen by phagocytotsis via Kupffer cells or macrophages. Nanoparticles of the size lower than 100 nm will stay in blood vessels, as they can escape from being engulfed by macrophages, or their identification by opsonins is limited (Stolnik *et al.*, 1995). Furthermore, the shape of the particle also plays a role in its circulation inside the body. Spherical particles can travel without restriction, whereas particles with uneven shape may fall or tangle in filtering organs (Champion *et al.*, 2014). Nanoparticles can be classified in various categories based on their surface structure, size, and chemistry (Wang and Wang, 2014). Some vital classes of nanoparticles include liposomes, polymeric nanoparticles, nanoparticles, gold nanoparticles, and solid lipid nanoparticles.

The last few years have observed unparalleled expansion in the application of nanotechnology in the domain of drug delivery (Jong and Bormm, 2008). Nanocarriers as potent drug delivery systems are one of the most recent applications out of the diverse application of nanoparticles (Murthy, 2007). Thus, exploration in the area of nanomedicine has not only turned into a leading movement but has also emerged as an innovative drug delivery field (Guo *et al.*, 2015). This chapter focuses on various novel drug delivery strategies employed currently by various researchers to improve the bioavailability and therapeutic index of nutraceuticals.

3.1.1 Nutraceuticals Documented for Chemotherapeutic Profile

In recent years, interest has shifted toward the exploration of nutraceuticals in the prevention of cancer. A large range of nutraceuticals are available; the most common are shown in Figure 3.1 and are discussed.

3.1.1.1 Curcumin

Curcumin from turmeric (*Curcuma longa*) is under exhaustive exploration not only due to its renowned anti-inflammatory potential that can propel the possibility of cancer, but also for the probability of managing various metabolic disorders such as diabetes, obesity, and atherosclerosis, stimulated by constant untreated inflammation (Bachmeier *et al.*, 2010). Various experiments conducted all over the world have demonstrated that curcumin can amend the expressions of pro-inflammatory cytokines along with its capacity to suppress an assemblage of signalling pathways convened by the nuclear transcription factors. Chemokines such as cycloxygenase-2, vascular endothelial growth factors, and various interleukins like IL-1, IL-6, and IGF are the proteins modulated by curcumin as stated in various reports (Kamat *et al.*, 2009; Kang *et al.*, 2009; Kunnumakkara *et al.*, 2009; Lin *et al.*, 2009; Sandur *et al.*, 2009).

3.1.1.2 Resveratrol

Resveratrol is a polyphenol (Signorelli and Ghidoni, 2005), predominantly found in grapes, peanuts, berries, turmeric, and other food products (Udenigwe *et al.*, 2008). Bharali *et al.* (2011) unravelled that it was Jang and his co-workers (1997) who first demonstrated the anticancer activity of resveratrol in the prevention of skin cancer development in mice. Since then, there has been extensive research conducted globally to study resveratrol as an chemotherapeutic agent against various cancer models, such as skin, lung, breast, gastric, colon, prostate, leukemia, brain (neuroblastoma), and pancreas (Bharali *et al.*, 2011).

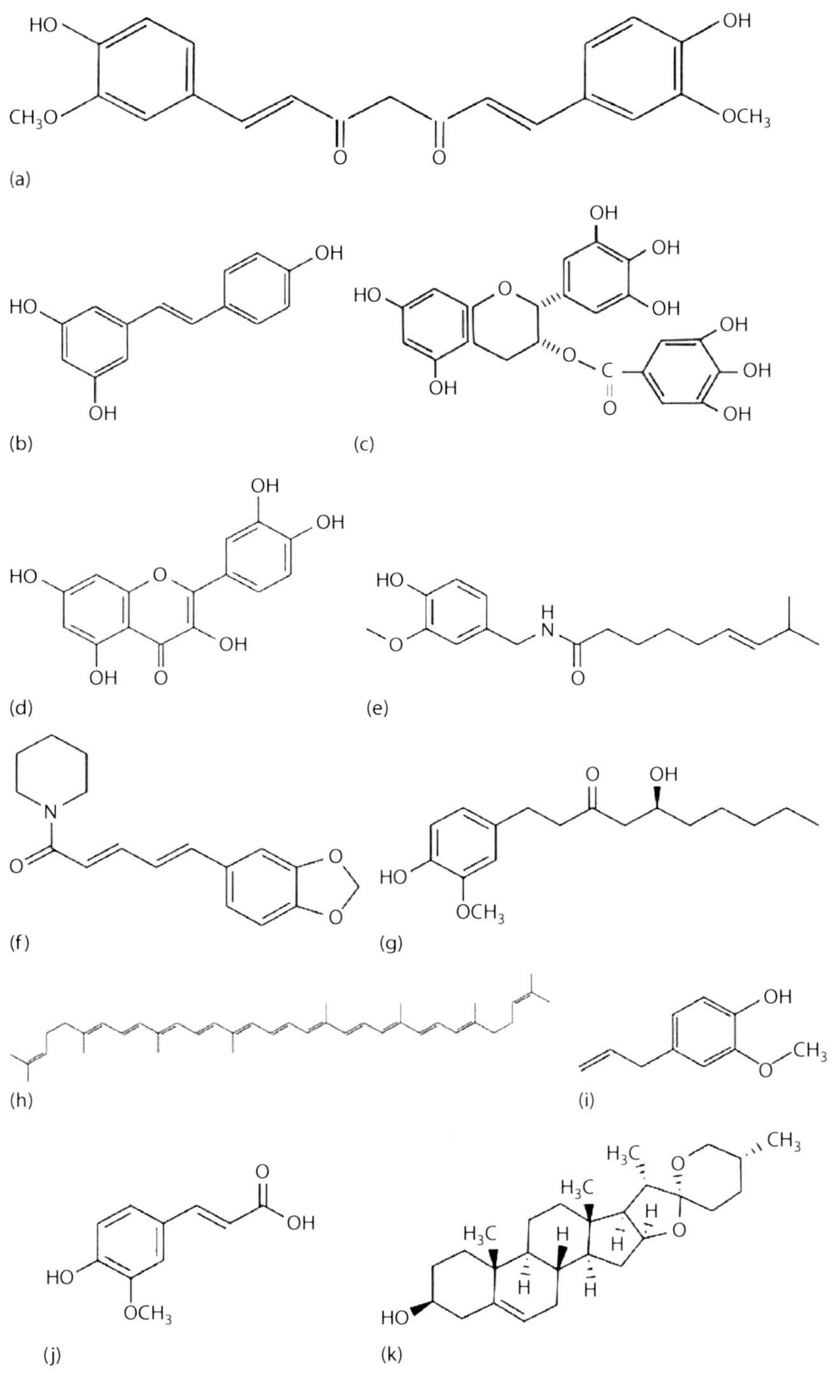

(a)

(b)

(c)

(d)

(e)

(f)

(g)

(h)

(i)

(j)

(k)

FIGURE 3.1 Chemical structures of Nutraceuticals: (a) curcumin, (b) resveratrol, (c) EGCG, (d) quercetin, (e) capsaicin, (f) piperene, (g) gingerol, (h) lycopene, (i) eugenol, (j) ferulic acid, (k) diosgenin.

3.1.1.3 Quercetin

Quercetin is a naturally occurring flavonoid present in abundance in plants (Niu *et al.*, 2010) and is believed to be an excellent free-radical scavenger (Zhang *et al.*, 2011). Quercetin is reported to inhibit A-549 human lung adenocarcinoma cancer cell line through induction of apoptosis by downregulating the expression of the anti-apoptotic gene bcl-2, and upregulating the expression of pro-apoptotic gene bax (Zheng *et al.*, 2012). This mode of action was also demonstrated by Niu and his coworkers (2010) in their study. Further, they also observed that quercetin induces apoptosis through caspase-3 dependant mitochondrial pathway by restricting the expression of the cyclooxygenase-2-mRNA and its protein in leukemia cells (Niu *et al.*, 2010).

3.1.1.4 Green Tea Polyphenols

The capability of green tea polyphenols to constrict the neovascularization and formation of new blood vessels is symptomatic, and thus may affect the growth of subsisting tumors too (Jung and Ellis, 2001). Epigallocatechin gallate (EGCG), found in green tea (*Camellia sinensis*), has been one of the most meticulously researched nutraceuticals (Boehm *et al.*, 2009; Butt and Sultan, 2009; Sagar *et al.*, 2006). EGCG not only restricts the growth of various tumors in culture but has also been known to reduce formation of new vasculature encouraged by VEGF in colon, breast, and head and neck cancer cell lines (Hazgui *et al.*, 2008; Kato *et al.*, 2008; Larsen and Dashwood, 2010). EGCG has been observed to diminish molecular signals such as nuclear factor kappa-β, matrix metallo proteinases, and Cyclooxygenase-2 proteins, all engaged in spreading and subsistence of tumor cells (Park *et al.*, 2009; Roomi *et al.*, 2010).

3.1.1.5 Lycopene

Lycopene has been documented for its strong antioxidant profile, which is also accountable for its beneficial effects (Johary *et al.*, 2012). It is the lead precursor of β-carotene in tomato, which accrues after the lycopene cyclase gene is decreased during the ripening stage. Experimental studies carried out by Ono *et al.* (2015) demonstrated that lycopene constricted the growth of chemically induced carcinogenesis *in vivo*. The pathway includes the regulation of growth factors, cell cycle arrest, and/or apoptosis induction, along with alteration in antioxidant and phase II detoxifying enzymes. Anti-inflammatory property of lycopene is also considered to play an important role in suppressing the promotion and development of carcinogenesis, inhibition of cell invasion, angiogenesis, and metastasis. A well-planned clinically involved trial is required to further elucidate the exact function of lycopene in cancer prevention (Ono *et al.*, 2015). Work done by Gajowik and Dobrzyńska (2014) further suggested that supplementation of lycopene might be useful in mitigating the negative side effects of cancer radiotherapy or in declining the poor condition caused by possible radiation accidents on human health.

3.1.1.6 Propolis

Propolis is a dense concoction of flavonoids, polyphenols, phenolic acids, phenolic aldehydes, aglycones, and ketones (Krocko, 2013). Premratanachai and Chanchao (2014) in their review article demonstrated that the water-soluble extracts of propolis, if given before the tumor exposure, exhibited superior *in vitro* efficacy. Also, there was significant difference between the sensitivity toward cancer cells and normal fibroblasts. Study also demonstrated the mechanism involved in cell death as apoptosis via activation of caspases (Premratanachai and Chanchao, 2014). In another study, it has been shown that the suppression of complexes of cyclins as well as cell cycle arrest is responsible for antiproliferative effects of propolis in cancer cells (Sawicka *et al.*, 2012). An experiment conducted with ethanolic extract of propolis indicated its cytotoxic properties. Further, its assistance synergistically improved the growth inhibiting activity of temozolomide against glioblastoma U87MG cell line and the process is partially mediated by a reduced activity of NF-κB (Renata *et al.*, 2013).

3.1.1.7 Ferulic Acid

Available in several fruits and cereals, ferulic acid (FA) is an extensively dispersed hydroxycinnamic acid that exhibits potent antioxidant and anticancer activities (Panwar *et al.*, 2016). FA along with *para*-coumaric acid (*p*-CA) has an effect on the metabolic activity, proliferation, cell cycle phase distribution, and kinetics of the colonic endothelial tumor cell line, Caco-2 (Janick *et al.*, 2005). Another study has documented the efficacy of FA against human cervical carcinoma cells ME-180 and HeLa and reported significant depletion of radiation surviving factor and increase in the lipid peroxidation indices in cancer cells (Subburayan *et al.*, 2011).

3.1.1.8 Piperine and Piplartine

A review article authored by Wargovich *et al.* (2010) has construed that piperene, a biologically active constituent in black pepper (*Piper nigrum*), helps to improve the absorption and availability index of curcumin and tea polyphenols, while modulating their isoforms during drug metabolism in the liver. Another alkaloid present in *Piper* species, piplartine, is also reported to improve the anticancer efficiency of 5-fluorouracil, both *in vitro* and *in vivo*. But, possibly, the most interesting activity documented about piperine is its role in hindering the production and development of mammary glands. Its efficacy to alter breast cancer stem cell biology still needs to be studied (Wargovich *et al.*, 2010).

3.1.1.9 Capsaicin

Capsaicin, the active compound present in Chile pepper (*Capsicum* sp.), is reported to hinder the cancer cells proliferation *in vitro* and curb the inflammatory response. Capsaicin may induce apoptosis by manipulating the GSKβ3 protein in beta-catenin pathway, altered in several cancers (Lee *et al.*, 2010; Yang *et al.*, 2009). Promising results have been shown by capsaicin in the preliminary studies against prostate cancer, lung cancer, and various leukemias (Chhabra *et al.*, 2012).

3.1.1.10 Eugenol

Tumor growth control potential of eugenol present in oils of clove (*Syzygium aromaticum*), basil, etc. was observed in human cancer cell lines and in the MNNG-rat model of gastric cancer (Manikandan *et al.*, 2009; Slamenova *et al.*, 2009). It was found that cell proliferation was subdued by suppressing nuclear factor kappa-β. Later, in another study, it was also demonstrated that eugenol has anti-inflammatory effects in the rat gastric cancer model by downregulating the expression of matrix metalloproteinases, vascular endothelial growth factors, tissue inhibitor of metalloproteinases, and RECK that all are associated with incursion (Manikandan *et al.*, 2010).

3.1.1.11 Gingerol

Gingerol (6-gingerol), present in ginger (*Zingiber officinale*), is one of the biologically active compounds (Kundu *et al.*, 2009). It is reported to have promising anti-inflammatory activity and has been stated to inhibit incursion, motility, and adhesion in human breast cancer cells, human hepatocarcinoma cells, and some other cell lines (Lee *et al.*, 2008; Kundu *et al.*, 2009; Dugasani *et al.*, 2010). Also, one study documented benefit of ginger is not due to its anticancer potential, but owing to its capability to counteract nausea and vomiting, that is, the negative impacts of chemotherapy (Hoffman, 2007).

3.1.1.12 Diosgenin

Diosgenin is a steroidal saponin present in fenugreek seeds (*Trigonella foenum-graecum*) and is observed to be quite harmless even at elevated doses when studied in rats (Taylor *et al.*, 2000). Diosgenin has been known to exhibit anti-proliferative potential against various cancer cell lines like human myelogenous leukemia, colon cancer, and breast cancer (Shishodia *et al.*, 2006; Raju and Bird, 2007; Srinivasan *et al.*, 2009). It restricts the activities of Nuclear Factor κappa-β, cyclooxgenase enzymes, and is presumed to have effect on 5-LOX also, though its effect on LOX suppression is yet to be confirmed (Raju and Mehta, 2009). Another study indicated that pure and impure diosgenin prevented telomerase activity by decreasing the hTERT gene expression in A549 lung cancer cell line (Rahmati-yamchi *et al.*, 2014).

3.1.1.13 Phycocyanin

Phycocyanin (Pc), a phycobilliprotein, is a potent therapeutic agent. In a study conducted by Subhashini and coworkers (2004), C-phycocyanin was demonstrated to induce apoptosis in human chronic myeloid leukemia K562 cell line. Another study exhibited that phycocyanin induced apoptosis in SKOV-3 ovary cancer cells by modulating protein expression (Pan *et al.*, 2015). A recent study demonstrated that phycocyanin induced apoptosis in pancreatic cancer cells too by upregulating p38 and JNK pathways, and stimulated autophagic cell death by restraining PI3/Akt/mTOR signaling pathways (Liao *et al.*, 2016).

3.1.2 Limitations

However, despite having such versatile profiles, therapeutical applications of nutraceuticals are limited because of their hydrophobicity, poor bioavailability, and instability. Sometimes, the optimum dose for the penetration and retention in the cells is quite high, and that can lead to toxic side effects. Further, shorter biological half-life and in certain cases immunotoxicity and reproductive toxicity also hindered their effective utilization in clinics.

3.1.3 Role of Nanotechnology in Nutraceuticals Delivery

Nanotechnology has given us many tools that can help in achieving new horizons for the application of this therapy. Nanotechnology possibly has a significant role in negating the drawbacks of otherwise potent nutraceuticals. Several nanocarriers have already been exploited for the efficient delivery of nutraceuticals. Table 3.1 is an assemblage of various nutraceuticals-loaded nanoparticles-mediated delivery systems, resulting in their enhanced bioavailability and biological activities, as compared to their counterparts.

3.2 CONCLUSION

Employing chemopreventive nutraceuticals in medicinal regime is restricted, as reported. To conquer these restrictions and to harness the potential to their best, the researchers have commenced the idea of exploiting nanotechnology to perk up their pharmacokinetic and pharmacodynamic potential. Further, in the perspectives of its utilization in the field of cancer therapy as well as in its diagnosis, nanotechnology has emerged as one of the most powerful approaches. Largely, this chapter outlines the promising position of diversified nanocarriers as potent nutraceutical delivery vehicles, in terms of their increased bioavailability and efficiency.

TABLE 3.1
Applications of Nanotechnology in Nutraceuticals Delivery

Nutraceuticals	Observations	Reference
Curcumin	**Polymeric NPs encapsulated curcumin evaluated against pancreatic tumor cells** • Apoptosis Induction • Inhibition of nuclear factor kappa B (NFκB) • Depletion of pro-inflammatory cytokines like IL-6, IL-8, and TNFα	Bisht *et al.*, 2007
	Curcumin-loaded biocompatible thermoresponsive polymeric NPs • Toxicity toward PC3 cancer cell line via apoptosis	Rejinold *et al.*, 2011
	Curcumin-Loaded Chitosan NPs • Cytotoxic activity against cancerous cells • Nontoxic against normal ones	Anitha *et al.*, 2012
	Curcumin loaded NPs against prostate cancer • Well internalization of nano particles and better tumor deterioration in xenograft mice • Reduction in expressions of nuclear β-catenin and androgen receptors • Regression in STAT3 and AKT phosphorylat • Restriction of Mcl-1 & Bcl-xL proteins • PARP cleavage • Oncogenic miR21 depletion and increase in miR-205 levels	Yallapu *et al.*, 2014
	Curcumin-loaded hydrogel NPs derived clusters against lung cancer cell line (A-549) • Increment in apoptosis rates	Teong *et al.*, 2015
	Curcumin NPs • Specificity toward PC3 cancer cells as compared to normal HEK cells • Hemolysis% for both (nanocurcumin and native) was almost similar	Adahoun *et al.*, 2016
Resveratrol	**Resveratrol NPs** • Better active target drug delivery system due to lack of surface amines	Nair, 2010
	Resveratrol-loaded gelatin NPs • Upregulation in cellular toxicity against NCI-H460 cells • Reactive oxygen species (ROS) generation and DNA damage • Enhanced bioavailablity and longer half-life	Karthikeyan *et al.*, 2013
	Resveratrol-bovine serum albumin NPs • Apoptosis in ovary cancer cells via caspase cascade	Guo *et al.*, 2015
	Colloidal mesoporous silica NPs of resveratrol • Superior cytoxicity than native resveratrol via PARP and cIAP1 pathways • Inhibition of lipopolysaccharide-induced NF-κB activation in macrophages	Natalie *et al.*, 2016
	Solid lipid NPs of resveratrol modified by adding Apolipo E protein • Better cellular uptake of resveratrol	Neves *et al.*, 2016
Green Tea Polyphenols	**PLGA NPs of tea polyphenols** • Enhanced protection of mouse skin tissue against DNA damage	Srivastava *et al.*, 2013
	Self-assembled micellar nanocomplexes of green tea catechins and herceptin • Adjuvant efficacy leading to better tumour specificity and anti-proliferative activity • Increased blood half-life	Chung *et al.*, 2014

(*Continued*)

TABLE 3.1 (CONTINUED)
Applications of Nanotechnology in Nutraceuticals Delivery

Nutraceuticals	Observations	Reference
	Gold-conjugated green tea NPs • Upregulation in anticancer activity and protection in liver tissues • Apoptosis indicated via increase in Bax levels, depletion in Bcl2 and Caspase 3 levels • Anti-oxidant activity and better cellular uptake	Mukherjee *et al.*, 2015
	Green tea polyphenols • Increased specific *in vitro* cytotoxicity in bladder cancer cells as compared to normal Vero cells • Induction of apoptosis via mitochondrial pathway • Downregulation of tumor cells when given orally to C3H/HeN mice	Granja *et al.*, 2016
	Epigallocatechin gallate (EGCG)-gold NPs • Increased apoptosis in B16F10 murine melanoma cells via intrinsic pathway • Enhanced biocompatibility and lesser injury to erythrophils • Depletion in tumor growth *in vivo*	Granja *et al.*, 2016
	EGCG-conjugated gold NPs • Higher stability and perseverance of antioxidant profile • Apoptosis in neuroblastoma cells	Granja *et al.*, 2016
Quercetin	**Protein-quercetin bioactive NPs** • Perseverance of antioxidant potential by encapsulation	Fang *et al.*, 2011
	Quercetin embedded NPs • Sustained release and enhanced anticancer efficacy	Pandey *et al.*, 2015
	Quercetin-loaded PLGA NPs • Targeted apoptosis in HepG2 cell line • Efficiently repressed tumor growth in solid tumor-bearing mouse model	Guan *et al.*, 2016
Ferulic acid	**Ferulic acid (FA) NPs** • Enhanced anticancer effect via increase in Reactive Oxygen Species (ROS) generation • Upregulation in mitochondrial polarization, DNA injury and TBARS	Merlin *et al.*, 2011
	FA-inorganic nanohybrids • Two times higher anticancer activity	Kim *et al.*. 2013
	FA-loaded chitosan NPs • Characteristics morphological changes of apoptosis in ME-180 cells	Panwar *et al.*, 2016
Lycopene	**Encapsulated lycopene against leukemia cell line (K562)** • Restriction in the cellular growth and enhancement in apoptosis • Inhibition of telomerase activity	Gharib and Faezizadeh, 2014
	Nanoemulsion delivering gold NPs and lycopene • Likely combined effects on HT-29 cell line	Huang *et al.*, 2015
Piperene	**Piperine-loaded PEG-PLGA NPs against breast cancer cells** • Adjuvant therapeutic effect with Paclitaxel along with conjugated aptamer NPs • Noteworthy decrease in required paclitaxel dosages, thus making nano scale piperine a safe adjuvant in cancer chemotherapy	Pachauri *et al.*, 2015
	Rapamycin- and piperine-loaded polymeric NPs • Better absorption profile with improved bioavailability of rapamycin in combination with piperene as chemo sensitizer	Katiyar *et al.*, 2015

(*Continued*)

TABLE 3.1 (CONTINUED)
Applications of Nanotechnology in Nutraceuticals Delivery

Nutraceuticals	Observations	Reference
	Piperine and piperlongumine-containing NPs • Decreased *in vitro* migration and incursion of triple-negative breast cancer cells • Bioactivity of phytochemicals was uncompromised	Ghassemi-Rad and Hoskin, 2015
Capsaicin	**Capsaicin-loaded trimethyl chitosan-based NPs** • Increase in Bax levels and depletion in BCL-2 as well as MDR-1 genes in human HepG2 hepatocarcinoma cells • Blank trimethyl-chitosan NPs had a little anti-apoptotic activity of their own too	Elkholi *et al.*, 2014
	Capsaicin-loaded NPs • Increased uptake in human glioblastoma U251 cells. • Enhanced anticancer potential due to their ability to cross the blood-brain barrier	Jiang *et al.*, 2015
Propolis	**Propolis nanofood** • Easy dispersion in aqueous media • Identical *in vitro* toxicity against a battery of human pancreatic cancer cells as of free propolis	Kim *et al.*, 2008
	Synthesized silver NPs using stingless bee propolis • Having enhanced anticancer potential against human lung cancer cell line (A-549) • Of greater compatibility, lesser costly, and ecofriendly	Kothai and Jayanthi, 2014
Eugenol	**Eugenol-loaded nanoemulsions** • Superior antiproliferative profile against colon cancer • ROS production leading to apoptosis	Majeed *et al.*, 2014
Gingerol	**Ginger-derived NPs** • Protective effect toward liver against alcohol-induced liver damage • Nuclear factor erythroid 2-related factor-2 activated, leading to the activation of antioxidant genes, which further decrease ROS generation	Zhuang *et al.*, 2015
Diosgenin	**Self-assembled NPs of diosgenin** • Optimum size, higher drug loading capacity, slow release of the drugs, and higher combined efficiency	Li *et al.*, 2015
Phycocyanin	**Silver nanoparticles fabricated utilising phycocyanin** • Noticeable growth inhibition of Gram +ve and Gram –ve bacteria • Promising cytotoxic effect against breast cancer (MCF-7) cell line as well as in ascites tumor possessing mice	El-Naggar *et al.*, 2017
	Selenium nanoparticles designed with phycocyanin • Protectivity by mitigating apoptotic effects of Palmitic acid in pancreatic β cells by hindering mitochondria depolarisation	Liu *et al.*, 2017
	Phycocyanin nanoformulation delivering paclitaxel • Better cellular uptake lead to enhanced efficiency of drug against U87MG cell line	Agrawal *et al.*, 2017
	Phycocyanin as a nanocargo for doxorubicin • Stimulation of ROS-intervening apoptosis • Controlling drug resistance	Huang *et al.*, 2017

Note: NPs: nanoparticles.

REFERENCES

Acosta, E. Bioavailability of nanoparticles in nutrient and nutraceutical delivery. *Curr Opin Colloid Interface Sci.* 2009; 14: 3–15.

Adahoun, M. A., Al-Akhras, M. H., Jaafar, M. S., and Bououdina, M. Enhanced anti-cancer and antimicrobial activities of curcumin nanoparticles. *Artif Cells Nanomed Biotechnol.* 2016; 8: 1–10.

Agrawal, M., Yadav, S. K., Agrawal, S. K., and Karmakar, S. Nutraceutical phycocyanin nanoformulation for efficient drug delivery of paclitaxel in human glioblastoma U87MG cell line. *J Nanoparticle Res.* 2017; 19: 272.

Anitha, A., Maya, S., Deepa, N., Chennazhi, K. P., Nair, S. V., and Jayakumar, R. Curcumin-loaded N,O-Carboxymethyl chitosan nanoparticles for cancer drug delivery. *J Biomater Sci Polym Ed.* 2012; 23(11): 1381–1400.

Bachmeier, B. E., Killian, P., Pfeffer, U., and Nerlich, A. G. Novel aspects for the application of Curcumin in chemoprevention of various cancers. *Front Biosci: A Journal and Virtual Library (Schol Ed)* 2010; 2: 697–717.

Bharali, D. J., Siddiqui, I. A., Adhami, V. M., Chamcheu, J. C., Aldahmash, A. M., Mukhtar, H. *et al.* Nanoparticle delivery of natural products in the prevention and treatment of cancers: Current status and future prospects. *Cancers.* 2011; 3: 4024–4045.

Bhatia, S. Nanoparticles types, classification, characterization, fabrication methods and drug delivery applications. *Natural polymer drug delivery systems.* 2016; Springer, Cham.

Bhuvaneswari, V., and Nagini, S. Lycopene: A review of its potential as an anticancer agent. *Curr Med Chem Anticancer Agents.* 2005; 5(6): 627–635.

Bisht, S., Feldmann, G., Soni, S., Ravi, R., Karikar, C., Maitra, A. *et al.* Polymeric nanoparticle-encapsulated curcumin ("nanocurcumin"): A novel strategy for human cancer therapy. *J NanoBiotech.* 2007; 5: 3.

Boehm, K., Borrelli, F., Ernst, E., Habacher, G., Hung, S. K., Milazzo, S. *et al.* Green tea (*Camellia sinensis*) for the prevention of cancer. *Cochrane Database Syst Rev.* 2009; CD005004.

Butt, M. S., and Sultan, M. T. Green tea: Nature's defense against malignancies. *Crit Rev Food Sci Nutr.* 2009; 49: 463–473.

Caldorera-Moore, M., Guimard, N., Shi, L., and Roy, K. Designer nanoparticles: Incorporating size, shape, and triggered release into nanoscale drug carriers. *Expert Opin Drug Deliv.* 2010; 7(4): 479–495.

Champion, J., Katare, Y., and Mitragotri, S. Particle shape: A new design parameter for micro- and nanoscale drug delivery carriers. *J Control Release.* 2014; 121(1–2): 3–9.

Chapman, J., Sullivan, T., and Regan, F. Nanoparticles in anti-microbial materials: Use and characterisation. 2012; 23, RSC NanoScience & Nanotechnology.

Chhabra, N., Aseri, M. L., Goyal, V., and Sankhla, S. Capsaicin: A promising therapy—A critical reappraisal. *Int J Nutr Pharmacol Neurol Dis.* 2012; 2(1): 8–15.

Chung, J. E., Tan, S., Gao, S. J., Yongvongsoontorn, N., Kim, S. H., Lee, J. H. *et al.* Self-assembled micellar nanocomplexes comprising green tea catechin derivatives and protein drugs for cancer therapy. *Nat Nanotechnol.* 2014; 9: 907–912.

Dugasani, S., Pichika, M. R., Nadarajah, V. D., Balijepalli, M. K., Tandra, S., Korlakunta, J. N. Comparative antioxidant and anti-inflammatory effects of [6]-gingerol, [8]-gingerol, [10]-gingerol and [6]-shogaol. *J Ethnopharmacol.* 2010; 127: 515–520.

Elkholi, I. E., Hazem, N. M., ElKashef, W. F., Sobh, M. A., Shaalan, D., Sobh, M. *et al.* Evaluation of anti-cancer potential of capsaicin-loaded trimethyl chitosan-based nanoparticles in HepG2 hepatocarcinoma cells. *J Nanomed Nanotechnol.* 2014; 5(6): 1–8.

El-Naggar, N. E., Hussein, M. H., El-Sawah, A. A. Bio-fabrication of silver nanoparticles by phycocyanin, characterization, *in vitro* anticancer activity against breast cancer cell line and *in vivo* cytotxicity. *Sci Rep.* 2017; 7(1): 10844.

Fang, R., Jing, H., Chai, Z., Zhao, G., Stoll, S., Ren, F. *et al.* Design and characterization of protein-quercetin bioactive nanoparticles. *J Nanobiotechnology.* 2011; 9: 19.

Gajowik, A., and Dobrzyńska, M. M. Lycopene - Antioxidant with radioprotective and anticancer properties. A review. *Rocz Panstw Zakl Hig.* 2014; 65(4): 263–271.

Gharib, A., and Faezizadeh, Z. *In vitro* anti-telomerase activity of novel lycopene-loaded nanospheres in the human leukemia cell line K562. *Pharmacogn Mag.* 2014; 10(Suppl 1): S157–S163.

Ghassemi-Rad, J., and Hoskin, D. W. Nanoparticle-encapsulated piperine and piperlongumine inhibit breast cancer cell growth and metastatic activity. *Cancer Res.* 2015; 75: 3684.

Granja, A., Pinheiro, M., and Reis, S. Epigallocatechin gallate nanodelivery systems for cancer therapy. *Nutrients.* 2016; 8(5): 307.

Guan, X., Gao, M., Xu, H., Zhang, C., Liu, H., Li, L. *et al.* Quercetin-loaded poly (lactic-*co*-glycolic acid)-d-α-tocopheryl polyethylene glycol 1000 succinate nanoparticles for the targeted treatment of liver cancer. *Drug Deliv.* 2016; 3: 1–12.

Guo, L., Peng, Y., Li, Y., Yao, J., Zhang, G., and Chen, J. Cell death pathway induced by resveratrol-bovine serum albumin nanoparticles in a human ovarian cell line. *Oncol Lett.* 2015; 9(3): 1359–1363.

Hazgui, S., Bonnomet, A., Nawrocki-Raby, B., Milliot, M., Terryn, C., Cutrona, J. *et al.* Epigallocatechin-3-gallate (EGCG) inhibits the migratory behavior of tumor bronchial epithelial cells. *Respiratory Res.* 2008; 9: 33.

Hoffman, T. Ginger: An ancient remedy and modern miracle drug. *Hawaii Med J.* 2007; 66: 326–327.

Huang, R. F. S., Wei, Y. J., Inbaraj, B. S., and Chen, B. H. Inhibition of colon cancer cell growth by nanoemulsion carrying gold nanoparticles and lycopene. *Int J Nanomedicine.* 2015; 8(10): 2823–2846.

Huang, Y., He, L., Song, Z., Chan, L., He, J., Huang, W., Zhou, B., and Chen. Phycocyanin-based nanocarrier as a new nanoplatform for efficient overcoming of cancer drug resistance. *J Mater Chem B.* 2017; 5: 3300.

Ikada, Y. Phagocytosis of polymer microspheres by macrophages. *Adv Polym Sci.* 1990; 94: 107–141.

Illum, L., Davis, S., Wilson, C., Thomas, N., Frier, M., and Hardy, J. Blood clearance and organ deposition of intravenously administered colloidal particles. The effects of particle size, nature and shape. *Int J Pharm.* 1982; 12(2–3): 135–146.

Jang, M., Cai, L., Udeani, G. O., Slowing, K. V., Thomas, C. F., Beecher, C. W. *et al.* Cancer chemopreventive activity of Resveratrol, A natural product derived from grapes. *Science.* 1997; 275: 218–220.

Janicke, B., Önning, G., and Oredsson, S. M. Differential effects of ferulic acid and *p*-coumaric acid on S phase distribution and length of S phase in the human colonic cell line caco-2. *J Agric Food Chem.* 2005; 53 (17): 6658–6665.

Jiang, Z., Wang, X., Zhang, Y., Zhao, P., Luo, Z., and Li, J. Effect of capsaicin-loading nanoparticles on gliomas. *J Nanosci Nanotechnol.* 2015; 15(12): 9834–9839.

Johary, A., Jain, V., and Misra, S. Role of lycopene in the prevention of cancer. *Int J Nutr Pharmacol Neurol Dis.* 2012; 2: 167–170.

Jong, W. H. D., and Borm, P. J. A. Drug delivery and nanoparticles: Applications and hazards. *Int J Nanomedicine.* 2008; 3(2): 133–149.

Jung, Y. D., and Ellis, L. M. Inhibition of tumour invasion and angiogenesis by epigallocatechin gallate (EGCG), a major component of green tea. *Int J Exp Pathol.* 2001; 82(6): 309–316.

Kamat, A. M., Tharakan, S. T., Sung, B., and Aggarwal, B. B. Curcumin potentiates the antitumor effects of Bacillus Calmette-Guerin against bladder cancer through the downregulation of NF-kappaB and upregulation of TRAIL receptors. *Cancer Res.* 2009; 69: 8958–8966.

Kang, H. J., Lee, S. H., Price, J. E., and Kim, L. S. Curcumin suppresses the paclitaxel-induced nuclear factor-kappa beta in breast cancer cells and potentiates the growth inhibitory effect of paclitaxel in a breast cancer nude mice model. *Breast J.* 2009; 15: 223–229.

Karthikeyan, S., Rajendra Prasad, N. R., Ganamani, A., and Balamurugan, E. Anticancer activity of resveratrol-loaded gelatin nanoparticles on NCI-H460 non-small cell lung cancer cells. *Biomed Preventive Nutr.* 2013; 3(1): 64–73.

Katiyar, S. S., Muntimadugu, E., Rafeeqi, T. A., Domb, A. J., and Khan, W. Co-delivery of rapamycin- and piperine-loaded polymeric nanoparticles for breast cancer treatment. *Drug Deliv.* 2015; 3:1–9.

Kato, K., Long, N. K., Makita, H., Toida, M., Yamashita, T., Hatakeyama, D. *et al.* Effects of green tea polyphenol on methylation status of RECK gene and cancer cell invasion in oral squamous cell carcinoma cells. *British J Cancer.* 2008; 99: 647–654.

Kim, D. M., Lee, G. D., Aum, S. H., and Kim, H. J. Preparation of propolis nanofood and application to human cancer. *Biol Pharm Bull.* 2008; 31(9): 1704–1710.

Kim, H. J., Ryu, K., Kang, J. H., Choi, A. J., Kim, T. I., and Oh, J. M. Anticancer activity of ferulic acid-inorganic nanohybrids synthesized via two different hybridization routes, reconstruction and exfoliation-reassembly. *Sci World J.* 2013; 2013: 9.

Kothai, S., and Jayanthi, B. Anti cancer activity of silver nano particles bio-synthesized using stingless bee propolis (*Tetragonula iridipennis*) of Tamilnadu. *Asian J Biomed Pharma Sci.* 2014; 04(40): 30–37.

Krocko, M. Effect of Propolis used in the broiler feed mixtures to the selected quality indicators of cooked Ham. *Scientific Papers Animal Science and Biotechnologies.* 2013; 46(2): 22–26.

Kundu, J. K., Na, H. K., and Surh, Y. J. Ginger-derived phenolic substances with cancer preventive and therapeutic potential. *Forum Nutr.* 2009; 61:182–192.

Kunnumakkara, A. B., Diagaradjane, P., Anand, P., Harikumar, K. B., Deorukhkar, A., Gelovani, J. *et al.* Curcumin sensitizes human colorectal cancer to capecitabine by modulation of cyclin D1, COX-2, MMP-9, VEGF and CXCR4 expression in an orthotopic mouse model. *Int J Cancer.* 2009; 125: 2187–2197.

Larsen, C. A., and Dashwood, R. H. Epigallocatechin-3-gallate inhibits Met signaling, proliferation, and invasiveness in human colon cancer cells. *Arch Biochem Biophys*. 2010; 501(1): 52–57.

Lee, H. S., Seo, E. Y., Kang, N. E., and Kim, W. K. [6]-Gingerol inhibits metastasis of MDA-MB-231 human breast cancer cells. *J Nutr Biochem*. 2008; 19: 313–319.

Lee, S. H., Krisanapun, C., and Baek, S. J. NSAID-activated gene-1 as a molecular target for capsaicin-induced apoptosis through a novel molecular mechanism involving GSK3beta, C/EBPbeta and ATF3. *Carcinogenesis*. 2010; 31: 719–728.

Li, C., Dai, L., Liu, K., Deng, L., Pei, T., and Lei J. A self-assembled nanoparticle platform based on poly (ethylene glycol)–diosgenin conjugates for co-delivery of anticancer drugs. *RSC Adv*. 2015; 5: 74828–74834.

Liao, G., Gao, B., Gao, Y., Yang, X., Cheng, X., and Ou, Y. Phycocyanin inhibits tumorigenic potential of pancreatic cancer cells: Role of apoptosis and autophagy. *Sci Rep*. 2016; 6: 34564.

Lin, S. S., Lai, K. C., Hsu, S. C., Yang, J. S., Kuo, C. L., Lin, J. P. *et al*. Curcumin inhibits the migration and invasion of human A549 lung cancer cells through the inhibition of matrix metalloproteinase-2 and -9 and vascular endothelial growth factor (VEGF). *Cancer Lett*. 2009; 285: 127–133.

Liu, C., Fu, Y., Li, C., Chen, T., and Li, X. Phycocyanin functionalized selenium nanoparticles reverse palmitic acid induced Pancreatic β cell apoptosisby enhancing cellular uptake and blocking reactive oxygen species (ROS) mediated mitrochondria dysfunction. *J Agric Food Chem*. 2017; 65(22): 4405–4413.

Lopes, C. M., Fernandes, J. R., and Martins-Lopes, P. Application of nanotechnology in the agro-food sector. Nanotechnology in agro-food. *Food Technol. Biotechnol*. 2013; 51(2): 183–197.

Majeed, H., Antoniou, J., and Fang, Z. Apoptotic effects of eugenol-loaded nanoemulsions in human colon and liver cancer cell lines. *Asian Pac J Cancer Prev*. 2014; 15: 9159–9164.

Manikandan, P., Murugan, R. S., Priyadarsini, R. V., Vinothini, G., and Nagini, S. Eugenol induces apoptosis and inhibits invasion and angiogenesis in a rat model of gastric carcinogenesis induced by MNNG. *Life Sci*. 2010; 86(25–26): 936–941.

Manikandan, P., Vinothini, G., Vidya Priyadarsini, R., Prathiba, D., and Nagini, S. Eugenol inhibits cell proliferation via NF-kappaB suppression in a rat model of gastric carcinogenesis induced by MNNG. *Invest New Drugs*. 2009; 23: 23.

Men, K., Duan, X., Wei, X. W., Gou, M. L., Huang, M. J., Chen, L. J. *et al*. Nanoparticle-delivered quercetin for cancer therapy. *Anticancer Agents Med Chem*. 2014; 14(6): 826–832.

Merlin, J. P. J., Prasad, N. R., Shibli, S. M. A., and Sebeela, M. Ferulic acid loaded Poly-d,l-lactide-co-glycolide nanoparticles: Systematic study of particle size, drug encapsulation efficiency and anticancer effect in non-small cell lung carcinoma cell line *in vitro*. *Biomed Preventive Nutr*. 2012; 2(1): 69–76.

Mukherjee, S., Ghosh, S., Das, D. K., Chakraborty, P., Choudhury, S., and Gupta, P. Gold-conjugated green tea nanoparticles for enhanced anti-tumor activities and hepatoprotection—Synthesis, characterization and *in vitro* evaluation. *J Nutr Biochem*. 2015; 26(11): 1283–1297.

Murthy, S. K. Nanoparticles in modern medicine: State of the art and future challenges. *Int J Nanomedicine*. 2007; 2(2): 129–141.

Nair, H. B., Sung, B., Yadav, V. R., Kannappan, R., Chaturvedi, M. M., and Aggarwal, B. B. Delivery of anti-inflammatory nutraceuticals by nanoparticles for the prevention and treatment of cancer. *Biochem Pharmacol*. 2010; 80(12): 1833–1843.

Natalie, S., Zhi, O., Naisarg, P., Yong, S., Siddharth, J., and Michael, M. G. Colloidal mesoporous silica nanoparticles enhance the biological activity of resveratrol. *Colloids Surf B Biointerfaces*. 2016; 144: 1–7.

Neves, A. R., Queiroz, J. F., and Reis, S. Brain-targeted delivery of resveratrol using solid lipid nanoparticles functionalized with apolipoprotein. *E J Nanobiotechnology*. 2016; 14: 27.

Niu, G., Yin, S., Xie, S., Li, Y., Nie, D., Ma, L. *et al*. Quercetin induces apoptosis by activating caspase-3 and regulating Bcl-2 and cyclooxygenase-2 pathways in human HL-60 cells. *Acta Biochim Biophys Sin*. 2010; 43(1): 30–37.

Ono, M., Takeshima, M., and Nakano, S. Mechanism of the anticancer effect of lycopene (*Tetraterpenoids*). *Enzymes*. 2015; 37: 139–166.

Pachauri, M., Gupta, E. D., and Ghosh, P. C. Piperine loaded PEG-PLGA nanoparticles: Preparation, characterization and targeted delivery for adjuvant breast cancer chemotherapy. *J Drug Deliv Sci Technol*. 2015; 29: 269–282.

Pan, R., Lu, R., Zhang, Y., Zhu, M., Zhu, W., and Yang, R. Spirulina phycocyanin induces differential protein expression and apoptosis in SKOV-3 cells. *Int J Biol Macromol*. 2015; 81: 951–959.

Pandey, S. K., Patel, D. K., Thakur, R., Mishra, D. P., Maiti, P., and Haldar, C. Anti-cancer evaluation of quercetin embedded PLA nanoparticles synthesized by emulsified nanoprecipitation. *Int J Biol Macromol*. 2015; 75: 521–529.

Panwar, R., Sharma, A. K., Kaloti, M., Dutt, D., and Pruthi, V. Characterization and anticancer potential of ferulic acid-loaded chitosan nanoparticles against ME-180 human cervical cancer cell lines. *Appl Nanosci.* 2016; 6: 803.

Park, I. J., Lee, Y. K., Hwang, J. T., Kwon, D. Y., Ha, J., and Park, O. J. Green tea catechin controls apoptosis in colon cancer cells by attenuation of H_2O_2-stimulated COX-2 expression via the AMPK signaling pathway at low-dose H_2O_2. *Ann N Y Acad Sci.* 2009; 1171: 538–544.

Premratanachai, P., and Chanchao, C. Review of the anticancer activities of bee products. *Asian Pac J Trop Biomed.* 2014; 4(5): 337–344.

Rahmati-Yamchi, M., Ghareghomi, S., Haddadchi, G., Milani, M., Aghazadeh, M., and Daroushnejad H. Fenugreek extract diosgenin and pure diosgenin inhibit the hTERT gene expression in A549 lung cancer cell line. *Mol Biol Rep.* 2014; 41(9): 6247–6252.

Raju, J., and Bird, R. P. Diosgenin, a naturally occurring steroid [corrected] saponin suppresses 3-hydroxy-3-methylglutaryl CoA reductase expression and induces apopto-sis in HCT-116 human colon carcinoma cells. *Cancer Lett.* 2007; 255: 194–204.

Raju, J., and Mehta, R. Cancer chemopreventive and therapeutic effects of diosgenin, a food saponin. *Nutr Cancer.* 2009; 61: 27–35.

Rejinold, S. N., Muthunarayanan, M., Divyarani, V. V., Sreerekha, P. R., Chennazhi, K. P., Nair, S. V. *et al.* Curcumin-loaded biocompatible thermoresponsive polymeric nanoparticles for cancer drug delivery. *J Colloid Interface Sci.* 2011; 360(1): 39–51.

Renata, M. Z., Borawska, M. H., Fiedorowicz, A., Naliwajko, S. K., Sawicka, D., and Car, H. Propolis changes the anticancer activity of temozolomide in U87MG human glioblastoma cell line. *BMC Complement Altern Med.* 2013; 13: 50–58.

Roomi, M. W., Monterrey, J. C., Kalinovsky, T., Rath, M., and Niedzwiecki, A. Comparative effects of EGCG, green tea and a nutrient mixture on the patterns of MMP-2 and MMP-9 expression in cancer cell lines. *Oncol Rep.* 2010; 24: 747–757.

Sagar, S. M., Yance, D., and Wong, R. K. Natural health products that inhibit angiogenesis: A potential source for investigational new agents to treat cancer—Part 1. *Curr Oncol.* 2006; 13: 14–26.

Salami, A., Seydi, E., Pourahmad, J. Use of nutraceuticals for prevention and treatment of cancer. *Iran J Pharm Res.* 2013; 12(3): 219–220.

Sandur, S. K., Deorukhkar, A., Pandey, M. K., Pabon, A. M., Shentu, S., Guha, S. *et al.* Curcumin modulates the radiosensitivity of colorectal cancer cells by suppressing constitutive and inducible NF-kappaB activity. *Int J Rad Oncol Biol Phys.* 2009; 75: 534–542.

Sawicka, D., Car, H., Borawska, M. H., and Nikliński, J. The anticancer activity of propolis. *Folia Histochem Cytobiol.* 2012; 50(1): 25–37.

Shahidi, F. Nutraceuticals, functional foods and dietary supplements in health and disease. *J Food Drug Anal.* 2012; 20(1): 226–230.

Shishodia, S., Aggarwal, B. B., Raju, J., Bird, R. P., Srinivasan, S., Koduru, S. *et al.* Diosgenin inhibits osteoclastogenesis, invasion, and proliferation through the downregulation of Akt, I kappa B kinase activation and NF-kappa B-regulated gene expression. *Oncogene.* 2006; 25: 1463–1473.

Signorelli, P., and Ghidoni, R. Resveratrol as an anticancer nutrient: Molecular basis, open questions and promises. *J Nutr Biochem.* 2005; 16: 449–466.

Singh, H. Nanotechnology applications in functional foods: Opportunities and challenges. *Prev Nutr Food Sci.* 2016; 21(1): 1–8.

Slamenova, D., Horvathova, E., Wsolova, L., Sramkova, M., Navarova, J., Manikandan, P. *et al.* Investigation of anti-oxidative, cytotoxic, DNA-damaging and DNA-protective effects of plant volatiles eugenol and borneol in human-derived HepG2, Caco-2 and VH10 cell lines. *Mutat Res.* 2009; 677: 46–52.

Srinivasan, S., Koduru, S., Kumar, R., Venguswamy, G., Kyprianou, N., and Damodaran, C. Diosgenin targets Akt-mediated prosurvival signaling in human breast cancer cells. *Int J Cancer.* 2009; 125: 961–967.

Srivastava, A. K., Bhatnagar, P., Singh, M., Mishra, S., Kumar, P., Shukla, Y. *et al.* Synthesis of PLGA nanoparticles of tea polyphenols and their strong *in vivo* protective effect against chemically induced DNA damage. *Int J Nanomedicine.* 2013; 8: 1451–1462.

Srivastava, A. K., Bhatnagar, P., Singh, M., Mishra, S., Kumar, P., Shukla, Y. *et al.* Synthesis of PLGA nanoparticles of tea polyphenols and their strong *in vivo* protective effect against chemically induced DNA damage. *J Nutr Biochem.* 2015; 26(11): 1283–1297.

Srividya, A. R., Venkatesh, N., Vishnuvarthan, V. J. Neutraceutical as medicine. *Int J Adv Pharma Sci.* 2010; 1(2): 132–133.

Stolnik, S., Illum, L., and Davis, S. Long circulating microparticulate drug carriers. *Adv Drug Deliv Rev.* 1995; 16(95): 195–214.

Subburayan, K., Govindhasamy, K., Nagarajan, R. P., and Rajendran, M. Radiosensitizing effect of ferulic acid on human cervical carcinoma cells *in vitro*. *Toxicol in Vitro*. 2011; 25: 1366–1375.
Subhashini, J., Mahipal, S. V., Reddy, M. C., Mallikarjuna Reddy, M., Rachamallu, A., and Reddanna, P. Molecular mechanisms in C-Phycocyanin induced apoptosis in human chronic myeloid leukemia cell line-K562. *Biochem Pharmacol*. 2004; 68(3): 453–462.
Taylor, W. G., Elder, J. L., Chang, P. R., and Richards, K. W. Microdetermination of diosgenin from fenugreek (Trigonella foenum-graecum) seeds. *J Agric Food Chem*. 2000; 48: 5206–5210.
Teong, B., Lin, C. Y., Chang, S. J., Niu, G. C. C., Yao, C. H., Chen, I. F. *et al*. Enhanced anti-cancer activity by curcumin-loaded hydrogel nanoparticle derived aggregates on A549 lung adenocarcinoma cells. *J Mater Sci Mater Med*. 2015; 26: 49.
Udenigwe, C. C., Ramprasath, V. R., Aluko, R. E., and Jones, P. J. Potential of resveratrol in anticancer and anti-inflammatory therapy. *Nutr Rev*. 2008; 66(8): 445–454.
Wang, E. C., and Wang, A. Z. Nanoparticles and their applications in cell and molecular biology. *Integr Biol (Camb)*. 2014; 6(1): 9–26.
Wargovich, M. J., Morris, J., Brown, V., Ellis, J., Logothetis, B., and Weber R. Nutraceutical use in late-stage cancer. *Cancer Metastasis Rev*. 2010; 29(3): 503–510.
Yallapu, M. M., Khan, S., Maher, D. M., Ebeling, M. C., Sundram, V., Chauhan, N. *et al*. Anti-cancer activity of curcumin loaded nanoparticles in prostate cancer. *Biomaterials*. 2014; 35(30): 8635–8648.
Yang, K. M., Pyo, J. O., Kim, G. Y., Yu, R., Han, I. S., Ju, S. A. *et al*. Capsaicin induces apoptosis by generating reactive oxygen species and disrupting mitochondrial transmembrane potential in human colon cancer cell lines. *Cell Mol Biol Lett*. 2009; 14: 497–510.
Zhang, M., Swarts, S. G., Yin, L., Liu, C., Tian, Y., Cao, Y. *et al*. Antioxidant properties of quercetin. *Adv Exp Med Biol*. 2011; 701: 283–289.
Zheng, S. Y., Li, Y., Jiang, D., Zhao, J., and Ge, J. F. Anticancer effect and apoptosis induction by quercetin in the human lung cancer cell line A-549. *Mol Med Rep*. 2012; 5(3): 822–826.
Zhuang, X., Deng, Z. B., Mu, J., Zhang, L., Yan, J., Miller, D. *et al*. Ginger-derived nanoparticles protect against alcohol-induced liver damage. *J Extracell Vesicles*. 2015; 4: 28713.

4 Harnessing Nanotechnology Using Nutraceuticals for Cancer Therapeutics and Intervention

Gargi Ghoshal, Ashay Jain, and O. P. Katare*

CONTENTS

4.1 INTRODUCTION

Cancer, a proliferative syndrome, is the leading cause of deaths worldwide. It is also known as malignant neoplasm, representing a group of diseases (WHO Cancer, 2016). According to a 2016 WHO report, about two million new cancer patients are identified each year, and out of them, one-third are expected to die in the United States from cancer (WHO Cancer, 2016). Nearly 90–95% of cancer cases arise owing to environmental factors, while 5–10% are due to inherent factors (Kumar and Kumar, 2014; WHO Cancer, 2016). The occurrence of cancer depends on the type of factors (internal: immune conditions, hormones, and inherited genetic mutations, or external: unhealthy diet, tobacco, X-rays, UV light, and infectious organisms) and mode of sequence of these factors (Chabner *et al.*, 1998). In the past few decades, momentous headway has been carried out on the perceptive, diagnostic, therapeutic, preventive, and management aspects of cancer therapy as next-generation sequencing, yet millions of people are still affected by cancer (Biemar and Foti, 2013; Vogelstein *et al.*, 2013). Recent evidence reports that cancer cells exhibit persistently high ROS levels as a consequence of metabolic, genetic, and microenvironment-associated alterations. High proportion of ROS, being toxic, can readily damage or mutate the DNA of cells by escaping from the respiratory chain (Poljsak *et al.*, 2013).

* Corresponding author.

Systematic management of cancer basically includes surgery, radiation, and chemotherapy, with chemotherapy out of these as most adopted. Administration of antioxidant(s), along with cancer therapy, has become an essential component of cancer therapy to overcome the toxicities responsible for the incompetence of drugs in cancer therapy (Wang *et al.*, 2010). The role of antioxidants in combating ROS-triggered damage in tumor cells has been implicated in chemoresistance with pitiable prognosis. Harnessing nanotechnology via nanocarriers in cancer biology has led to the evolution of highly efficient delivery systems with minimum toxicity (Socinski *et al.*, 2012). This chapter unravels the potential use of antioxidants together with chemotherapeutic agents (Poljsak *et al.*, 2013).

4.2 REACTIVE OXYGEN SPECIES AND CANCER

Reactive oxygen species (ROS) represent a group of oxidants capable of generating free radicals (Jain *et al.*, 2015b) and non-free radical oxygen comprising molecules such as hydroxyl radical (OH), singlet oxygen ($1/2O_2$), superoxide ($O_2\cdot-$), and hydrogen peroxide (H_2O_2) (Pacher and Szabo, 2008), derived from the single electron reductions of molecular oxygen and generated as a normal byproduct of aerobic metabolism (Equation 4.4) (Taverne *et al.*, 2013). These ROS also play a vital role in regular biochemical functions such as gene expression (Figure 4.1), cellular development, and basic defense mechanism of body at low physiological concentration (Poljsak *et al.*, 2013; Stone and Smith, 2004).

$$O_2^{*-} + e^- + 2H^+ \rightarrow H_2O_2H_2O_2 \quad \text{(Eq. 4.1)}$$

$$H_2O_2 + e^- + H^+ \rightarrow H_2O + OH^* \quad \text{(Eq. 4.2)}$$

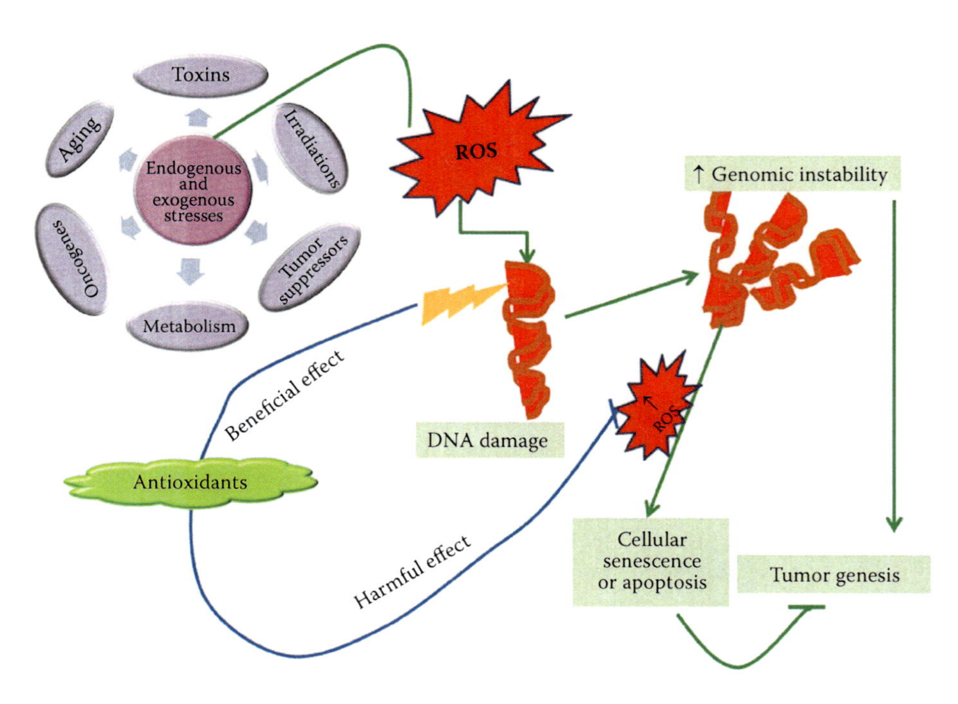

FIGURE 4.1 Role of ROS in cancer.

$$OH^{*} + e^{-} + H^{+} \rightarrow H_2O \qquad \text{(Eq. 4.3)}$$

$$O_2^{*-} + NO \rightarrow ONOO^{-} \qquad \text{(Eq. 4.4)}$$

Free radicals are generated from endogenous interaction (auto-oxidation of lipid and enzymatic inactivation of small molecule) and exogenously from external environment (include pesticides, hyperoxic environments, anesthetics, certain organic solvents and pollutants, and tobacco smoke). However, after metabolism and conversion into intermediate molecules, they cause tissue damage as excessive generation of ROS leads to oxidative stress (Machlin and Bendich, 1987). The optimum concentration of ROS at the cellular level is required to maintain an equilibrium state between rate of production and clearance by various antioxidants. Numerous reducing agents are present within the cellular microenvironment that determines the redox condition of cell and ensures limited generation of ROS (Kohen and Nyska, 2002). ROS play a noteworthy role in ordinary cell signaling as an arbitrator that normalizes vascular function, that is, physiological tone of vascular smooth muscle (Lakshmi *et al.*, 2009). However, deregulated generation of ROS leads to increased oxidative stress on cellular environment and causes change in molecular pathways, through which human disorders, such as cancer, physiological aging, neurological diseases, and heart diseases (like hypertension, atherosclerosis, and myocardial infarction) tend to develop (Pacher and Szabo, 2008; Lushchak, 2011a, 2011b).

In a mitochondrial paradigm of cancer, Wallace (2005) illustrated that rate and extent of generation of ROS increased on mutations in nuclear or mitochondrial genes of the mitochondrial electron transport chain (ETC). Owing to partial inhibition of electron transfer, the electrons get accumulated at ETC sites and uptaken by O_2, thus leading to generation of superoxide. Further, the superoxide dismutase aggravates the radicals and yield H_2O_2 (hydrogen peroxide), which contributes to genetic instability after diffusion to the nucleus and subsequent attack on DNA. A number of evidences favor the idea that enhancement of ROS may cause transformation and pave the way for cancer progression by augmenting genomic instability. Earlier reports also indicate that cancer cells usually generate more ROS than normal cells (Verma, 2009).

4.3 NUTRACEUTICALS

The term *nutraceutical* is derived by adjoining the terms *nutrition* and *pharmaceutical* (Figure 4.2). According to Health Canada, the nutraceutical is "a product isolated or purified from foods that is generally sold in medicinal forms not usually associated with food." A nutraceutical is regarded for the medicinal value or to safeguard from chronic disease (Esther *et al.*, 2000). Initially, in 1989, Stephen DeFelice coined this term. According to him, "a nutraceutical is any substance that is a food or a part of food and provides medical or health benefits, including the prevention and treatment of disease" (Brower, 1998; Pandey *et al.*, 2010). Today nutraceuticals are widely used in the treatment of many diseases (Figure 4.3). Lack of side effects coupled with beneficial effects during the course of disease, but not merely just ameliorating symptoms, are the key attractions of nutraceutical therapy.

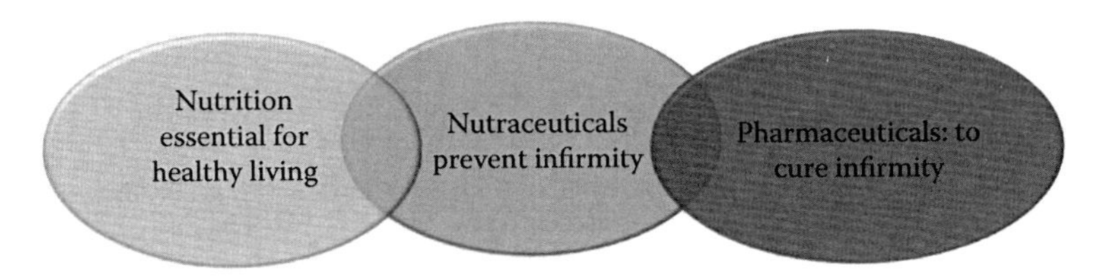

FIGURE 4.2 Nutraceutical term coined from *nutrition* and *pharmaceuticals*.

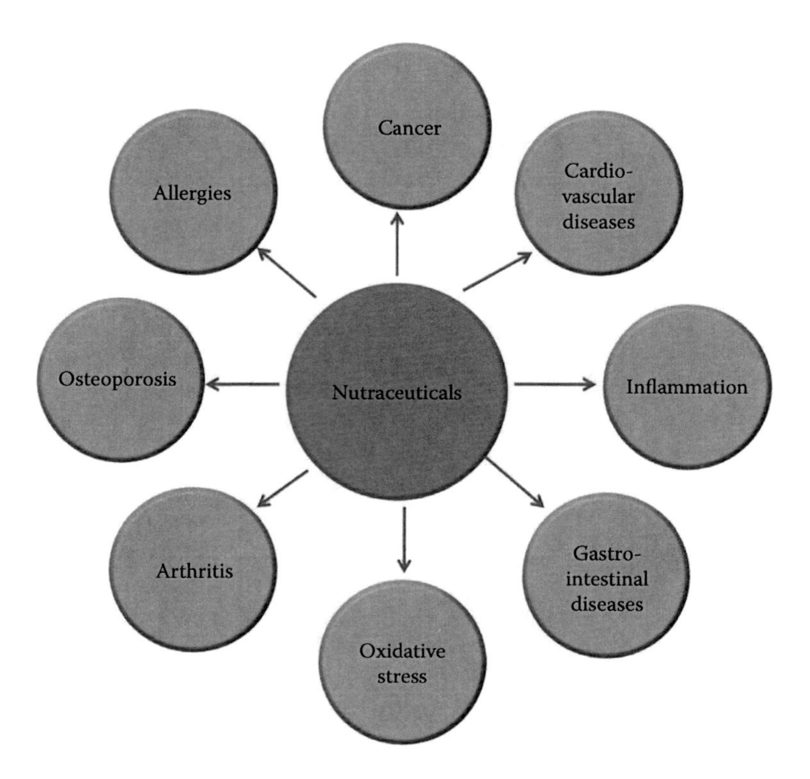

FIGURE 4.3 Role of nutraceuticals in various ailments.

4.4 LIMITATIONS OF NUTRACEUTICALS

The applications of nutraceutical in healthcare are increasing owing to their natural abundance without any recommendation from a medical practitioner. Natural products are attaining interest because, in some disease, standard therapeutic agents result in no effect. They are regarded to have health benefits above their usual nutritive value, for example, ω-3 oils, oil-soluble vitamins, carotenoids, and polyphenols. There are a number of challenges in incorporating many of these nutraceuticals into commercial food products, such as low water-miscibility, high melting point, low physicochemical stability, and poor bio-accessibility (Garti, 2008; McClements, 2012).

4.5 CANCER NANOTECHNOLOGY

Of late, nanotechnology has shown remarkable potential to facilitate the development of novel strategies for the diagnosis, treatment, and effective suppression of carcinoma of various types. Various scientific reports provide evidences for large and far-reaching influence of cancer nanotechnology (Garg *et al.*, 2016). Moreover, the physiochemical behavior of bulk materials get changed after incorporation in the nanocarrier systems, and sometimes the differences are unexpected (Garg *et al.*, 2017). These differences are certainly accredited to their high surface area to mass ratio, quantum or physical confinement effects, or associated lesser known phenomena that appear prevalent when a substance becomes ultra-small. There is evidence suggesting that the subtle-sized particulate substances may possess an intense influence on physiological environments.

A complex dynamic cascade of biotic/abiotic physicochemical interactions results after administration of nanobiomaterials to the biological systems. These physicochemical interactions also describe the disposition and fate in physiological systems. The nanomaterial and biological systems, together or separate, should be well-comprehended to make strides in cancer nanotechnology. Additionally, a strong correlation between effectively categorized nanomaterial attributes and trustworthy biological end-points are indeed prerequisite to guide the researchers toward clinical nanotechnology-based elucidations for cancer. The clinical applications of nanotechnology and nanotoxicology are associated with complex challenges that need multidisciplinary expertise from areas such as materials science, chemical engineering, toxicology, chemistry, and medicine to emerge with an adequate solution.

Since the vital factors that influence the biological performance of nanobiomaterials may differ extensively and are mostly strange, it is strongly suggested that characterization of the test materials should be as comprehensive and broad as possible (Tervonen *et al.*, 2009). Further, it is also recommended that a systematic and interactive approach to material characterization should be attempted to explicate the biological interactions with nanomaterials. Since the biological milieu is extremely vigorous, the nanoparticulate interactions within the biological surroundings will likely be reasonably dynamic. These biological interactions should be cautiously extrapolated or generalized to that of different nanomaterials and/or organisms until a key understanding of the mechanisms prevailing for these phenomena is attained.

4.5.1 Nutraceuticals in Cancer

Despite tremendous advancements in cancer therapy, cancer remains a major health problem worldwide. Of late, dozens of plant extracts are reported to be effective in the prevention and treatment of various types of cancer. Various reports advocated that nutritional modulation can prove to be advantageous for patients suffering from cancer (Weisburger, 1999). Previous reports also indicate that food items with high fiber, moderately low carbohydrates with enough quantity of high-quality protein, and reduced amount of fat (specifically fats of the omega-3 fatty acid group) are favorable for cancer patients (Tuomisto *et al.*, 2004).

Nutraceuticals may prove beneficial in minimizing the adverse effects allied with the chemotherapy and radiation therapy, thus possibly resulting in better health by minimizing cancer cachexia (Grimble, 2003). Moreover, at several stages of cellular level, the phytochemicals displayed their action via various mechanisms (Table 4.1). A number of phytochemicals are also regarded as versatile sources of antioxidants and influence the signaling pathway associated with the redox-mediated transcription factors. In addition, antioxidants modulate the endocrine system, immunological cascade, and enzymes linked with inflammation. Ironically, some of these impose an effect upon DNA repair and cleavage process. Even in a tangible amount, the dietary bioactive compound may have a possible influence on the regulation of gene expression. A systematic investigation on the effects of nutraceuticals on gene expression is indeed warranted to understand the disease preventing mechanism of dietary supplements in various ailments regarding, for example, hypertension, atherosclerosis, diabetes, obesity, and cancers of various types. Additionally, it was reported that phytochemicals can defend against lipid peroxidation and modulate inflammatory responses (Issa *et al.*, 2006). These beneficial effects of bioactive substances, along with lack of toxicity, make them potential candidates to be used alone or in combination with chemotherapeutic agent(s) while treating cancer. Though the mechanism of action and efficacy of each bioactive substance in any type of cancer should be taken into account, the correct compound can be applied in specific medical condition only. Despite the beneficiary outcomes of functional food and nutraceuticals in the prevention and control of cancer progress, a food supplement-based add-on therapy is certainly warranted to offer a quality life to cancer patients (Ogilvie, 1998). Nutritive interference can prove to be an influential device to modulate the malignant disease and diminish the chemotherapy and radiation therapy associated toxicity (McCullough and Giovannucci, 2004). Furthermore, nutraceuticals can remarkably increase the natural killers (NK) cells' function and tumor necrosis factor-α (TNF-α) in patients with late stage cancer (See *et al.*, 2002).

TABLE 4.1
Nutraceuticals in Cancer: Mechanism of Actions at Different Cellular Levels

Nutraceuticals Studied for Cancer	Effectiveness and Mechanism of Action	References
Curcumin	Inhibits neurotensin mediated interleukin 8 production and migration of HCT116 human colon cancer cells.	(Wang *et al.*, 2006)
Polyphenols	Important for their capacity for blocking initiation of carcinogenic process and to suppress cancer progression.	(Lecour and T. Lamont, 2011)
Epigallocatechin-3-gallate (EGCG)	Studies determining the anticancer drug properties of EGCG are preclinical. The antitumor properties of EGCG have been recognized on multiple cancer cell lines, including less common tumors such as anaplastic thyroid carcinoma and malignant mesothelioma.	(Ranzato *et al.*, 2012; De Amicis *et al.*, 2013)
Resveratrol	It possesses a natural anti-proliferative activity due to its role as a phytoalexin (plant antibiotic). It is believed to have also multiple bioactivities including anticancer, anti-carcinogenesis, and anti-inflammatory effects. Resveratrol down regulates survivin and induces apoptosis in human multi-drug-resistant SPCA1/CDDP cells. Suppression of 7,12dimethylbenz(a)anthracene induced mammary carcinogenesis in rats by resveratrol: role of nuclear factor kappa B, cyclooxygenase 2, and matrix metalloprotease 9.	(Banerjee *et al.*, 2002; Singh *et al.*, 2013; Zhao *et al.*, 2016)
Quercetin	Quercetin are considered to be related to the induction of cell apoptosis through multiple mechanisms. Quercetin can inhibit melanoma growth, invasion, and metastatic potential.	(Zhang *et al.*, 2012)
Ascorbate (Vitamin C)	Ascorbate toxicity reported the induction of apoptosis through cell cycle arrest, activation of the apoptosis factors, and interference with iron uptake in cells.	(Ranzato *et al.*, 2011)
Astaxanthin	Inhibits tumor invasion by decreasing extracellular matrix production and induces apoptosis in experimental rat colon carcinogenesis by modulating the expressions of ERK2, NF-κB, and COX2.	(Nagendraprabhu and Sudhandiran, 2011)
Indole-3-carbinol –(I3C)	Increases apoptosis, represses growth of cancer cells, and enhances adenovirus mediated oncolysis.	(Chen *et al.*, 2014)

4.5.2 Nanonutraceuticals for Cancer

Nanoscience has shown tremendous utilities in various science disciplines including food, medicine, biotechnology, microbiology, energy production, communication, and electronics. Nanotechnology has sizeable potential to transform agriculture and food systems. However, nutraceuticals are being used in the treatment of various ailments as natural bioactive substances, exhibiting tremendous health benefits. Carotenoids, resveratrol, and curcumin have gained increasing research interest owing to their inherent antioxidant activity and health benefits, while improving the immunity during and after the treatment of various diseases including cancers, cardiovascular, and neurodegenerative disorders like Alzheimer's disease. However, low solubility and limited bioavailability allied with the bioactive compounds results in poor absorption and reduced biological activity and are the major concerns today. In the food and agriculture sector, effective loading of bioactive in the carrier, sustained release, and efficient delivery are few of the emerging challenges. These issues can be addressed efficiently by using the concept of nanotechnology for expert delivery of nutraceuticals and to improve the biological performance.

4.5.3 Nanocarriers in Delivery of Nutraceuticals to Cancer

Nanoparticle innovation has been generally employed in medicine, including for tumor treatment. As medication nanocarriers, nanoparticles have several meritorious elements: (i) enhanced solubilization of therapeutic agent, (ii) high surface area to volume proportions, (iii) sustained release of drugs or solubilize drugs for systemic delivery, (iv) improved bioavailability, (iv) unrivalled pharmacokinetics, (v) provided targeted (cellular or tissue) delivery of drugs, (vi) protected therapeutic agents against enzymatic degradation (i.e., nucleases and proteases), and (vii) biocompatibility. A number of nanotechnology-based anticancer formulations are now available for clinical purposes or preclinical trials. The subsequent sections illustrate an outline of nanocarriers for nutraceuticals and general recommendations that can be used to define the role of nutraceuticals and nanotechnology for cancer therapeutics.

4.5.3.1 Liposomes

Liposomes illustrate a dynamic and innovative nanotechnology-based delivery system for a variety of bioactive agents (Suntres, 2011). The loading of the bioactives, both lipophilic and hydrophilic, into the liposomes bypasses the toxicity usually associated with the drug/bioactive as generally often seen in anticancer agents. Liposomes provide an impressive opportunity to improve the efficacy of clinical agent. They have been regarded as an excellent prototype of cell membranes and widely applied for estimation of the antioxidant activity of various hydrophilic and hydrophobic antioxidant compounds against oxidant insults. Liposomes provide ease for various therapeutic agents to deliver the active agent at cellular compartment via diffusion through bilayer lipidic membrane, and receptor-mediated endocytosis (Suntres, 2011).

During the last a couple of decades, various reports confirmed the efficacy of curcumin while treating neurological disorders, AIDS, cancer, cardiovascular disease, and various inflammatory disorders. The key challenge related with the therapeutic efficacy of curcumin is its poor bioavailability owing to its low aqueous solubility and stability in the digestive tract (Anand *et al.*, 2008). Various emerging delivery systems, including self-emulsifying drug delivery systems, polymeric nanoparticles, solid dispersion, microspheres, solid lipid nanoparticles, liposomes, nanoemulsions, and microemulsions, have been today widely employed to modulate the bioaccessibility and tissue-targeting potential of curcumin *in vitro* and *in vivo* (Padhye *et al.*, 2010). In a report, the bioavailability and antioxidant activity of curcumin was found to be enhanced following oral intake of curcumin-loaded liposomes (Takahashi *et al.*, 2009). A recent study also advocated the enhanced antineoplastic potential of curcumin, while loaded in liposomes against human pancreatic carcinoma cells. Study results revealed that curcumin-loaded liposomes obstruct pancreatic carcinoma growth and, additionally, unveil anti-angiogenic effects (Mach *et al.*, 2009). Another study revealed that curcumin-loaded phospholipid vesicles or lipid-nanospheres showed ability to scavenge ROS as antioxidants in dispersions. Following 6 h intravenous injection in rats (2 mg/kg), confocal microscopic images of tissue sections displayed the massive distribution of curcumin in tissues macrophages, specifically bone marrow and splenic macrophages. Results merely suggested the therapeutic potential of curcumin-loaded liposomes as an antioxidant and anti-inflammatory agent (Sou *et al.*, 2008).

Quercetin, usually abundant in berries, onions, apples, and red wine, is a polyphenolic flavonoid accounted for its strong antioxidant and anti-inflammatory characteristics. The clinical usage of quercetin is hindered by its poor aqueous solubility and high metabolism rate (Graf *et al.*, 2005). A study demonstrated that the quercetin-loaded PEGylated liposomes remarkably increased the solubility and bioaccessibility of quercetin and improved its antineoplastic efficiency in immunocompetent C57BL/6N mice bearing LL/2 Lewis lung cancer and in BALB/c mice bearing CT26 colon adenocarcinoma (Yuan *et al.*, 2006).

The anticancer activity of curcumin and resveratrol was found to be improved in prostate-specific PTEN knockout mice, when they were co-administered after being incorporated into liposomes. The treatment was found to be far more effective than that seen when each agent was given alone, either in their free form or after incorporation into the liposomes (Narayanan *et al.*, 2009).

4.5.3.2 Polymeric Nanoparticles

Polymer nanoparticles (NPs) play dynamic roles in various discipline of science, including biotechnology, medicine, environmental technology, pollution control, photonics, electronics, material science, and so forth (Rao and Geckeler, 2011). In the biomedical field of science, biodegradable nanoparticles are extensively used for effective delivery of therapeutic compounds owing to their improved bioavailability, higher encapsulation efficiency, controlled release, and nontoxic properties.

Epigallocatechin-3-gallate (EGCG), the most abundant tea catechin in green tea, is accounted for its strong antioxidant property. Various epidemiological and preclinical reports advocate that EGCG can decrease the risk of cancer, which is mainly attributed to its inhibitory effects on enzyme activities and signal transduction pathways, which led to the suppression of cell proliferation and enhancement and apoptosis. However, EGCG is unstable in plasma and intestinal juices, leading to its low bioavailability. Polymer nanoparticles assembled from gallic acid (GA) grafted chitosan (CS, GA-g-CS for GA grafted CS) and caseinophosphopeptides (CPP) were developed to deliver EGCG as novel functional foods. The GA-g-CS-CPP nanoparticles showed intense antioxidant activity and toxicity against Caco-2 colon cancer cells. The EGCG-loaded GA-g-CS-CPP nanoparticles (84–90% for encapsulation efficiency) showed improved delivery property, controlled release of EGCG under simulated gastrointestinal environments, preventing its degradation under neutral and alkaline conditions, and amplifying its antineoplastic potential against Caco-2 cells (Hu *et al.*, 2015).

Ferulic acid (FA), a widely distributed hydroxycinnamic acid, is abundant generally in cereals and fruits. FA exhibits strong antioxidant and anticancer activities, though the bioavailability of FA is quite less owing to its low solubility and permeability. In a report presented by Panwar *et al.* (2016), nontoxic chitosan-tripolyphosphate pentasodium (CS-TPP) nanoparticles (NPs) were used to entrap FA to increase its availability in the biological system. FA-loaded CS-TPP nanoparticles were found to enhance the cytotoxicity, solubility, and anticancer potential against ME-180 cell lines, thus presenting it as a potent therapeutic agent for medicinal and clinical purposes (Panwar *et al.*, 2016).

4.5.3.3 Solid Lipid Nanoparticles

Solid lipid nanoparticles (SLNs) are the rapidly developing field of nanotechnology and are extensively used systems for delivery of drug, diagnostic purposes, clinical medicine and research, etc. (Jain *et al.*, 2010). SLNs were effectively developed to overwhelm the disadvantages of other colloidal carriers, such as polymeric nanoparticles, liposomes, microemulsions, etc. SLNs show artistic release profile, targeted delivery with excellent physical stability, when used as drug delivery vehicles (Jain *et al.*, 2015a).

Umbelliferone manifests low hydrophilicity and minimal absorption in gastrointestinal tract, which seriously restricted its use in biomedical applications. As a result, supplementation of functional food with natural antioxidant formulations such as lipid nanoparticles loaded with umbelliferone inhibits the formation of free radicals thus leading to prevention of some diseases. At this juncture, Lacatusu *et al.* (2013) synthesized SLNs loaded with umbelliferone by a modified-high shear homogenization technique (HSH) and analyzed the effect of umbelliferone encapsulation into SLNs on its antioxidant activity (Lacatusu *et al.*, 2013). Physically stable SLNs loaded with umbelliferone were produced with promising antioxidant properties. By encapsulation of umbelliferone in lipid nanoparticles an improvement of antioxidant activity with almost 25% was observed in the case of SLNs loaded with 1.13 μM umbelliferone. The study enlightens the major impact of the encapsulation process on the antioxidant properties, as it could induce an enhancement of this property due to a coupled effect of umbelliferone presence and lipid matrix, by generating other structures from lipid matrix with scavenger properties (Lacatusu *et al.*, 2013).

Triptolide (TP) is regarded to have anti-inflammatory, immunosuppressive, anti-fertility, and anti-neoplastic properties. The clinical applications of FA are limited to some extent owing to its poor aqueous solubility and few adverse effects. An innovative delivery vehicle like SLNs is warranted to administer TP and lessen its limitations (Mei *et al.*, 2003). With TP, the occurrence of adverse drug reaction (ADRs) is considerably greater than other bioactive compounds. The body systems affected by ADRs of TP include bone marrow, circulatory system, cardiovascular, urogenital, gastrointestinal, and skin. The SLNs of TP were developed with the aim to deliver TP and to overcome the solubility issue and toxicity associated with its usage (Jenning *et al.*, 2000). The developed systems could provide significant advantages in the clinical use of the TP. In another study, the utility of SLNs dispersions and microemulsions as carriers for topical delivery of TP were exploited. The results of the study indicated that these SLN dispersions and microemulsions can function as effective promoters for TP to penetrate into skin (Mei *et al.*, 2003).

Out of the 600 identified naturally abundant carotenoids, lycopene (LYC) and beta carotene (BC) have potent antioxidant properties and anticancer activity, particularly against prostate cancer (Jain *et al.*, 2016b; Thakur *et al.*, 2017). Intake of carotenoids has also been concomitant with decreased risk of chronic diseases including cancer and cardiovascular diseases (Jain *et al.*, 2015c; Jain *et al.*, 2016a). Numerous epidemiological studies suggested the anti-carcinogenic and anti-atherogenic potential of LYC and BC, which have been accredited predominantly to their antioxidant properties.

Jain and coworkers (2017) examined the anticancer efficacy of lycopene encompassed lipidic particles (LYC-SLNs), while co-administered with the clinically relevant anticancer drug methotrexate (MTX). The *in vitro* cell culture experiments showed a remarkably improved cellular uptake of LYC-SLNs in MCF-7 cells, as compared to the free LYC. The survival of MCF-7 cells was significantly inhibited after co-administration of MTX with LYC-SLNs, as compared to MTX alone. Co-delivery of MTX with LYC-SLNs described here as a worthwhile approach to elucidate the effectiveness of combination therapy in an array of natural antioxidant and cancer therapeutics (Jain *et al.*, 2017b).

Jain and associates (2017) also conducted a study intended to investigate the effect of co-administration of beta carotene (BC) and methotrexate (MTX) in breast cancer therapeutics. The study also described the possible protective role of BC on MTX-induced toxicity. Herein, BC and MTX co-loaded lipid-polymeric hybrid nanoparticles (F-BC-MTX-LPHNPs) was developed. Further, periphery of the system was functionalized with fructose. Results of *in vitro* cellular study indicated that F-BC-MTX-LPHNPs induced the highest apoptosis index (0.89 level) against MCF-7 cells, which was far greater than that observed with free MTX (0.46 level). Outcomes of the *in vivo* study suggested that the residual tumor progression was low (~32%) in the animals treated with both the molecules, while loaded into nanostructured system. Besides, BC ameliorated the MTX-induced hepatic and renal toxicity (Jain *et al.*, 2017a).

4.5.3.4 Self-Nano Emulsifying Drug Delivery Systems (SNEDDS)

SNEDDS are nanostructured isotropic blends of drug, lipids, and surfactants, generally consisting of one or more hydrophilic co-solvents or co-emulsifiers. Upon gentle mixing and subsequent dilution with aqueous media, they form fine (oil in water) emulsion instantly. SNEDDS is a broad term, usually constituting the emulsions with a globule size ranging from a few nanometers to several-microns. Self micro-emulsifying drug delivery systems (SMEDDS), on the other hand, designate the formulations developing transparent microemulsions with globule size between 100 and 250 nm. SNEDDS may be a potential candidate in place of orally administered emulsions due of their relatively high physical stability and ability to be delivered in standard soft gelatin capsules (Tripathi *et al.*, 2016; Sandhu *et al.*, 2017).

Ellagic acid has been reported to have many beneficial properties that help in the treatment of several oxidation-linked chronic diseases such as cardiovascular disease, prostate cancer, neurodegenerative diseases, and breast cancer (Larrosa *et al.*, 2010; Landete, 2011). Despite having several beneficial health effects, EA has limited potential for therapeutic use due to its poor absorption because of low aqueous solubility (Lei *et al.*, 2003), extensive metabolic transformation (Seeram *et al.*, 2004), and rapid elimination, leading to ineffective drug plasma concentration. The present study has undoubtedly proved the potential effectiveness of SNEDDS for formulating EA with improved release profile and permeability. *In vitro* dissolution and *ex vivo* permeability studies revealed improved release profile and permeation of EA, respectively, from EAPL complex SNEDDS compared to EA suspension. Moreover, EAPL complex SNEDDS overcame food effect on release pattern of EA (Avachat and Patel, 2015).

Co-administration of quercetin (QT) with tamoxifen (Tmx), owing to its antioxidant and antiproliferative action, could increase the therapeutic efficacy vis-à-vis reducing the hepatotoxicity. Herein, Tmx-QT-SNEDDS was developed by Jain *et al.* (2013) for improving the therapeutic efficacy in synergistic manner as well as for reducing the hepatotoxicity of Tmx. Ultimately, complete eradication of hepatotoxicity was observed after repetitive oral administration of s-Tmx-QT-SNEDDS whereas Tmx citrate showed hepatotoxicity when estimated in terms of levels of liver toxicity markers. The results clearly indicate that co-encapsulation of QT along with Tmx in solid SNEDDS can be employed as potential strategy for improving the therapeutic efficacy of Tmx. Co-incorporation of an antioxidant, QT along with clinically available anticancer drug Tmx in single delivery vehicle, bring out the noticeable increment in anticancer activity as compared to marketed Tmx formulation (Jain *et al.*, 2014).

In their report, Sandhu *et al.* (2017) anticipated tamoxifen (TMX) and naringenin (NG) co-entrapped self-emulsifying nano-miceller drug delivery system (TMX-NG-SNEDDS) to increase the effectiveness of TMX therapy while counteracting the toxic effects of TMX. The developed formulation has shown considerable increments in cellular mortality, when incubated with human adenocarcinoma cells (MCF-7). Quantitative apoptosis assay illustrated that the percent apoptosis was found maximum with TMX-NG–SNEDDS (84.7%). The pharmacodynamic model exhibited the improved efficacy in a synergistic way by reducing tumor volume, and by increased survival rate of the animals (Sandhu *et al.*, 2017).

4.6 CONCLUSIONS

Cancer is the major cause of increased rate of morbidity and mortality across the globe. The effective pharmacological treatments may efficiently diminish cancer associated deaths. Although various challenges obstruct the application of conventional therapies against cancer, nano-constructed carriers may prove to be a promising therapeutic tool for targeted treatment in cancer therapeutics. The increasing numbers of investigations demonstrate the pivotal role played by nutraceuticals against cancer. There is a renewed attention in the expansion and application of nanocarriers, such as polymeric NPs, liposomes, self-emulsifying drug delivery systems, and SLNs, to deliver bioactive for cancer. These formulation approaches are suggestive of sustained/controlled drug release, improved bioavailability, target specificity, and improved drug stability. In this chapter, efforts have been made to present appropriate content of diverse studies highlighting the intrinsic worth and potentials of various nutraceuticals using nanocarriers to optimize the pharmacotherapy against cancer. It is firmly believed that increasing perception on *in vivo* fate of antioxidant-loaded nanocarriers will prove to be a boon in facilitating the future prospects of improved functioning and efficient delivery system for anticancer therapeutics.

REFERENCES

Anand, P., Thomas, S. G., Kunnumakkara, A. B., Sundaram, C., Harikumar, K. B., Sung, B. *et al.* Biological activities of curcumin and its analogues (Congeners) made by man and mother nature. *Biochem Pharmacol.* 2008; 76: 1590–1611.

Avachat, A. M., and Patel, V. G. Self nanoemulsifying drug delivery system of stabilized ellagic acid–phospholipid complex with improved dissolution and permeability. *Saudi Pharm J.* 2015; 23: 276–289.

Banerjee, S., Bueso-Ramos, C., and Aggarwal, B. B. Suppression of 7,12-dimethylbenz(a)anthracene-induced mammary carcinogenesis in rats by resveratrol: Role of nuclear factor-kappaB, cyclooxygenase 2, and matrix metalloprotease 9. *Cancer Res.* 2002; 62: 4945–4954.

Biemar, F., and Foti, M. Global progress against cancer-challenges and opportunities. *Cancer Biol Med.* 2013; 10: 183–186.

Brower, V. Nutraceuticals: Poised for a healthy slice of the healthcare market? *Nat Biotechnol.* 1998; 16: 728–731.

Chabner, B. A., Boral, A. L., and Multani, P. Translational research: Walking the bridge between idea and cure. *Cancer Res.* 1998; 58: 4211–4216.

Chen, L., Cheng, P.-H., Rao, X.-M., McMasters, K. M., and Zhou, H. S. Indole-3-carbinol (I3C) increases apoptosis, represses growth of cancer cells, and enhances adenovirus-mediated oncolysis. *Cancer Biol Ther.* 2014; 15: 1256–1267.

De Amicis, F., Perri, A., Vizza, D., Russo, A., Panno, M. L., Bonofiglio, D. *et al.* Epigallocatechin gallate inhibits growth and epithelial-to-mesenchymal transition in human thyroid carcinoma cell lines. *J Cell Physiol.* 2013; 228: 2054–2062.

Esther, B., Rapport, L., and Lockwood, B. What is nutraceutical? *Pharm J.* 2000; 265 (7104): 57–58.

Garg, N. K., Singh, B., Jain, A., Nirbhavane, P., Sharma, R., Tyagi, R. K., Kushwah, V., Jain, S., and Katare, O. P. Fucose decorated solid-lipid nanocarriers mediate efficient delivery of methotrexate in breast cancer therapeutics. *Colloids Surf B Biointerfaces.* 2016; 146: 114–126.

Garg, N. K., Tyagi, R. K., Sharma, G., Jain, A., Singh, B., Jain, S., and Katare, O. P. Functionalized lipid polymer hybrid nanoparticles mediated co-delivery of methotrexate & aceclofenac: A synergistic effect in breast cancer with improved pharmacokinetics attributes. *Mol Pharm.* 2017; 14: 1883–1897.

Garti, N. *Delivery and controlled release of bioactives in foods and nutraceuticals.* CRC Press; 2008; p. 478.

Graf, B. A., Mullen, W., Caldwell, S. T., Hartley, R. C., Duthie, G. G., Lean, M. E. J. *et al.* Disposition and metabolism of [2-14C]quercetin-4'-glucoside in rats. *Drug Metab Dispos.* 2005; 33: 1036–1043.

Grimble, R. F. Nutritional therapy for cancer cachexia. *Gut.* 2003; 52: 1391–1392.

Hu, B., Wang, Y., Xie, M., Hu, G., Ma, F., and Zeng, X. Polymer nanoparticles composed with gallic acid grafted chitosan and bioactive peptides combined antioxidant, anticancer activities and improved delivery property for labile polyphenols. *J Funct Foods.* 2015; 15: 593–603.

Issa, A. Y., Volate, S. R., and Wargovich, M. J. The role of phytochemicals in inhibition of cancer and inflammation: New directions and perspectives. *J Food Compos Anal.* 2006; 19: 405–419.

Jain, A., Agarwal, A., Majumder, S., Lariya, N., Khaya, A., Agrawal, H., Majumdar, S., and Agrawal, G. P. Mannosylated solid lipid nanoparticles as vectors for site-specific delivery of an anti-cancer drug. *J Control Release.* 2010; 148: 359–367.

Jain, A., Kesharwani, P., Garg, N. K., Jain, A., Jain, S. A., Jain, A. K., Nirbhavane, P., Ghanghoria, R., Tyagi, R. K., and Katare, O. P. Galactose engineered solid lipid nanoparticles for targeted delivery of doxorubicin. *Colloids Surf B Biointerfaces.* 2015a; 134: 47–58.

Jain, A., Kesharwani, P., Garg, N. K., Jain, A., Nirbhavane, P., Ghanghoria, R., Banerjee, S., Iyer, A. K., and Mohd Amin, M. C. I. Nano-constructed carriers loaded with antioxidant: Boon for cardiovascular system. *Curr Pharm Des.* 2015b; 21: 4456–4464.

Jain, A., Sharma, G., Kushwah, V., Garg, N. K., Kesharwani, P., Ghoshal, G., Singh, B., Shivhare, U. S., Jain, S., and Katare, O. P. Methotrexate and beta-carotene loaded-lipid polymer hybrid nanoparticles: A preclinical study for breast cancer. *Nanomedicine.* 2017a; 12: 1851–1872.

Jain, A., Sharma, G., Kushwah, V., Thakur, K., Ghoshal, G., Singh, B., Jain, S., Shivhare, U. S., and Katare, O. P. Fabrication and functional attributes of lipidic nanoconstructs of lycopene: An innovative endeavour for enhanced cytotoxicity in MCF-7 breast cancer cells. *Colloids Surf B Biointerfaces.* 2017b; 152: 482–491.

Jain, A., Thakur, D., Ghoshal, G., Katare, O. P., and Shivhare, U. S. Characterization of microcapsulated β-carotene formed by complex coacervation using casein and gum tragacanth. *Int J Biol Macromol.* 2016a; 87: 101–113.

Jain, A., Thakur, D., Ghoshal, G., Katare, O. P., and Shivhare, U. S. Microencapsulation by complex coacervation using whey protein isolates and gum acacia: An approach to preserve the functionality and controlled release of β-carotene. *Food Bioprocess Technol.* 2015c; 8: 1635–1644.

Jain, A., Thakur, D., Ghoshal, G., Katare, O. P., Singh, B., and Shivhare, U. S. Formation and functional attributes of electrostatic complexes involving casein and anionic polysaccharides: An approach to enhance oral absorption of lycopene in rats *in vivo. Int J Biol Macromol.* 2016b; 93: 746–756.

Jain, A. K., Thanki, K., and Jain, S. Co-encapsulation of tamoxifen and quercetin in polymeric nanoparticles: Implications on oral bioavailability, antitumor efficacy and drug-induced toxicity. *Mol Pharm.* 2013; 10: 3459–3474.

Jain, A. K., Thanki, K., and Jain, S. Solidified self-nanoemulsifying formulation for oral delivery of combinatorial therapeutic regimen: Part I. formulation development, statistical optimization and *in vitro* characterization. *Pharm Res.* 2014; 31: 923–945.

Jenning, V., Schafer-Korting, M., and Gohla S. Vitamin A-loaded solid lipid nanoparticles for topical use: Drug release properties. *J Control Release.* 2000; 66: 115–126.

Kohen, R., and Nyska, A. Oxidation of biological systems: Oxidative stress phenomena, antioxidants, redox reactions, and methods for their quantification. *Toxicol Pathol.* 2002; 30: 620–650.

Kumar, N., and Kumar, R. Nanomedicine for cancer treatment. In: *Nanotechnology and nanomaterials in the treatment of life-threatening diseases*; 2014; pp. 177–246.

Lacatusu, I., Badea, N., Murariu, A., Oprea, O., Bojin, D., and Meghea, A. Antioxidant activity of solid lipid nanoparticles loaded with umbelliferone. *Soft Mater.* 2013; 11: 75–84.

Lakshmi, S. V., Padmaja, G., Kuppusamy, P., and Kutala, V. K. Oxidative stress in cardiovascular tissue. *Indian J Biochem Biophys.* 2009; 46: 421–440.

Landete, J. M. Ellagitannins, ellagic acid and their derived metabolites: A review about source, metabolism, functions and health. *Food Res Int.* 2011; 44: 1150–1160.

Larrosa, M., García-Conesa, M. T., Espín, J. C., and Tomás-Barberán, F. A. Ellagitannins, ellagic acid and vascular health. *Mol Aspects Med.* 2010; 31: 513–539.

Lecour, S., and Lamont, K. T. Natural polyphenols and cardioprotection. *Mini Rev Med Chem.* 2011; 14: 1191–1199.

Lei, F., Xing, D.-M., Xiang, L., Zhao, Y.-N., Wang, W., Zhang, L.-J. *et al.* Pharmacokinetic study of ellagic acid in rat after oral administration of pomegranate leaf extract. *J Chromatogr B.* 2003; 796: 189–194.

Lushchak, V. I. Adaptive response to oxidative stress: Bacteria, fungi, plants and animals. *Comp Biochem Physiol C Toxicol Pharmacol.* 2011a; 153: 175–190.

Lushchak, V. I. Environmentally induced oxidative stress in aquatic animals. *Aquat Toxicol.* 2011b; 101: 13–30.

Mach, C. M., Mathew, L., Mosley, S. A., Kurzrock, R., and Smith, J. A. Determination of minimum effective dose and optimal dosing schedule for liposomal curcumin in a xenograft human pancreatic cancer model. *Anticancer Res.* 2009; 29: 1895–1899.

Machlin, L. J., and Bendich, A. Free radical tissue damage: Protective role of antioxidant nutrients. *Clin Nutr.* 1987; 1: 441–445.

McClements, D. J. Advances in fabrication of emulsions with enhanced functionality using structural design principles. *Curr Opin Colloid Interface Sci.* 2012; 17: 235–245.

McCullough, M. L., and Giovannucci, E. L. Diet and cancer prevention. *Oncogene.* 2004; 23: 6349–6364.

Mei, Z., Chen, H., Weng, T., Yang, Y., and Yang, X. Solid lipid nanoparticle and microemulsion for topical delivery of triptolide. *Eur J Pharm Biopharm.* 2003; 56: 189–196.

Nagendraprabhu, P., and Sudhandiran, G. Astaxanthin inhibits tumor invasion by decreasing extracellular matrix production and induces apoptosis in experimental rat colon carcinogenesis by modulating the expressions of ERK-2, NFkB and COX-2. *Invest New Drugs.* 2011; 29: 207–224.

Narayanan, N. K., Nargi, D., Randolph, C., and Narayanan, B. A. Liposome encapsulation of curcumin and resveratrol in combination reduces prostate cancer incidence in PTEN knockout mice. *Int J Cancer.* 2009; 125: 1–8.

Ogilvie, G. K. Interventional nutrition for the cancer patient. *Clin Tech Small Anim Pract.* 1998; 13: 224–231.

Pacher, P., and Szabo, C. Role of the peroxynitrite-poly(ADP-ribose) polymerase pathway in human disease. *Am J Pathol.* 2008; 173: 2–13.

Padhye, S., Chavan, D., Pandey, S., Deshpande, J., Swamy, K. V., and Sarkar, F. H. Perspectives on chemopreventive and therapeutic potential of curcumin analogs in medicinal chemistry. *Mini Rev Med Chem.* 2010; 10: 372–387.

Pandey, M., Verma, R. K., and Saraf, S. A. Nutraceuticals: New era of medicine and health. *Asian J Pharm Clin Res.* 2010; 3: 11–15.

Panwar, R., Sharma, A. K., Kaloti, M., Dutt, D., and Pruthi, V. Characterization and anticancer potential of ferulic acid-loaded chitosan nanoparticles against ME-180 human cervical cancer cell lines. *Appl Nanosci.* 2016; 6: 803–813.

Poljsak, B., Šuput, D., Milisav, I., and Milisav, I. Achieving the balance between ROS and antioxidants: When to use the synthetic antioxidants. *Oxid Med Cell Longev.* 2013; 2013: 956792.

Ranzato, E., Biffo, S., and Burlando, B. Selective ascorbate toxicity in malignant mesothelioma. *American Thoracic Society.* 2011; 44.

Ranzato, E., Martinotti, S., Magnelli, V., Murer, B., Biffo, S., Mutti, L. *et al.* Epigallocatechin-3-gallate induces mesothelioma cell death via H_2O_2 –dependent T-type Ca 2+ channel opening. *J Cell Mol Med.* 2012; 16: 2667–2678.

Rao, J. P., and Geckeler, K. E. Polymer nanoparticles: Preparation techniques and size-control parameters. *Prog Polym Sci.* 2011; 36: 887–913.

Sandhu, P. S., Kumar, R., Beg, S., Jain, S., Kushwah, V., Katare, O. P., and Singh, B. Natural lipids enriched self-nano-emulsifying systems for effective co-delivery of tamoxifen and naringenin: Systematic approach for improved breast cancer therapeutics. *Nanomedicine Nanotechnology Biol Med.* 2017; 13: 1703–1713.

See, D., Mason, S., and Roshan, R. Increased tumor necrosis factor alpha (Tnf-A) and natural killer cell (Nk) function using an integrative approach in late stage cancers. *Immunol Invest.* 2002; 31: 137–153.

Seeram, N. P., Lee, R., and Heber, D. Bioavailability of ellagic acid in human plasma after consumption of ellagitannins from pomegranate (*Punica granatum L.*) juice. *Clin Chim Acta.* 2004; 348: 63–68.

Singh, C. K., George, J., and Ahmad, N. Resveratrol-based combinatorial strategies for cancer management. *Ann N Y Acad Sci.* 2013; 1290: 113–121.

Socinski, M. A., Bondarenko, I., Karaseva, N. A., Makhson, A. M., Vynnychenko, I., Okamoto, I. *et al.* Weekly nab -Paclitaxel in combination with carboplatin versus solvent-based paclitaxel plus carboplatin as first-line therapy in patients with advanced non-small-cell lung cancer: Final results of a phase III trial. *J Clin Oncol.* 2012; 30: 2055–2062.

Sou, K., Inenaga, S., Takeoka, S., and Tsuchida, E. Loading of curcumin into macrophages using lipid-based nanoparticles. *Int J Pharm.* 2008; 352: 287–293.

Stone, W. L., and Smith, M. Therapeutic uses of antioxidant liposomes. *Mol Biotechnol.* 2004; 27: 217–230.

Suntres, Z. E. Liposomal antioxidants for protection against oxidant-induced damage. *J Toxicol.* 2011; 2011: 152474.

Takahashi, M., Uechi, S., Takara, K., Asikin, Y., and Wada, K. Evaluation of an oral carrier system in rats: Bioavailability and antioxidant properties of liposome-encapsulated curcumin. *J Agric Food Chem.* 2009; 57: 9141–9146.

Taverne, Y. J. H. J., Bogers, A. J. J. C., Duncker, D. J., and Merkus, D. Reactive oxygen species and the cardiovascular system. *Oxid Med Cell Longev.* 2013; 86: 4–23.

Tervonen, T., Linkov, I., Rui, J., Ae, F., Steevens, J., Chappell, M. *et al.* Risk-based classification system of nanomaterials. *J Nanopart Res.* 2009; 11: 757–766.

Thakur, D., Jain, A., Ghoshal, G., Shivhare, U. S., and Katare, O. P. Microencapsulation of β-carotene based on casein/guar gum blend using zeta potential-yield stress phenomenon: An approach to enhance photostability and retention of functionality. *AAPS PharmSciTech.* 2017; 18: 1447–1459.

Tripathi, S., Kushwah, V., Thanki, K., and Jain, S. Triple antioxidant SNEDDS formulation with enhanced oral bioavailability: Implication of chemoprevention of breast cancer. *Nanomedicine Nanotechnology Biol Med.* 2016; 12: 1–13.

Tuomisto, J. T., Tuomisto, J., Tainio, M., Niittynen, M., Verkasalo, P., Vartiainen, T. *et al.* Risk-benefit analysis of eating farmed salmon. *Science.* 2004; 305(5683): 476–477.

Verma, M. *Cancer epidemiology.* Volume 2 Modif Factors. ISBN 978-1-60327-492-0, 2009; 1930: 467–475.

Vogelstein, B., Papadopoulos, N., Velculescu, V. E., Zhou, S., Diaz, L., and Kinzler, K. W. Cancer genome landscapes. *Science.* 2013; 339: 1546–1558.

Wallace, D. C. A mitochondrial paradigm of metabolic and degenerative diseases, aging, and cancer: A dawn for evolutionary medicine. *Annu Rev Genet.* 2005; 39: 359–407.

Wang, Q., Zhang, L., Hu, W., Hu, Z. H., Bei, Y. Y., Xu, J. Y. *et al.* Norcantharidin-associated galactosylated chitosan nanoparticles for hepatocyte-targeted delivery. *Nanomedicine.* 2010; 6: 371–381.

Wang, X., Wang, Q., Ives, K. L., and Evers, B. M. Curcumin inhibits neurotensin-mediated interleukin-8 production and migration of HCT116 human colon cancer cells. *Clin Cancer Res.* 2006; 12: 5346–5355.

Weisburger, J. H. Antimutagens, anticarcinogens, and effective worldwide cancer prevention. *J Environ Pathol Toxicol Oncol.* 1999; 18: 85–93.

WHO Cancer, 2016. WHO.

Yuan, Z., Chen, L., Fan, L., Tang, M., Yang, G., Yang, H. *et al.* Liposomal quercetin efficiently suppresses growth of solid tumors in murine models. *Clin Cancer Res.* 2006; 12: 3193–3199.

Zhang, X., Chen, L.-X., Ouyang, L., Cheng, Y., and Liu, B. Plant natural compounds: Targeting pathways of autophagy as anti-cancer therapeutic agents. *Cell Prolif.* 2012; 45: 466–476.

Zhao, W., Bao, P., Qi, H., and You, H. Resveratrol down-regulates survivin and induces apoptosis in human multidrug-resistant SPC-A-1/CDDP cells. *Oncol Rep.* 2016; 23: 279–286.

5 Nanostructured Drug Delivery of Nutraceuticals for Counteracting Oxidative Stress

*Shobhit Kumar, Bharti Gaba, Jasjeet K. Narang, Javed Ali, and Sanjula Baboota**

CONTENTS

* Corresponding author.

5.1 INTRODUCTION

Age progression is an inevitable characteristic of life. The susceptibility of degenerative disorders increases with aging and has been affecting millions of people each year, thus becoming a major problem of modern society. Free radicals are the major contributing reason for aging (loss of muscle strength and stamina). Free radicals contain unpaired electron(s), usually present in the outer orbit, and are responsible for changing the chemical reactivity behavior of a molecule. They make the unstable molecule highly reactive and electron acceptor, wherein these molecules accept electrons from other molecules, leading to damage to cells and their cellular components such as DNA and proteins (Sies *et al.*, 2017). Both peroxynitrite ($ONOO^-$) and hydrogen peroxide (H_2O_2) are responsible for free radicals' formation, as they are highly reactive (Aprioku, 2013). Generally, these molecules and free radicals are known as reactive oxygen species (ROS), which cause harmful effects, including low cellular energy production, impaired cell reproduction leading to cancer, and arterial clogging and blockage, eventually leading to heart attacks and strokes (Parker *et al.*, 2017; Sultana *et al.*, 2017).

ROS decreases the activity of different endogenous antioxidant enzymes like glutathione, catalase, and mitochondrial succinate dehydrogenase enzyme. Furthermore, it also damages glucose transport and results in induction of membrane lipoperoxidation, mutation in the gene encoding for antioxidant enzymes, deposition of mitochondrial ion and amyloid beta peptide, thus precipitating various neurodegenerative diseases. Figure 5.1 illustrates oxidative stress-induced diseases. Table 5.1 enlists different harmful effects of free radicals on various organs of the human body.

Diet containing high level of phenolic compounds (caffeic and chlorogenic acids), electromagnetic radiation (X-rays and gamma rays), inorganic particles (asbestos, quartz, silica), ozone, industrial chemicals, certain drugs (e.g., nitrofurantoin, bleomycin, anthracyclines), and pesticides and air pollutants (e.g., cigarette smoke comprised of oxides of nitrogen) are some of the major exogenous sources for free radicals. Among endogenous sources of free radicals is aerobic respiration, which is the main natural source responsible for adding free radicals in the body. Mitochondria consume oxygen for adenosine triphosphate (ATP) production, and during this process they generate production of H_2O_2 and hydroxyl species as by-products. Our body destroys and evades the bacterial or viral attack by phagocytosis mechanism resulting in generation of O_2^-, $NO^\bullet$, H_2O_2, and OCl^- radicals. This phagocytosis mechanism represents the second endogenous source of free radicals. The third source includes peroxisomes, an organelle that is responsible for fatty acid the degradation by

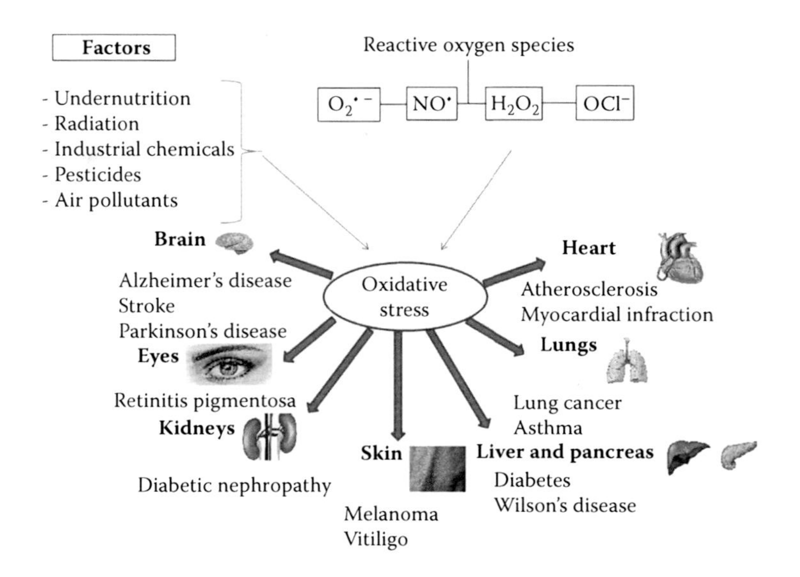

FIGURE 5.1 Oxidative stress-induced diseases.

TABLE 5.1
Detrimental Effects of Oxidative Stress on Various Body Organs

Organ	Detrimental Effect	Reference
Brain	Alzheimer's disease (suppression of mitochondrial succinate dehydrogenase enzyme activity, impairing glucose transport, induction of membrane lipoperoxidation, mitochondrial ion and amyloid beta peptide deposition)	Meng *et al.*, 2015
	Stroke (vasodilatation via potassium channels opening and altered vascular reactivity, breakdown of BBB)	Ozkul *et al.*, 2007
	Parkinson's disease (breakdown in the regulation of dopamine)	Jenner, 2003
Eyes	Retinitis pigmentosa (death of rods and cone cell)	Usui *et al.*, 2009
	Cataract (opacity of the lens)	Zoric, 2003
Heart	Atherosclerosis (increased oxidized low-density lipoproteins particles are associated with plaque instability and thrombosis)	Steinberg, 1997
	Myocardial infarction (activation of metalloproteinases that are responsible for degradation of extracellular matrix, leading to ventricular dilatation)	Di Filippo *et al.*, 2006
Kidney	Diabetic nephropathy (polyol pathway become deficient and stimulation of renin-angiotensin system)	Prabhakar *et al.*, 2007
Lungs	Lung cancer (stimulation of transcription factors like Nrf2 and NF-κB, altered gene expression patterns causing carcinogenesis)	Szabo *et al.*, 1996
Skin	Melanoma (DNA damage and mutations in genes)	Liu-Smith *et al.*, 2015
	Vitiligo (lipid peroxidation and increase antimelanogenic cytokines formation)	Jimbow *et al.*, 2001
Liver and pancreas	Diabetes (non-enzymatic glycation of proteins)	Kalita *et al.*, 2014
	Wilson's disease (increased level of cytokines and glutamate)	Kalita *et al.*, 2014
Joints	Systemic lupus erythematosus (redox-dependent stimulation of serine/threonine-protein kinase)	Nuttall *et al.*, 2003

producing H_2O_2 as by-product. The fourth endogenous source of free radicals is cytochrome P450 enzyme, which acts as a defense system against toxic substances. But a few oxidants as byproducts are produced during detoxification of reactive species that have the capability to damage DNA and other cellular components (Lobo *et al.*, 2010). For instance, a superoxide molecule may react with a fatty acid and take out one of its electrons. In this way, fatty acid gets converted into a free radical, which may further react with other fatty acids. These generated free radicals alter the permeability of cell membranes and structure of receptor proteins, which, in turn, change or stop their function.

5.1.1 The Body's Defense

To quench the excess amount of free radicals present in our body, the human system is equipped with free radical detoxifying enzymes and antioxidant chemicals, which act as defence mechanism to combat them. But when there is an overload of free radicals, oxidative stress arises, and the body needs to be supplemented with exogenous supply of antioxidants as endogenous levels are not sufficient to provide defence to reduce their impact. Free radical detoxifying enzyme systems are responsible for protecting the cell organelles from ROS damage. An antioxidant is a molecule that prevents free radicals from stealing electrons.

5.1.2 Free Radical Detoxifying Enzymes

5.1.2.1 Superoxide Dismutase (SOD)

This enzyme comprised of metals is commonly present in extracellular fluids. They are responsible for the catalysis of two superoxides into H_2O_2 and oxygen. This generated H_2O_2 is still considered a ROS; however, it is less reactive than a superoxide. Among fast acting enzymes, SOD is one of them.

5.1.2.2 Catalase

They use iron as a cofactor and are responsible for converting H_2O_2 to water, completing, in this manner, the detoxification reaction initiated by SOD. Catalase is highly efficient and is capable of destroying millions of H_2O_2 molecules per second.

5.1.2.3 Glutathione Peroxidases

They have selenium-cofactors, which are responsible for the catalysis of breakdown of H_2O_2. They are mostly found in hepatic tissues.

5.1.3 Antioxidant Chemicals

Antioxidants are broadly classified as either hydrophilic or hydrophobic, and this categorization evaluates their site of action in the body. Hydrophilic antioxidants act in the extracellular fluids. On the other hand, hydrophobic antioxidants act on cell membranes and protect them from ROS damage. The body can synthesize several antioxidants; however, some of them must be taken from the diet.

5.1.4 Antioxidant Chemicals the Body Synthesizes

5.1.4.1 Glutathione

In cells, glutathione is maintained in the reduced form by the enzyme glutathione reductase. Due to its high concentration and central role in maintaining the cell's redox state, glutathione is one of the most important cellular antioxidants.

5.1.4.2 Uric Acid

It is the main metabolic intermediate during the breakdown of nucleotides, such as adenine. It circulates in high concentrations in the blood and disables circulating free radicals.

5.1.4.3 Melatonin

Unlike other antioxidants, melatonin does not undergo redox cycling, which is the ability of a molecule to undergo repeated reduction and oxidation. Melatonin, once oxidized, cannot be reduced to its former state, because it forms several stable end-products upon reacting with free radicals. Therefore, it has been referred to as a terminal (or suicidal) antioxidant.

5.1.5 Antioxidant Chemicals Obtained from the Diet

A number of important antioxidants are obtained from food, including selenium and vitamins, A, C, and E.

5.1.6 Beneficial Effects of Antioxidants on Body

Antioxidants are the main ingredients of breakfast cereals, and many more processed foods products, to provide energy. These are promoted as additives that can prevent and treat various health-related problems such as cancer, cataract, memory loss, neurological diseases, immune dysfunction, and heart disease. Figure 5.2 illustrates the antioxidants' protective pathway. Moreover, by protecting skin from ROS, antioxidant prevents erythema and skin cancers. Table 5.2 lists the role of the various nutraceutical compounds on different organs of the body.

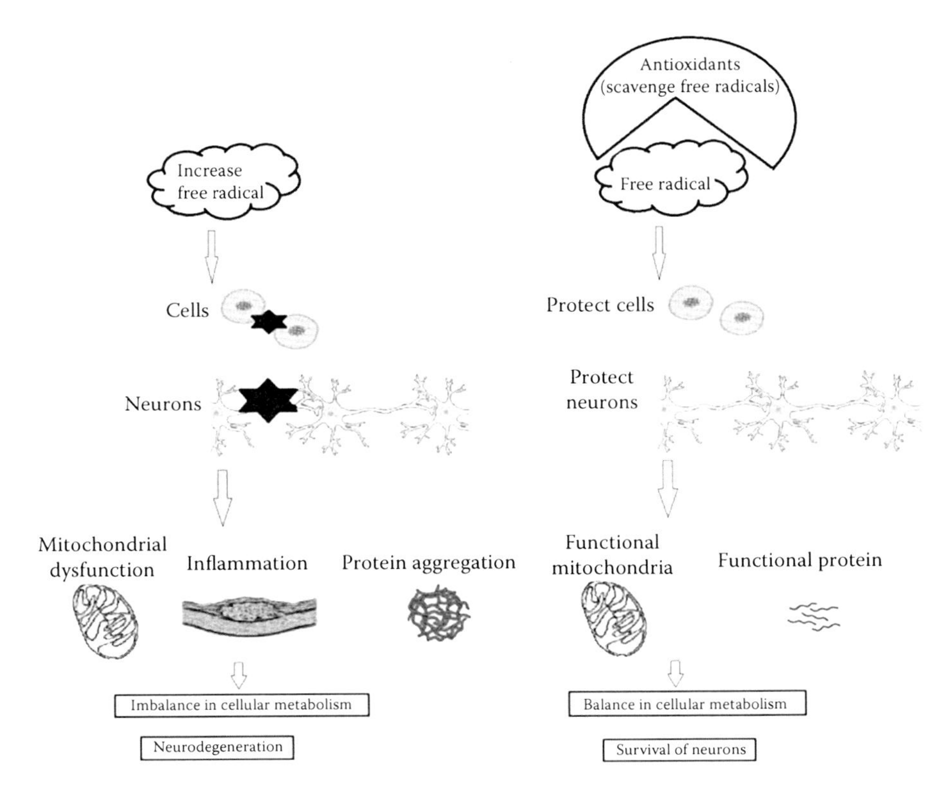

FIGURE 5.2 Various protective pathways of antioxidants.

5.1.7 Drawbacks Associated with Antioxidants

Antioxidants such as ascorbic acid and curcumin are known to possess effective anti-inflammatory activity that prevents skin damage, but their protective efficacy is reduced because of their instability and low skin penetration. Coenzyme Q10 (CoQ10) has poor topical bioavailability due to large molecular weight and high lipophilicity. Another important antioxidant, vitamin E, has high elimination rate and low cutaneous delivery, because of less water solubility. On the other hand, resveratrol, curcumin, naringenin, CoQ10, and silymarin have poor bioavailability because of less water solubility. To overcome these problems, it is required to formulate delivery systems that can facilitate delivery of antioxidants effectively to the target site in the body and reduce oxidative stress, thereby helping in prevention and treatment of various diseases. Table 5.3 summarizes the drawbacks of nutraceuticals that are hampering their clinical use and the novel delivery approaches to overcome the problems.

5.1.8 Delivery Systems for Different Antioxidants

5.1.8.1 Ascorbic Acid

TABLE 5.2
Antioxidant Role of the Various Nutraceutical Compounds on Different Organs of the Body

Nutraceutical Compound	Beneficial Effects on Different Organs of the Body	References
Ascorbic acid	Eye (reduced oxidative modification of lens proteins observed with aging)	McCusker *et al.*, 2016
	Antitumor activity (enhancing the immune system and prevent metastasis)	Pires *et al*, 2016
	Skin (reduces DNA destruction and lipid peroxidation, decreased the release of pro-inflammatory cytokines, and protects against apoptosis)	Serrano *et al.*, 2015
Resveratrol	Skin (detoxification of ROS)	Fujimura *et al.*, 2016
	Cardioprotective effect (upregulation of nitric oxide production)	Abbas *et al.*, 2016
	Eye (inhibit angiogenesis and suppressed retinal neovascularization)	Lee *et al.*, 2015
Curcumin	Skin (quench ROS and decreases inflammation through nuclear factor-KB [NF-kB] inhibition. Enhance deposition of collagen and fibroblast in injury thus improving wound-healing)	El-Refaie *et al.*, 2015; Sintov, 2015
	Brain (decrease cyclooxygenase, phospholipase, lipidperoxidation and proliferation of neuroglial cell, reduction in beta amylora plaques. Increase action of SOD and oligodentrocytes)	Mishra and Palanivelu, 2008
	Eye (decreses NF-kB activation, expression of inflammatory genes)	Liu *et al.*, 2015
Naringenin	Eye (increase the blood flow in iris, ciliary body and choroid)	Zhang *et al.*, 2016; Tsai *et al.*, 2015
	Skin (reducing superoxide anion production)	Chlapanidas *et al.*, 2014
	Antitumor activity (inhibit cell proliferation)	Sulfikkarali *et al.*, 2013
	Liver (restore protein levels)	Yen *et al.*, 2009
Vitamin E	Eye (maintain health of cell membranes, DNA repair)	Paradiso *et al.*, 2016
	Skin (neutralizes ROS)	Brownlow *et al.*, 2015
Silymarin	Brain (decreases protein nitrosylation)	Hirayama *et al.*, 2016
	Skin (anti-inflammatory action as well as inhibition of mitogenic signaling)	Mady *et al*, 2016
	Antitumor activity (decrease activation of NFκB)	Zheng *et al.*, 2016
	Liver (increases glutathione level, which increases liver capacity for detoxification)	Chaudhary *et al.*, 2015
CoQ10	Skin (reduces ROS damage and increases skin repair and regeneration)	Yadav *et al.*, 2016
	Brain (decreases deposition of destructive amyloid β-peptide proteins in brain cells, helps in maintaining neurotransmitter level and act as a free radical scavenger in lipid and mitochondrial membranes)	Belhaj *et al.*, 2012; Young *et al.*, 2007
Quercetin	Brain (increasing the expression of antioxidant and altering inflammation)	Dajas *et al.*, 2015
	Eye (anti-inflammatory effects on the lacrimal functional unit)	Oh *et al.*, 2015
	Antitumor activity (induces of cell-cycle arrest and/or apoptosis)	Gibellini *et al.*, 2011
Rutin	Brain (decreases pro-inflammatory cytokines in neuroblastoma cells)	Qu *et al.*, 2014
	Antitumor activity (prevents apoptotic cell death)	Perk *et al.*, 2014

Ascorbic acid is one of the forms of vitamin C, which was initially called as L-hexuronic acid, later renamed to ascorbic acid. Ascorbic acid is referred to as one of the most popular scavengers of ROS that is valuable in preventing the harmful effects of superoxides and peroxynitrites (Leonard *et al.*, 2002). It is commonly used in prevention of scurvy and cataracts. It helps in boosting the immune system and accelerates the process of healing wounds. The main drawbacks associated with ascorbic acid are less stability and poor permeability that consequently result in low bioavailability.

5.1.8.1.1 Enhancement of Dermal Delivery

Vitamin C is commonly known for antioxidant action that aids photo-damage of the keratinocytes, but its effectiveness is minimized because of its variability and lesser infiltration capabilities into

TABLE 5.3
Drawbacks of Nutraceuticals Hampering their Clinical Use, and Novel Approaches Used to Overcome Them

Nutraceuticals	Limitations	Novel Approaches	References
Ascorbic acid	Less stability and poor permeability	Liposomes	Serrano *et al.*, 2015
		Gel	Wang *et al.*, 2013
		SLNs	Guney *et al.*, 2014
Resveratrol	Low bioavailability due to poor aqueous solubility	Nanoemulsion	Pangeni *et al.*, 2014
		SNEDDS	Singh *et al.*, 2014
		SMEDDS	Seljak *et al.*, 2014
		Liposomes	Csiszar *et al.*, 2015
		Nanotube	Vergaro *et al.*, 2012
Curcumin	Low stability, permeability and bioavailability	Nanoemulsion	Naz and Ahmad, 2015
		Microemulsion	Sintov, 2015
		NLC	Meng *et al.*, 2015
		Microsponge	Sareen *et al.*, 2014
Naringenin	Low bioavailability due to poor aqueous solubility	SNEDDS	Khan *et al.*, 2015
		Nanoparticles	Sulfikkarali *et al.*, 2013; Ban *et al.*, 2015
		Liposomes	Wang *et al.*, 2017
Vitamin E	High elimination rate	Microsphere	Prado *et al.*, 2012
	Low cutaneous delivery due to poor aqueous solubility	Nanoemulsion	Brownlow *et al.*, 2015
		PVA and sodium alginate loaded vitamin E film	Garrastazu *et al.*, 2014
Silymarin	Low bioavailability due to poor aqueous solubility and degradation in gastric fluid	Proliposome	Wang *et al.*, 2015
		NLCs	Chaudhary *et al.*, 2015
		SNEDDS	Chen *et al.*, 2015
		Nanoparticles	Yang *et al.*, 2013; El-Nahas *et al.*, 2017
		Nanoemulsion	Nagi *et al.*, 2017
CoQ10	Poor bioavailability due to poor aqueous solubility.	Liposomes	Shao *et al.*, 2015
		Chitosan-coated liposome	Zhao *et al.*, 2017
	Poor topical bioavailability due to large molecular weight and high lipophilicity.	Micelles	Qin *et al.*, 2017
		Proniosomal gel	Yadav *et al.*, 2016
	Low stability	NLC	Keck *et al.*, 2014
		Micelle	Liu *et al.*, 2016a
Quercitin	Poor bioavailability due to poor aqueous solubility	Nanoparticles	Anwer *et al.*, 2016
		Complex	Kellici *et al.*, 2016
		Mesoporous SBA-15 silica carrier	Trendafiloval *et al.*, 2017
Rutin	Poor bioavailability due to poor aqueous solubility	2-hydroxypropyl-beta-cyclodextrin complex	Miyake *et al.*, 2000
		NLCs	Kamel *et al.*, 2015

the skin. To overcome this problem, Serrano *et al.* (2015) prepared liposomes loaded with ascorbate to surmount the obstacle of the stratum corneum. Diffusion of these liposomes was tested on complete skin, epidermis, and dermis in 10 patients. Phosphatidylcholine-based liposomal formulation enhanced skin permeability of fluorescein and ascorbate. Fluorescein phosphatidylcholine liposomes improved circulation into epidermis more than than dermis, whereas ascorbate liposomes improved the same vice versa.

Pepe and coworkers (2012) developed sugar-based microemulsions loaded with two antioxidants (lycopene and ascorbic acid) to enhance the cutaneous delivery of both molecules and promote antioxidant activity. Formulation was prepared using different oils such as isopropyl myristate and mixed with monocaprylin and monoolein. Observations revealed that formulation was capable to enhance lycopene penetration into porcine skin approximately 3.3–8.0 times. In the case of ascorbic acid, it was found that the cutaneous penetration was 1.5–3.0 times more than the pure drug solution. The study illustrated the potential of drug delivery systems in increasing the tissue antioxidant activity by showing rapid cutaneous penetration.

5.1.8.1.2 Enhancement of Stability

To increase the thermal stability of vitamin C, Gao *et al.* (2014) designed a drug delivery system composed of Ca/Al layered double hydroxide. The designed system was also evaluated for the controlled release of vitamin C. In their work co-precipitation technique was used for the intercalation of vitamin in the gallery of Ca/Al layered double hydroxide. According to them, about 36.4 wt% of vitamin C was loaded in layered double hydroxide, and it showed increased thermal stability of vitamin. It was found that the release of vitamin C was extended and showed 80% maximum release at pH 7.4.

5.1.8.1.3 Drug Delivery or Targeting

Experimental studies have revealed that ascorbic acid may reduce neuronal harm caused by free radicals and also facilitate its repair. Guney *et al.* (2014) explored the use of ascorbic acid loaded solid lipid nanoparticles (SLNs) for efficient cellular uptake and accumulation in cancer cell. They used hot homogenization techniques for incorporation of ascorbic acid in SLNs. Formulation showed high drug entrapment efficiency, along with particle size of less than 250 nm. They concluded that SLNs were capable of efficiently delivering ascorbic acid to cancer cells and showed sustained release in comparison to pure drug (Guney *et al.*, 2014).

5.1.8.1.4 Sustained Release

Ascorbic acid also suffers with the drawback of sustaining the activity. To overcome this problem, Pierucci and associates (2006) evaluated potential use of microparticles in sustaining the activity of ascorbic acid. Spray-drying method was employed for the preparation of formulation using pea protein as wall-forming agent. The prepared formulations had irregular and rough surfaces and showed average particle size below 8 microns. The authors suggested that pea protein could be used for fabrication of microparticles to obtain sustained release of drug (Pierucci *et al.*, 2006).

5.1.8.2 Resveratrol

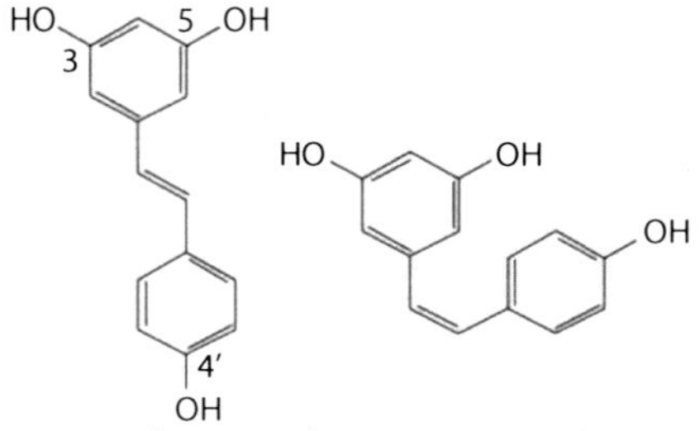

HO 3 5 OH 4′ OH HO OH OH

trans-Resveratrol *cis*-Resveratrol

Peanuts, soy beans, pomegranates, berries, and grapes are the vital sources of resveratrol. It acts as a direct scavenger to free radicals showing antioxidative property and is helpful in ameliorating oxidative stress and delaying the risk of neurodegenerative and associated diseases. This compound has poor bioavailability because of low water solubility.

5.1.8.2.1 *Enhancement of Aqueous Solubility and Bioavailability*

Singh and Pai (2014) prepared and evaluated resveratrol self-nanoemulsifying drug delivery system (SNEDDS) with the aim to improve oral drug bioavailability. The pharmacokinetic study results revealed that the prepared formulation showed 1.33 times increment in area under curve (AUC). The authors concluded that the preparation showed effective delivery and enhanced oral bioavailability of resveratrol.

Seljak and associates (2014) prepared resveratrol-loaded self-microemulsifying drug delivery system (SMEDDS) with the objective to improve the pharmacokinetic properties of resveratrol. Results showed that the formulation showed excellent self-emulsification along with an increased release of drug and with an appreciable decrease in drug metabolite efflux, which, in turn, increased oral bioavailability of drug.

5.1.8.2.2 *Drug Delivery or Targeting*

To attain therapeutic doses of resveratrol in the body tissues, Csiszar and coworkers (2015) prepared resveratrol-loaded fusogenic liposomes and examined its efficiency for the increased uptake of drug by the aged cells. The findings showed that short-term incubation of aged cerebromicrovascular endothelial cells with fusogenic liposomes lead to increase in intra-cellular resveratrol level, whereas resveratrol loaded in conventional liposomes was not effective in increasing the intra-cellular drug content. According to the investigators, their formulation (resveratrol-loaded fusogenic liposomes) increased the uptake of drug in the cells and was helpful in rapidly attenuating oxidative stress associated problems.

Vergaro and associates (2012) developed resveratrol loaded clay nanotubes fabricated with halloysite, wherein drug-loaded clay showed resveratrol release of 48 h and enhanced cytotoxicity activity, causing rapid cell apoptosis. Pangeni *et al.* (2014) prepared a kinetically stable resveratrol-loaded nanoemulsion with the aim to deliver resveratrol directly to the brain through intranasally. The incorporated resveratrol into nanoemulsion was protected from degradation into the nasal cavity, which further enhanced the availability in brain and blood. The optimized nanoemulsion formulation showed high free radical scavenging potential. Moreover, the formulation was capable of delivering resveratrol to the brain following intranasal administration as indicated by the results of pharmacokinetic studies.

5.1.8.3 Curcumin

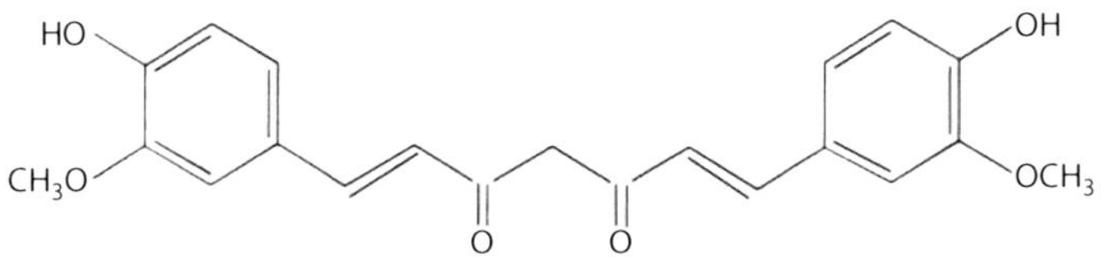

Curcumin is a phenolic compound obtained from turmeric, a member of family Zingiberaceae. It is a potent antioxidant and anti-inflammatory molecule. The phenolic hydroxyl group is responsible for its antioxidant activity. Additionally, it has ferric ion reducing potential as well as ferrous ions chelating action. Besides the numerous activities, curcumin suffers from various drawbacks such as low stability, permeability, and bioavailability.

5.1.8.3.1 *Enhancement of Stability*

To overcome the problem of instability, Hazzah and co-workers (2015) formulated curcumin-loaded SLNs. The experimental results indicated that prepared formulation had high entrapment efficiency and showed minimum inhibitory concentration of 0.187, 1.5, 3, 0.75, 0.093, and 0.185 mg/mL, against *Candida albicans, Lactobacillus acidophilus, Escherichia coli, Viridans streptococci, Streptococcus mutans*, and *Staphylococcus aureus*, respectively. The findings suggested that the antimicrobial action and stability of curcumin were maintained by loading it in lipid-based nanocarrier.

5.1.8.3.2 Enhancement of Bioavailability

Bioavailability of the curcumin was enhanced by formulating their nanoemulsions by Naz and Ahmad (2015). They incorporated this nanoemulsion into a gel by using Carbopol-980 to improve topical bioavailability of curcumin. The results revealed that prepared nanoemulsion gel improved the permeation (4 times) of drug due to nanodroplet size (less than 70.0 ± 2.7 nm). Furthermore, the gel system reversed the arthritic symptoms in rats in comparison to curcumin solution in oil.

Li and associates (2012) prepared curcumin-loaded flexible liposomes and found that there was a 2.35- and 7.76-fold increase in bioavailability of curcumin, respectively, in comparison to the suspension of curcumin. This indicated the efficacy of prepared silica-coated formulation to act as a carrier that improved the oral bioavailability of curcumin.

5.1.8.3.3 Enhancement of Dermal Delivery

El-Refaie and associates (2015) designed a curcumin loaded gel-core hyaluosome employing film hydration technique with the aim to increase drug penetration through the skin. The burn-wound healing test results illustrated that developed formulation showed recovery at seventh day of treatment and lead to normal skin at 11th day with no sign of scar. This may have been due to increased (5-fold) skin deposition of curcumin from gel core hyaluosome as compared to simple curcumin transfersomal gel.

Sintov (2015) formulated and characterized curcumin-loaded microemulsion and determined its potential for transdermal delivery. The analyst found that the permeability coefficient of curcumin in microemulsion comprising 10% water was 2 times more than the preparation consisting of 5% and 20% aqueous segment. All these changes may have been due to the decrease in the globule size and concentration in preparation as the water content decreased from 20% to 5%. The study showed that the infusion improvement ratio between curcumin microemulsion to curcumin solution was 55, establishing the capacity of curcumin-laden microemulsion for efficient transdermal delivery of curcumin.

5.1.8.3.4 Drug Delivery or Targeting

Targeted delivery of curcumin to the specific organs is also an important factor to be considered for ensuring complete release. Meng and associates (2015) explained the preparation of curcumin-loaded NLCs by solvent evaporation method. By electrostatic interaction, surfaces of prepared NLCs were modified with lactoferrin so that it mimicked as low density lipoprotein, which ensured its easy permeation across blood–brain barrier (BBB). Experimental results indicated that surface-modified NLCs achieved 2.78 times better drug accumulation in brain vis-à-vis unmodified NLCs. The authors concluded that lactoferrin surface-modified NLCs can act as a promising carrier for brain targeting.

Sareen *et al.* (2014) investigated the potential of curcumin microsponges for colon delivery. The results exhibited that curcumin release from the delivery system was observed at alkaline pH, but not at upper GIT pH, suggesting its promising role in drug delivery to colon. Experimental findings portrayed reduction in edema by microsponges as compared with simple curcumin. The present work illustrated the potential of microsponge dosage form in improving biopharmaceutical behavior of curcumin.

5.1.8.3.5 Enhancement of Ocular Delivery

Liu and colleagues (2016b) developed heat-sensitive *in situ* nanogel loaded with curcumin cationic nanostructured lipid system for enhancement of ocular reservation and curcumin infiltration. They compared the permeability coefficient and AUC of curcumin solution with nanogel-based formulation and found that the nanogel formulation showed 9.24-fold and 1.56-fold increase in AUC and permeability coefficient, respectively. They also observed that the gel-based system showed controlled release of curcumin due to prolonged ocular mean residence time.

5.1.8.4 Naringenin

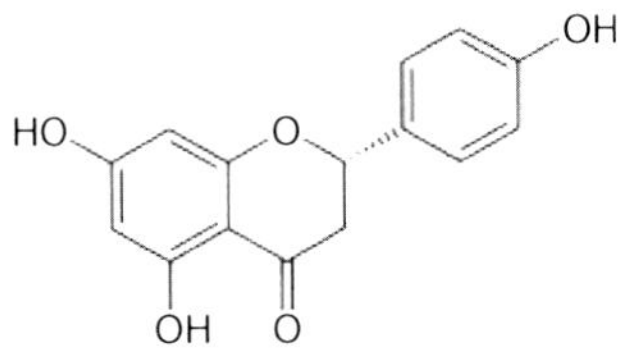

Naringenin is a flavonoid (i.e., trihydroxy flavanone). It scavenges the harmful $OH^{\cdot}$ and superoxide radicals, thus preventing oxidation of lipids, lipoproteins, and glutathione. The main drawback associated with its oral administration is poor aqueous solubility, which limits its oral bioavailability to around 5.8%. To prevail upon such difficulties, strategies such as complexation, salt formation, nanoparticles, etc. have been employed.

5.1.8.4.1 Enhancement of Aqueous Solubility and Bioavailability

In a study, Yen and colleagues (2009) designed nanoparticles loaded with naringenin to increase oral bioavailability. Prepared nanoparticles exhibited enhanced solubility of naringenin and showed significantly better release rate in comparison to pure naringenin. Obtained data suggested that the developed nanoparticles could effectively improve naringenin release and show more oral bioavailability.

Ban and associates (2015) designed oil-based naringenin loaded lipid nanoparticles. The developed lipid nanoparticles exhibited improved bioaccessibility (71%) of naringenin vis-à-vis pure naringenin, which showed only 15% of bioaccessibility. According to the authors, their work established lipid nanoparticles to be a possible alternative drug delivery system for improving bioavailability of naringenin.

Recently, Wang and associates (2017) developed naringenin-loaded liposomes to enhance bioavailability. Results showed that liposomes exhibited particle size of 70.53 ± 1.71 nm. They compared the drug release from naringenin loaded liposome with free naringenin. Liposomes showed better *in vitro* release than the free drug due to decreased particle size. They reported that after oral administration of liposomes to mice, there was 13.44 times increment in relative bioavailability.

5.1.8.4.2 Enhancement of Dermal Delivery

Naringenin can increase the level of melanin in skin and prevent oxidative skin damage. In several studies, naringenin formulation has been developed for topical application. Trai and coworkers (2015) prepared and evaluated naringenin-loaded liposomes for topical application. Experimental data showed that by using liposomes formulation the deposition content of drug in skin was increased about 7.3–11.8 times when compared with drug solution-treated groups. Naringenin-loaded liposomes caused less skin irritation in comparison to the standard (paraformaldehyde) group. They concluded that liposomes can act as potential carrier for naringenin for topical application.

Chlapanidas and colleagues (2014) investigated the action of naringenin-encapsulated sericin microparticles in the management and prevention of psoriasis. Microparticles lead to cytotoxic effect at a dose level of 200 μg/mL, and action was more in comparison to pure naringenin. Moreover, the prepared microparticles resulted in downregulation of cytokine, making them a suitable drug delivery carrier for the prevention and treatment of psoriasis.

5.1.8.4.3 Enhancement of Ocular Delivery

In recent years, it has been confirmed that naringenin could be used for oculopathy, as it has various action on eyes like retinal pigment epithelium degeneration, choroidal neovascularisation, retinal function recovery, etc. But the ocular use of naringenin is low due to less water solubility. Zhang *et al.* (2016) developed naringenin-loaded nanoparticles and evaluated for topical ophthalmic delivery. In their study the drug was complexed with sulfobutylether-cyclodextrin to increase its solubility. Release study indicated that naringenin-loaded nanoparticles showed sustained release characteristics of drug. *In vivo* studies showed that prepared nanoparticles lead to enhanced bioavailability of drug in aqueous humor. From the study, it was suggested that the nanoparticles could effectively be used for ocular administration of naringenin.

5.1.8.5 Vitamin E

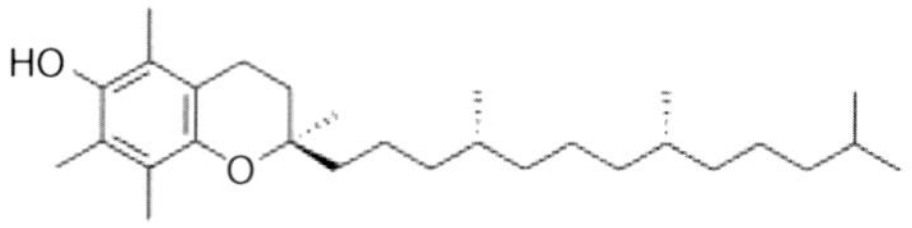

HO O

Vitamin E is well known for its protective action against oxidative stress. It is usually found in many foods such as cereals, meat, poultry, eggs, fruits, and vegetables. It helps in the prevention of digestive problems, cystic fibrosis, sunburn, uveitis, dementia, stroke, rheumatoid arthritis, dysmenorrheal, anemia, cataracts, and many other diseased conditions. The systemic availability of vitamin E depends on plasma lipoproteins, which are responsible for its transportation into the circulation. Blood level of this vitamin is controlled by hepatic system, wherein the transfer proteins play a key role.

5.1.8.5.1 Enhancement of Dermal Delivery

Vitamin E neutralizes ROS on exposure to sunlight, thus preventing oxidative skin injury, and is used for a variety of cutaneous disorders. A nanoemulsion system for improving the dermal delivery of vitamin E was developed by Brownlow *et al.* (2015). Prepared formulation showed good photostability and characteristic features (polydispersity index less than 0.25, droplet size less than 150 nm and –30 mV of surface charge). Vitamin E enriched nanoemulsion provided protection to cutaneous pathologies against UV radiations and exhibited excellent biocompatibility. The study revealed that formulation of vitamin E into nanoemulsion could improve vitamin E aqueous solubility and increase cutaneous delivery.

Vitamin E helps in curing burn injuries, especially due to its antioxidant activity in reperfusion of ischemic tissues. Garrastazu and associates (2014) developed a polymer film composed of polyvinyl alcohol (PVA) and sodium alginate loaded with vitamin E. The film released up to 30% of vitamin in 12 h. Also, the prepared film had the potential to deposit more amount of it in skin in comparison to conventional cream.

5.1.8.5.2 Targeted Delivery

Various studies have revealed that the combination of anticancer drug and vitamin E may be an effective strategy for enhancing antitumor efficacy and for reducing cytotoxicity to normal cells. Kutty and associates (2015) designed a micellar system of vitamin E loaded with docetaxel, with the aim to evaluate the superiority of cetuximab conjugated vitamin E micelle system for improving the potential of delivering docetaxel, which could be helpful in cancer treatment. The purpose of adding cetuximab ligand was to target the receptors present on breast tumor cells (MDA-MB-231). On direct intravenous injection, the micellar system deposited in cancer cells and retained themselves for one day, providing successful delivery of anticancer drug to cancer cells and by their action of anti-proliferation inhibiting the growth of cancer cells in much better manner as compared to Taxotere®. Investigators concluded that these targeting micellar systems effectively avoided receptor overexpression of MDA-MB-231 cancers.

5.1.8.6 Silymarin

Silymarin is a bioflavonoid having anticancer, antioxidant, anti-lipid-peroxidative, antifibrotic, immunoregulatory, anti-inflammatory, and liver regenerating activity. Milk thistle, *Silybum marianum*, is the basic source of silymarin.

5.1.8.6.1 Improvement of Aqueous Solubility and Bioavailability

Silymarin has low oral bioavailability because of poor solubility and degradation by gastric fluid. To overcome such problems, various strategies have been exploited including encapsulating it into SLNs, proliposomes, and solubilizing it in micellar drug delivery systems, SNEDDS, etc. Wang *et al.* (2015) developed silymarin proliposome to enhance silymarin's poor bioavailability and hepatoprotective activity. Drug-loaded liposomes exhibited multilayer structure and showed higher intestinal transport of drug due to small particle size. *In vivo* studies carried out using injured liver showed that drug-loaded liposomes exhibited more hepatoprotective activity by elevating SOD and glutathione levels, suggesting the superiority of liposomes over silymarin tablets. Yang and coworkers (2013) fabricated silymarin nanoparticles to achieve better hepatoprotective activity. The designed dosage form showed 1,300 times enhancement in silymarin solubility. Findings demonstrated that formulation exhibited high C_{max} and more AUC (about 1.3 times increase) in comparison to the marketed formulation of silymarin. Additionally, the prepared formulation displayed reduction in carbon tetrachloride-induced liver toxicity implying its advantage to deliver silymarin.

5.1.8.6.2 Drug Targeting

Chaudhary *et al.* (2015) designed NLCs as a nanolipid carrier system of silymarin for targeting liver. The prepared NLCs exhibited 84.60% release of silymarinin over 24 h. The *in vivo* experiments illustrated that the absorption of silymarin took place via lymphatic transport mechanism that was confirmed by C_{max} of 5.008 μg/mL in plasma. Additionally, there was 2-times increase in drug bioavailability from NLCs vis-à-vis its suspension. In liver, about 19.26 μg of silymarin reached in 2 h, but in other organs negligible amounts of drug were found. This study suggested that direct targeting of silymarin to liver via lymphatic pathway could easily be obtained by utilizing NLCs.

5.1.8.6.3 Controlled Release

Controlled release systems become attractive strategy to obtain drug action for an extended period of time. Xie *et al.* (2013) developed silymarin monolithic osmotic tablets by compressing solid dispersion of drug. The tablets were composed of cellulose acetate coating, which acted as a semipermeable membrane. In 20 h, the tablets showed zero-order release, which indicated that the formulation was capable of releasing the drug in a controlled manner and provided prolonged availability of the drug in the body.

5.1.8.6.4 Parenteral Delivery

To avoid gastrointestinal degradation and to provide the highest availability and fast effect of silybin, parenteral route is a good option. Different nano-drug delivery carriers such as nanoparticles, nanoemulsions, and nanomicelles have potential to improve parenteral delivery of lipophilic compounds. Jia *et al.* (2010) explored the use of NLCs for parenteral delivery of silybin. Emulsion evaporation technique was used for the production of NLCs. Silybin-NLCs in comparison to drug solution exhibited more AUC and remained for a prolonged time in circulation. The tissue distribution study conducted in New Zealand rabbits and Kunming mice showed that higher amounts of silybin-NLCs reached RES organs, predominantly the liver. The findings suggested that NLC are promising carriers to attain sustained release of silybin.

5.1.8.7 CoQ10

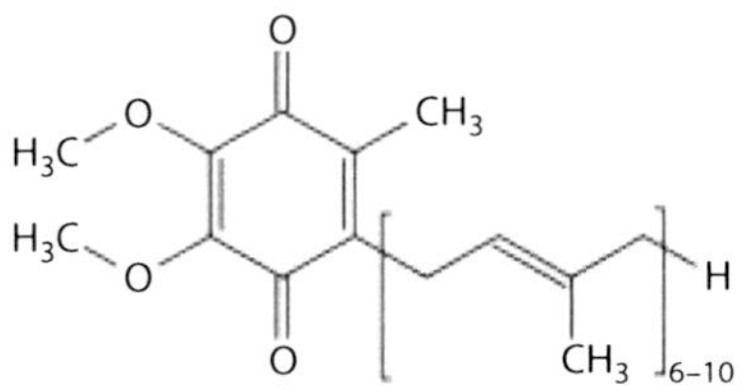

CoQ10 is a well-known endogenous antioxidant and also has anti-amyloidogenic activity. Both of these properties help in the treatment of disorders such as diabetes mellitus, CNS pathologies, and cancer. Generally, it is involved in energy production where it acts as a cofactor. It has the potential to block apoptosis mechanism of cell death and could be useful in the management and prevention of neuroprotective disorders (Young *et al.*, 2007; Belhaj *et al.*, 2012). Furthermore, it was investigated that it also delayed the amyloid beta-protein toxic effects, which makes its role useful in the prevention of Alzheimer's disease (Young *et al.*, 2007). In spite of its usefulness, poor oral bioavailability is a problem in clinical development of CoQ10, which is mainly caused by its poor aqueous solubility (<0.25 µg/mL). There is need to prepare carriers capable of improving CoQ10 solubility and stability. Different strategies have been applied for increasing CoQ10 solubility, such as liposomes, SNEDDS, and NLCs.

5.1.8.7.1 Enhancement of Stability

Liu and associates (2016a) improved the stability of coenzyme Q10 by using micelle system composed of Kolliphor® HS 15. The amorphous form of drug was entrapped in the formulation, which was analyzed and proved by differential scanning calorimetry study. It was examined that CoQ10 followed first order release pattern. Moreover, the formulation showed good stability at a temperature below 25°C, making it a potential carrier for drug delivery.

5.1.8.7.2 Enhancement of Aqueous Solubility and Bioavailability

Shao and coworkers (2015) evaluated the use of liposomes as a drug delivery carrier for CoQ10 with the aim to increase its oral bioavailability. In their study, they formulated chitosomes, tocopheryl polyethylene glycol 1000 succinate (TPGS)-chitosomes, and TPGS-liposomes. Among all formulations, chitosan-coated liposomes exhibited more mucoadhesive strength. In contrast to pure drug, the designed formulations, especially TPGS-chitosomes, showed 30 times more uptake of drug in Caco-2 cells. Moreover, it showed sustained release of drug in rats, which was up to 24 h, due to which it showed enhanced antioxidant activity of CoQ10. The experimental findings implied that the prepared TPGS-chitosomes had the potential to increase the oral bioavailability of drug and showed extended drug release, thus it can be used for effective oral drug delivery.

5.1.8.7.3 Enhancement of Dermal Delivery

It is well known that CoQ10 effectively minimizes UV radiation-mediated oxidative damage by thiol depletion, stimulation of phosphotyrosine kinases, and collagenase suppression. It is also responsible for reducing wrinkle depth. CoQ10 has been used in many cosmetics, meant for skin to obtain healthy skin and for antiaging. A problem associated with CoQ10 skin application is its poor topical bioavailability, which is mainly due to its large molecular weight and high lipophilic nature. However, a few attempts have been made in the past to improve the topical bioavailability and patient compliance using niosomes and NLC preparations. Based on these considerations and to overcome the problem associated with poor CoQ10 bioavailability through topical route, Yadav and associates (2016) developed CoQ10-loaded proniosomal gel. Biochemical estimations conducted in mice indicated that the developed proniosomal gel was capable of restoring the levels of antioxidant enzymes. Histopathological observation results indicated that CoQ10-loaded proniosomal gel had good efficiency to treat UV radiation damaged skin as compared to free drug, and was also safe for skin application as it had negligible effect on normal histology.

Keck and associates (2014) explored the use of ultra-small NLC as a drug carrier for CoQ10 to increase its release as well as its skin penetration. Experimental results revealed the usefulness of designed formulation for promising delivery of drug for enhanced antioxidant activity. On topical application of NLCs, occlusion is obtained through the formation of film, which in turn reduces water loss and results in enhancement of skin hydration and penetration of drug to skin. On the basis of these facts, using lipid nanoparticles for dermal application of CoQ10 could be beneficial.

5.1.8.8 Quercetin

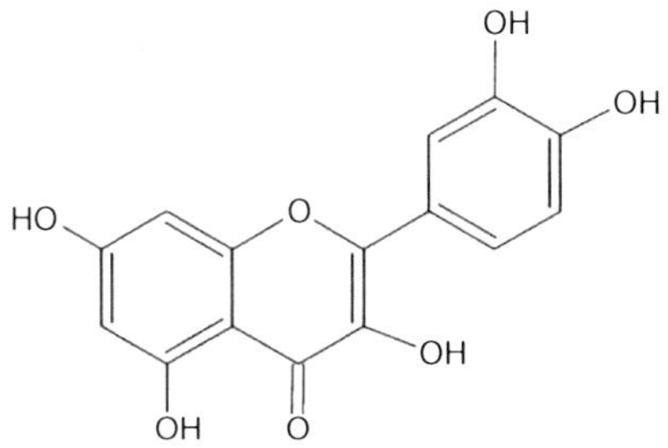

Quercetin is a plant pigment extracted from several leaves, fruits, grains, and vegetables such as red wine, buckwheat, green tea, berries, and onions. It is used as an antioxidant, as it fights damage against free radical. It also demonstrates anticancer properties. Because of the antioxidant properties of quercetin, it exhibits several benefits such as lowering inflammation (which is the root cause of the major diseases), fighting allergies (including food allergies and skin reactions as quercetin is able to stabilize the release of histamines, which are accountable for the allergic symptoms), maintenance of heart health (quercetin helps in reducing the risk of atherosclerosis, lowers cholesterol levels), as an anticancer (has chemopreventive activities and may show an exclusive antiproliferative effect on cancerous cell) and also protects skin health (quercetin blocks the release of many pro-inflammatory cytokines and helps in protecting skin from disorders like photosensitivity and dermatitis).

5.1.8.8.1 Enhancement of Aqueous Solubility and Bioavailability

The major problem associated with quercetin is its low aqueous solubility, which hampers its bioavailability. Approaches like co-solvency, size reduction, solid dispersion, hydrotrophy, complexation, nanonization, and many others have been employed to improve solubility of quercetin. Anwer and coworkers (2016) formulated polymeric nanoparticles of quercetin to overcome this shortcoming. Physicochemical characterization was conducted, which indicated that the drug was adequately entrapped in the polymer and showed a sustained drug release profile and a significant antioxidant activity.

5.1.8.8.2 Enhancement of Antioxidant Activity

Antioxidant potential of quercetin is well known and the methods to enhance the use are in progress. Kumar *et al.* (2016) used *Rhizopus oryzae* lipase to prepare quercetin ferulate (acylated quercetin). This synthetic technique is an efficient method for the production of the flavonoid derivation in comparison to the conventional chemical methods as the production may be scaled up to 25 grams. Results show an enhanced antioxidant activity of quercetin ferulate.

5.1.8.9 Rutin

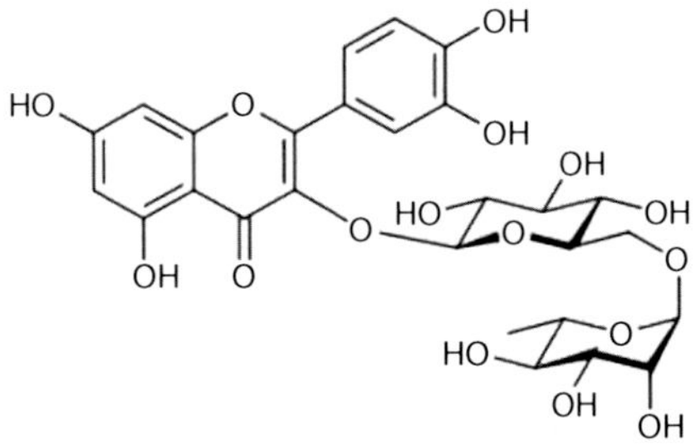

Rutin comes from the plant *R*uta *graveolens*, thus the genesis of the name, rutin. It is also believed to prevent mucositis, a side effect of cancer treatment. It may also be used for osteoarthritis in combination with proteins such as trypsin and bromelain. It is generally regarded as safe, though it may cause some side effects such as flushing, rashes, headache, or stomach upset. It has low oral bioavailability, which is its major disadvantage.

5.1.8.9.1 Enhancement of Aqueous Solubility and Bioavailability

Kumar and associates (2012) prepared SEDDS of rutin with the aim to enhance rutin bioavailability. The results demonstrated that higher amount of drug release was obtained after 30 min (93%) for optimized SEDDS. Nanocrystals of rutin by Mauludin and coworkers (2009) for improving its dissolution behavior and bioavailability were prepared. The dried drug nanocrystals were compressed into tablets by direct compression. Small sized particles were seen to be completely redispersed in aqueous phase. Results revealed that nanocrystals showed drug release after 30 min with complete dissolution in water. Prepared nanocrystals showed dissolution of 71% in comparison to 55% obtained from the marketed formulation.

5.1.8.9.2 Enhancement of Antioxidant Activity

The antioxidant potential of rutin was evaluated by Ahmad and co-workers (2016). Drug encapsulated-chitosan nanoparticles (RUT-CS-NPs) were prepared and analyzed for different parameters to investigate its antioxidant ability to treat ischemic diseases. Results revealed an increased content of rutin in the brain. The authors concluded that these nanoparticles were highly effective in targeting rutin to the brain.

5.1.8.9.3 Enhancement of Dermal Activity

Kamel *et al.* (2015) developed rutin NLCs to explore its antioxidant properties. The formulation prepared had good occlusivity, high drug encapsulation, and release efficiencies. In comparison to the standard antioxidant, NLCs showed two times higher antioxidant effect.

5.2 CONCLUSIONS

Due to age progression and increased exposure to environmental damage, genetic defects, and immunological factors, body endogenous free radical scavenging mechanism is not totally active to defend against oxidative stress, which is responsible for many diseases. Using approaches that can reduce or prevent oxidative damage could offer a management approach for diseases associated with oxidative stress. An opportunity that could be focused in the future is to alter antioxidants for enhancement of their bioavailability. Also, the developed formulations of antioxidants should be subjected to clinical trials so that companies developing them could enter the global market. In the future, investigators would focus on the use of antioxidants to maintain redox balance in body. Moreover, there is a need to understand the physiological alterations caused by free radicals because it offers exciting possibilities for curing diseases.

REFERENCES

Abbas, A. M. Cardioprotective effect of resveratrol analogue isorhapontigenin versus omega-3 fatty acids in isoproterenol-induced myocardial infarction in rats. *J Physiol Biochem*. 2016; 72: 469–484.

Ahmad, N., Ahmad, R., Naqvi, A. A., Alam, M. A., Ashafaq, M., Samim, M. *et al.* Rutin-encapsulated chitosan nanoparticles targeted to the brain in the treatment of Cerebral Ischemia. *Int J BiolMacromol*. 2016; 91: 640–655.

Anwer, M. K., Al-Mansoor, M. A., Jamil, S., Al-Shdefat, R. I., Ansari, M. N., Shakeel, F. *et al.* Development and evaluation of PLGA polymer based nanoparticles of quercetin. *Int J BiolMacromol*. 2016; pii: s0141-8130(16)30702-4.

Aprioku, J. S. Pharmacology of free radicals and the impact of reactive oxygen species on the testis. *J Reprod Infertil*. 2013; 14: 158–172.

Ban, C., Park, S. J., Lim, S., Choi, S. J., and Choi, Y. J. Improving flavonoid bioaccessibility using an edible oil-based lipid nanoparticle for oral delivery. *J Agric Food Chem*. 2015; 63: 5266–5272.

Belhaj, N., Dupuis, F., Arab-Tehrany, E., Denis, F. M., Paris, C., Lartaud, I. *et al.* Formulation, characterization and pharmacokinetic studies of coenzyme Q10 PUFA's nanoemulsions. *Eur J Pharm Sci*. 2012; 47: 305–312.

Brownlow, B., Nagaraj, V. J., Nayel, A., Joshi, M., and Elbayoumi, T. Development and *in vitro* evaluation of vitamin E-enriched nanoemulsion vehicles loaded with genistein for chemoprevention against UVB-induced skin damage. *J Pharm Sci*. 2015; 104: 3510–3523.

Chaudhary, S., Garg, T., Murthy, R. S., Rath, G., and Goyal, A. K. Development, optimization and evaluation of long chain nanolipid carrier for hepatic delivery of silymarin through lymphatic transport pathway. *Int J Pharm*. 2015; 485: 108–121.

Chen, C. H., Chang, C. C., Shih, T. H., Aljuffali, I. A., Yeh, T. S., and Fang, J. Y. Self-nanoemulsifying drug delivery systems ameliorate the oral delivery of silymarin in rats with Roux-en-Y gastric bypass surgery. *Int J Nanomedicine*. 2015; 10: 2403–2416.

Chlapanidas, T., Perteghella, S., Leoni, F., Farago, S., Marazzi, M., Rossi, D. *et al.* TNF-α blocker effect of naringenin-loaded sericin microparticles that are potentially useful in the treatment of psoriasis. *Int J Mol Sci*. 2014; 15: 13624–13636.

Csiszar, A., Csiszar, A., Pinto, J. T., Gautam, T., Kleusch, C., Hoffmann, B. *et al.* Resveratrol encapsulated in novel fusogenic liposomes activates Nrf2 and attenuates oxidative stress in cerebromicrovascular endothelial cells from aged rats. *J Gerontol A Biol Sci Med Sci*. 2015; 70: 303–313.

Dajas, F., Abin-Carriquiry, J. A., Arredondo, F., Blasina, F., Echeverry, C., Martínez M. *et al.* Quercetin in brain diseases: Potential and limits. *Neurochem Int*. 2015; 89: 140–148.

Di Filippo, C., Cuzzocrea, S., Rossi, F., Marfella, R., and D'Amico, M. Oxidative stress as the leading cause of acute myocardial infarction in diabetics. *Cardiovasc Drug Rev*. 2006; 24: 77–87.

El-Nahas, A. E., Allam, A. N., El-Kamel, A. H. Mucoadhesive buccal tablets containing silymarin Eudragit-loaded nanoparticles: Formulation, characterisation and *ex vivo* permeation. *J Microencapsul*. 2017: 1–12.

El-Refaie, W. M., Elnaggar, Y. S., El-Massik, M. A., and Abdallah, O. Y. Novel curcumin-loaded gel-core hyaluosomes with promising burn-wound healing potential: Development, *in-vitro* appraisal and *in-vivo* studies. *Int J Pharm*. 2015; 486: 88–98.

Fujimura, A. T., Martinez, R. M., Pinho-Ribeiro, F. A., Lopes Dias da Silva, A. M., Baracat, M. M., Georgetti, S. R. *et al.* Resveratrol-loaded liquid-crystalline system inhibits UVB-induced skin inflammation and oxidative stress in mice. *J Nat Prod*. 2016; 79: 1329–1338.

Gao, X., Chen, L., Xie, J., Yin, Y., Chang, T., Duan, Y. *et al. In vitro* controlled release of vitamin C from Ca/Al layered double hydroxide drug delivery system. *Mater Sci Eng C Mater Biol Appl*. 2014; 39: 56–60.

Garrastazu, G., Stanisçuaki, G. S., Balducci, A. G., Colombo, P., Sonvico, F. Polymeric films loaded with vitamin E and Aloe vera for topical application in the treatment of burn wounds. *BioMed Res Int*. 2014; 2014: 1–9.

Gibellini, L., Pinti, M., Nasi, M., Montagna, J. P., De Biasi, S., Roat, E. *et al.* Quercetin and Cancer Chemoprevention. *Evid Based Complement Alternat Med*. 2011; 2011: 591356.

Guney, G., Kutlu, H. M., and Genc, L. Preparation and characterization of ascorbic acid loaded solid lipid nanoparticles and investigation of their apoptotic effects. *Colloids Surf B Biointerfaces*. 2014; 121: 270–280.

Hazzah, H. A., Farid, R. M., Nasra, M. M., Hazzah, W. A., El-Massik, M. A., and Abdallah, O. Y. Gelucire-based nanoparticles for curcumin targeting to oral mucosa: Preparation, characterization, and antimicrobial activity assessment. *J Pharm Sci*. 2015; 104: 3913–3924.

Hirayama, K., Oshima, H., Yamashita, A., Sakatani, K., Yoshino, A., and Katayama, Y. Neuroprotective effects of silymarin on ischemia-induced delayed neuronal cell death in rat hippocampus. *Brain Res.* 2016; 1646: 297–303.

Jenner, P. Oxidative stress in Parkinson's disease. *Ann Neurol.* 2003; 53: S26–S36.

Jia, L., Zhang, D., Li, Z., Duan, C., Wang, Y., Feng, F. *et al.* Nanostructured lipid carriers for parenteral delivery of silybin: Biodistribution and pharmacokinetic studies. *Colloids Surf B Biointerfaces.* 2010; 80: 213–218.

Jimbow, K., Chen, H., Park, J. S., and Thomas, P. D. Increased sensitivity of melanocytes to oxidative stress and abnormal expression of tyrosinase-related protein in vitiligo. *Br J Dermatol.* 2001; 144: 55–65.

Kalita, J., Kumar, V., Misra, U. K., Ranjan, A., Khan, H., and Konwar, R. A study of oxidative stress, cytokines and glutamate in Wilson disease and their asymptomatic siblings. *J Neuroimmunol.* 2014; 274: 141–148.

Kamel, R., and Mostafa, D. M. Rutin nanostructured lipid cosmeceutical preparation with sun protective potential. *J PhotochemPhotobiol B.* 2015; 153: 59–66.

Keck, C. M., Baisaeng, N., Durand, P., Prost, M., Meinke, M. C., and Müller, R. H. Oil-enriched, ultra-small nanostructured lipid carriers (usNLC): A novel delivery system based on flip-flop structure. *Int J Pharm.* 2014; 477: 227–235.

Kellici, T. F., Chatziathanasiadou, M. V., Diamantis, D., Chatzikonstantinou, A., Andreadelis, I., Christodoulou, E. *et al.* Mapping the interactions and bioactivity of quercetin–(2-hydroxypropyl β–)-clodextrin complex. *Int J Pharm.* 2016; 511(1): 303–311.

Khan, A. W., Kotta, S., Ansari, S. H., Sharma, R. K., and Ali, J. Self-nanoemulsifying drug delivery system (SNEDDS) of the poorly water-soluble grapefruit flavonoid Naringenin: Design, characterization, *in vitro* and *in vivo* evaluation. *Drug Deliv.* 2015; 22: 552–561.

Kumar, V., Jahan, F., Mahajan, R. V., and Saxena, R. K. Efficient regioselective acylation of quercetin using *Rhizopusoryzae* lipase and its potential as antioxidant. *Bioresour Technol.* 2016; 218: 1246–1248.

Kumar, V. P., and Khan, A. Formulation design and evaluation of rutin loaded self emulsifying drug delivery system (SEDDs) using edible oil. *Asia J Pharm Clin Res.* 2012; 5: 76–78.

Kutty, R. V., Chia, S. L., Setyawati, M. I., Muthu, M. S., Feng, S. S., and Leong, D. T. *In vivo* and *ex vivo* proofs of concept that cetuximab conjugated vitamin E TPGS micelles increases efficacy of delivered docetaxel against triple negative breast cancer. *Biomaterials.* 2015; 63: 58–69.

Lee, C. S., Choi, E. Y., Lee, S. C., Koh, H. J., Lee, J. H., and Chung, J. H. Resveratrol inhibits hypoxia-induced vascular endothelial growth factor expression and pathological neovascularization. *Yonsei Med J.* 2015; 56: 1678–1685.

Leonard, S. S., Cutler, D., Ding, M., Vallyathan, V., Castranova, V., and Shi, X. Antioxidant properties of fruit and vegetable juices: More to the story than ascorbic acid. *Ann Clin Lab Sci.* 2002; 32: 193–200.

Li, C., Zhang, Y., Su, T., Feng, L., Long, Y., and Chen, Z. Silica-coated flexible liposomes as a nanohybrid delivery system for enhanced oral bioavailability of curcumin. *Int J Nanomedicine.* 2012; 7: 5995–6002.

Liu, L., Mao, K., Wang, W., Pan, H., Wang, F., Yang, M., and Liu, H. Kolliphor® HS 15 micelles for the delivery of coenzyme Q10: Preparation, characterization, and stability. *AAPS PharmSciTech.* 2016a; 17: 757–766.

Liu, R., Sun, L., Fang, S., Wang, S., Chen, J., Xiao, X., and Liu, C. Thermosensitive *in situ* nanogel as ophthalmic delivery system of curcumin: Development, characterization, *in vitro* permeation and *in vivo* pharmacokinetic studies. *Pharm Dev Technol.* 2016b; 21: 576–582.

Liu-Smith, F., Poe, C., Farmer, P. J., and Meyskens, F. L., Jr. Amyloids, melanins and oxidative stress in melanomagenesis. *Exp Dermatol.* 2015; 24: 171–174.

Lobo, V., Patil, A., Phatak, A., and Chandra, N. Free radicals, antioxidants and functional foods: Impact on human health. *Pharmacogn Rev.* 2010; 4: 118–126.

Mady, F. M., Essa, H., El-Ammawi, T., Abdelkader, H., and Hussein, A. K. Formulation and clinical evaluation of silymarin pluronic-lecithin organogels for treatment of atopic dermatitis. *Drug Des Devel Ther.* 2016; 10: 1101–1110.

Mauludin, R., Muller, R. H., and Keck, C. M. Development of an oral rutin nanocrystal formulation. *Int J Pharm.* 2009; 370: 202–209.

McCusker, M. M., Durrani, K., Payette, M. J., and Suchecki, J. An eye on nutrition: The role of vitamins, essential fatty acids, and antioxidants in age-related macular degeneration, dry eye syndrome, and cataract. *Clin Dermatol.* 2016; 34: 276–285.

Meng, F., Asghar, S., Gao, S., Su, Z., Song, J., Huo, M. *et al.* A novel LDL-mimic nanocarrier for the targeted delivery of curcumin into the brain to treat Alzheimer's disease. *Colloids Surf B Biointerfaces.* 2015; 134: 88–97.

Mishra, S., and Palanivelu, K. The effect of curcumin (turmeric) on Alzheimer's disease: An overview. *Ann Indian Acad Neurol.* 2008; 11: 13–19.

Miyake, K., Arima, H., Hirayama, F., Yamamoto, M., Horikawa, T., Sumiyoshi, H. *et al.* Improvement of solubility and oral bioavailability of rutin by complexation with 2-hydroxypropyl-beta-cyclodextrin. *Pharm Dev Technol.* 2000; 5: 399–407.

Nagi, A., Iqbal, B., Kumar, S., Sharma, S., Ali, J., and Baboota, S. Quality by design based silymarin nanoemulsion for enhancement of oral bioavailability. *J Drug Deliv Sci Techno.* 2017; 40: 35–44.

Naz, Z., and Ahmad, F. J. Curcumin-loaded colloidal carrier system: Formulation optimization, mechanistic insight, *ex vivo* and *in vivo* evaluation. *Int J Nanomedicine.* 2015; 10: 4293–4307.

Nuttall, S. L., Heaton, S., Piper, M. K., Martin, U., and Gordon, C. Cardiovascular risk in systemic lupus erythematosus-evidence of increased oxidative stress and dyslipidaemia. *Rheumatology.* 2003; 42: 758–762.

Oh, H. N., Kim, C. E., Lee, J. H., and Yang, J. W. Effects of quercetin in a mouse model of experimental dry eye. *Cornea.* 2015; 34: 1130–1136.

Ozkul, A., Akyol, A., Yenisey, C., Arpaci, E., Kiylioglu, N., and Tataroglu, C. Oxidative stress in acute ischemic stroke. *J Clin Neurosci.* 2007; 14: 1062–1066.

Pangeni, R., Sharma, S., Mustafa, G., Ali, J., and Baboota, S. Vitamin E loaded resveratrol nanoemulsion for brain targeting for the treatment of Parkinson's disease by reducing oxidative stress. *Nanotechnology.* 2014; 25: 485102.

Paradiso, P., Serro, A. P., Saramago, B., Colaço, R., and Chauhan, A. Controlled release of antibiotics from vitamin E-loaded silicone-hydrogel contact lenses. *J Pharm Sci.* 2016; 105: 1164–1172.

Parker, W., Hornik, C. D., Bilbo, S., Holzknecht, Z. E., Gentry, L., Rao, R. *et al.* The role of oxidative stress, inflammation and acetaminophen exposure from birth to early childhood in the induction of autism. *J Int Med Res.* 2017; 45: 407–438.

Pepe, D., Phelps, J., Lewis, K., Dujack, J., Scarlett, K., Jahan, S. *et al.* Decylglucoside-based microemulsions for cutaneous localization of lycopene and ascorbic acid. *Int J Pharm.* 2012; 434: 420–428.

Perk, A. A., Shatynska-Mytsyk, I., Gerçek, Y. C., Boztaş, K., Yazgan, M., Fayyaz, S. *et al.* Rutin mediated targeting of signaling machinery in cancer cells. *Cancer Cell Int.* 2014; 14: 124.

Pierucci, A. P., Andrade, L. R., Baptista, E. B., Volpato, N. M., and Rocha-Leão, M. H. New microencapsulation system for ascorbic acid using pea protein concentrate as coat protector. *J Microencapsul.* 2006; 23: 654–662.

Pires, A. S., Marques, C. R., Encarnacão, J. C., Abrantes, A. M., Mamede, A. C., Laranjo, M. *et al.* Ascorbic acid and colon cancer: An oxidative stimulus to cell death depending on cell profile. *Eur J Cell Biol.* 2016; 95: 208–218.

Prabhakar, S., Starnes, J., Shi, S., Lonis, B., and Tran, R. Diabetic nephropathy is associated with oxidative stress and decreased renal nitric oxide production. *J Am Soc Nephrol.* 2007; 18: 2945–2952.

Prado, A. G., Santos, A. L., Nunes, A. R., Tavares, G. W., and de Almeida, C. M. Designed formulation based on α-tocopherol anchored on chitosan microspheres for pH-controlled gastrointestinal controlled release. *Colloids Surf B Biointerfaces.* 2012; 96: 8–13.

Qin, B., Liu, L., Pan, Y., Zhu, Y., Wu, X., Song, S. *et al.* PEGylated Solanesol for oral delivery of coenzyme Q10. *J Agric Food Chem.* 2017; 65: 3360–3367.

Qu, J., Zhou, Q., Du, Y., Zhang, W., Bai, M., Zhang, Z. *et al.* Rutin protects against cognitive deficits and brain damage in rats with chronic cerebral hypoperfusion. *Br J Pharmacol.* 2014; 171: 3702–3715.

Sareen, R., Nath, K., Jain, N., Dhar, K. L. Curcumin loaded microsponges for colon targeting in inflammatory bowel disease: Fabrication, optimization, and *in vitro* and pharmacodynamic evaluation. *Biomed Res Int.* 2014; 2014: 340701.

Seljak, K. B., Berginc, K., Trontelj, J., Zvonar, A., Kristl, A., and Gašperlin, M. A self-microemulsifying drug delivery system to overcome intestinal resveratrol toxicity and presystemic metabolism. *J Pharm Sci.* 2014; 103: 3491–3500.

Serrano, G., Almudéver, P., Serrano, J. M., Milara, J., Torrens, A., Expósito, I. *et al.* Phosphatidylcholine liposomes as carriers to improve topical ascorbic acid treatment of skin disorders. *Clin Cosmet Investig Dermatol.* 2015; 8: 591–599.

Shao, Y., Yang, L., and Han, H. K. TPGS-chitosome as an effective oral delivery system for improving the bioavailability of Coenzyme Q10. *Eur J Pharm Biopharm.* 2015; 89: 339–346.

Sies, H., Berndt, C., and Jones, D. P. Oxidative Stress. *Annu Rev Biochem.* 2017; 86: 715–748.

Singh, G., and Pai, R. S. *In vitro* and *in vivo* performance of supersaturable self-nanoemulsifying system of trans-resveratrol. *Artif Cells Nanomed Biotechnol.* 2014; 21: 1–7.

Sintov, A. C. Transdermal delivery of curcumin *via* microemulsion. *Int J Pharm.* 2015; 481: 97–103.

Steinberg, D. Low density lipoprotein oxidation and its pathobiological significance. *J Biol Chem.* 1997; 272: 20963–20966.

Sulfikkarali, N., Krishnakumar, N., Manoharan, S., and Nirmal, R. M. Chemopreventive efficacy of naringenin-loaded nanoparticles in 7,12-dimethylbenz(a)anthracene induced experimental oral carcinogenesis. *Pathol Oncol Res.* 2013; 19: 287–296.

Sultana, Z., Maiti, K., Aitken, J., Morris, J., Dedman, L., and Smith R. Oxidative stress, placental ageing-related pathologies and adverse pregnancy outcomes. *Am J Reprod Immunol.* 2017; 77.

Szabo, E., Riffe, M. E., Steinberg, S. M., Birrer, M. J., and Linnoila, R. I. Altered cJUN expression: An early event in human lung carcinogenesis. *Cancer Res.* 1996; 56: 305–315.

Trendafilova, I., Szegedi, A., Mihaly, J., Momekov, G., Lihareva, N., and Popova, M. Preparation of efficient quercetin delivery system on Zn-modified mesoporous SBA-15 silica carrier. *Mater Sci Eng C Mater Biol Appl.* 2017; 73: 285–292.

Tsai, M. J., Huang, Y. B., Fang, J. W., Fu, Y. S., and Wu, P. C. Preparation and characterization of naringenin-loaded elastic liposomes for topical application. *PLoS One.* 2015; 10: e0131026.

Usui, S., Ovesson, B. C., Lee, S. Y., Jo, Y. J., Yoshida, T., Miki, A. *et al.* NADPH oxidase plays a central role in cone cell death in retinitis pigmentosa. *J Neurochem.* 2009; 110: 1028–1037.

Vergaro, V., Lvov, Y. M., and Leporatti, S. Halloysite clay nanotubes for resveratrol delivery to cancer cells. *Macromol Biosci.* 2012; 12: 1265–1271.

Wang, M., Xie, T., Chang, Z., Wang, L., Xie, X., Kou, Y. *et al.* A new type of liquid silymarin proliposome containing bile salts: Its preparation and improved hepatoprotective effects. *PLoS One.* 2015; 10: e0143625.

Wang, P. C., Huang, Y. L., Hou, S. S., Chou, C. H., and Tsai, J. C. Lauroyl/palmitoyl glycol chitosan gels enhance skin delivery of magnesium ascorbyl phosphate. *J Cosmet Sci.* 2013; 64: 273–286.

Wang, Y., Wang, S., Firempong, C. K., Zhang, H., Wang, M., Zhang, Y. *et al.* Enhanced solubility and bioavailability of naringenin *via* liposomal nanoformulation: Preparation and *in vitro* and *in vivo* evaluations. *AAPS PharmSciTech.* 2017; 18: 586–594.

Xie, Y., Lu, Y., Qi, J., Li, X., Zhang, X., Han, J. *et al.* Synchronized and controlled release of multiple components in silymarin achieved by the osmotic release strategy. *Int J Pharm.* 2013; 441: 111–120.

Yadav, N. K., Nanda, S., Sharma, G., and Katare, O. P. Systematically optimized coenzyme q10-loaded novel proniosomal formulation for treatment of photo-induced aging in mice: Characterization, biocompatibility studies, biochemical estimations and anti-aging evaluation. *J Drug Target.* 2016; 24: 257–271.

Yang, K. Y., Hwang du, H., Yousaf, A. M., Kim, D. W., Shin, Y. J., Bae, O. N. *et al.* Silymarin-loaded solid nanoparticles provide excellent hepatic protection: Physicochemical characterization and *in vivo* evaluation. *Int J Nanomedicine.* 2013; 8: 3333–3343.

Yen, F. L., Wu, T. H., Lin, L. T., Cham, T. M., and Lin, C. C. Naringenin-loaded nanoparticles improve the physicochemical properties and the hepatoprotective effects of naringenin in orally-administered rats with CCl(4)-induced acute liver failure. *Pharm Res.* 2009; 26: 893–902.

Young, A. J., Johnson, S., Steffens, D. C., and Doraiswamy, P. M. Coenzyme Q10: A review of its promise as a neuroprotectant. *CNS Spectr.* 2007; 12: 62–68.

Zhang, P., Liu, X., Hu, W., Bai, Y., and Zhang, L. Preparation and evaluation of naringenin-loaded sulfobutylether-β-cyclodextrin/chitosan nanoparticles for ocular drug delivery. *Carbohydr Polym.* 2016; 149: 224–230.

Zhao, G., Hu, C., and Xue, Y. *In vitro* evaluation of chitosan-coated liposome containing both coenzyme Q10 and alpha-lipoic acid: Cytotoxicity, antioxidant activity, and antimicrobial activity. *J Cosmet Dermatol.* 2017: 1–5.

Zheng, N., Liu, L., Liu, W., Zhang, P., Huang, H., Zang, L. *et al.* ERβ up-regulation was involved in silibinin-induced growth inhibition of human breast cancer MCF-7 cells. *Arch Biochem Biophys.* 2016; 591: 141–149.

Zoric, L. Parameters of oxidative stress in the lens. Aqueous humor and blood in patients with diabetes and senile cataracts. *Srp Arh Celok Lek.* 2003; 131: 137–142.

6 Nanonutraceuticals in Central Nervous System Disorders

Amita Sarwal, Nisha Rawat, Gurpreet Singh, V. R. Sinha, Sumit Sharma, and Dinesh Kumar*

CONTENTS

6.1 INTRODUCTION TO NUTRACEUTICALS

Recently, many bioactive constituents such as extracts of food or phytochemicals have been marketed as pharmaceutical products (e.g., tablets, capsules, powders, solutions, gels, liquors, etc.), which demonstrate physiological benefit directly or indirectly. As such these products cannot be accurately termed as "food nutrients" or "pharmaceuticals," so the cross term *nutraceuticals* has been devised to delegate them (Figure 6.1). In Canada, they have been given a new category as "natural health products" that promote health.

* Corresponding author.

FIGURE 6.1 The precept of nutraceutical.

Until now, there has been no legitimate definition of the term *nutraceutical*, but today it is commonly used to elucidate a wide range of products that are marketed, labeled as food constituents, and are said to have benefits such as healing and mitigation of disease and enhancement of strength and welfare of an individual (Kalra, 2003).

Hippocrates, the well-recognized father of modern medicine, emphasizing the principle "Let food be the medicine, and medicine be the food," conceptualized the relationship between the use of nutritional food and its therapeutic advantages. According to this, nutraceuticals are foodstuff and ingredients that provide therapeutic benefits and blur the line among foodstuff and drugs, although not comfortably falling into legal classifications of food or drug per se (Chauhan *et al.*, 2013). As originally designated, Stephen Defelice, founder and chair of Foundation for Innovation in Medicine, defined *nutraceutical* as any commodity, including a food or a part of foodstuff, that provides therapeutic or health benefits, with the ability to prevent and treat any disease (Pandey *et al.*, 2010). These tend to cover a broad list of substances, from conventional isolated constituents, specific foods, probiotic microbes and enzymes, to genetically engineered and fortified ones, as shown in Figure 6.2.

Nutraceuticals, sometimes also termed *medical foods*, *nutritional supplements*, or *dietary supplements*, have encountered considerable volume of interest due to their supposed safety, and ostensible nutritional and therapeutic effects. Health can be improved by supplementation or by consuming foods either traditional or fortified.

Nutraceuticals have received enough popularity as being advantageous in cancer, osteoporosis, diabetes, coronary heart disease, obesity, besides other chronic neurodegenerative diseases such as Parkinson's disease (PD) and Alzheimer's disease (AD). Research findings elaborate a wide number of biological processes playing a vital role in defense and resistance to pathologies of changes

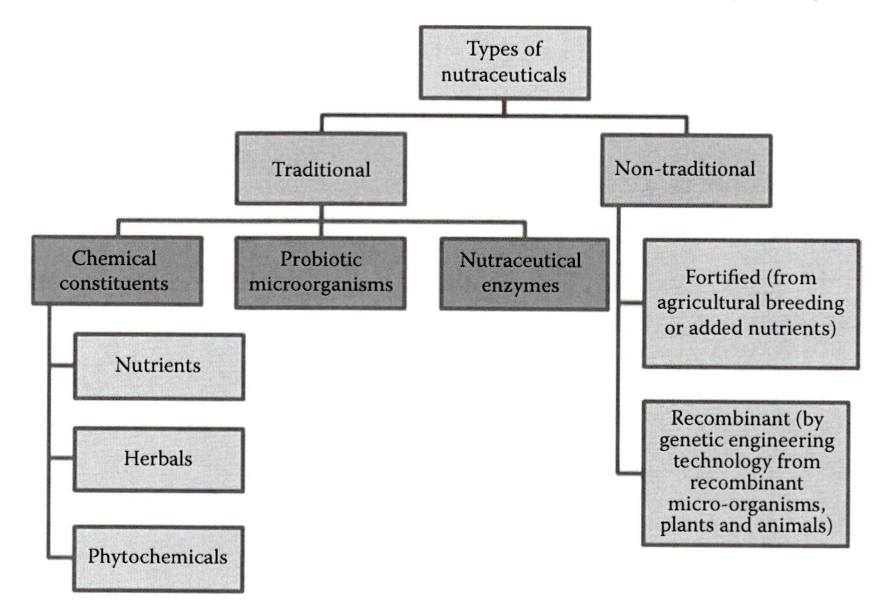

FIGURE 6.2 Classification of nutraceuticals.

associated with age or chronic disorders by natural compounds that include preservation of mitochondrial integrity, signal transduction pathways, trigger of antioxidant defenses, fluctuations in cell proliferation and differentiation, and expression of genes associated with cell survival.

Currently, approximately 480 nutraceuticals and phytochemicals are available, possessing documented health advantage(s), many of them originated from plants. This is because several plants synthesize secondary plant metabolites, like alkaloids, which have shown the capability to defend people from infection and different ailments, and therefore can also be used in preventing human infections and different ailments. However, nutraceuticals can also be acquired from diverse sources, including marine supply, animal sources, and so forth (Sarin *et al.*, 2012).

Alarming rise in frequency of occurrence of neurotic disorders, accompanied by lack of powerful modern remedies, demand increased global sense of distinctive lifestyle, which could significantly affect and initiate prevention of brain disorders affecting cognitive performance. Phytochemicals in nutraceuticals are bioactive compounds that sustain or promote fitness. The former are widely defined as phytoestrogens, terpenoids, limonoids, carotenoids, phytosterols, glucosinolates, polyphenols, flavonoids, isoflavonoids, and anthocyanidins (Prakash and Gupta, 2009). They cause extraordinary influence on healthcare systems and offer health benefits like prophylaxis or therapeutic management of various diseases and/or physiological disorders due to of toxic hazards of drug (Prakash *et al.*, 2012). Consumers are actually turning toward food dietary supplements to improve health fitness, wherein prescribed drugs tend to fail. Consequently, in today's situation, the demand of nutraceuticals is steeply growing because of its safer use.

Epidemiological and animal studies propose that the everyday intake of fruits, greens, and entire grains minimizes threats of long-term diseases caused by the oxidative damage (Scalbert *et al.*, 2005). Carotenoids, ascorbates, tocopherols, lipoic acids, and polyphenols are sturdy herbal antioxidants documented with free radical scavenging properties (Kondratyuk and Pezzuto, 2004).

6.1.1 Current Scenario of Nutraceuticals

Generations of nutraceuticals are found in pharmaceutical and biotech organizations. The idea of nutraceutical began in Germany, and France, which moved into the United States in the mid-1900s. The current nutraceutical market started off in the 1980s through Japan. The nutraceutical business in the United States is currently over $100 billion, the figure being somewhat higher in Europe. Japan involves roughly one-fourth of their $6 billion yearly food sale on nutraceuticals, with nearly 47% of the Japanese population consuming nutraceuticals regularly. In India, nutraceuticals are being significantly promoted by a few leading brands through companies like GlaxoSmithKline, Dabur, Cadila Healthcare, Zandu, Himalaya, Amway, Sami, Elder, and Sun Pharma (Gupta *et al.*, 2010). The U.S. Department of Agriculture certified organic Wellements Probiotic Gripe Water, which is used to treat stomach discomfort in infants. Some of the marketed formulations are listed in Table 6.1 (Dureja *et al.*, 2003).

TABLE 6.1
Different Formulation of Nutraceuticals for Brain Disorders

Formulation	Active Ingredient(s)	Route of Administration	Reference
Nanoemulsion	Curcumin, resveratrol, lutein	Oral	Swati *et al.*, 2016
Solid Lipid Nanoparticles	Thujone, pinene, camphor	Oral	Campos *et al.*, 2015
Liposomes	Follic acid	Intranasal	Ravouru *et al.*, 2013
Nanoemulsion	Huperzine A	Transdermal	Patel *et al.*, 2013
Liposomes	Rivastigmine	Intranasal	Yang *et al.*, 2013
Solid Lipid Nanoparticles	Curcumin	Intranasal	Sood *et al.*, 2013
Nanoemulsion	*G. biloba* extract	Oral	Jin *et al.*, 2013
Solid Lipid Nanoparticles	Resveratrol	Oral and intraperitoneal	Frozza *et al.*, 2010
Polymeric nanoparticles	Tacrine	Intravenous	Wilson *et al.*, 2008

6.2 NEUROLOGICAL DISORDERS

The brain controls the body. It is a part of the sensory system, which additionally incorporates the spinal cord and an extensive system of nerves and neurons, which controls five senses and all the muscles in the body. When the brain is injured, it can harm a wide range of activities, including memory, sensation, and even identity. It includes conditions that have an impact on the cerebrum, including states generated by disease, gene acquired attributes, or traumatic destruction. Various disorders associated with the brain are multiple sclerosis, AD, PD, stroke, depression, migraine, mania, anxiety, seizers, epilepsy, trauma, and so on. A report distributed by the World Health Organization Neurological Disorders, Public Health Challenges states that around one billion individuals experience the ill-effects of neurological issue, including epilepsy, stroke, cerebral pain, AD ailment and different dementias, PD infection, various sclerosis, cerebrum wounds, and neuro-infections.

6.2.1 Events and Obstacles of CNS Delivery

Psychotic disorders certainly occupy an integral and crucial region in the world of pharmaceutical sciences, seeking requisite treatment due to unmet medical demands. Drug development resulting in inferior outcomes, combined with lengthy development stages, has lately acquired attention in the educational and commercial sectors. Numerous bio-therapeutics, including peptides, monoclonal antibodies, nucleic acids, and proteins, are being investigated for diverse CNS therapies.

6.2.2 Dominant Hurdles to CNS Delivery

- Blood–brain barrier (BBB): High molecular weight compounds are unable to permeate through the tight junctions of endothelial cells at the BBB, intercepting 98% of effective molecules from passing through.
- Blood-Cerebrospinal Spinal Fluid (CSF) barrier: Though the blood-CSF barrier does hold tight junctions of epithelial cells in the choroid plexus, it is likely to cause lesser obstruction than the BBB due to its lesser surface area.
- Systemic distribution and clearance: A significant part of thinning and metabolizing of drug molecules in the systemic circulation takes place, also for lipophilic compounds, which are developed to enhance brain permeation. (Mansoor *et al.*, 2013)

6.3 NANOTECHNOLOGY IN NUTRACEUTICALS

Nanoscience and nanotechnology have created massive influence in the fields of medicine and science (Figure 6.3). It has also shown considerable application in the areas of nutraceuticals or phytochemicals (Mansoor *et al.*, 2013). Various phytochemicals/nutraceuticals with potential health and physiological effects such as curcumin, resveratrol, fatty acids, etc. have gained interest in the last few years. However, a majority of these phytochemicals/nutraceuticals have exhibited low bioavailability, attributing to its poor solubility in the gastrointestinal fluids and causing insufficient absorption from the gastrointestinal tract or high first-pass metabolism, resulting in unsatisfactory biological activity. Hence, to overcome such issues, nanotechnology came to the rescue to enhance the biological activity and delivery of nutraceuticals. Poor water solubility can also be enhanced using nanotechnology-enabled techniques for targeted delivery, to penetrate through tight epithelial and endothelial barriers, release of high molecular weight molecules, combined delivery of two or more drugs, and surveillance of sites of drug delivery by co-loading herbal drugs with imaging modalities. There are various nanoformulations that can be developed to enhance the bioavailability and site-specific targeting of the nutraceuticals and provide sustained delivery to the effective site, such as nanoemulsions, micelles, nanoparticles, nanocapsules, and nanocrystals (Sahni, 2012).

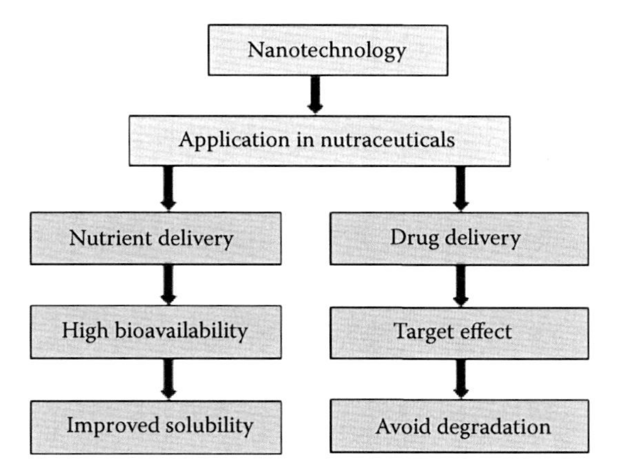

FIGURE 6.3 Application of nanotechnology in nutraceuticals.

6.4 VARIOUS FORMS OF NUTRACEUTICALS USED IN BRAIN-ASSOCIATED DISEASES

6.4.1 Flavonoid Polyphenols

Fruits and vegetables containing polyphenols, which include blueberries, cranberries, strawberries, and spinach, are useful for brain health and its functioning (Lau *et al.*, 2005). They are comprised of bioactives with high antioxidant properties that can be estimated by the changed oxygen radical absorbance ability (ORAC) assay (Wu *et al.*, 2004).

These are useful in brain-associated depreciation in neuronal signal transduction, in addition to the cognitive and motor deficits (Joseph *et al.*, 2005). Aging leads to declining motor and cognitive functions, which are escalated in age-associated neurodegenerative diseases, consisting of amyotrophic lateral sclerosis (ALS), AD, and PD (Esposito *et al.*, 2002). The neurogenesis function of the hippocampus region to produce neurons dwindles throughout aging, with the consequent decline in cognitive functions (Drapeau *et al.*, 2003). Polyphenolic supplements enhance neuronal signaling by strengthening the performance of motor nerves, for example, tea-derived epigallocatechin 3-gallate (EGCG) and quercetin extracts from apples and other fruits (Henning *et al.*, 2012).

6.4.1.1 Epigallocatechin 3-Gallate (EGCG)

EGCG is a principle antioxidant compound of polyphenolic flavonoid obtained from green tea. EGCG inhibits Aβ (amyloid beta) fibril formation and oligomerization and subsequently minimizes β-amyloid instigated toxicity in hippocampal neurons. EGCG also protects SH-SY5Y neuroblastoma (human derived cell lines used in scientific research) in humans from the toxicity caused by amyloid precursor protein (APP), 3-hydroxykynurenine or 6-hydroxydopamine (6-OHDA) (Schroeder *et al.*, 2009). Researchers have also demonstrated the preventive action of dopamine neurons from 1-methyl-4-phenylpyridinium (MPP+) toxicity. It has also been discovered that co-administration of EGCG prevented MPTP (1methyl-4phenyl-1236-tetrahydropyridine)-associated loss of dopamine neurons from the substantia nigra pars compacta.

Nanolipidic particles of EGCG were prepared to report higher oral bioavailability than EGCG in its free form and they were found to be advantageous in the treatment of AD. These nanolipidic particles enhanced bioavailability as well as α-secretase inducing ability of EGCG, benefitting the treatment of AD. EGCG was altered to avoid the need of both, co-administered with an additional drug and encapsulation into a micelle. Smaller particle sizes with stable carrier properties were successfully achieved due to the avoidance of micelle encapsulation (Smith *et al.*, 2010).

6.4.1.2 Quercetin

Quercetin is a flavonoid compound, obtained commonly from capers and apples having anti-oxidative belongings (Kelsey *et al.*, 2010), and assumed to have defensive outcomes against ischemia-induced brain injury and may be used to enhance cognitive impairment (Kumar *et al.*, 2008). Furthermore, earlier research has shown that quercetin prevents tissue harm through numerous mechanisms, for instance, with the aid of inhibiting neuronal apoptosis inside the brain. Similarly, research work is being undertaken for exploring quercetin as a goal for anti-epileptic therapy (Nieoczym *et al.*, 2014). It is quite likely that the circulating flavonoid metabolites may also indirectly affect brain function and cognitive performance by crossing the BBB or by modulating nitric oxide-dependent cerebrovascular function at the levels of cerebral endothelium, while some flavanoid metabolites can also act centrally by modulating the neuronal receptors, signaling kinases and neurotrophins, thus bringing about alterations in the synaptic function, as depicted in Figure 6.4.

Despite observance of reformed learning and memory ability upon oral administration of quercetin in mouse, its high metabolism, low intestinal absorption, and rapid elimination leads to its reduced efficacy. To overcome these problems, liposomes were chosen as the carrier to encapsulate quercetin for nasal delivery, with results demonstrating improved antianxiety activity and cognitive effect. The ability of liposomes to act as semi-lipophilic particles facilitated CSF absorption of the quercetin liposomes. Furthermore, the same group depicted retardation in degeneration of the hippocampal neurons by depletion of oxidative stress (Tong-un *et al.*, 2010). Encapsulation of quercetin to form nanoparticles containing sodium alginate and chitosan showed improved *in vitro* neuroprotective effect in oxidative stress model (Aluani *et al.*, 2017).

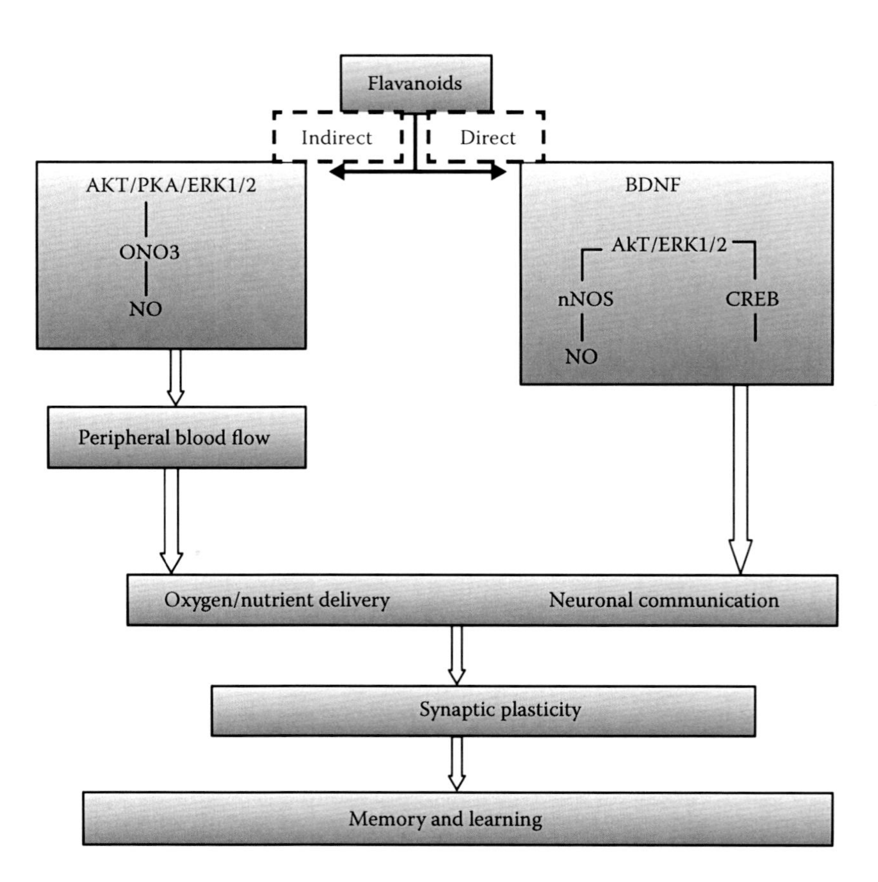

FIGURE 6.4 Proposed mechanisms of dietary flavonoids on memory and learning.

6.4.2 Non-flavonoid Polyphenols

6.4.2.1 Resveratrol

Resveratrol is a polyphenolic compound that exhibits antioxidant property and is found in abundance in grapes. It is thought to have cardiovascular benefits, but it also possesses neuroprotective activity (Bertelli and Das, 2009; Sadruddin and Arora, 2009). Resveratrol protects organotypic hippocampal slices from oxygen-glucose deprivation (Zamin *et al.*, 2006), embryonic rat mesencephalic cultures from tert-butyl hydroperoxide (Karlsson *et al.*, 2001), and cerebellar granule neurons from MPP (+)-brought toxicity (Alvira *et al.*, 2007).

Orally administered trans-resveratrol is considerably metabolized in the enterocyte prior to entry in the blood or target organs. It is photosensitive, effortlessly oxidized, and displays unfavourable pharmacokinetics. Despite suitable absorption from the gastrointestinal lumen, it shows low bioavailability due to its quick metabolism and clearance. Therefore, it is important to counterbalance trans-resveratrol to obtain satisfactory therapeutic activities and bioavailability in the brain. Trans-resveratrol was encapsulated into lipid-core nanocapsules and assessed for various physicochemical parameters to exhibit increased concentration in brain tissue, compared to its free form (Frozza *et al.*, 2010).

6.4.2.2 Curcumin

In accordance with the recently published reports, curcumin is less regularly used than resveratrol. Curcumin is found to be highly effective against Japanese encephalitis (JE), an arboviral disease. Neuro2a cell lines were taken and infected with JE virus, the contaminated cells were investigated with different portions of curcumin where the cell viability, reactive oxygen species (ROS) production inside the cells was evaluated, and any alteration in cell membrane integrity was calculated, which may be responsible for its effectiveness (Dutta *et al.*, 2009). It was also found to be noteworthy in cerebral ischemic (Yang *et al.*, 2009). Curcumin is neuroprotective toward the MPTP that could be a prodrug of neurotoxin MPP+, and due to the reason for therapeutic management of everlasting signs of Parkinson's sickness. Curcumin guards the nigrostriatal tract of mice from MPTP-precipitated neurodegeneration. Curcumin also guards the striatum and midbrain of MPTP dealt with mice from catalase and superoxide dismutase (SOD) and additionally saves glutathione depletion and lipid peroxidation resulting from this toxin (Rajeswari, 2006). The therapeutic implementation of curcumin is limited owing to its hydrophobic nature and insolubility in aqueous media. To circumvent such limitations, curcumin nanoparticles were prepared that were water-soluble. Due to the increase in the water solubility, there will be increase in the biodistribution and bioavailability of curcumin as well, along with retardation of its metabolism and systemic elimination (Mathew *et al.*, 2012). Several drug delivery systems have been tested for improved targeting of curcumin in solid lipid nanoparticles (SLNs), liposomes, polymeric nanoparticles with poly(lactic-co-glycolic acid) (PLGA), nanogels, micelles, and complexes with dendrimer/dimer. SLNs of curcumin showed great recovery in membrane lipids, as well as acetylcholinesterase (AChE) activity in aluminium chloride-treated mice, an effect that was comparable to rivastigmine, the most prescribed drug for AD (Bagli *et al.*, 2016). Microemulsion of curcumin, along with DHA, is used for brain targeted delivery, which is suitable for IV as well as intranasal administration, and had shown better efficacy and stability. DHA itself is reported to have antitumor activity, and when given in combination with curcumin, enhances the activity (Shinde and Devarajan, 2017).

6.4.3 Phenolic Diterpenes

Rosmarinic and Carnosic acids are observed in rosemary and are found in high concentrations in this herb.

6.4.3.1 Rosmarinic Acid

Rosmarinic acid has a marvelous scavenging asset and scavenges reactive nitrogen species, reactive oxygen species, and other peroxynitrite protecting activity (Choi *et al.*, 2002). The oxidative pressure on SH-SY5Y human neuroblastoma cells caused by hydrogen peroxide can also be prevented (Lee *et al.*, 2008). In the research studies of AD, within animal models, it was observed that rosmarinic acid may treat Aβ neurotoxicity and drastically retard the onset of ailment due to its scavenging of peroxynitrite (Alcamo *et al.*, 2007). Studies have also shown the preventive effect of rosmarinic acid in the animal model of ALS (Shampoo *et al.*, 2010).

SLNs of rosmarinic acid were prepared for intranasal administration, which leads to optimal brain drug levels revealing the effectiveness of nasal route over the IV route. Nasal delivery also offers advantages such as intercepting undesired distribution and metabolism in the body and strengthening the treatment of Huntington's disease (Bhatt *et al.*, 2015).

6.4.3.2 Carnosic Acid

Within the *in vitro* studies, it was observed that carnosic acid activates the nuclear factor 2 (Nrf2) transcription pathways that help in shielding the neurons from oxidative strain. However, in *in vivo* research it was determined to go through the BBB and normalize the reduced glutathione level inside the brain, thereby reducing the harm induced due to center cerebral artery ischemia or reperfusion (Satoh *et al.*, 2008).

The carnosic acid-chitosan nanoparticles were formulated, which were proven to be favorable for upregulation of neurotrophin levels in the brain tissue through the intranasal route. They exhibited a rise in the intranasal residence time with defense against enzymatic degradation, showcasing probabilities of minimized frequency of administration (Vaka *et al.*, 2013).

6.4.4 Polyunsaturated Fatty Acids (PUFAs)

The n-3 PUFAs and n-6 PUFAs are known as vital fatty acids, which might be generally obtained from weight loss, as they are not capable of being synthesized in the human body. These are also referred to as omega-3 and omega -6 fatty acids, respectively (Figure 6.5). Requisite uptake of omega-3 and omega-6 polyunsaturated fatty acids, n-3/n-6 PUFAs provide neurological benefits (Barros *et al.*, 2014).

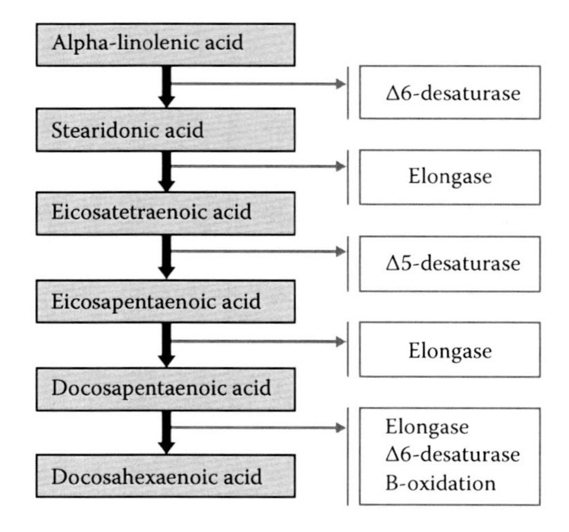

FIGURE 6.5 Biosynthesis of (n-3) PUFA.

6.4.4.1 n-3 PUFA

It is also known as α-Linolenic acid (ALA), transformed into eicosapentaenoic acid (EPA) and eventually into docosahexanoic acid (DHA). Even as n-6 PUFA is called linoleic acid, it endogenously gets converted into arachidonic acid (AA), which is a precursor of pro-inflammatory and eicosanoids compounds (Stillwell and Wassall, 2003). n-3 PUFAs antioxidant property is useful for cognitive improvement, synaptogenesis, and synaptic transmission.

6.4.4.2 n-6 PUFA

This acts as second messengers for pro-apoptotic and inflammatory occasions. As a result, if there is an imbalance in n-3/n-6 PUFA ratio, it may lead to neuronal harm, as occurred in neurodegenerative disease (Yehuda *et al.*, 2005). The powerful n-3/n-6 PUFA ratio is assumed to decide function of membrane-bound proteins and membrane fluidity (Uauy and Dangour, 2006). With increase in the age or in older age, and particularly with patients having neurodegenerative disease, brain DHA levels decrease, which shows a decline in the n-3/n-6 PUFAs ratio in brain tissues, followed by loss in memory and other cognitive functions. As a person gets older, there is an exchange in morphology and body structure of the brain, resulting in higher levels of ROS/RNS and decreased antioxidant ability. Imbalanced ratio of n-3/n-6 PUFAs has been confirmed in augmentation of brain-related disorders (Hardy and Higgins, 1992). DHA and EPA are especially located in fish oil (FO) along with trout, salmon, and tuna. Vegetable oils along with soya and linseed oils are rich sources of ALA (Whelan and Rust, 2006).

Various studies have proved that administration of DHA helps in maintaining DHA concentration in body and thus in maintaining brain functioning (Calon and Cole, 2007). In an observation, Barberger-Gateau *et al.* (2007) determined a cohort (n = 8085) without any dementia. Following a treatment period of four years, it was found that weekly fish consumption results in decreased risk of AD. A double-blinded, placebo-controlled study confirmed that patients suffering from PD, when administered fish oil, without or with antidepressants, exhibited decreased depressive signs and symptoms, thus indicating that n-3 PUFAs had assumed to have antidepressant impact. Hence, it can be used as a supportive treatment along with other medicament for PD (Da Silva *et al.*, 2008; De Lau *et al.*, 2005).

Nanoparticles of low density lipoproteins (LDL), together with pulsed focused ultrasound (FUS), **lead to** localized delivery of DHA in the brain through systemic administration. The concentration of DHA in the brain, derived from dietary supplements, is comparatively more time-consuming. Hence, in states expressing low levels of brain DHA content, it would take a lot of time to restore requisite DHA levels sourcing from dietary intake. In such cases, introduction of FUS/LDL-DHA nanoparticles help two-fold DHA concentrations in brain. Hence, according to investigations on the role of omega-3 fatty acids in the CNS, LDL-mediated delivery of DHA to the brain enhance cognition, decrease neuroinflammation, and safeguard from seizures and strokes (Mulik *et al.*, 2016).

6.4.5 St. John's Wort

St. John's Wort (SJW), *Hypericum perforatum* (family Hypericaceae), is used in the treatment of mild to moderate forms of depression. It is also known to have antitumoral, antiviral, anti-angiogenic, antibacterial (gram positive), and anti-inflammatory properties. SJW flowers and leaves contain hyperforin and hypericum as the major lipophilic constituent, but it is unsteady and liable to oxidative degradation that causes modifications in the constituents of extracts. Hyperforin showed antidepressant activity in various animal models, and the amount of hypoforin determines the efficacy of SJW (Vazzana *et al.*, 2014).

In vitro studies revealed that hyperforin inhibits the intake of various neurotransmitters by elevating free cytosolic concentrations of Na+, whereas *in vivo* it elevates the extracellular levels of glutamate, noradrenaline dopamine, and serotonin except GABA in the rat brain. Hatanaka *et al.* (2011)

prepared several formulations of SJW comprising of cyclodextrin inclusion (SJW-CD), solid dispersion (SJW-SD), dry-emulsion (SJW-DE), and nanoemulsion (SJW-NE) along with characterization studies. SJW-NE approach may prove to be constructive in refining the oral bioavailability and anti-nociceptive effect of SJW extract, lipid nanoparticles upgrade the chemical stability of compounds and reveal a vast applicability as drug delivery system, thus lipidic nanoparticles were found to helpful in the delivery of SJW extract. Prakash *et al.* (2010) have studied the protective role of *H. perforatum* gold nanoparticles (HPGNPs) for restraint stress-induced behavior in mice and given an outcome of positive impact on antioxidant enzyme activities such as superoxide dismutase, catalase, glutathione peroxidase, and reduced glutathione in the stress-treated animals. Both the studies unraveled an impressive behavior and neuroprotective effect.

6.4.6 Ginsenosides

Ginseng is the dried root of Panax ginseng. It is a popular medicinal herb with potentials of strengthening energy, vitality, and immune system, and it removes unwanted free radicals. Ginsenoside is the main bioactive constituent of the herb. It has scavenging property and thus scavenging of hydrogen peroxide free radicals leads to reduction of oxidative stress in the liver, brain, and other organs, which extends cell survival of a PD model (Palanivel *et al.*, 2015). Ginsenoside intranasal delivery of oil-in-water microemulsion shows enhanced brain bioavailability with improved protective effect vis-à-vis the conventional methods.

Absorption, followed by blood and tissues levels of ginsenoside, was higher from the intranasal formulation than that by the orally administered form (Li *et al.*, 2015). Liposomal nanovesicular systems were prepared of ginseng extract (GE) to augment the intracellular antioxidant activity of GE. Liposomes protect the incorporated drug against hydrolytic and metabolic decomposition, deriving in extended therapeutic drug levels. Nanoliposomes encapsulating ginseng extract abundant in ginsenosides exhibited protective effect against hydrogen peroxide-induced oxidative stress. The study confirmed that liposomal nanocarriers productively reduced hydrogen peroxide-induced oxidative stress (Tsai *et al.*, 2012).

6.4.7 Lycopene

Lycopene is sourced from watermelon, tomato, and pink guava and is the most potent antioxidant among various common carotenoids (Jain *et al.*, 2017). It has the potential to scavenge free radicals and shows protective behavior in cerebral ischemia-reperfusion injury. Lycopene supplementation reduces oxidative stress in rotenone-induced PD and restores antioxidant levels by reversing complex inhibition. Further calculations indicated therapeutic prospective of lycopene in neurodegenerative diseases associated with oxidative stress (Kaur *et al.*, 2011). Recently, it has been reported that lycopene has protective effect in 3-nitropropionic acid (3-NP)-induced model for Huntington's disease. Lycopene is a familiar lipophilic molecule with sensitivity toward heat and light; it exhibits low stability and low bioavailability. The desired therapeutic effect of lycopene, ingested through food, can be magnified by intercepting its degradation and boosting its solubility through the integration of the oil phase within oil-in-water nanoemulsions. Lycopene, developed in the form of a self-nanoemulsifying system or nanoemulsion, results in enhanced bioavailability of lycopene. Nanosized lycopene, prepared using a nanoemulsion delivery technique, enhances antioxidant activity with increased bioavailability (Ha *et al.*, 2015). Nano-lycopene, developed using a nanostructured lipid carrier delivery system, shows less chemical degradation of lycopene and improve *in vitro* antioxidant activity (Okonogi and Riangjanapatee, 2015). Using various aforementioned techniques, the stabilization study was conducted on lycopene. Study results show that lycopene is potentially bioavailable for protecting against oxidative stress that leads to PD.

6.5 FUTURE PROSPECTS

Nanotechnology in drug delivery has manifested into nanoformulations showing unique pharmacokinetic and pharmacodynamic properties, especially in targeted delivery. Nutraceuticals are a rich source of compounds possessing antioxidants and constituents that can be therapeutically used. Integration of research among the traditional "nutraceuticals" and of novel nanocarriers has established attractive therapies in the pharmaceutical area, thus improving the health of people. It is expected that the useful and valuable relevance of the natural products and natural remedies using nanocarriers will enhance the significance of current drug delivery systems. Numerous publications report the excellent *in vitro* bioactivity of available phytocomponents. Major drawbacks such as larger molecular size, poor aqueous solubility, extensive metabolisms, degradation during gastric emptying, etc. are mainly responsible for the limited clinical utility of these phytocomponents or phytonutrients. Application of nanotechnology tools results in improved bioavailability and bioactivity of phytomedicines by reducing particle size, surface modification, entrapping the nutraceuticals with different polymers. Nanomaterials not only help in targeted delivery, sustained delivery and improvement of the pharmacokinetic profile, but also enhanced diffusion of drugs into various organs by crossing various biological barriers including the blood–brain barrier.

Nevertheless, the researchers need to recognize that the proposed clinical applications of any nanoformulation require approval of federal agencies such as Food and Drug Administration (FDA). Accordingly, extensive clinical studies must be applied to prove the safety and efficacy of such new nanoformulations. In the future, nanonutraceuticals are certainly anticipated to alter the human body functioning in ways that we cannot imagine, yet consideration of their plausible efficacy-to-safety issues is pivotal too.

6.6 CONCLUSIONS

Today nutraceuticals are appealing because they're handy for today's lifestyle. Research is being carried out to prove their efficacy as therapeutic agents or as a supplement with other therapeutic agents, and many of them had proved their effectiveness in various psychotropic ailments. Also, the nanotechnology had enhanced its effectiveness, which provides a greater interest toward them. The combination of nanotechnology with traditional herbal medicine may provide a useful tool in designing future nanoneutraceutical formulation with improved bioavailability profile and less toxicity. Besides this, scientists and researchers need to pay heed to the toxicity and safety particulars of these nanoneutraceuticals related to both ecosystem and biological bodies, alongside cost-effeciency and enduring safety of the nano materials.

6.7 ACKNOWLEDGMENTS

The authors express no conflicts of interest. The authors accept responsibility for all contents of the chapter.

REFERENCES

Aluani, D., Tzankova, V., Yordanov, Y., Kondeva-Burdina, M., and Yoncheva, K. *In vitro* protective effects of encapsulated quercetin in neuronal models of oxidative stress injury. *Biotechnol Biotechnol Equip.* 2017; 1–10.

Alvira, D., Yeste-Velasco, M., Folch, J., Vardaguer, E., Canudas, A. M., and Pallas, M. Comparative analysis of the effects of resveratrol in two apoptotic models: Inhibition of complex I and potassium deprivation in cerebellar neurons. *Neuroscience.* 2007; 147: 746–756.

Barberger-Gateau, P., Raffaitin, C., Letenneur, L., Berr, C., Tzourio, C., Dartigues, J. F. *et al.* Dietary patterns and risk of dementia: The Three-City cohort study. *Neurology.* 2007; 69: 1921–1930.

Bagli, E., Goussia, A., Moschos, M. M., Agnantis, N., and Kitsos, G. Natural compounds and neuroprotection: Mechanisms of action and novel delivery systems. *In Vivo*. 2016; 30(5): 535–547.

Bertelli, A. A., and Das, D. K. Grapes, wines, resveratrol, and heart health. *J Cardiovasc Pharmacol*. 2009; 54: 468–476.

Bhatt, R., Singh, D., Prakash, A., and Mishra, N. Development, characterization and nasal delivery of rosmarinic acid-loaded solid lipid nanoparticles for the effective management of Huntington's disease. *Drug Deliv*. 2015; 22(7): 931–939.

Calon, F., and Cole, G. Neuroprotective action of omega-3 polyunsaturated fatty acids against neurodegenerative diseases: Evidence from animal studies. *Prostaglandins Leukot Essent Fatty Acids*. 2007; 77: 287–293.

Campos, D. A., Madureira, A. R., Sarmento, B., Gomes, A. M., and Pintado, M. M. Stability of bioactive solid lipid nanoparticles loaded with herbal extracts when exposed to simulated gastrointestinal tract conditions. *Food Res Int*. 2015; 78: 131–140.

Chauhan, B., Kumar, G., Kalam, N., and Ansari, H. S. Current concepts and prospects of herbal nutraceutical: A review. *J Adv Pharm Technol Res* 2013; 4: 4–8.

Choi, H. R., Choi, J. S., Han, Y. N., Bae, S. J., and Chung, H. Y. Peroxynitrite scavenging activity of herb extracts. *Phytother Res*. 2002; 16: 364–367.

Da Silva, T. M., Munhoz, R. P., Alvarez, C., Naliwaiko, K., Kiss, Á., Andreatini, R. *et al.* Depression in Parkinson's disease: A double-blind, randomized, placebo-controlled pilot study of omega-3 fatty-acid supplementation. *J Affect Disord*. 2008; 111: 351–359.

De Lau, L. M., Schipper, C. M., Hofman, A., Koudstaal, P. J., and Breteler, M. M. Prognosis of Parkinson disease: Risk of dementia and mortality: The Rotterdam Study. *Arch Neurol*. 2005; 62: 1265–1269.

Drapeau, E., Mayo, W., Aurousseau, C., Le Moal, M., Piazza, P. V., and Abrous, D. N. Spatial memory performances of aged rats in the water maze predict levels of hippocampal neurogenesis. *Proc Natl Acad Sci USA*. 2003; 100: 4385–4390.

Dureja, H., Kaushik, D., and Kumar, V. Developments in nutraceuticals. *Indian J Pharmacol*. 2003; 35: 363–372.

Dutta, K., Ghosh, D., and Basu, A. Curcumin protects neuronal cells from Japanese encephalitis virus-mediated cell death and also inhibits infective viral particle formation by dysregulation of ubiquitin-proteasome system. *J Neuroimmune Pharmacol*. 2009; 4: 328–337.

Esposito, E., Rotilio, D., Di Matteo, V., Di Giulio, C., Cacchio, M., and Algeri, S. A review of specific dietary antioxidants and the effects on biochemical mechanisms related to neurodegenerative processes. *Neurobiol Aging*. 2002; 23: 719–735.

Frozza, R. L., Bernardi, A., Paese, K., Hoppe, J. B., da Silva, T., Battastini, A. M., Pohlmann, A. R., Guterres, S. S., and Salbego, C. Characterization of trans-resveratrol-loaded lipid-core nanocapsules and tissue distribution studies in rats. *J Biomed Nanotechnol*. 2010; 6(6): 694–703.

Gupta, S., Chauhan, D., Mehla, K., and Nair, A. An overview of nutraceuticals: Current scenario. *J Basic Clin Pharm*. 2010; 1: 55–62.

Ha, T. V. A., Kim, S., Choi, Y., Kwak, H. S., Lee, S. J., Wen, J. *et al.* Antioxidant activity and bioaccessibility of size-different nanoemulsions for lycopene-enriched tomato extract. *Food Chem*. 2015; 178: 115–121.

Hardy, J. A., and Higgins, G. A. Alzheimer's disease: The amyloid cascade hypothesis. *Science*. 1992; 256: 184–185.

Hatanaka, J., Shinme, Y., Kuriyama, K., Uchida, A., Kou, K., Uchida, S. *et al. In vitro* and *in vivo* characterization of new formulations of St. John's Wort extract with improved pharmacokinetics and anti-nociceptive effect. *Drug Metab Pharmacokinet*. 2011; (6): 551–558.

Henning, S. M., Wang, P., and Heber, D. Quercetin increased the antiproliferative activity of green tea polyphenol (–)-epigallocatechin gallate in prostate cancer cells. *Nutr Cancer*. 2012; 64(4): 580–587.

Jain, A., Sharma, G., Kushwah, V., Thakur, K., Ghoshal, G., Singh, B. *et al.* Fabrication and functional attributes of lipidic nanoconstructs of lycopene: An innovative endeavour for enhanced cytotoxicity in MCF-7 breast cancer cells. *Colloids Surf B Biointerfaces*. 2017; 152: 482–491.

Jin, Y., Wen, J., Garg, S., Liu, D., Zhou, Y., Teng, L. *et al.* Development of a novel niosomal system for oral delivery of Ginkgo biloba extract. *Int J Nanomedicine*. 2013; 8: 421–430.

Joseph, J. A., Shukitt, H. B., and Casadesus, G. Reversing the deleterious effects of aging on neuronal communication and behavior: Beneficial properties of fruit polyphenolic compounds. *Am J Clin Nutr*. 2005; 81: 313S–316S.

Kalra, E. K. Nutraceutical—Definition and Introduction. *AAPS PharmSci*. 2003; 5: 27–28.

Karlsson, J., Emgard, M., Brundin, P., and Burkitt, M. J. Trans-resveratrol protects embryonic mesencephalic cells from tert-butyl hydroperoxide: Electron paramagnetic resonance spin trapping evidence for a radical scavenging mechanism. *J Neurochem*. 2001; 75: 141–150.

Kaur, H., Chauhan, S., and Sandhir, R. Protective effect of lycopene on oxidative stress and cognitive decline in rotenone induced model of Parkinson's disease. *Neurochem Res.* 2011; 36: 1435–1443.

Kelsey, N. A., Wilkins, H. M., and Linseman, D. A. Nutraceutical antioxidants as novel neuroprotective agents. *Molecules.* 2010; 15: 7792–7814.

Kondratyuk, T. P., and Pezzuto, J. M. Natural product polyphenols of relevance to human health. *Pharm Biol.* 2004; 42: 46–63.

Kumar, A., Sehgal, N., Kumar, P., Padi, S. S., and Naidu, P. S. Protective effect of quercetin against ICV colchicine-induced cognitive dysfunctions and oxidative damage in rats. *Phytother Res.* 2008; 22: 1563–1569.

Lau, F. C., Shukitt-Hale, B., and Joseph, J. A. The beneficial effects of fruit polyphenols on brain aging. *Neurobiol Aging.* 2005; 26: 128–132.

Lee, H. J., Cho, H. S., Park, E., Kim, S., Lee, S. Y., Kim, C. S. *et al.* Rosmarinic acid protects human dopaminergic neuronal cells against hydrogen peroxide-induced apoptosis. *Toxicology.* 2008; 250: 109–115.

Li, T., Shu, Y. J., Cheng, J. Y., Liang, R. C., Dian, S. N., Lv, X. X. *et al.* Pharmacokinetics and efficiency of brain targeting of ginsenosides Rgl and Rbl given as Nao-Qing microemulsion. *Drug Dev Ind Pharm.* 2015; 41(2): 224–231.

Mansoor, A., Shah, L., and Yadav, S. Nanotechnology for CNS delivery of bio-therapeutic agents. *Drug Deliv Transl Res.* 2013; 3(4): 336–351.

Mathew, A., Fukuda, T., Nagaoka, Y., Hasumura, T., Morimoto, H., Yoshida, Y. *et al.* Curcumin loaded-PLGA nanoparticles conjugated with Tet-1 peptide for potential use in Alzheimer's disease. *PLoS One.* 2012; 7(3): e32616.

Mulik, R. S., Bing, C., Wodzak, M. L., Munaweera, I., Chopra, R., and Corbin, I. R. Localized delivery of low-density lipoprotein docosahexaenoic acid nanoparticles to the rat brain using focused ultrasound. *Biomaterials.* 2016; 83: 257–268.

Nieoczym, D., Socała, K., Raszewski, G., and Wlaź, P. Effect of quercetin and rutin in some acute seizure models in mice. *Prog Neuropsychopharmacol Biol Psychiatry.* 2014; 54: 50–58.

Okonogi, S., and Riangjanapatee, P. Physicochemical characterization of lycopene-loaded nanostructured lipid carrier formulations for topical administration. *Int J Pharm.* 2015; 478(2): 726–735.

Pandey, M., Verma, R. K., and Saraf, S. A. Nutraceuticals: New era of medicine and health. *Asian J Pharm Clin Res.* 2010; 3: 11–15.

Patel, P. A., Patil, S. C., Kalaria, D. R., Kalia, Y. N., Patravale, V. B. Comparative *in vitro* and *in vivo* evaluation of lipid based nanocarriers of Huperzine A. *Int J Pharm.* 2013; 446(1–2): 16–23.

Prakash, D. J., Arulkumar, S., and Sabesan, M. Effect of nanohypericum (Hypericum perforatum gold nanoparticles) treatment on restraint stress induced behavioral and biochemical alteration in male albino mice. *Pharmacognosy Res.* 2010; 2(6): 330–334.

Prakash, D., Gupta, C., and Sharma, G. Importance of phytochemicals in nutraceuticals. *J Tradit Chin Med.* 2012; 1: 70–78.

Prakash, D., and Gupta, K. R. The antioxidant phytochemicals of nutraceutical importance. *Open Nutraceuticals J.* 2009; 2: 20–35.

Rajeswari, A. Curcumin protects mouse brain from oxidative stress caused by 1-methyl-4-phenyl-1,2,3,6-tetrahydropyridine. *Eur Rev Med Pharmacol Sci.* 2006; 10: 157–161.

Ravouru, N., Kondreddy, P., Korakanchi, D., and Haritha, M. Formulation and evaluation of niosomal nasal drug delivery system of folic acid for brain targeting. *Curr Drug Discov Technol.* 2013; 10(4): 270–282.

Sadruddin, S., and Arora, R. Resveratrol: Biologic and therapeutic implications. *J Cardiometab Syndr.* 2009; 4: 102–106.

Sahni, J. K. Exploring delivery of nutraceuticals using nanotechnology. *Int J Pharm Investig.* 2012; 2(2): 53.

Sarin, R., Sharma, M., Singh, R., and Kumar, S. Nutraceuticals: A review. *International Research Journal of Pharmacy.* 2012; 3: 95–99.

Satoh, T., Kosaka, K., Itoh, K., Kobayashi, A., Yamamoto, M., Shimojo, Y. *et al.* Carnosic acid, a catechol-type electrophilic compound, protects neurons both *in vitro* and *in vivo* through activation of the Keapl/Nrf2 pathway via S-alkylation of targeted cysteines on Keapl. *J Neurochem.* 2008; 104: 1116–1131.

Scalbert, A., Manach, C., Morand, C., and Remesy, C. Dietary polyphenols and the prevention of diseases. *Crit Rev Food Sci Nutr.* 2005; 45: 287–306.

Schroeder, E. K., Kelsey, N. F., Doyle, J., Breed, E., Bouchard, R. J., Loucks, F. A. *et al.* Green tea epigallocatechin 3-gallate accumulates in mitochondria and displays a selective antiapoptotic effect against inducers of mitochondrial oxidative stress in neurons. *Antioxid Redox Signal.* 2009; 11: 469–480.

Shinde, R. L., and Devarajan, P. V. Docosahexaenoic acid–mediated, targeted and sustained brain delivery of curcumin microemulsion. *Drug Deliv.* 2017; 24(1): 152–161.

Smith, A., Giunta, B., Bickford, P. C., Fountain, M., Tan, J., and Shytle, R. D. Nanolipidic particles improve the bioavailability and alpha-secretase inducing ability of epigallocatechin-3-gallate (EGCG) for the treatment of Alzheimer's disease. *Int J Pharm.* 2010; 389: 207–212.

Sood, S., Jain, K., and Gowthamarajan, K. Curcumin-donepezil–loaded nanostructured lipic carriers for intranasal delivery in an Alzheimer's disease model. *Alzheimers Dement.* 2013; 9(4): P299.

Stillwell, W., and Wassall, S. R. Docosahexaenoic acid: Membrane properties of a unique fatty acid. *Chem Phys Lipids.* 2003; 126: 1–2.

Tong-un, T., Muchimapura, S., Wattanathorn, J., and Phachonpai, W. Nasal administration of quercetin liposomes improves memory impairment and neurodegeneration in animal model of Alzheimer's disease. *Am J Agric Biol Sci.* 2010; 5(3): 286–293.

Tsai, W. C., Li, W. C., Yin, H. Y., Yu, M. C., and Wen, H. W. Constructing liposomal nanovesicles of ginseng extract against hydrogen peroxide-induced oxidative damage to L929 cells. *Food Chem.* 2012; 132(2): 744–775.

Uauy, R., and Dangour, A. D. Nutrition in brain development and aging: Role of essential fatty acids. *Nutr Rev.* 2006; 64: S24–S33.

Vaka, S. R., Shivakumar, H. N., Repka, M. A., and Murthy, S. N. Formulation and evaluation of carnosic acid nanoparticulate system for upregulation of neurotrophins in the brain upon intranasal administration. *J Drug Target.* 2013; 21(1): 44–53.

Vazzana, M., Macedo, A. S., Santini, A., Faggio, C., and Souto, E. B. Novel neuroprotective formulations based on St. John's Wort. *J Food Res.* 2014; 3: 1927–0895.

Whelan, J., and Rust, C. Innovative dietary sources of n-3 fatty acids. *Annu Rev Nutr.* 2006; 26: 75–103.

Wilson, B., Samanta, M. K., Santhi, K., Kumar, K. P., Paramakrishnan, N., and Suresh, B. Targeted delivery of tacrine into the brain with polysorbate 80-coated poly(n-butylcyanoacrylate) nanoparticles. *Eur J Pharm Biopharm.* 2008; 70(1): 75–84.

Wu, X., Beecher, G. R., Holden, J. M., Haytowitz, D. B., Gebhardt, S. E., and Prior, R. L. Lipophilic and hydrophilic antioxidant capacities of common foods in the United States. *J Agric Food Chem.* 2004; 52: 4026–4037.

Yang, C., Zhang, X., Fan, H., and Liu Y. Curcumin upregulates transcription factor Nrf2, HO-1 expression and protects rat brains against focal ischemia. *Brain Res.* 2009; 1282: 133–114.

Yang, Z. Z., Zhang, Y. Q., Wang, Z. Z., Wu, K., Lou, J. N., and Qi, X. R. Enhanced brain distribution and pharmacodynamics of rivastigmine by liposomes following intranasal administration. *Int J Pharm.* 2013; 452(1–2): 344–354.

Yehuda, S., Rabinovitz, S., and Mostofsky, D. I. Essential fatty acids and the brain: From infancy to aging. *Neurobiol Aging.* 2005; 26: 98–102.

Zamin, L. L., Dillenburg-Pilla, P., Argenta-Comiran, R., Horn, A. P., Simão, F., Nassif, M. *et al.* Protective effect of resveratrol against oxygen-glucose deprivation in organotypic hippocampal slice cultures: Involvement of PI3-K pathway. *Neurobiol Dis.* 2006; 24: 170–182.

7 Alzheimer's Disease: Potential of Nanotailored Nutraceuticals

Vandita Kakkar, Komal Saini,*
Suneera Adlakha, and Indu Pal Kaur

CONTENTS

7.1 INTRODUCTION

Alzheimer's disease (AD) is a chronic and progressive neurodegenerative disease that constitutes the key cause of loss of memory (dementia) in the elderly (Kawas *et al.*, 2000). Reports by World Health Organization (WHO) suggest that 5% of men and 6% of women ages 60 years and up are affected with Alzheimer's type dementia worldwide (Chow *et al.*, 2015). With an increase in the life period and aging population in the developed countries, the patients afflicted with AD will rise if no useful therapy is developed in the near future. The patients in most cases preserve motor functions but demonstrate the cognitive impairment that begins with loss of memory and that strengthens with the progression until dementia to death. The damage of nerve cells initiates with the cells involved in learning and memory, which gradually then spreads to the cells that control other aspects of thinking, judgment, and behavior. This ultimately affects the cells that are involved in controlling and coordinating movements (Chiappelli *et al.*, 2006). The statement below is the state of mind of the patients suffering from AD.

> "Those with dementia are still people and they still have stories and they still have character and they are all individuals and they are all unique. And they just need to be interacted with on a human level."
> —Carey Mulligan

There are seven stages that lucidly depict the concepts of mild, moderate, moderately severe, and severe AD (Neugroschl and Wang, 2011) (Table 7.1).

The pathological hallmarks of AD are the senile neuritic plaques (SNPs) and neurofibrillary tangles (NFTs), which engross the amyloid-β peptides and hyperphosphorylated tau, respectively (Pietronigro *et al.*, 2017). There are increasing evidences that link these neurological hallmarks to brain cholesterol (Reid *et al.*, 2007). The clearance of amyloid-β, in addition to the formation of neurofibrillary tangles, is affected by the amount of brain cholesterol via action at the lipid rafts

* Corresponding author.

TABLE 7.1
Stages of Alzheimer's Disease

Sr. No.	Stages	General Features	Duration of Stage
1.	No Cognitive Impairment	Mentally healthy person	–
2.	Very Mild Decline	Memory lapses and changes in thinking are rarely detected by friends, family or medical personnel, especially as about half of all people over 65 begin noticing problems in concentration and word recall.	–
3.	Mild Cognitive Decline	1. Difficulty with retrieving words, planning, organization, misplacing objects, and forgetting recent learning, which can affect life at home and work. 2. Depression and other changes in mood.	2–7 years
4.	Moderate Cognitive Decline (Mild or Early-stage)	1. Problems handling finances result from mathematical challenges. 2. Recent events and conversations are increasingly forgotten, although most people in this stage still know themselves and their family. 3. Problems carrying out sequential tasks, including cooking, driving, ordering food at restaurants, and shopping. (Accurate diagnosis of AD is possible at this stage.)	Lasts roughly 2 years
5.	Moderately Severe Cognitive Decline (Moderate or Mild-stage)	1. Decline is more severe and requires assistance. 2. No longer one is able to manage independently, recall personal history details, and contact information. 3. People in this stage experience severe decline in numerical abilities and judgment skills. 4. Basic daily living tasks like eating and dressing require increased supervision.	1.5 years
6.	Severe Cognitive Decline (Moderately severe or mid-stage)	1. Total lack of awareness of present events and inability to accurately remember the past. 2. People in this stage progressively lose the ability to take care of daily living activities like dressing, toileting, and eating, but are still able to respond to nonverbal stimuli. 3. Agitation and hallucinations often show up in the late afternoon or evening. Many can't remember close family members, but know they are familiar.	2.5 years
7.	Very Severe Cognitive Decline (Severe or late-stage)	1. In this final stage, speech becomes severely limited, as well as the ability to walk or sit. 2. Individuals lose the ability to smile, and the ability to hold their head up. 3. Reflexes become abnormal and muscles grow rigid. 4. Swallowing is impaired.	1 to 2.5 years

located in neuronal membranes. The presence of intravascular cerebral cholesterol also contributes to atherosclerosis and is a significant risk factor for ischemic cerebrovascular disease. These cerebral infarcts further enhance the chances of dementia by 20 times (Kapogiannis *et al.*, 2017). Figure 7.1 presents a comparison of cellular changes in the normal brain against an advanced AD patient's crosswise "slice" through the brain.

In the Alzheimer's brain:

- The shriveling up of the cortex damages the brain areas characteristically involved in thinking, planning, and remembering.
- Shrinkage of the cortical area, that is, the hippocampus, which plays a key function in development of new memories, occurs.
- Large-sized ventricles (fluid-filled spaces within the brain) are developed (Islam *et al.*, 2010).

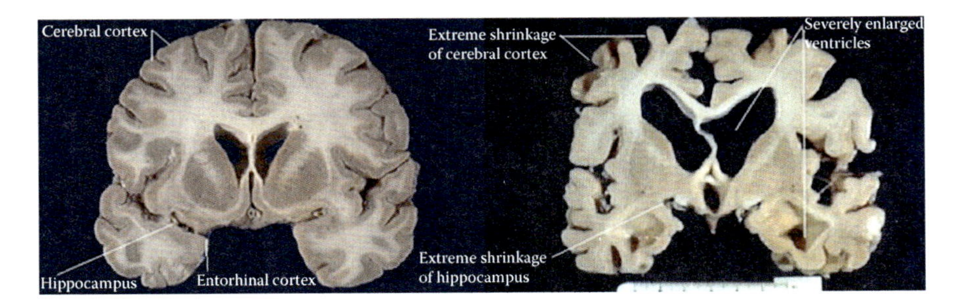

FIGURE 7.1 Comparison of a normal aged brain (left) and the brain of person with AD (right).

Current treatment options for AD, which include cholinesterase inhibitors or NMDA antagonists, help to cope with the symptoms and do not provide a complete cure, thereby obviating the need for new approaches to deal with the underlying mechanisms of AD. Nanotailored formulations are now considered as the smart and novel drug delivery systems for remote regions such as the brain, so as to protect various therapeutic agents while delivering them efficiently into the damaged areas. Several nanotechnology-enabled formulations are reported in the literature, which are administered via different routes with or without ligands for enhancing their ability to cross the blood–brain barrier (BBB). This chapter thus describes the usefulness of nanocoutured nutraceuticals for the treatment of AD.

7.1.1 Epidemiology

AD is the most common cause of dementia, which affects approximately 18 million cases, and thus displays a dramatic epidemic owing to the phenomenal immense growth of the aged population around the world (Figure 7.2). In the United States, an estimated 5.2 million Americans of all ages have AD, as reported in 2014, and approximately 200,000 individuals with early onset of AD are under the age of 65 (Seth and Kakkar, 2015). In comparison to Africa, Asia, and Europe, the prevalence of AD is much higher in the United States, which can also be related to the superior methods of detection. AD is the sixth foremost reason of death in the United States, as detailed by the National Center for Health Statistics of the Centers for Disease Control and Prevention (CDC), and with the mortality rate being 27 deaths per 100,000 people (Murphy *et al.*, 2016). Alzheimer's and

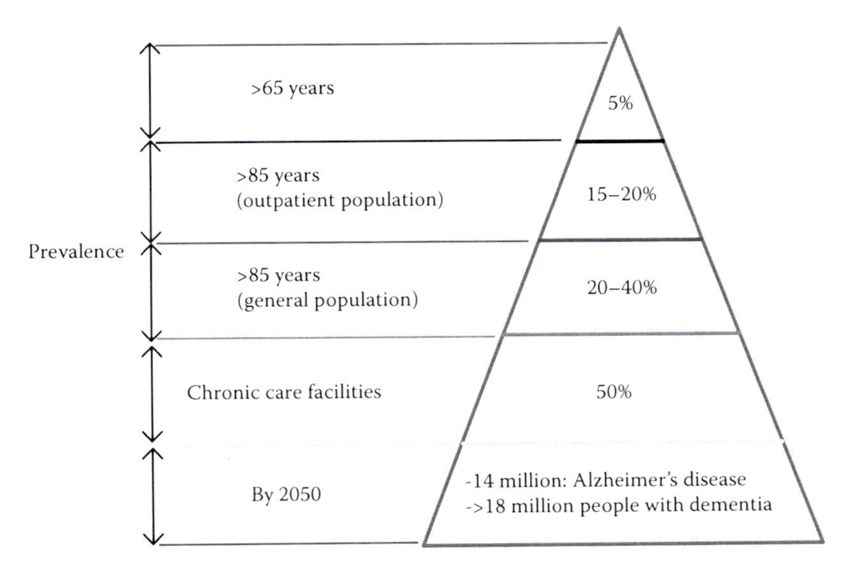

FIGURE 7.2 Prevalence of dementia with advancing age and medical facilities.

Related Disorders Society of India (ARDSI) have reported that approximately 3.7 million people (2.1 million women and 1.5 million men) of all communities are affected by AD in India alone, with an estimated two-folds increase of AD patients by 2030 (The Dementia India Report, 2010). Such distressing circumstances on the exponential rise in AD cases, both in developed and other developing nations, pose a huge socioeconomic burden on families, caregivers, and the whole society.

7.1.2 Pathogenesis

The pathogenesis of AD is a complex linkage of neuroinflammation, oxidative stress, amplified cholesterol levels, and modulation in acetylcholinesterase activity (Mirshafiey *et al.*, 2016). With pathophysiology of AD, the question goes back to the Alzheimer's time (1907) when it was examined that the neuropathological features of this disease are amyloidal plaques as well as hyperphosphorylated NFTs. Several hypotheses, including the amyloid beta (Aβ), cholinergic and tau hypothesis (implication of oxidative stress and neuroinflammation), have been put forth on the basis of various causative factors to explain this multifactorial disorder (Kurz and Perneczky, 2011).

Amyloid cascade hypothesis theorizes that the cortical plaques in the brain affected with AD mainly are composed of Aβ protein, which is made via amyloid precursor protein (APP) processing. On chromosome 21, gene encoding exists with APP. The physiologic role of APP per se is not clear, yet its impact on neuronal function as well as cerebral development have been accounted for (Zheng *et al.*, 1995). The imbalance between production as well as clearance of Aβ peptide hypothesis has been quite convoluted (Salomone *et al.*, 2012). Latter causes its deposition as diffuse senile plaques, once it is thrashed by the secretases, which results in its aggregation into soluble oligomers and finally as an insoluble beta-sheet conformation. Some studies have revealed that Aβ 42 oligomers, which are formed by combined activities of neurons and its associated astrocytes, stimulate the oxidative impairment and promote tau hyperphosphorylation, resulting in toxic effects on the synapses and the mitochondria (Kurz and Perneczky, 2011). Aβ 42 plaques have been made known to appear during late stages of AD, ensuing microglial activation causing production of IL-1β, TNF-α, and IFN-γ (Rosenmann, 2013). The cytokines stimulate the nearby astrocyte-neuron to further produce Aβ 42 oligomers, thus activating more Aβ 42 production and their dispersal (Dal Pra *et al.*, 2014). Furthermore, oligodendroglias (OLGs) are associated with neurons–astrocyte complex and the Aβ oligomers cause its destruction (Lakatos *et al.*, 2016). These Aβ oligomer aggregates are responsible factors for neuronal as well as vascular degeneration with AD brains resulting in oxidative stress. This results in impaired capability to the scavenge oxygen radicals (Roth *et al.*, 2005). It has been shown that Aβ 42 oligomers possess capability of destroying the membranes rich in cholesterol, particularly the ones found in OLGs and myelin (Subasinghe *et al.*, 2003).

The tau hypothesis categorically discusses the abnormal phosphorylation, which occurs under the pathological conditions of AD to form aggregates, which get converted to filamentous brain inclusions referred to as NFTs (Stoothoff and Johnson, 2005). Dynamics of neurons in the central nervous system (CNS) is regulated by microtubule-associated protein, that is, Tau. Although tau composition and the structure of NFTs are quite well-characterized, the process of tangle formation is still unknown (Pooler *et al.*, 2013). Tau is abnormally hyperphosphorylated (in AD and related disorders) and gets accumulated as intraneuronal tangles of paired helical filaments (PHF), twisted ribbons, and/or straight filaments (Iqbal *et al.*, 2005). This characteristic brain lesion directly associates with dementia in these patients and other neurodegenerative disorders such as Parkinsonism, Down syndrome, Pick disease, etc. (Arrigada *et al.*, 1992). A comparative study on adults and fetal brain biopsies has shown that the phosphorylation of the analogous positions is known to occur in PHF-tau (Goedert, 1996). One of the other important factors that contribute to the initiation and progression of AD is oxidative stress. The undue production of reactive oxygen species (ROS) results in mitochondrial dysfunction and/or aberrant accumulation of transition metals. Growing evidence has revealed that extensive OS is a characteristic biomarker of AD brains with established pathology of senile plaques as well as NFT. Nonetheless, the mechanisms that direct to redox balance disturbance and creation of free radicals are elusive (Kim *et al.*, 2015).

With the accretion of free radicals, the damage and alterations affecting the expression of antioxidant (AO) enzymes, like superoxide dismutase (SOD) and catalase, has been found in the central nervous system as well as peripheral tissues of AD patients (Padurariu *et al.*, 2010). In AD and mild cognitive impairment (MCI), the augmented oxidative damage to lipids and proteins, and depleted levels of glutathione and AO enzyme activities, are more limited to a small area to the synapses. The latter correlates with the severity of the disease, signifying an involvement of oxidative stress in AD-related synaptic loss (Ansari and Scheff, 2010). Increase in the levels of oxidative stress has been implicated in MCI, an intermediate state between normal aging and dementia, thus highlighting the OS-induced deterioration in AD. These outcomes entail that OS may be the earliest alterations that occur during the initiation and growth of AD.

In the first stage, ROS excites the c-Jun N-terminal kinases (JNK) and p38, and deactivates protein phosphatase 2A (PP2A). JNK and p38 then uphold the expression of Tau, which is then repressed by PP2A. Also, activation of JNK and p38 further stimulates AβPP cleaving enzyme 1 (BACE1), causing Aβ1-42 accumulation, leading eventually to the activation of NADPH oxidase (Nox) to attain the production of additional $O_2\bullet^-$, which further results in Ca^{2+} influx to extract the excitatory neurotoxicity as well (Figure 7.3).

The ROS formation may slow down the progress of AD by multiple mechanisms, comprising but not limited to decline in OS-mediated neuronal toxicity and inhibition of Aβ, reduction in tau

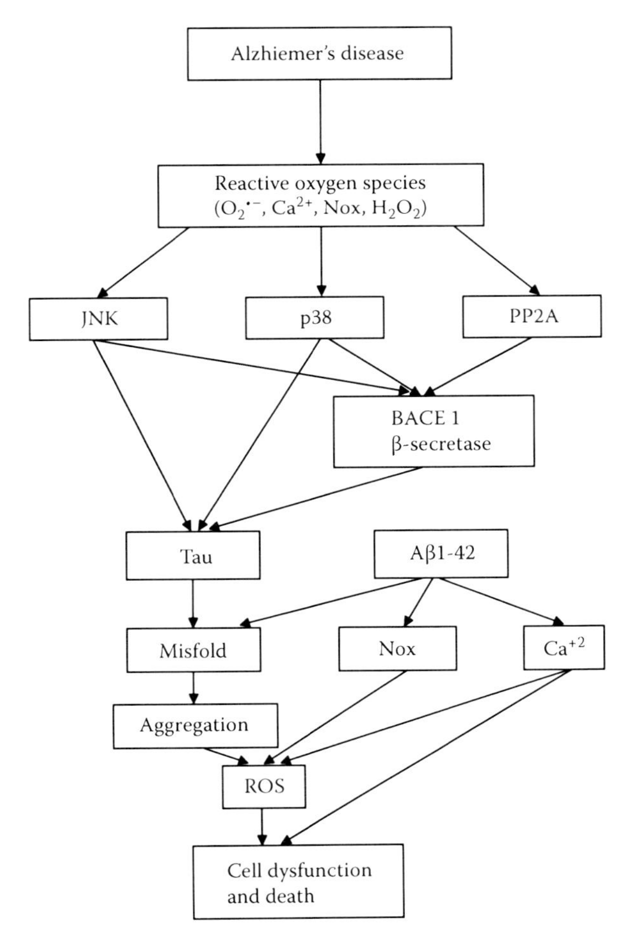

FIGURE 7.3 Causal relationship between ROS and misfolded proteins, underlying AD.

polymerization and phosphorylation, and reestablishment of the mitochondria role as well as metal homeostasis. Consequently, the treatment of AD with some natural antioxidants has the capability to target several molecular events involved during the pathogenesis of AD (Zhao and Zhao, 2013).

Neuroinflammation, represented in terms of plaques of Aβ protein and tangles of tau protein, is a characteristic trait of AD, which is different from the peripheral inflammation (Tuppo and Arias, 2005). Brain inflammation is difficult to detect owing to lack of pain fibers. In addition, the classic marks of inflammation, which include redness, swelling, heat, and pain, are typically not seen in the CNS. Furthermore, the presence of BBB, composed of tight junction within the capillaries in the CNS, impedes the access of inflammatory cells, pathogens, and some macromolecules into the subarachnoid space. It protects the sensitive post-mitotic neurons from the inflammatory destructions (Akiyama *et al.*, 2000).

The inflammatory mechanisms involved in the cascade of neuroinflammation are related to the involvement of brain cells (e.g., microglia and astrocytes), the classic and alternate pathways of the complement system, the pentraxin acute-phase proteins, neuronal-type nicotinic acetylcholine receptors (AChRs), peroxisomal proliferators-activated receptors (PPARs), and cytokines as well as chemokines (Figure 7.4). This signifies the setting in of the neuroinflammation to be an extremely interactive process, wherein the first to counteract the neuronal injuries are the microglial cells, which are the brain macrophages (Dheen *et al.*, 2007). In AD, a cleavage of the APP and the aggregation of Aβ peptides (Aβ1-42 and Aβ1-40) trigger the activation of microglials and astrocytes, following activation of transcription factors (i.e., nuclear factor-kappa B [NF-kB] and activator protein [AP]-1), persuading the prolongation of ROS and several other proinflammatory mediators (Dheen *et al.*, 2007). The liberation of these proinflammatory cytokines and ROS further cause the neuronal destruction or neurotoxicity, resulting in apoptosis as well as necrosis. These pro-inflammatory mediators from the microglials and astrocytes can also stimulate each other to strengthen the signs of inflammatory signals to the neurons (Pan *et al.*, 2010).

A meager involvement of lymphocytes or monocytes beyond the normal surveillance of brain, and an increasing effect of certain cytokines in neurodegeneration and AD, have been reported (Frank-Cannon *et al.*, 2009) (Table 7.2). Establishing correlation between the neuroinflammation levels and the amount of release of cytokine can further help to portray a clear picture of the incurred neuroinflammation (Frank-Cannon *et al.*, 2009).

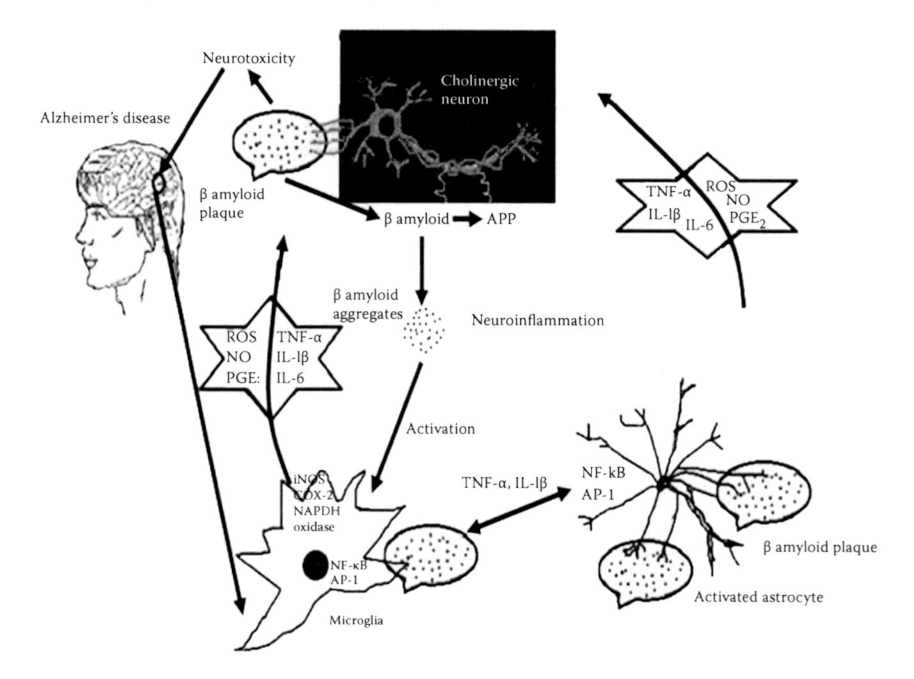

FIGURE 7.4 Mechanism of occurrence of neuroinflammation in AD.

TABLE 7.2
Pro-Inflammatory Elements during the Process of Neuroinflammation

Pro-Inflammatory Elements	Effect
1. Chemokines	Dysfunction, apoptosis and necrosis of neuron, microglia and astrocytes.
2. IL-1β, IL-6, IL-12, INF-γ, TNF-α	Astrocytes and microglia activation, INF-γ, TNF-α dysfunction, apoptosis and necrosis of neuron, microglia and astrocytes.
3. NO, ROS	Oxidative stress in cells; dysfunction, apoptosis and necrosis of neuron, microglia and astrocytes.

7.1.3 Current Treatment Strategies

The treatment options for AD aim to manage the symptoms by acting centrally in the brain. These include the use of anti-amyloid drugs, which are under active investigation (O'Brien and Wong, 2011). Additionally, the therapeutic utility of passive transmission of Aβ monoclonal antibodies from vaccinated mice to the AD model in mice to reducing the cerebral amyloidosis (Lemere, 2009) has been accounted. A member of the pepsin family, β-secretase, is a membrane-anchored aspartyl protease (Dislich and Lichtenthaler, 2012). Aβ generation of APP is assisted by β and γ-secretase-mediated cleavage. Further, β-secretase inhibitors reduce the levels of Aβ, but mechanism-based side effects occur owing to the inhibition of β-cleavage of non-amyloid substrates like neuregulin are also observed to create a therapeutic effect in AD (Panza *et al.*, 2010). Moreover, the utility of statins, also known as 3-hydroxy-3-methylglutaryl co-enzyme A (HMG-CoA) reductase inhibitors, which recovers the blood-flow by decreasing the coagulation for AD, has been predicted. They are generally used as cholesterol-reducing drugs and have been revealed to lower the incidence of primary and secondary coronary heart disease in the clinical studies (Kotyla, 2010). The latter acts by blocking the enzyme essential for the manufacturing of L-mevalonate, an intermediary product in the production of cholesterol. A correlation between a high cholesterol level during midlife and AD has been reported, and it has been shown that the consumption of statins, surprisingly, lower the risk of development of AD by inhibiting the synthesis of cholesterol (Endo, 2010). Reduction of cholesterol levels by statins helps the cells to upregulate the low-density lipoprotein (LDL) receptor, while enhancing the LDL-receptor degradation, such that the surface expression of the receptors is unaffected (Lagace, 2014). These drugs are broadly used and are secure for long-term therapy, and thus might delay or lower the onset of AD (Son *et al.*, 2015). NMDA receptor antagonist Memantine has been approved for the cure of moderate to severe AD by the Food and Drug Administration (FDA) (Parsons *et al.*, 2007). Cognitive decline in patients with AD is connected with neuronal destruction from the excitatory toxicity, which is caused by the persistent over-activation of NMDA (N-methyl-D-aspartate) receptors by glutamate. Both Aβ and tau protein expression triggers the activation of NMDA receptors, thus resulting in excitatory toxic trail that directs to the cell death (Rammes, 2009; Revett *et al.*, 2013).

Further, application of cholinesterase inhibitors (CIs) chiefly as long-term symptomatic therapy for AD, which act by reducing the acetylcholinesterase-assisted destruction of acetylcholine in the synaptic cleft, has been reported (Coste *et al.*, 2017).

After the CIs, the role of neurotrophins to promote the cell survival has been documented. They act by signaling via specific tyrosine kinase receptors, by engaging the internal cellular machinery to efficiently block the process of apoptosis to take place in damaged neurons. Neurotrophins are measured prospective therapeutic agents for neurodegenerative disorder (Nagahara and Tuszynski, 2011). Function of nerve growth factor (NGF) in sustaining the neuronal reliability, as well as survival in reaction to injury of the basal cholinergic forebrain neurons, has been indicated. Therefore, the NGF and related agents are purported to have neurorestorative as well as neuroprotective properties (Aloe *et al.*, 2012).

Apart from the various approaches discussed above, the utility of antioxidant (AO) therapy for AD has also been well-realized and evidenced. AO is a compound that, even in lower concentration, delays or prevents the oxidation of the substrate. It acts by lowering oxidative stress, in addition to DNA mutations and other malignant transformations, along with the other factors of cell-damaging. Several studies have depicted that the activity of AO on ROS activity resulted in decreased prevalence of neurodegenerative and other diseases (Nita and Grzybowski, 2016).

The first identified kinds of AO defense systems developed against oxidative damage are the ones that prevent ROS occurrence, and/or block or capture the radicals formed thereof (Sharma *et al.*, 2012). They mainly act by a repair process that eliminates the damaged biomolecules, earlier than their aggregation, thus enabling changes in the cell metabolism. The mechanisms involved are (a) removal of oxidized proteins using proteolytic systems; (b) renovation of lipids to be oxidized through phospholipases, peroxidases, or acyl transferases; and (c) repairing of nucleic acids that are oxidatively damaged by using the specific enzymes (Lien *et al.*, 2012). It has been indicated that the decomposition of the repair systems directs more to the age-linked diseases than modest alteration in the AO defense's prospective against ROS occurrence (Radak *et al.*, 2013).

The AOs, which are typically deficient or are in little amount, are referred to as the exogenous antioxidants. These mainly include the use of dietary nutrient supplements (e.g., vitamins, minerals, fibers, fatty acids, or amino acids), lycopene, glutathione, polyphenols (i.e., flavonoids and non-flavonoids), resveratrol, and the N-acetylcysteine (Orihuela-Campos *et al.*, 2015).

7.1.4 Intervention of Nutraceuticals for the Treatment of AD

The word *nutraceutical* is known to be a blend of the words *nutrition* and *pharmaceutical* (Kalra, 2003). It is a big umbrella term that explicitly defines any product derivatives of food sources accompanied with health advantages in addition to the basic nutritional value found in foods. There is no official definition of nutraceuticals, yet their advertisement(s) as food and dietary nutritional supplements has been flourishing (Kakkar *et al.*, 2016).

Phytochemicals of nutraceutical signify the bioactive ingredients that promote health activities and occur at the intersection of food and pharmaceutical industries (Karwande and Borade, 2015). They are principally termed as the polyphenols, phytoestrogens, terpenoids, carotenoids, phytosterols, flavonoids, isoflavonoids, and anthocyanidins. They have a significant impact on health in terms of disease prevention and immunity boosters. Several foods, such as whole grains, fruits and vegetables, and herbs, are composed of phytochemicals of nutraceutical significance, and several such phytochemical constituents have been reported to be advantageous for neurodegenerative diseases (Kim *et al.*, 2015). These particular nutritional food groups are rich in micronutrients and vitamins for their nutritional quality and advantages in health (Pandey *et al.*, 2010). The global nutraceuticals product market is anticipated to reach $285.0 billion in 2021 at a compound annual growth rate of 7.5% from 2016 to 2021 (BCC Research, 2017).

Abundant literature supports the usefulness of polyphenolics from phytochemicals in curing AD (Lim *et al.*, 2013). However, clinical translation of these molecules is often met with failure. This is mainly assigned to a compromised bioavailability attributed to their poor aqueous solubility, poor permeability, eventual incomplete absorption, fast metabolism, instability at physiological pH, and light-assisted degradation. Accordingly, to improve their biological activity, the principles of nanotechnology have been successfully employed for proficient delivery of these nutraceuticals.

7.1.4.1 Nanotransformed Phytochemicals

Advances in nanotechnology have provided excellent opportunities for management of AD. Loading a drug in a suitably formulated nanocarrier system helps to increase the concentration of the drug in the brain cells in comparison to the drug per se. Moreover, the small particle size will help in crossing the BBB and reach the target neuronal cell. These systems agree to transport of non-transportable drugs or diagnostic agents crosswise in the BBB by disguising their physicochemical properties by enveloping bioactives in these nanostructured systems. Moreover, they may trim down the leakage of the drug in the brain tissue and thus decrease the peripheral toxicity (De Rosa *et al.*, 2012). Nanotailored nutraceuticals with immense potential to modify the physicochemical properties of drugs can be examined as magic bullets for neurodegenerative diseases (Mazzarino *et al.*, 2012). Besides targeted drug delivery, the sustained release drug profiles and intracellular asylum to protect the therapeutics from degradation are all achieved with fruition employing such nanostructured drug delivery system (Bozdağ Pehlivan, 2013). Nanocarriers for polyphenols can be used for achieving bioenhanced activity of any treatment (Yang *et al.*, 2012), owing to their effect on the pharmacokinetics as well as pharmacodynamics of drugs, while reducing the incidence of their plausible side effects. Many useful nanostructured approaches, including solid lipid nanoparticles (SLNs), nano-lipidic carriers (NLCs), polymer-based nanoparticles, liposomes, nanoemulsions, microemulsions (MEs), and liquid crystals (LCs) (Das *et al.*, 2010; Rejinold *et al.*, 2011; Gangwar *et al.*, 2012; Mazzarino *et al.*, 2012; Omidfar *et al.*, 2013) are explicitly illustrated sequentially in Table 7.3.

Doggui *et al.* (2012) studied polymer-based nanoparticles to deliver curcumin to neuronal cells enclosed in biodegradable poly (lactide-co-glycolide) (PLGA) based nanoparticulate formulations. The latter was estimated for the cellular damage and the biological action. Cytotoxicity assays established that the void PLGA-nanoparticles (NPs) and curcumin-loaded PLGA nanoparticles (NPs-Cur) were safe for human neuroblastoma SK-N-SH cells. Furthermore, the NPs-Cur was proficient to protect the SK-N-SH cells against H_2O_2 and check the elevation of ROS and the expenditure of glutathione provoked by H_2O_2. Interestingly, NPs-Cur also could check the generation of

TABLE 7.3
Various Nanostructured Nutraceuticals for Treatment of AD

Nanotechnology-Enabled Systems	Drug or Bioactive Molecule	Route(s) of Administration	Reference(s)
1. Polymeric nanoparticles	-Curcumin	*In vitro*	Doggui *et al.*, 2012
	-Curcumin	*In vitro*	Mulik *et al.*, 2012
	-Quercetin	*In vitro*	Sun *et al.*, 2016
2. Solid lipid nanoparticles and nano-lipid carriers	-Resveratrol	Oral	Reis *et al.*, 2016
	-Piperine	Intraperitoneal	Yusuf *et al.*, 2013
	-Curcumin	Oral route	Kakkar *et al.*, 2013
	-Curcumin and donepezil	Intranasal	Sood *et al.*, 2013
	-Curcumin	Oral	Kakkar and Kaur, 2011
	-Quercetin	Intravenous	Dhawan *et al.*, 2011
	-Resveratrol	Oral and intraperitoneal	Figueiró *et al.*, 2011
	-Vinpocetine	Oral route	Zhuang *et al.*, 2010
3. Liposomes	-Curcumin-PEG derivative	*In vitro*	Mourtas *et al.*, 2014
	-Curcumin	*In vitro*	Sancini *et al.*, 2013
	-Lipid curcumin derivatives	*In vitro*	Mourtas *et al.*, 2011
	-*Ginkgo biloba* extract	Oral route	Naik *et al.*, 2006
4. Nanoemulsions	-β-Asarone	Intranasal	Zhang *et al.*, 2014
	-Huperzine A	Transdermal	Patel *et al.*, 2013
	-*Tabernaemontana divaricate*	Transdermal	Chaiyana *et al.*, 2013
5. Microemulsions	-Huperzine A and ligustrazine phosphate	Transdermal	Shi *et al.*, 2012
6. Liquid crystals	-*Tabernaemontana divaricate*	Transdermal	Chaiyana *et al.*, 2013

the redox-sensitive transcription factor Nrf2 in the existence of H_2O_2. Their findings suggested that NPs-Cur could be a potential drug delivery strategy to protect the neurons against oxidative damage, as observed in AD.

Another strategy reported the employment of nanoparticles (NPs) embellished with suitable ligands for the brain delivery of curcumin. This strategy was based on the development of curcumin-loaded PnBCA poly(n-butyl cyanoacrylate) NPs tagged with ApoE3 ligands to take the advantage of LDL-r-mediated transcytosis crossways of the BBB and through SHSY5Y neuroblastoma cells as well. A comparison of effect of free curcumin versus nanocarrier inhibitory activity by ApoE3-functionalized nanocarriers on Aβ1-42-mediated toxicity was calculated on SH-SY5Y cells. The results indicated significant reduction (40% compared with free drug at 100 nM Aβ) of Aβ1-42-related toxicity on cells that take care of functionalized nanospheres beside the decline in ROS development (Mulik *et al.*, 2012).

Sun *et al.* (2016) investigated the PLGA-functionalized quercetin (PLGA@QT) NPs inhibitory effects on A42 fibrils. PLGA@QT NPs illustrated lower cytotoxicity when tested on SH-SY5Y cells *in vitro*. PLGA@QT NPs cytotoxicity studies lead to a concentration-related behavior on the SH-SY5Y human neuroblastoma cells and established inhibitory signals on the neurotoxicity of Zn2+-A42 system and better viability of the neuron cells. Overall, quercetin-based nanoscale drug delivery system showed higher therapeutic index and lower incidence of side effects.

Reis *et al.* (2016) formulated resveratrol-loaded SLNs, which were functionalized with apolipoprotein E and recognized by the LDL receptors, over-expressed on the BBB. Their results revealed enhanced permeation through hCMEC/D3 cell line monolayers (1.8-fold higher) for functionalized SLNs vis-à-vis the non-functionalized ones. The authors highlighted the promise of nanosystems for resveratrol delivery into the brain, while protecting the bioactive from degradation in the bloodstream.

Yusuf *et al.* (2013) proposed the formulation of piperine SLNs with polysorbate 80 coating using emulsification-solvent diffusion technique. Their results revealed an increased acetylcholinesterase activity and augmentation in cognitive activities superior to those of donepezil in ibotenic acid-induced AD in mice model. Further, the histopathological studies confirmed decline in plaques and tangles, thus emphasizing the task of nanotechnology for distributing drug across the BBB.

Kakkar *et al.* (2013) demonstrated that curcumin-SLNs (C-SLNs) improved the action of AChE vis-à-vis free drug. Further, the concentration of curcumin in the brain was two-fold as compared to the free drug upon oral administration in an animal model cerebral ischemia. Another study by our group was conducted to confirm the delivery of C-SLNs in an aluminum chloride induced model of AD in mice. C-SLNs showed significantly better results (97.46 and 73% recovery in LPO and AChE) at a dose of 50 mg/kg ($p < 0.001$), which were comparable to those achieved with rivastigmine (standard drug for AD). Histopathology of the brain sections of C-SLNs treated groups further confirmed the potential of C-SLNs for AD (Kakkar and Kaur, 2011).

In another study, Sood *et al.* (2013) developed curcumin/donepezil-loaded NLCs for intranasal delivery to the brain. A higher brain drug concentration using intranasal delivery was confirmed vis-à-vis intravenous administration. *In vivo* study in mouse model illustrated the improvement in the memory and learning processes in comparison free drug group. On the other hand, the intensity of acetylcholine was enhanced and oxidation damage decreased in the groups treated with the NLCs.

Dhawan *et al.* (2011) reported systematic development of SLNs of a plant bioactive, quercetin, for its improved CNS delivery as potential treatment for Alzheimer's disease. Quercetin SLNs showed markedly superior memory retention in rats with aluminium-induced dementia vis-à-vis pure quercetin during behavioral studies ($p < 0.0001$). Highly significant reversal of aluminum-induced neurotoxicity was achieved employing SLNs ($p < 0.00001$), along with maintenance of lipid peroxidation, glutathione, and nitrite levels in brain homogenates, corroborating successful targeting of quercetin-SLNs to the CNS.

Figueiró *et al.* (2011) developed NLCs loaded with resveratrol to enhance their cerebral bioavailability. Authors showed approximately three times, seven times, and three times higher drug concentrations in the brain, liver, and kidneys in mice treated with the NLCs vis-à-vis free resveratrol.

Zhuang *et al.* (2010) showed that vinpocetine-loaded NLCs illustrated a two-fold increase, three-fold increase, and 0.35-fold lowering in the maximum concentration, maximum time, and elimination constant in plasma, respectively, in comparison with the vinpocetine suspension post-oral administration in rats. Moreover, the drug bioavailability was augmented to 322% in rats after oral administration of NLCs in comparison to free drug, thus highlighting the significance of incorporating the poorly soluble drugs (Zhuang *et al.*, 2010).

Sancini *et al.* (2013) formulated the transactivating transcriptional activator (TAT) functionalized curcumin derivative nanoliposomes (NL) to enhance their permeability across the BBB. Further, they displayed a high affinity for Aβ peptide, thus proving it to be a promising tool for delivery of drug and contrast agents to AD brain.

Mourtas and coworkers (2011, 2014) reported high affinity of curcumin analog- and curcumin-loaded liposomes for senile plaques, when evaluated on postmortem brain tissues of AD patients. Further, they showed that the tagging of curcumin-derivative liposomes with monoclonal antibodies (MAbs) extensively enhanced the intake by the cellular model of BBB. These findings confirmed the importance of such multifunctional liposomes for AD treatment.

Naik *et al.* (2006) evaluated the AO activity of *Ginkgo biloba* (*G. biloba*) phytosomes and its possible mechanism(s) of action by assaying AO enzymes in rats with sodium nitrite induced hypoxia as the oxidative stress.

Patel *et al.* (2013) reported the comparison between huperzine A-loaded SLNs (HupA-SLNs) and the NLCs. NLCs confirmed enhanced permeability via abdominal rat skin direct to the SLNs in *ex vivo* permeation studies. The *in vivo* studies revealed substantial development of cognition in scopolamine-induced amnesia of mice model post-treatment with the nanocarrier-based formulations vis-à-vis the control group. These nanocarriers were demonstrated as efficient promoters for transport of drug and being nonirritating across the skin of rat. They exhibited superior rate of permeation across skin in comparison to SLNs and NLCs. The sustained and controlled release characteristics of drug from nanocarriers specified the decrease in transfer latency time over the period of 3 days.

Zhang *et al.* (2014) demonstrated that administration of β-asarone-loaded MEs via intranasal route resulted in a ratio of AUC_{brain}/AUC_{plasma} considerably superior to intravenous administration. In another similar study, Chaiyana *et al.* (2013) developed ME loaded with anticholinesterase alkaloidal extract from *Tabernaemontana divaricata* (*T. divaricata*) exhibiting improved skin penetration and better formulation retention for 24 hours upon transdermal application. Furthermore, they examined the stability of the MEs and AChE activity of more than 80% by 180 days.

Shi *et al.* (2012) illustrated an inventive ME-based patch for concurrent transdermal delivery of huperzine A (HA) and ligustrazine phosphate (LP) for prevention of AD. The permeation studies demonstrated that the MEs improved the permeation rates of HA and LP vis-à-vis the control, and the penetration kinetics of the transdermal patch indicating it to be a zero-order process. Furthermore, the pharmacodynamic studies signified the role of combination therapy of HA and LP for treating amnesia in contrast to monotherapy. They demonstrated a dose-dependent antiamnesic effect when estimated in scopolamine-induced amnesia model in rats post-transdermal administration. The potential of liquid crystal (LC) of an alkaloidal extract from *T. divaricata* for enhancing the acetylcholine level has also been successfully reported (Chaiyana *et al.*, 2013).

Although ample literature have confirmed the effective transport of nanotailored nutraceuticals across the BBB, still there is a requirement to optimize this strategy in terms of efficiency and targetability. Safety and toxicity aspects of the NPs are important concerns that have to be taken sincerely, understood as well as resolved, before extensive clinical use of these formulations for the treatment of AD.

Thus, it can be deciphered that nanocoutured drug delivery systems hold immense prospective for the treatment and cure of AD, once the issues pertaining to its safety are carefully catered.

7.2 CONCLUSION AND FUTURE PERSPECTIVE

Role of phytochemicals of nutraceutical significance in Alzheimer's disease (AD) is well-supported with definitive evidence. Looking onto the drastic imbalance between the rising number of AD patients and the available drug therapies, it seems a breakthrough technological revolution is on its way. Pharmaceutical nanotailoring of these nutraceuticals is predicted to act as targeted delivery system for alleviating the pathological events during AD. Pleiotropic action of these nanotransformed phytochemicals decreases oxidative stress (OS) and neuroinflammation, possesses the ability to modulate the cholinergic system, and finally depletes the synthesis of cholesterol. Moreover, success in *in vivo* clinical translation is expected as a result of nanotechnology-based interventions. Future studies with nanostructured drug delivery of nutraceuticals per se, or in combination with available synthetic drugs, after having established safety and toxicity, can open newer avenues in the treatment of AD in the coming years.

REFERENCES

Akiyama, H., Barger, S., Barnum, S., Bradt, B., Bauer, J., Cole, G. M. *et al.* Inflammation and Alzheimer's disease. *Neurobiol Aging.* 2000; 21(3): 383–421.

Aloe, L., Rocco, M. L., Bianchi, P., and Manni, L. Nerve growth factor: From the early discoveries to the potential clinical use. *J Transl Med.* 2012; 10: 1–15.

Ansari, M. A., and Scheff, S. W. Oxidative stress in the progression of Alzheimer disease in the frontal cortex. *J Neuropathol Exp Neurol.* 2010; 69(2): 155–167.

Arrigada, P. A., Growdon, J. H., Hedley-White, E. T., and Hyman, B. T. Neurofibrillary tangles but not senile plaques parallel duration and severity of Alzheimer's disease. *Neurology* 1992; 42: 631–639.

BCC Research. Available online at: www.bccresearch.com/market-research/food-and-beverage (accessed on February, 2018).

Bozdağ Pehlivan, S. Nanotechnology-based drug delivery systems for targeting, imaging and diagnosis of neurodegenerative diseases. *Pharm Res.* 2013; 30: 2499–2511.

Chaiyana, W., Rades, T., Okonogi, S. Characterization and *in vitro* permeation study of microemulsions and liquid crystalline systems containing the anticholinesterase alkaloidal extract from *Tabernaemontana divaricata. Int J Pharm.* 2013; 452(1–2): 201–210.

Chiappelli, F., Navarro, A. M., Moradi, D. R., Manfrini, E., and Prolo, P. Evidence-based research in complementary and alternative medicine III: Treatment of patients with Alzheimer's disease. Evid based complement. *Alternat Med.* 2006; 3: 411–424.

Chow, A. H. L., Cheng, K. K., Chan, P. S., Fan, S., Kwan, S. M., Yeung, K. L. *et al.* Curcumin-conjugated magnetic nanoparticles for detecting amyloid plaques in Alzheimer's disease mice using magnetic resonance imaging (MRI). *Biomaterials.* 2015; 44: 155–172.

Coste, F., Bent-Ennakhil, N., Xie, L., Aigbogun, M. S., Wang, Y., Kariburyo, F. *et al.* A real-world analysis of treatment patterns for cholinesterase inhibitors and memantine among newly-diagnosed Alzheimer's disease patients. *Neurol Ther.* 2017; 6(1): 131–144.

Dal Prà, I., Armato, U., Chioffi, F., Pacchiana, R., Whitfield, J. F., Chakravarthy, B. *et al.* The Aβ peptides-activated calcium-sensing receptor stimulates the production and secretion of vascular endothelial growth factor-A by normoxic adult human cortical astrocytes. *Neuromolecular Med.* 2014; 16(4): 645–657.

Das, R. K., Kasoju, N., and Bora, U. Encapsulation of curcumin in alginate-chitosan-pluronic composite nanoparticles for delivery to cancer cells. *Nanomedicine.* 2010; 6: e153–e60.

De Rosa, G., Salzano, G., Caraglia, M., and Abbruzzese A. Nanotechnologies: A strategy to overcome blood–brain barrier. *Curr Drug Metab.* 2012; 13(1): 61–69.

Dhawan, S., Kapil, R., and Singh, B. Formulation development and systematic optimization of solid lipid nanoparticles of quercetin for improved brain delivery. *J Pharm Pharmac.* 2011; 63(3): 342–351.

Dheen, S. T., Kaur, C., Ling, E. A. Microglial activation and its implications in the brain diseases. *Med Chem.* 2007; 14: 1189–1197.

Dislich, B., and Lichtenthaler, S. F. The membrane-bound aspartyl protease BACE1: Molecular and functional properties in Alzheimer's disease and beyond. *Front Physiol.* 2012; 3: 8.

Doggui, S., Sahni, J. K., Arseneault, M., Dao, L., and Ramassamy, C. Neuronal uptake and neuroprotective effect of curcumin-loaded PLGA nanoparticles on the human SK-N-SH cell line. *J Alzheimers Dis.* 2012; 30: 377–392.

Endo, A. A historical perspective on the discovery of statins. *Proc Jpn Acad Ser B Phys Biol Sci.* 2010; 86: 484–493.

Figueiró, F., Bernardi, A., Frozza, R. L., Terroso, T., Zanotto-Filho, A., Jandrey, E. H. F. *et al.* Resveratrol-loaded lipid-core nanocapsules treatment reduces *in vitro* and *in vivo* glioma growth. *J Biomed Nanotechnol.* 2011; 9: 516–526.

Frank-Cannon, T. C., Alto, L. T., McAlpine, F. E., and Tansey, M. G. Does neuroinflammation fan the flame in neurodegenerative diseases? *Mol Neurodegener.* 2009; 4: 47.

Gangwar, R. K., Dhumale, V. A., Kumari, D., Nakate, U. T., Gosavi, S. W., and Sharma, R. B. Conjugation of curcumin with PVP capped gold nanoparticles for improving bioavailability. *Mat Sci Eng.* 2012; 32: 2659–2663.

Goedert, M. Tau protein and the neurofibrillary pathology of Alzheimer's disease. *Ann NY Acad Sci.* 1996; 777: 121–131.

Iqbal, K., Alejandra, A. D., Chen, S., Chohan, M. O., El-Akkad, E., Gong, C. *et al.* Tau pathology in Alzheimer disease and other tauopathies. *Biochim Biophys Acta.* 2005; 1739: 198–210.

Islam, S., Alam, S. B., Ferdousy, R., and Chowdhury, E. H. Analysis of morphological brain change of Alzheimer disease (AD) patients. *Appl Phys Res.* 2010; 2(2): 148–155.

Kakkar, V., and Kaur, I. P. Evaluating potential of curcumin loaded solid lipid nanoparticles in aluminium induced behavioural, biochemical and histopathological alterations in mice brain. *Food Chem Toxicol.* 2011; 49(11): 2906–2913.

Kakkar, V., Modgill, N., Kumar, M. From Nutraceuticals to nanoceuticals. *Nanoscience in Food and Agriculture 3, Sustainable Agriculture Reviews.* 2016; 23: 183–198.

Kakkar, V., Muppu, S. K., Chopra, K., and Kaur, I. P. Curcumin loaded solid lipid nanoparticles: An efficient formulation approach for cerebral ischemic reperfusion injury in rats. *Eur J Pharm Biopharm.* 2013; 85: 339–345.

Kalra, E. K. Nutraceutical-definition and introduction. *AAPS PharmSci.* 2003; 5: 27–28.

Kapogiannis, D., Mullins, R. J., Diehl, T. C., and Chia, C. W. Insulin resistance as a link between amyloid-beta and tau pathologies in Alzheimer's disease. *Front Aging Neurosci.* 2017; 9: 118.

Karwande, V., and Borade, R. (2015). Phytochemicals of nutraceutical importance. Vol. 4, Nutraceuticals. *In*: A M Grumezescu, *ed.*, Nanotechnology in the Agri-Food Industry Scitus Academics LLC, USA.

Kawas, C., Gray, S., Brookmeyer, R., Fozard, J., and Zonderman, A. Age-specific incidence rates of Alzheimer's disease: The Baltimore Longitudinal Study of Aging. *Neurology.* 2000; 54: 2072–2077.

Kim, G. H., Kim, J. E., Rhie, S. J., and Yoon, S. The role of oxidative stress in neurodegenerative diseases. *Exp Neurobiol.* 2015; 24: 325–340.

Kim, S. Y., Venkatesan, R., and Ji, E. Phytochemicals that regulate neurodegenerative disease by targeting neurotrophins: A comprehensive review. *Biomed Res Int.* 2015; 1–22.

Kotyla, P. The role of 3-hydroxy-3-methylglutaryl coenzyme a reductase inhibitors (statins) in modern rheumatology. *The Adv Musculoskelet Dis.* 2010; 2: 257–269.

Kurz, A., and Perneczky, R. Novel insights for the treatment of Alzheimer's disease. *Prog Neuropsychopharmacol Biol Psychiatry.* 2011; 35(2): 373–379.

Lagace, T. A. PCSK9 and LDLR degradation: Regulatory mechanisms in circulation and in cells. *Curr Opin Lipidol.* 2014; 25: 387–393.

Lakatos, A., Tyzack, G., and Patani, R. Human stem cell-derived astrocytes: Specification and relevance for neurological disorders. *Curr Stem Cell Rep.* 2016; 2(3): 236–247.

Lemere, C. A. Developing novel immunogens for a safe and effective Alzheimer's disease vaccine. *Prog Brain Res.* 2009; 175: 83.

Lien, Y. C., Feinstein, S. I., Dodia, C., and Fisher, A. B. The roles of peroxidase and phospholipase A(2) activities of peroxiredoxin 6 in protecting pulmonary microvascular endothelial cells against peroxidative stress. *Antioxid Redox Signal.* 2012; 16: 440–451.

Lim, H. J., Shim, S. B., Jee, S. W., Lee, S. H., Lim, C. J., Hong, J. T. *et al.* Green tea catechin leads to global improvement among Alzheimer's disease-related phenotypes in NSE/hAPP-C105 Tg mice. *J Nutr Biochem.* 2013; 24: 1302–1313.

Mazzarino, L., Travelet, C., Ortega-Murillo, S., Otsuka, I., Pignot-Paintrand, I., and Lemos-Senna, E. Elaboration of chitosan-coated nanoparticles loaded with curcumin for mucoadhesive applications. *J Colloid Interface Sci.* 2012; 370: 58–66.

Mirshafiey, A., Azizi, G., Navabi, S. S., Al-Shukaili, A., Seyedzadeh, M. H., and Yazdani, R. The role of inflammatory mediators in the pathogenesis of Alzheimer's disease. *Sultan Qaboos Univ Med J.* 2015; 15(3): e305–e316.

Mourtas, S., Canovi, M., Zona, C., Aurilia, D., Niarakis, A., La Ferla, B. *et al.* Curcumin-decorated nanoliposomes with very high affinity for amyloid-β1-42 peptide. *Biomaterials.* 2011; 32: 1635–1645.

Mourtas, S., Lazar, A. N., Markoutsa, E., Duyckaerts, C., and Antimisiaris, S. G. Multifunctional nanoliposomes with curcumin–lipid derivative and brain targeting functionality with potential applications for Alzheimer disease. *Eur J Med Chem.* 2014; 80: 175–183.

Mulik, R. S., Mönkkönen, J., Juvonen, R. O., Mahadik, K. R., and Paradkar, A. R. ApoE3 mediated polymeric nanoparticles containing curcumin: Apoptosis induced *in vitro* anticancer activity against neuroblastoma cells. *Int J Pharm.* 2012; 437: 29–41.

Murphy, S. L., Xu, J. Q., and Kochanek, K. D. Deaths: Final data for 2014. National Vital Statistics Reports. Hyattsville, MD. *Nat Cent Health Stat.* 2016; 61(4): 1–121.

Nagahara, A. H., and Tuszynski, M. H. Potential therapeutic uses of BDNF in neurological and psychiatric disorders. *Nat Rev Drug Discov.* 2011; 10: 209–219.

Naik, S. R., Pilgaonkar, V. W., and Panda, V. S. Evaluation of antioxidant activity of Ginkgo biloba phytosomes in rat brain. *Phytother Res.* 2006; 20: 1013–1016.

Neugroschl, J., and Wang, S. Alzheimer's disease: Diagnosis and treatment across the spectrum of disease severity. *Mt Sinai J Med.* 2011; 78(4): 596–612.

Nita, M., and Grzybowski, A. The role of the reactive oxygen species and oxidative stress in the pathomechanism of the age-related ocular diseases and other pathologies of the anterior and posterior eye segments in adults. *Oxi Med Cell Longe.* 2016; 2016: 3164734.

O'Brien, R. J., and Wong, P. C. Amyloid precursor protein processing and Alzheimer's disease. *Annu Rev Neurosci.* 2011; 34: 185–204.

Omidfar, K., Khorsand, F., and Darziani Azizi, M. New analytical applications of gold nanoparticles as label in antibody based sensors. *Biosens Bioelectron.* 2013; 43: 336–347.

Orihuela-Campos, R. C., Tamaki, N., Mukai, R., Fukui, M., Miki, K., and Terao, J. Biological impacts of resveratrol, quercetin, and N-acetylcysteine on oxidative stress in human gingival fibroblasts. *J Clin Biochem Nutr.* 2015; 56: 220–227.

Padurariu, M., Ciobica, A., Hritcu, L., Stoica, B., Bild, W., and Stefanescu, C. Changes of some oxidative stress markers in the serum of patients with mild cognitive impairment and Alzheimer's disease. *Neurosci Lett.* 2010; 469(1): 6–10.

Pan, M., Laia, C., and Ho, C. Anti-inflammatory activity of natural dietary flavonoids. *Food Funct.* 2010; 1, 15–31.

Pandey, M., Verma, R. K., and Saraf, S. A. Nutraceuticals: New era of medicine and health. *Asian J Pharm Clin Res.* 2010; 3(1): 11–15.

Panza, F., Frisardi, V., Imbimbo, B. P., Capurso, C., Logroscino, G., Sancarlo, D., Seripa, D., Vendemiale, G., Pilotto, A., and Solfrizzi, V. 2010. γ-Secretase inhibitors for the treatment of Alzheimer's disease: The current state. *CNS Neuro Ther.* 2010; 16: 272–284.

Parsons, C. G., Stöffler, A., and Danysz, W. Memantine: A NMDA receptor antagonist that improves memory by restoration of homeostasis in the glutamatergic system - too little activation is bad, too much is even worse. *Neuropharmacol.* 2007; 53: 699–723.

Patel, P. A., Patil, S. C., Kalaria, D. R., Kalia, Y. N., Patravale, V. B. Comparative *in vitro* and *in vivo* evaluation of lipid based nanocarriers of Huperzine A. *Int J Pharm.* 2013; 446(1–2): 16–23.

Pietronigro, E. C., Della Bianca, V., Zenaro, E., and Constantin, G. NETosis in Alzheimer's disease. *Front Immunol.* 2017; 8: 211.

Pooler, A. M., Polydoro, M., Wegmann, S., Nicholls, S. B., Spires-Jones, T. L., and Hyman, B. T. Propagation of tau pathology in Alzheimer's disease: Identification of novel therapeutic targets. *Alzheimer's Res Ther.* 2013; 5: 49.

Radak, Z., Zhao, Z., Koltai, E., Ohno, H., and Atalay, M. Oxygen consumption and usage during physical exercise: The balance between oxidative stress and ROS-dependent adaptive signaling. *Antioxid Redox Signal.* 2013; 18: 1208–1246.

Rammes, G. Neramexane: A moderate-affinity NMDA receptor channel blocker: New prospects and indications. *Exp Rev Clin Pharmacol.* 2009; 2: 231–238.

Reid, P. C., Urano, Y., Kodama, T., and Hamakubo, T. Alzheimer's disease: Cholesterol, membrane rafts, isoprenoids and statins. *J Cell Mol Med.* 2007; 11(3): 383–392.

Reis, S., Neves, A. R., and Queiroz, J. F. Brain-targeted delivery of resveratrol using solid lipid nanoparticles functionalized with apolipoprotein E. *J Nanobiotechnol.* 2016; 14: 1–11.

Rejinold, N. S., Sreerekha, P. R., Chennazhi, K. P., Nair, S. V., and Jayakumar, R. Biocompatible, biodegradable and thermo-sensitive chitosan-g-poly (N-isopropylacrylamide) nanocarrier for curcumin drug delivery. *Int J Bio Macromol.* 2011; 49: 161–172.

Revett, T. J., Baker, G. B., Jhamandas, J., and Kar, S. Glutamate system, amyloid β peptides and tau protein: Functional interrelationships and relevance to Alzheimer disease pathology. *J Psychiatry Neurosci.* 2013; 38: 6–23.

Rosenmann, H. Immunotherapy for targeting tau pathology in Alzheimers disease and tauopathies. *Curr Alzheimer Res.* 2013; 10(3): 217–228.

Roth, A. D., Ramı́rez, G., Alarco´n, R., and Von Bernhardi, R. Oligodendrocytes damage in Alzheimer's disease: Beta amyloid toxicity and inflammation. *Biol Res.* 2005; 38(4): 381–387.

Salomone, S., Caraci, F., Leggio, G. M., Fedotova, J., and Drago, F. New pharmacological strategies for treatment of Alzheimer's disease: Focus on disease modifying drugs. *Br J Clin Pharmacol.* 2012; 73(4): 504–517.

Sancini, G., Gregori, M., Salvati, E., Cambianica, I., Re, F., Ornaghi, F. *et al.* Functionalization with TAT-peptide enhances blood-brain barrier crossing *in vitro* of nanoliposomes carrying a curcumin-derivative to bind amyloid-b peptide. *J Nanomed Nanotechol.* 2013; 4: 3.

Seth, S., and Kakkar, V. Past and present therapeutics for Alzheimer's disease. *Int J Life Sci.* 2015; 4(3): 131–142.

Sharma, P., Jha, A. B., Dubey, R. S., and Pessarakli, M. Reactive oxygen species, oxidative damage, and anti-oxidative defense mechanism in plants under stressful conditions. *J Botany.* 2012; 2012: 26.

Shi, J., Cong, W., Wang, Y., Liu, Q., and Luo, G. Microemulsion-based patch for transdermal delivery of huperzine A and ligustrazine phosphate in treatment of Alzheimer's disease. *Drug Dev Ind Pharm.* 2012; 38: 752–761.

Son, S. M., Kang, S., Choi, H., Mook-Jung, I. Statins induce insulin-degrading enzyme secretion from astrocytes via an autophagy-based unconventional secretory pathway. *Mol Neurodeg.* 2015; 10: 56.

Sood, S., Jain, K., and Gowthamarajan, K. Curcumin-donepezil loaded nanostructured lipid carriers for intranasal delivery in an Alzheimer's disease model. *Alzheimers Dement.* 2013; 9: P299.

Stoothoff, W. H., and Johnson, G. V. W. Tau phosphorylation: Physiological and pathological consequences. *Biochim Biophys Acta.* 2005; 1739: 280–297.

Subasinghe, S., Unabia, S., Barrow, C. J., Mok, S. S., Aguilar, M. I., and Small, D. H. Cholesterol is necessary both for the toxic effect of Ab peptides on vascular smooth muscle cells and for Ab binding to vascular smooth muscle cell membranes. *J Neurochem.* 2003; 84(3): 471–479.

Sun, D., Li, N., Zhang, W., Zhao, Z., Mou, Z., Huang, D. *et al.* Design of PLGA-functionalized quercetin nanoparticles for potential use in Alzheimer's disease. *Colloids Surf B.* 2016; 148: 116–29.

The Dementia India Report, 2010. Available online at: http://ardsi.org/downloads/ExecutiveSummary.pdf. (Accessed on: February, 2018).

Tuppo, E. E., and Arias, H. R. The role of inflammation in Alzheimer's disease. *Int J Biochem Cell Biol.* 2005; 37: 289–305.

Yang, R., Zhang, S., Kong, D., Gao, X., Zhao, Y., and Wang, Z. Biodegradable polymer-curcumin conjugate micelles enhance the loading and delivery of low-potency curcumin. *Pharm Res.* 2012; 29: 3512–25.

Yusuf, M., Khan, M., Khan, R. A., and Ahmed, B. Preparation, characterization, *in vivo* and biochemical evaluation of brain targeted Piperine solid lipid nanoparticles in an experimentally induced Alzheimer's disease model. *J Drug Target.* 2013; 21: 300–11.

Zhao, Y., and Zhao, B. Oxidative Stress and the Pathogenesis of Alzheimer's disease. *Oxi Med Cell Longe.* 2013; 1–10.

Zhang, L. K., Xu, R. X., and Jiang, M. Evaluation of brain-targeting of β-asarone microemulsion by intranasal administration. *Chin Trad Herbal Drugs.* 2014; 45(1): 86–89.

Zheng, H., Jiang, M., and Trumbauer, M. E. Beta-amyloid precursor protein-deficient mice show reactive gliosis and decreased locomoter activity. *Cell.* 1995; 81: 525–31.

Zhuang, C. Y., Li, N., Wang, M., Zhang, X. N., Pan, W. S., Peng, J. J. *et al.* Preparation and characterization of vinpocetine loaded nanostructured lipid carriers (NLC) for improved oral bioavailability. *Int J Pharm.* 2010; 394: 179–85.

8 Nanoemulsions: A New Application in Nutraceutical and Food Industry

*Silki Chandel, Priyanka Jain, Saket Asati, and Vandana Soni**

CONTENTS

* Corresponding author.

8.1 INTRODUCTION

The nutraceuticals and food sector is a vital part of the global industry. Therefore, all the major industries are constantly looking for possible approaches to improve quality, safety, and production efficiency. They are using different approaches of advanced nanotechnology for production of good quality food as well as nutraceuticals with their functionalized properties. Nanoencapsulation of bioactives, food quality monitoring using biosensors (which are proficient for bacterial identification), active and smart food packaging systems, etc., are some instances where nanotechnology can be applied in the food industry.

Nutraceuticals can be described in a better way as the products originated from the food sources having additional health benefits with basic nutritional value. They are considered as unspecific biological medicines used to control symptoms and prevent malignant processes. Functional compounds are generally meant to provide health benefits to prevent or treat diseases (Chu *et al.*, 2007a,b). Numerous forms of tablets and capsules of functional compounds are already available in the market, but they may not be able to sustain the health benefits due to low bioavailability of lipophilic compounds (Chen *et al.*, 2006; Spernath and Aserin, 2006). The major lipophilic functional compounds are frequently required to be modified using various nanotechnological approaches and can be categorized as follows:

- Fatty acids
- Antioxidants
- Carotenoids
- Phytosterols

Presently, the main focus of scientists is to work with the systems that ensure the long kinetic stability of the nutraceuticals and foods for commercial application. Nanoemulsions can be used as carriers for nutraceuticals, drugs, flavors, antioxidant and antimicrobial agents, overcoming the problems associated with these compounds like their solubility, bioavailability etc. (McClements *et al.*, 2007a; Weiss *et al.*, 2008; Dasgupta *et al.*, 2016). Nanoemulsions are the preparations, where the bioactive compounds are in small droplets of nanometric (or submicron) size range. It consists of two immiscible phases, oil phase and aqueous phase, the oil phase being dispersed throughout the aqueous phase with the help of emulsifiers (Tadros *et al.*, 2004; McClements *et al.*, 2007a; Acosta, 2009).

Nanoemulsions can be classified into three categories:

- Oil in water (o/w) nanoemulsions—oil phase dispersed in aqueous phase
- Water in oil (w/o) nanoemulsions—aqueous phase dispersed in oil phase
- Bicontinuous nanoemulsions—interdispersion of both oil and water micro droplets within a system

Emulsions are considered a quite thermodynamically unstable system. The nanoemulsion is considerably different from conventional emulsions due to the difference in their droplet sizes as well as in their degradation rate, which may be due to the effect of Brownian motion by the globules (McClements, 2005). Therefore, nanoemulsions are highly stable against coalescence, flocculation, and gravitational separation, but still they are unstable due to Ostwald ripening (Dickinson, 1992).

Nanoemulsions possess many interesting physical properties, which are different from the conventional emulsions like relatively transparent and rheological responses to mechanical shear. They appear to be slightly turbid or transparent because they have relatively smaller globule size than the wavelength of light ($r << \lambda$).

8.1.1 Advantages of Nanoemulsion (Chen *et al.*, 2011)

- Enhance the absorption rate
- Eliminate absorption variability
- Help in solubilizing the lipophilic moieties
- Provide an aqueous dosage form for lipophilic drugs and food materials
- Increase bioavailability
- Are able to be delivered by various routes like topical, oral, and intravenous, in the form of nanoemulsion
- Mask the unpleasant taste of drugs and food molecules
- Protect the encapsulated compound from hydrolysis
- Are easy to fabricate, that is, require less amount of energy for preparation
- Can carry both lipophilic and hydrophilic drugs as well as food products

8.1.2 Limitations and Challenges of Nanoemulsion-Based Systems (Chen *et al.*, 2011)

Limitations

- These systems require large amounts of surfactant and co-surfactant for stability.
- Stability of nanoemulsions is influenced by the pH, temperature, etc.

Challenges

Nanoemulsions have enormous applicability as a carrier for various bioactive food ingredients and nutraceuticals, though there are some major challenges associated with the nanoemulsion system:

- Preparation of nanoemulsion requires some specialized instruments and different processing techniques, due to which, the cost of production increases. For instance,

arrangement of high-pressure homogenizer for nanoemulsions preparation is expensive. Moreover, micro-fluidization and ultrasonication are also expensive techniques employed for the preparation of nanoemulsions.
- Although nanoemulsions are stable for many years, Ostwald ripening decreases the stability of nanoemulsion; therefore, they are necessary to be prepared again prior to use.
- The exact mechanism of their formation is not clearly known.
- The interfacial chemistry involved in the production of nanoemulsions is also not clear.

8.1.3 Physicochemical Properties of Nanoemulsions

The physicochemical properties of nanoemulsion play an important role in the designing of stable food and nutraceuticals-based products. In this section, various physicochemical properties that influence the formulation and performance of nanoemulsion-based food and nutraceutical products are being discussed (McClements, 2011).

8.1.3.1 Optical Properties

Overall appearance of nanoemulsion-based products is an important criterion that needs to be considered while designing them. Usually, a colloidal dispersion has optical properties that are characterized by their opacity and color. The optical properties of nanoemulsions depend on various parameters such as size and size distribution, droplet concentration, as well as relative refractive index.

8.1.3.2 Rheological Properties

Nanoemulsions are considerably different from conventional emulsions in respect to their rheological properties as they have relatively smaller particle size. Certain types of nanoemulsion-based food products, containing oil droplets, may exhibit different rheological properties such as plastics, elastic and viscoelastic solids, and viscous and viscoelastic liquids, depending on their composition and structure.

8.1.3.3 Stability

The small droplet exhibits Brownian motion within nanoemulsions, which provide greater stability toward sedimentation (or creaming). Table 8.1 shows various properties and their effects on formulation of emulsion.

8.2 COMPONENTS OF NANOEMULSION

The major components that can be used to formulate nanoemulsions are discussed below.

8.2.1 Oil Phase

Various nonpolar components (i.e., constituting the oil phase) are used to formulate nanoemulsions. These include oil-soluble vitamins, flavor oils, mineral oils, essential oils, waxes, free fatty acids, triacylglycerols, diacylglycerols and monoacylglycerols. Other components such as co-enzyme Q, curcumin, carotenoids, phytosterols and phytostanols, etc., are used as the formulation ingredients for nanoemulsions. The physicochemical characteristics of the oil phase, interfacial tension, water-solubility, polarity, density, viscosity, phase behavior, refractive index, and chemical stability, should also be considered to formulate the nanoemulsion, as they may affect the performance and stability of the formulation (Anton *et al.*, 2007, 2009). Triacylglycerol, sunflower, soybean, olive, safflower, corn, flax seed, fish oils, and algae, are most frequently used in the food industry for the nanoemulsion preparation owing to their abundance, low cost, and nutritional or

TABLE 8.1
Bulk Physiochemical Properties of Phases—Oil and Water That Affect the Stability and Performance of Nanoemulsion

Property	Stability	Performance
Viscosity (η)	The rate of creaming decreases with increasing aqueous phase viscosity	Affect the product formation, texture, shelf life, and release pattern
Interfacial tension (γ)	Low γ leads to poor emulsifier affinity for droplet surface and promotes droplet coalescence	Affects product preparation and shelf life
Solubility	High water dispersion of an oil phase promotes Ostwald ripening	Affects product shelf life
Polarity	It is a measure for the partitioning of components between oil and aqueous phase	Affects release pattern
Density (ρ)	Creaming rate increases with increasing density constant	Affects product shelf life
Refractive index	–	Turbidity increases with increasing refractive index contrast

functional characteristics. Among the oils of short-chain triacylglycerols (SCT), medium-chain triacylglycerols (MCT), or long-chain triacylglycerols (LCT), use of the MCT and LCT oils for preparing nanoemulsions is considered difficult due to their high viscosity, low polarity, and high tension between the two immiscible layers vis-à-vis SCT. In high pressure homogenization as one of the methods for the preparation of nanoemulsion, the droplet disruption within the homogenizer is found to be difficult due to the high viscosity of MCT and LCT oils (Wooster *et al.*, 2008; Qian *et al.*, 2012; Raikos *et al.*, 2017). The preparation of nanoemulsion via phase-inversion temperature (PIT) method is also difficult due to the hydrophobic nature of these oils (Witthayapanyanon *et al.*, 2006). But once the nanoemulsions are formed successfully using MCT and LCT oils, they are physically more stable for prolonged periods of time than those prepared using SCT oils. At the same time, some edible oils can also be used for the preparation of nanoemulsions, but they are not effective for stabilizing the droplets owing to their low interfacial tension, which causes Ostwald ripening. Therefore, ripening inhibitors are incorporated in the formulation to improve the long-term stability.

8.2.2 Aqueous Phase

Nanoemulsions usually consist of water as the aqueous phase, but some other polar components such as carbohydrates, minerals, proteins, alcohols, polyols, acids, and bases are also used. The formation and type of nanoemulsions, and their physicochemical properties such as pH, polarity, refractive index, interfacial tension, density, ionic strength, rheology, phase behavior, and stability depend on the composition and type of aqueous phase. In case of high pressure homogenization methods, the cosolvents like alcohol or polyols are added into the aqueous phase to get the desired ratio of dispersed-to-continuous phase. In phase inversion, cosolvents may be used to improve the surfactants behavior to form nanoemulsions.

8.2.3 Stabilizers

In nanoemulsion formulation, addition of stabilizers plays a vital role. It is well known that oil and an aqueous phase are not easily homogenized together, because of their different physicochemical behavior. As a result, coalescence, droplet flocculation, gravitational separation, and Ostwald ripening types of instability issues tend to crop in. Stabilizers, therefore, are considered essential to enhance the long-term stability of nanoemulsions.

8.2.4 Emulsifiers/Co-Emulsifiers

An emulsifier or surface-active agent has the ability to reduce interfacial tension between two immiscible phases. Therefore, selection of a suitable emulsifier, either single or as a combination, will affect the formulation of a nanoemulsion. The stability of nanoemulsions toward change in environmental conditions such as ionic strength, pH, etc., also affects the selection of emulsifier. The emulsifiers and surfactants that are used in the food and nutraceutical industries include polysaccharides, proteins, and phospholipids. Selection of emulsifying agents depends on the method of preparation, that is, type of homogenization techniques. Small molecule surfactants, for instance, are suitable for low and high pressure homogenization techniques, whereas polysaccharides and proteins are usually considered unsuitable at low pressure. At the same time, proteins and polysaccharides constitute the area of interest in the production of nanoemulsions due to their natural availability. Classification of surfactants on the basis of their electrical properties is discussed below (McClements, 2005).

8.2.4.1 Ionic Surfactants

These surfactants are either negatively or positively charged, for example, CITREM, DATEM, SLS, and lauric. They are suitable for both low and high energy techniques. But when used in high concentrations, they tend to cause irritations, a property that limits their use in food products (Solè *et al.*, 2006a,b).

8.2.4.2 Nonionic Surfactants

Nonionic surfactants are commonly used for the preparation of nanoemulsions because of their nontoxicity and nonirritability, and also suitability for both low and high energy techniques. Examples of nonionic surfactants that are commonly used include ethoxylated sorbitan esters (e.g., Spans, Tweens), polyoxyethylene ether (POE) (e.g., Brij 97), and sugar ester (e.g., sorbitanmonooleate, sucrose monopalmitate) (Chiu, 2006; Jafari *et al.*, 2007; Shakeel *et al.*, 2007; Henry *et al.*, 2009).

8.2.4.3 Zwitter-Ionic Surfactants

Two oppositely charged groups, when present in the same surfactant molecule, are known as zwitter-ionic surfactants. The most common examples of zwitter-ionic surfactants are phospholipids with generally regarded as safe (GRAS) status that also allows their use in food and nutraceuticals, for example, lecithin. On the other hand, several natural phospholipids, when used alone, are not suitable either to form or stabilize the nanoemulsions, but their effectiveness increases with co-surfactants (Morais *et al.*, 2006; Hoeller *et al.*, 2009; Xue and Zhong, 2014). Co-surfactants are amphiphilic molecules that work with other surfactants by enhancing their properties. Co-surfactants optimizes the viscosity of the formulation by changing the continuous phase ratio, causing fluidization of the oil-water interface, reducing the electrical repulsion between head groups of ionic surfactants, and inducing appropriate interfacial curvature (Shafiq-un-Nabi *et al.*, 2007).

8.2.5 Texture Modifier

Texture modifier is the substance that thickens the aqueous phase and increases its viscosity (Imeson, 2010). Texture modifiers are commercially added into the emulsion type food products to hinder the droplet movement subsequently enhancing the stability of an emulsion. They are also used to improve smoothness, thickness, and appeal or aesthetic attraction of the texture. Most frequently used instances in food industries are polysaccharides such as starch, guar gum, xanthan, alginate, pectin, and proteins such as milk, eggs, and vegetable proteins.

8.2.6 Weighting Agent

Weighting agents are the substances generally used to match the density of one phase with the surrounding phase. They work either by coating or partial crystallization of the lipid core (McClements, 2005). The main reason behind the usage of these agents is to prevent the sedimentation or creaming by decreasing the driving force of gravitational separation. In food and beverage industries, the weighting agents are one of the important ingredients. Damar gum, ester gum, brominated vegetable oils, and sucrose-acetate iso-butyrate are some of the common examples of weighting agents used in food industries (McClements, 2005).

8.2.7 Ripening Inhibitor

Ostwald ripening is a phenomenon where large droplets are formed due to diffusion of small droplets dispersed in a continuous phase. Therefore, the concept of using a ripening inhibitor in the food and nutraceuticals industry is required for stabilizing certain types of nanoemulsions, which contain short-chain triacylglycerols, essential and flavor oils. Ripening inhibitors are highly hydrophobic in nature and can be encapsulated into oil droplets; they generate entropy of mixing and inhibit Ostwald ripening. Examples of inhibitors are mineral oil, ester gum, and long-chain triglyceride (Moghimi *et al.*, 2016; Zhang *et al.*, 2016).

8.3 METHODS OF PREPARATION

The main objective of the nanoemulsion preparation is to obtain the droplet size of 100 nm or lower for the delivery of nutraceuticals or functional food (Ravi *et al.*, 2011; Salvia-Trujillo *et al.*, 2016). To achieve that size, there is a need of either high energy or high concentration of surfactants, or both. As a result, mainly two methods, high and low energy methods used for nanoemulsion preparation, are described below (Anton *et al.*, 2009).

8.3.1 High Energy Emulsification Methods

In these methods, high speed homogenizers are used to disrupt the large droplets into very small oil droplets and then intermingling the oil and aqueous phases. Commonly, sonication, high pressure valve homogenizers and micro-fluidizers are used (Gutierrez *et al.*, 2008; Leong *et al.*, 2009; de Oca-Ávalos *et al.*, 2017). Nowadays, high energy emulsification methods are most commonly used to prepare nanoemulsions on a large-scale. Such methods are also suitable for a variety of starting materials, though they are quite costly and require an extremely high intensive source of energy.

High energy emulsification methods include ultrasonication and high-pressure homogenization methods, described as under:

8.3.1.1 Ultrasonication Emulsification

Ultrasonication emulsification method uses the ultrasonic waves emitted via probe to disintegrate the conventional emulsions by the formation of cavitation forces. The cavitation forces are generated from the tip of the probe sonicator inserted in the oil and water phases being homogenized (O'Sullivan *et al.*, 2017). By varying the intensity and the time of the ultrasonic waves, nanoemulsion of the desired properties can be prepared. This method is used in small-scale production of nanoemulsions.

Preparation of Nanoemulsions by High Amplitude Ultrasound – High amplitude ultrasound waves possess numerous advantages over high pressure homogenizers. By using high amplitude ultrasound waves, attainment of the highest quality nanoemulsion at commercial scale has become simpler. The advantages, such as low equipment cost, ease to use (i.e., have high-shear mixer to prepare a preliminary emulsion and eliminate separate rotor-stator), and ease to clean (less number of wetted parts), make them a viable alternative to high pressure homogenizer.

8.3.1.2 High Pressure Homogenization

This method is most commonly used for the fabrication of nanoemulsions at large industrial scale. It can be performed either by using micro-fluidizers or homogenizers.

8.3.1.2.1 Microfluidization

A micro-fluidizer, used in microfluidization technology, is a device comprising of a high pressure positive displacement pump, which can be operated at a pressure of up to 20,000 psi. The generated high pressure forces the product through a series of micro-sized channels present in an interaction chamber, wherein the product collides with high velocity to an impingement area, resulting in the formation of nanoemulsions, as shown in Figure 8.1. The process can be repeated many times to get the nanoemulsion of desired size. Microfluidization method can be used for fabrication of nanoemulsions at both laboratory and industrial scales (Anton *et al.*, 2008; Constantinides *et al.*, 2008; Bai *et al.*, 2016; García-Márquez *et al.*, 2017). With the help of micro-fluidizers, nutraceutical emulsions, flavor emulsions, and homogenized milk can be prepared at industrial scale.

8.3.1.2.2 High Pressure Homogenizer

High pressure homogenizers are most commonly used in the food industries for the production of both conventional emulsion and nanoemulsions. In this method, high pressure is applied with the help of homogenizer to the system containing oil phase, aqueous phase, and surfactant and co-surfactant. Due to high pressure, the macroemulsion, when passed through a small orifice, forms nanoemulsion, as shown in Figure 8.2. During this process, several forces such as intense turbulence, hydraulic shear, and cavitation forces works and results in tiny droplets. This method is also suitable for the preparation of nanoemulsions at laboratory as well as at industrial scale (Constantinides *et al.*, 2008; Ruiz-Montañez *et al.*, 2017; Xu *et al.*, 2017). Poor productivity and component degradation at high temperature due to the heat generated

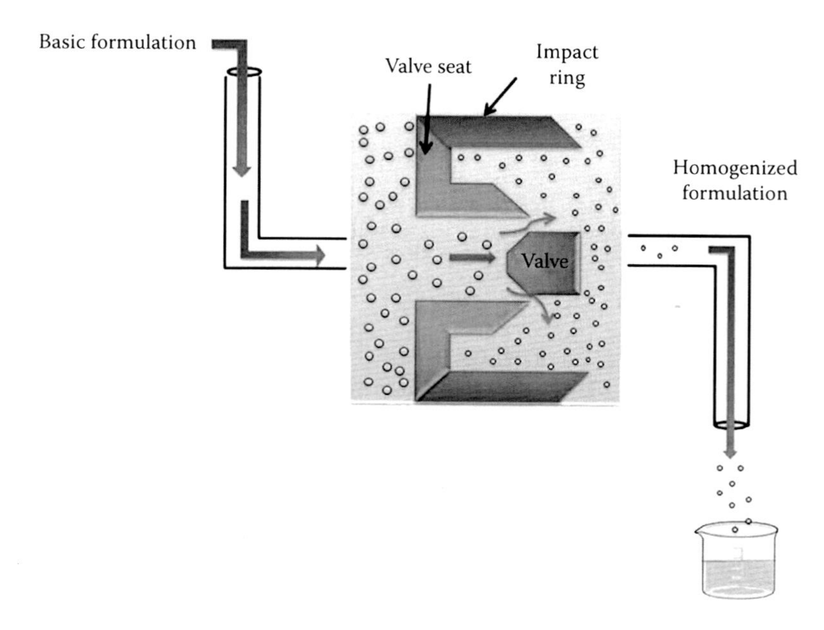

FIGURE 8.1 Micro-fluidization.

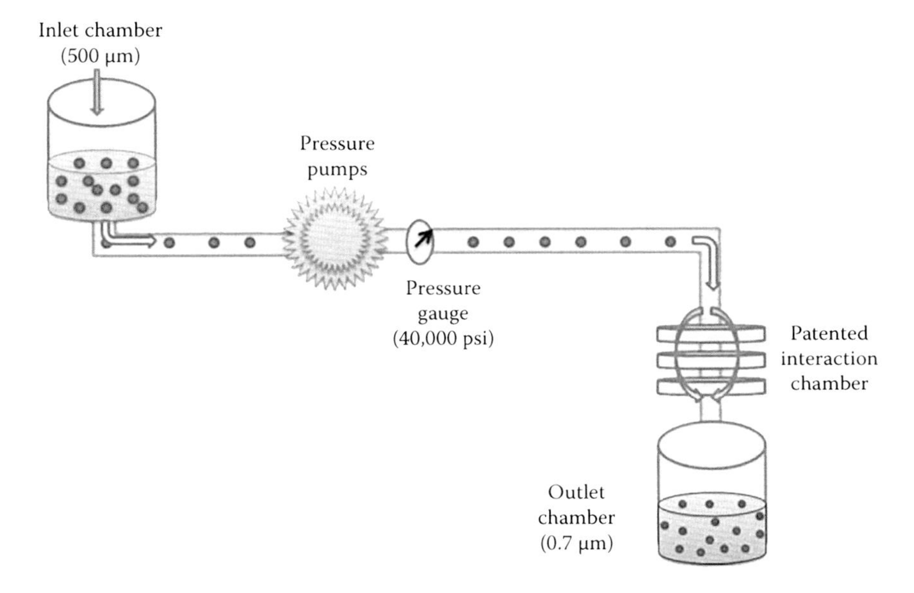

FIGURE 8.2 High pressure homogenization method.

during the process are the main problems associated with this method. Therefore, this method is not applicable for thermolabile drugs, such as proteins, nucleic acid, macromolecules, and enzymes. Also, the need of high intense energy by the sophisticated instruments increases the production cost of nanoemulsions.

8.3.2 Low Energy Emulsification Methods

As the name indicates, this method required the least amount of energy vis-à-vis the method discussed above for the preparation of nanoemulsions. This method totally depends on interfacial phenomenon or phase transitions, the inherent characteristic of the surfactants, co-surfactants/ co-emulsifiers and oil in order to yield emulsion in nanosized range. In this method, the internal chemical energy of the system is used for the emulsification. This method is classified on the basis of changes in the composition of the components or changes in the temperature (Morales *et al.*, 2003; Dasgupta *et al.*, 2016; Singh *et al.*, 2017).

Low energy emulsification based methods are discussed in the following section.

8.3.2.1 Phase Inversion Method

Structural changes of the surfactant is a prime factor responsible for the preparation of nanoemulsions via phase inversion method. On the basis of emulsification initiation factor, this method of phase inversion can be classified as (a) phase inversion temperature (PIT) method, when emulsification is triggered by a change in temperature, or (b) phase inversion composition (PIC) method when emulsification is triggered by a change in composition.

8.3.2.1.1 Phase Inversion Temperature (PIT) Method

PIT method was first described by Shinoda *et al.* (1968), wherein the nanoemulsions are formed by change in temperature. Changes in temperature lead to change in the physicochemical properties of the used nonionic surfactants, usually in the form of curvature of formed monolayer or molecular geometry or in solubility of surfactants. As a result of these inversions, type of emulsion gets converted from o/w to w/o, or vice versa. This can be easily understood by packing parameter, which describes the molecular geometry of a surfactant. It is denoted by p and is described by the formula as below (Israelachvili, 1992):

$$p = \frac{aT}{aH}$$

where, aT is cross-sectional area of the lipophilic tail-group and aH is the hydrophilic head-group.

Surfactants form monolayer at oil-water interface. The curvature of formed monolayer depends on the packing parameter of the surfactant and it decides the type of emulsion.

- When the value of p is less than one, the curvature is convex-shaped and it favors the formation of o/w emulsion.
- When value of p equals 1, it implies zero curvature, which leads to the formation of intermediate bi-continuous system.
- When value of p is greater than 1, then the curvature is concave-shaped and it favors the formation of w/o emulsion (Figure 8.3).

Effect of Temperature on the Solubility of the Surfactant – Relative solubility of surfactant in oil and water phases changes with temperature (Anton *et al.*, 2007, 2009). At low temperature, the head groups of nonionic surfactant are highly hydrated, and that results in the high solubility in water phase. As the temperature increases, the water of head group decreases gradually, causing a decrease in water solubility. At a certain temperature, the surfactant solubility becomes equal in both oil and water phases. This temperature dependent solubility of nonionic surfactant molecules is responsible for o/w or w/o nanoemulsion formation (Anton *et al.*, 2009).

On the basis of the above paragraph, the temperature at which o/w type emulsion changes to w/o emulsion type, or vice versa, is known as phase inversion temperature (PIT). At PIT less than 30°C, the surfactant has solubility in water and has less than one p value, thereby favoring the formation of o/w emulsions. But at high temperature, the water solubility of surfactant gradually decreases and the p value moves toward one. When the p value becomes equal to one the

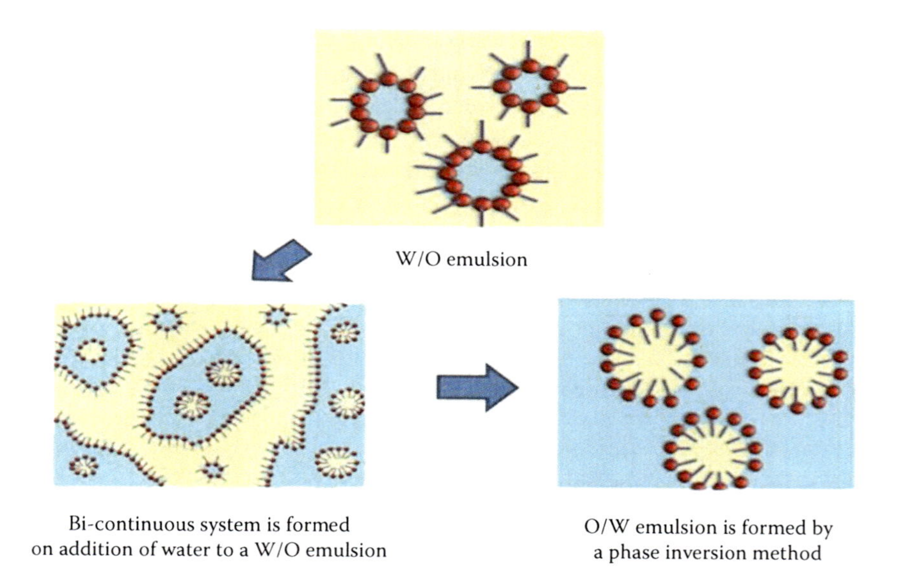

FIGURE 8.3 Phase inversion method.

emulsion starts breaking down due to low interfacial tension between oil and water phases, which further leads to the coalescence of droplets, followed by the formation of a bi-continuous system (Esquena *et al.*, 2003). At temperatures above PIT, the value of p is more than one, favoring the formation of a w/o emulsion.

Initially, PIT method was used for the preparation of o/w nanoemulsions. Now, it has also been modified and established for fabricating w/o nanoemulsions (Esquena *et al.*, 2003; Peng *et al.*, 2010). The PIT method is not suitable for the thermolabile substances such as peptides and tretinoin.

8.3.2.1.2 Phase Inversion Composition Method (PIC)

In PIC method, the molecular geometry of the surfactants is changed by modifying the contents of the formulations, instead of changing temperature as in PIT (Anton *et al.*, 2009). For instance, an o/w emulsion can be converted into w/o emulsion by the addition of salt. On addition of the salt, the packing parameter of the system changes from $p < 1$ to $p > 1$, because salt ions have the ability to change the charge on the polar group of the surfactant. On the other hand, a w/o emulsion with high concentration of salt can be changed into an o/w emulsion by increasing the quantity of water, because the dilutions reduce the ionic strength below its critical level.

8.3.2.2 Emulsion Inversion Point Method

In this method, one type of emulsion changes to another type (e.g., w/o type to o/w type or vice versa) by catastrophic phase inversion (Borrin *et al.*, 2016; Zhang *et al.*, 2017). Any changes in the ratio of the dispersed phase, either above or below to some critical ratio, lead the catastrophic-phase inversion. For instance, in a w/o emulsion (formed by using a surfactant), the aqueous phase is the dispersed phaseand oil phase is the continuous phase (Sajjadi, 2006). With the gradual addition of water in the system with continuous stirring, a critical stage is reached where water content exceeds than the oil phase, the rate of coalescence of water droplets exceeds the rate of coalescence of oil, which leads the phase inversion from a w/o type to o/w type system. Formulation variable like emulsifier concentration and process variables like rate of water addition and stirring speed affect the critical amount of water causing phase inversion. These variables also affect the size of the oil droplets formed (Thakur *et al.*, 2008). The small molecule surfactants that can stabilize the both w/o and o/w emulsions are used in this method for long time stability.

8.3.2.3 Spontaneous Emulsification Method

In this method, an emulsion is formed spontaneously during the mixing of the constituent phases together at a specified temperature. When the constituent phases are mixed together, the solvent molecules and/or surfactant molecules rapidly diffuse to the continuous phase from the dispersed phase without changing surfactant curvature. In the pharmaceutical industries, the drug delivery systems prepared by using this method are referred to as either self nano-emulsifying drug delivery systems or self-emulsifying drug-delivery systems. In this method, emulsions can be formed easily either by varying the compositions of the two phasesor by changing environmental parameters, for example, ionic strength, pH, temperature, or by changing the stirring speed as well as rate and order of mixing (Bouchemal *et al.*, 2004).

For instance, in one such method, an oil phase, which consists of hydrophobic oil, hydrophilic surfactant, and an organic solvent that is miscible to water, is added into the aqueous phase (Anton *et al.*, 2009), and, in another case, the aqueous phase is added to the oil phase consisting of the above mentioned ingredients. In most of the cases, ethanol or acetone is used as water-miscible organic solvent (Anton *et al.*, 2008). Today, various theories have been proposed for the formation of emulsions. According to these theories, two phases, oil and aqueous phase, are immiscible with each other. Out of the two phases, one phase should contain some other component that is partially miscible with both phases (e.g., an amphiphilic alcohol or surfactant). When both of the phases are mixed together, the partially miscible component will diffuse from its original phase to another

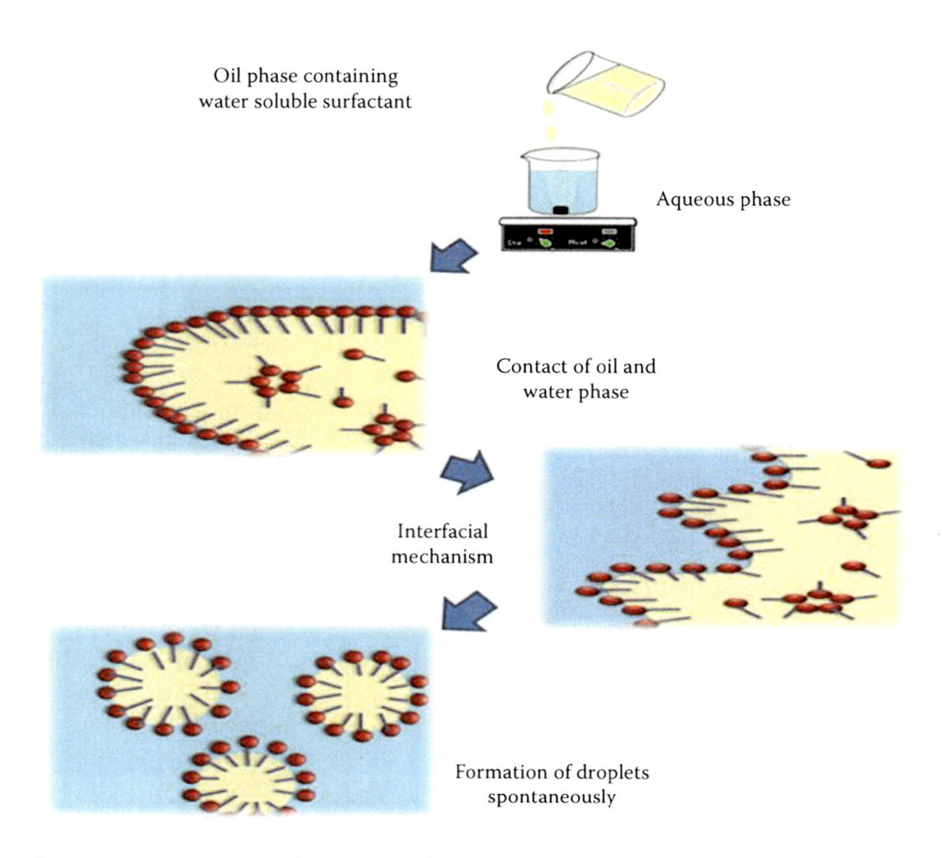

FIGURE 8.4 Spontaneous emulsification method.

phase that causes an increase in the interfacial area between oil and water, causing an increase in interfacial turbulence. The resultant increase in interfacial turbulence facilitates the spontaneous formation of droplets (Figure 8.4). In such cases, the droplet size can be controlled by varying both the formulation variables (i.e., compositions of the phases) and the process variables (i.e., ionic strength, pH, temperature, and stirring speed) (Bouchemal *et al.*, 2004). Overall, this method is an attractive and low-cost method. But the nanoemulsions prepared by this method are thermodynamically unstable and also require high concentration of surfactants.

8.4 SELF NANOEMULSIFYING DRUG DELIVERY SYSTEMS (SNEDDS)

The SNEDDS are prepared by the addition of the isotropic mixtures of oil, surfactant, and co-surfactant in the aqueous phase with gentle agitation. When these systems enter in the gastrointestinal tract, they convert into nanoemulsion by self-emulsification mechanism. The droplet size of the nanoemulsions formed by this mechanism is less than 100 nm and therefore, these systems are clear and transparent (Balakumar *et al.*, 2013). The successful formulation of these self nanoemulsifying drug systems depends on the physical, chemical, and biological properties of the ingredients used. The factors influencing the process of self nanoemulsification are (Date *et al.*, 2010)

- The physical and chemical propertiesof the components
- The ratio of the components
- The environmental conditions (i.e., pH and temperature of the aqueous phase)
- Physicochemical properties of the drug (i.e., polarity, solubility and partition coefficient, etc.)

8.5 FINISHING TECHNIQUES

Finishing techniques are referred as the possible changes made in droplet characteristics by using different methods after a nanoemulsion has been formed. The droplet size of the nanoemulsion is based on the physical and chemical characteristics of the formulation (food products and nutraceuticals), for example, release characteristics, rheological behavior, optical qualities, digestibility, and stability.

After the formation of the nanoemulsion, altering the emulsifier to change the properties of the interfacial layer (e.g., electrical charge, content, rheology, or thickness) enveloping the oil droplets may be beneficial for various applications. Single emulsifiers, for example, are mostly effective to produce small droplets (nano-size range) during the preparation of nanoemulsion, but it may not stabilize the droplets effectively for the long duration. Similarly, in case of PIT method, the nonionic surfactants are not effective to prevent the coalescence of the droplets at higher temperatures (Rao and McClements, 2010). Therefore, different approaches are required to change the properties of the interfacial layer. Different approaches for alteration of interfacial properties after nanoemulsion formation are discussed below.

8.5.1 Interfacial Displacement

In this process, the preformed nanoemulsion is mixed with an emulsifier solution, which displaces some or all of the emulsifier in a competitive manner at the oil water interface. The interfacial layer will depend completely on the surfactants available in the medium (Rao and McClements, 2010; Solans *et al.*, 2016).

8.5.2 Interfacial Deposition

Interfacial deposition is the phenomenon whereupon addition of a substance in nanoemulsions is adsorbed on the surface of the available emulsifier layer. For instance, when an emulsion is electrically charged due to the presence of emulsifier, the oppositely charged substances are attracted and deposited on the surfaces (Teo *et al.*, 2016).

8.5.3 Interfacial Cross-Linking

When the emulsifiers are adsorbed physically, chemically, or enzymatically on the oil-water interface, it is known as interfacial cross linking process. This process is used for the cross-linking of proteins and polysaccharides, which are covalently adsorbed on the interface layer (Littoz and McClements, 2008; Sandra *et al.*, 2008; Manchun *et al.*, 2014).

8.5.4 Solvent Displacement/Evaporation

In the solvent displacement method, an aqueous phase is first saturated with the partially miscible amphiphilic organic solvent. Then, water is added to the preformed emulsion, which results the transfer of organic solvent from the oil to the aqueous phase, and because of this reason the oil droplets are shrunk. After this, the organic solvent is evaporated at higher temperature. The droplet size of the nanoemulsion by this method depends on the water miscibility of the organic solvent, the amount of organic solvent, and the initial droplet size in the original emulsion (Tan *et al.*, 2016). For example, o/w emulsion was prepared that contained protein emulsifier and water as aqueous phase and ethyl acetate (an organic solvent) and corn oil as oil phase. The emulsion was diluted, by which the droplet size was minimized because of solvent displacement or evaporation of organic solvent at higher temperature. Recently, such approaches have been used for preparation of nanoemulsions containing protein-coated oil droplets. Nanoemulsions can also be prepared by using the various food-based emulsifiers including small molecule surfactants and whey proteins (Chu *et al.*, 2007a).

8.5.5 Lipid Phase Exchange

After the formation of o/w nanoemulsion, a mixture of oil phase containing the droplets may be altered by means of mixing with the microemulsion system. Oil phase has limited miscibility with the water phase; therefore, they will move between the lipid droplets by molecular diffusion (Pena and Miller, 2002; Capek, 2004). The surfactants may also be helpful in the formation of micelles by solubilizing the oil moieties (Weiss *et al.*, 2000). In lipid exchange method, these processes may be used for the preparation of nanoemulsion. Due to these compositional ripening effects, the movement of oil between droplets takes place until it reaches to the stable condition (Komaiko and McClements, 2016).

8.5.6 Lipid Crystallization

In this method, oil phase crystallization is performed by controlling the temperature and its constituents. The lipids are partially and fully crystallized after the formation of nanoemulsion. Crystallization of lipid phase of nanoemulsion has been accomplished by heating above its melting point. The temperature should be maintained during the preparation of nanoemulsion for crystallization of lipids. After the formulation of nanoemulsion, the lipid droplets are crystallized at cold temperature, which promotes their stability. The functional characteristics of nanoemulsion depend on the selection of oils. Therefore, the oils must be selected carefully, as they control the location, type, and concentration of the crystals (Kesisoglou *et al.*, 2007; Weiss *et al.*, 2007; Salminen *et al.*, 2016).

8.6 PACKAGING

The main targets of nutraceutical and food packaging are the protection and preservation of the products in a cost-effective manner, which satisfies the industry and consumer requirements, minimizes environmental impact, as well as provides information to the consumers regarding ingredients and nutritional values (Coles, 2003).

8.6.1 Protection/Preservation

Packaging should protect food and nutraceuticals from deterioration and maintain the advantageous effects of processing, nutritional values, extended shelf-life, and maintain or enhance the quality and safety of both. The packaging gives protection against chemical, biological, and physical factors that may be responsible for the deterioration. Different environmental factors (oxygen, humidity, radiations) can cause the chemical changes into the food product, and therefore chemical protection of the product is important. Various chemical barriers such as glass and metals are used for the protection of food products from the environmental factors.

The closure devices should contain such materials that allow minimum permeability. It is well-known that the plastic caps are quite permeable to various gases and vapors. Thus, the plastic closure used with gasket facilitates impermeable closing to allow sealing after filling. Plastic materials are also a good option for packaging, which provides a wide variety of barrier properties, but comparatively have higher permeability than metal and glass.

Microbes, insects, or rodents are the causative agents for the biological spoilage of food products; biological packaging is needed to develop the protection from spoilage. Additionally, this biological packaging also maintains the conditions to control ripening and aging (Galus and Kadzińska, 2015). Physical protection prevents food products from mechanical damage such as shock and vibrations, hence provide a cushioning effect during transportation.

8.6.2 Active and Intelligent Food Composite Materials (FCMs)

Active and intelligent food composite materials are useful for the enhancement of shelf life or sensory characteristics of the food during storage, because they are capable of secreting some nanoscale chemicals (such as flavors, antioxidants, or antimicrobial agents).

The active compounds are incorporated into food packaging materials, where these active compounds, known as food additives, are designed to release nanoparticulate chemicals into the food. Therefore, these materials should have required safety data to confirm their position in the list of food additives. Today, the NanoAg embedded baby bottle is an example of a FCM-modified active nanoparticulate material that is being used commercially (Alfadul and Elneshwy, 2010; Ntim and Noonan, 2017).

In spite of these composite materials, some nanosensors can also be mixed with packaging materials. These nanosensors have the capacity to detect environmental conditions, and chemical or microbial contaminates (Neethirajan and Jayas, 2011). This technology is based on the response to a stimulus, which is due to change in the environment. These advanced FCMs release chemical substances in response to any contamination, which can oppose or diminish the contamination in the food product (Pereira de Abreu *et al.*, 2012). These responses include specific pressure or pH or the presence of microbial products that act as the stimulator for the active release form the FCMs (Otles and Yalcin, 2008). Literature reports have shown that most of these agents are at development stages in the European countries.

8.6.2.1 Antimicrobial Nanopackaging (Active)

The antimicrobial nanopackaging is the new concept for the protection of food products from the microbial contamination. In this method, the packaging materials and FCMs are used in combination, and the latter tend to prolong shelf-life of the product, improve food quality and safety, and minimize the waste due to spoilage (Ozer *et al.*, 2014).

8.6.2.2 Time Temperature Logs (Intelligent)

These systems provide advantages such as traceability and biosecurity of the product. With the product having an expiration date, the recommended storage conditions are also mentioned on the product under definitive environmental conditions of moisture, temperature, presence of air, etc. As the storage conditions are not always the same during the storage period, these could affect the food product quality via premature degradation of the food product. The degradation may also be harmful to the consumer because of possible toxins or microbes. Some of the nanosensor substances incorporated in the packaging materials may act as specific indicator for the microbial contamination and also provide history about the effect of environmental factors. This advancement in packaging leads to elimination of need of expiration dates on some products and also gives more accurate data about the spoilage of food. The main aim of the manufacturers is to protect the product during transportation and storage, so that it may reach the consumer in a suitable condition. Nachay (2007) showed that some of the nanosensors are under the developmental stage and some of them have been commercialized. Due to the high cost and low uptake from the retailer and companies, such sensors are not much used in the European countries. As some chemicals are used in such nanosensors, some legislative restrictions may also be applicable (Pavelkova *et al.*, 2013).

8.7 POTENTIAL APPLICATION OF NANOEMULSION IN FOOD AND NUTRACEUTICALS

Applications of nanoemulsions in nutraceutical and food industries have received enormous attention due to increased consumers' requirements for healthier and safer food products. The demand is completed by using edible systems that will be able to encapsulate, protect, and release function compounds and also that are significant for food processing to maximize the nutrient content and will be ultimately beneficial for consumers. The applications of nanoemulsions in the nutraceutical and food industries are discussed below.

8.7.1 Nanoencapsulation of Probiotics

Probiotics are intended to deliver the live bacterial cells to the gut of humans, which is beneficial for their health. An example is Dr. Kim's probiotic nanofood combination with calcium, which reduces the risk of osteoporosis. Probiotics are also available in capsules, liquids, and powders. Powder forms should not be used at warmer places than room temperature, as the heat will kill the bacteria (Wani *et al.*, 2016). Probiotic-containing foods are the best way to get their beneficial effects by adding it into your diet, for example, kimchi (a Korean-fermented and salted vegetable dish), tempeh (a fermented soy product), kefir (a fermented milk drink), and yogurt (bacterial fermentation of milk).

8.7.2 Nanoencapsulation of Nutraceuticals

Nutraceuticals are the products obtained from food sources having additional health benefits along with basic nutrition provided to the consumer. Stability of the formulation is the main difficulty always associated with the production process at the nanoscale. Hence, it is desired to produce the formulation with improved stability (Katouzian and Jafari, 2016).

NutraLease, a well-established company, is performing research to enhance the action of functional compounds. Their beverages and food products bearing encapsulated functional (active) compounds like omega-3, β-carotene, lutein, lycopene, coenzyme Q10, phytosterols, isoflavones, and vitamins A, D, and E are already available on the market (NutraLease, 2011a). NutraLease's technology is based on self-assembled nanoemulsions, which are used to enhance the encapsulation rate and bioavailability of functional compounds in the living body (NutraLease, 2011b).

Certain other nutraceuticals such as curcumin, licorice, lycopene, β–carotene, etc. have also been encapsulated in nanoemulsions. Nanoemulsions may be a good choice to improve on nutraceuticals and nutritional value of processed foods by enhancing the water solubility, and absorption of nutrients and subsequently reducing their dose. Another example of nanoemulsion in the field of the food industry is "NovaSol" (beverages solutions) introduced by Aquanova. Aquanova affirms improved stability (in contexts of both temperature and pH) of entrapped functional materials and regulated additive amounts in these solutions.

NovaSol collection is categorized into two classes:

- Functional compounds: DL-α-tocopherol acetate, fat soluble vitamins, coenzyme Q10, and omega fatty acids possess some health benefits.
- Natural colorants: curcumin, chlorophyll, apocarotenal, sweet pepper extract, β-carotene, and lutein (AquaNova, 2017).

8.7.3 Nanoencapsulation of Flavors

Nanoemulsions are also used to encapsulate the flavors, thus providing the protection from the environmental conditions such as hydrolysis, enzymatic reactions, oxidation, and temperature. Also, they are thermodynamically stable at a wide range of pH. NutraLease's nanoemulsions protect the flavoring agents during the manufacturing processes and also during the beverage shelf-life (NutraLease, 2011c).

8.7.4 Nanoemulsions in Mayonnaise, Spreads, and Ice Creams

Nanoemulsions can also facilitate the use of lower fat content in food products without compromising their creaminess, which offers a healthier option to consumers. This is possible due to nanoemulsions, which have the small droplet size, providing unique textural characteristics for improved transparency and pleasant touch. Due to these two unique characteristics, there is wide applicability of nanoemulsion in food industries (Chaudhry *et al.*, 2008; Turner, 2016). Unilever has been using the nanoemulsion formulation for the preparation of healthy ice cream, without affecting its taste, with the aim to develop ice cream with minimum fat content, that is, by reducing fat level from 16% to 1%.

8.7.5 Other Applications of Nanoemulsion

A number of the food industries such as Nestlé and Unilever have also used nanotechnology in their food products. Nestlé has made w/o emulsion (10 to 500 nm), in which polysorbate or other micelle-forming substances were added to get easier and faster thawing cycle, thus helping in complete thawing of frozen food. Hence, Nestle got a patent on this approach for easier thawing (Möller *et al.*, 2009). Nanoemulsions also minimize the use of stabilizers in the product. During the formation of nanoemulsion, nanoemulsification reduces the size of the droplets and subsequently reducing the chances to break down and separate.

Some nanoemulsions, which contain soyabean oil, nonionic surfactant, and tributyl phosphate, are used as the antimicrobial substances and are active against a variety of microbes and pathogens. Other nanoemulsions have also been developed for the protection of food packaging equipment and food product from microbes (Sekhon, 2010).

8.8 BIOLOGICAL FATE AND TOXIC EFFECT OF NANOEMULSIONS

Application of nanoformulations in nutraceutical and food industries is gradually increasing; therefore, potential toxic effects associated with their small globule size became a major concern (Chaudhry *et al.*, 2008). Globule size of the nanoemulsion plays a critical role in the biological fate of the formulation. As the size of globule is reduced to nanosize range, it may affect its pharmacokinetic fate and toxicity. Such biological effects of nanoformulations also depend on their constituents and physicochemical characteristics, such as interfacial characteristics, particle size, concentration, and zeta potential (Hu *et al.*, 2009).

The composition of nanoemulsion includes different materials, minerals, surfactants, phospholipids, polysaccharides, and proteins. Due to the nature of these materials the GI tract shows different rate and extent of digestion affecting the biological fate of nanoemulsion. For instance, we know that dietary fibers may not be digested, while the proteins and phospholipids got digested in the stomach and small intestine with the help of protease and phospholipase enzymes. Therefore, there is the need of much more extensive research for understanding the effect of particle size and properties on the fate of nanoemulsion in the human body and to determine their adverse effect and toxicity (Bouwmeester *et al.*, 2009).

8.8.1 Biological Fate

Effect and toxicity of nanoemulsion in the human body can be understood by the proper study of physiological and physiochemical reactions, which occurs between the nanoemulsion and human body environment (Singh *et al.*, 2009).

8.8.1.1 Digestion

In the proceeding paragraph, a brief discussion on the digestion of nanoemulsions within the GI tract will be discussed. GI tract shows the complex reactions, hence, these reactions affect in a complex manner on the fate of nanoemulsion formulation (Singh *et al.*, 2009; van Aken, 2010b, Yao *et al.*, 2017).

Mouth: As the nanoemulsions are administered orally, first these come into the contact of the environment of the mouth cavity, which possesses saliva and several enzymes. This preexisting environment will change the ionic strength, pH, and temperature of the formulation (van Aken, 2010a). The nanoemulsion droplets are coated with mucin present in saliva and may adhere to each other to form aggregates.

Stomach: Nanoemulsion droplets enter into the stomach after crossing the esophagus and experience the extreme acidic environment (pH = 1 to 3). The nanoemulsion mixes with gastric enzymes and other substances such as proteins phospholipids, which may alter the fate of the droplets. Gastric enzymes perform different functions like proteases start protein digestion and lipases start lipid digestion. Hence, some of the proteins and lipids are digested in the stomach (Chen *et al.*, 2017).

Small Intestine: Partially digested bolus comes into contact with the higher pH medium of the small intestine, where it is mixed with the alkaline digestive enzymes. The digestion of surfactant in the nanoemulsion is affected by the surface active substances present in the small intestine. Various digestive enzymes (such as pancreatic lipase/co-lipase complex) may hydrolyze the components present within nanoemulsion droplet. These enzymes facilitate the conversion of diglycerides and triglycerides into free fatty acids and monoglycerides. In another case, proteins and phospholipids are digested by protease and phospholipase enzymes to peptides, amino acids, and to free fatty acids, respectively. Lipophilic or lipid components, like pharmaceuticals or nutraceuticals, encapsulated into the mixed micelles and vesicles, will also be moved into the enterocytes from the mucous layer, where they are absorbed. The behavior and properties of the nanoemulsions will be changed due to its exposure at various physiological conditions (McClements *et al.*, 2007b). The possible changes may be in the form of either changes in the composition of oil, water or surfactant or in aggregation state, size distribution, charges, interfacial characteristics, and physical state of the nanoemulsions. All these alterations are considered as the major factor for understanding the potential toxicity and fate of nanoemulsion formulation in the human body (Chen *et al.*, 2017).

8.8.1.2 Absorption

After the digestion, nanoemulsion and its digested products are absorbed by the GI tract (Acosta, 2009; Hu *et al.*, 2009). For their absorption, the two types of cells, epithelium and M-cell, present on the lining of GI tract are responsible (des Rieux *et al.*, 2006). Most of the lining of GI tract is made up of epithelium cells, but they are less effective for absorption of the globules. However, fewer numbers of M-cells are present on the epithelium lining (<1% of epithelium surface), but they are more efficient in absorbing the nanosize globules.

The fate of the nanoemulsions may be as follows:

- Nanoemulsions may be absorbed in the digested form primarily due to the presence of cellular enzymes.
- Nanoemulsions may directly be moved into the lymphatic or blood system via cells.
- They may be deposited at the particular site within the cells (Hu *et al.*, 2009; Sahay *et al.*, 2010a,b).

8.8.1.3 Metabolism, Distribution, and Excretion

These processes depend on the nature of the substances absorbed, as the nanoemulsion droplet or as the digested constituent parts of the formulation (Bouwmeester *et al.*, 2009). Some studies suggested that nanoparticles made up from indigestible materials (like inorganic or hydrocarbons) may be deposited in the different parts of the body for a different retention time, which depends on the size of the particles (Feliu *et al.*, 2016). Metallic nanoparticles (like gold) have smaller size and accumulate in the various organs when compared with the higher size particles. Therefore, researchers are investigating edible nanoformulations for direct absorption to delineate their fate in the human body.

8.8.2 Biological Fate of Bioactive or Functional Components Delivered through Nanoemulsion

Nanoemulsion delivery systems are commonly used to encapsulate lipophilic or hydrophobic constituents, like water-insoluble drugs, oil-soluble vitamins, phytosterols, ω-3 oils, and carotenoids (McClements *et al.*, 2007; Acosta, 2009). Therefore, there is acute need to understand the characteristics of nanoemulsions, which affect the absorption and release of the entrapped constituents for the globules. Hence, various physiological and physicochemical factors of the droplets in the formulation tend to significantly affect the bioavailability of poorly water-soluble components of nanoemulsion, following biological processes in the given manner (Acosta, 2009).

Transit Time: The rate and extent of enzymatic hydrolysis of digestible lipid depend on the transit time in GI tract. The absorption of the lipophilic materials also depends on it. Nanoemulsion formulation in the transit time of the particles can be compared to the conventional emulsion using various processes (Bouwmeester *et al.*, 2009):

- Nanoemulsion particles are smaller in size, which are able to cross the mucous layer of enterocytes, but large particles cannot.
- Nanoemulsion droplets with positive charge may also be entrapped in the mucous layer; hence, they are anionic in nature.

Digestion: Nanoemulsion contains lipid at the oil-water interfaces, the lipid digestion depends on the size of the droplets, as the smaller size droplets are more prone to digestion due to larger surface area available for the action of digestive enzymes (Lundin *et al.*, 2008). Therefore, the nanoemulsion formulations show faster release of entrapped components, when compared with the conventional formulations.

The interfacial layer lipid composition of the oil droplets may also affect the rate and extent of lipid digestion (Chu *et al.*, 2009). For example, if the interfacial layer on the droplet surface inhibits the absorption of digestive enzymes, the digestion may also be inhibited. Some lipid components are digested in the stomach and small intestine, so the entrapped components are easily released out from the nanoemulsion and vice versa. Molecular characteristics of the lipid components also affect the digestion like low molecular weight, and lipids having medium chain triglycerides are more susceptible to digest as compared with the higher molecular weight triglycerides (corn oil) (Porter *et al.*, 2007).

Solubilization: Thesolubilization of lipohilic contents of the nanoemulsion in the small intestine depends on presence of bile salts and surfactant (Porter *et al.*, 2007). The bile salts are released from the liver and reached to the small intestine bile duct. The bile salts increase the solubilization of the lipophilic components. Surfactant materials present in the formulation may also be helpful in the solubilization process.

Absorption: The absorption of the components affects the bioavailability of the formulation. The smaller size particles (< 500 nm) show better absorption due to various mechanisms that are given below (Acosta, 2009):

- As the particle size decreases, the solubility of lipidic components increases.
- The retention times of the smaller particles are also increased: the particles can enter the mucous layer and thus also increase the interaction with the lining of the GI tract.
- Smaller particles may show faster absorption from the mucous lining of the intestine by the paracellular or the passive or active transcellular pathway.

Metabolism: The metabolism and excretion of the entrapped materials highly depend on the direct absorption of the nanoemulsion droplets (Bouwmeester *et al.*, 2009).

8.8.3 Potential Toxicity of Nanoemulsions

Presently, there is limited information available regarding the potential toxicity of food based nano-emulsion. In spite of this, there are various physiological and physicochemical processes related to the particle size of the formulations that may cause the toxicity. But there is not any standardized protocol available for the determination of toxic effect of the nanoemulsions used for the nutraceutical and food products (Martins *et al.*, 2015).

8.8.3.1 Increased Bioavailability of Bioactive Components That Are Toxic at High Levels

As discussed before, as the particle size (100–1000 nm) of the droplets decrease, then the absorption of the entrapped content increases (Acosta, 2009). Ideally, bioactive components have higher bioavailability with no side effect in the human body when delivered through nanoemulsion. The increase in bioavailability should be considered particularly in those nutraceutical and food components that show toxic effects when used at large amount. Some active nutraceutical components may show toxic effects even at low bioavailability. The toxic effect of these formulations may be present due to their continuous consumption for a longer period of time.

8.8.3.2 Direct Absorption of Nanoemulsion Droplets

As described earlier, the indigestible component of nanoformulations, like inorganic and metallic components, can directly reach to the epithelium layer by transcellular or paracellular pathway. After the absorption, these formulations may be transported, metabolized, or excreted, or they may also be accumulated in some tissues. The physicochemical properties of the formulation affect these processes (Hu *et al.*, 2009). These properties include particle size, zeta potential, and nature of the various components in the formulation (Bouwmeester *et al.*, 2009).

8.8.3.3 Interference with Normal GI-Tract Function

The size of the droplets in the nanoformulations affects the normal functioning of the various parts of the GI tract (Chaudhry *et al.*, 2008). The smaller droplets provide high surface area and affect the activity of various digestive enzymes, which may alter the rate of digestion of nanoformulations. Some proteins are adsorbed by the surfaces of the nanoformulation, which changes the normal physiology of the body and leads to toxicity. The nanoformulations also affect the cell membrane structure by binding with the cellular receptors (Jiang *et al.*, 2009).

8.9 FUTURE TRENDS

In nutraceutical and food industries, applications of nanoemulsions are increasing progressively. It is also an interesting topic for various scientific as well as industrial communities due to their wide applicability, that is, encapsulation of bioactives, flavors, etc., masking taste and numerous benefits, such as enhanced bioavailability and improved physical stability over the conventional emulsion. On the other hand, there are certain limitations, which in turn become challenges for the researchers to overcome them to facilitate their better applicability in nutraceutical and food industries. The most challenging step for the researchers is to identify the appropriate food-based ingredients for the formulation of nanoemulsions, which can be used not only on the laboratory scale but can also be applicable for bulk production in the nutraceutical and food industries. Presently, the ingredients used to formulate nanoemulsions such as synthetic oils, organic solvents, synthetic polymers, or synthetic surfactants are considered unsuitable for large scale production in the industries. Ideally, the food industry would like to prepare nanoemulsions from most acceptable and widely used food product compositions such as triglyceride oils, various polysaccharides, proteins, and flavor oils.

Furthermore, there is an ardent need for an appropriate processing method, which can be economically and robustly used for the formulation of food grade products on the industrial scale. Recently, a lot of research has been undertaken to investigate the development of effective techniques for their scale-up at industrial level with special emphasis on the safety parameters for consumption of nanoparticlulate nutraceutical products.

8.10 CONCLUSIONS

A nanoemulsion system is the most encouraging system to enhance the solubility and functional bioavailability of the encapsulated lipophilic components. These have shown great application in the food and nutraceuticals industries, and therefore, these industries are seeking to utilize these systems for the encapsulation of lipophilic components. Various methods are available to prepare and characterize nanoemulsions, and some of them are more suitable than other methods. As nanoemulsions are non-equilibrated systems, their preparation requires the input of either surfactants or large amount of energy, or both. Recently, high energy methods such as ultrasonication and high-pressure homogenization methods have been employed for the purpose, which use various mechanical instruments. On the other hand, the low-energy methods use the internal chemical energy of the system for emulsification, each method having its distinct advantages and limitations. Most of these methods are in their development stage, these facilitate to prepare extremely fine particulate nanodispersions, allowing their continuous and controllable production. Nevertheless, the nanoemulsions in food industries pose several unaddressed challenges that are still needed to be resolved in terms of the process of production, production cost, product safety, and acceptance. On the other side, nanoemulsions with small size droplets are able to encapsulate large amount of nutraceuticals; therefore, they may enhance the bioavailability of such compounds. Hence, nanoemulsions may find extensive applications in food, beverages, and nutraceuticals industry to produce a transparent product with superior oral bioavailability.

REFERENCES

Acosta, E. Bioavailability of nanoparticles in nutrient and nutraceutical delivery. *Curr Opin Coll Inter Sci.* 2009; 14 (1): 3–15.

Alfadul, S. M., and Elneshwy, A. A. Use of nanotechnology in food processing, packaging and safety—Review. *Afr J Food Agr Nut Dev.* 2010; 10: 6.

Anton, N., Benoit, J. P., and Saulnier, P. Design and production of nanoparticles formulated from nanoemulsion templates—A review. *J Control Rel.* 2008; 128 (3): 185–199.

Anton, N., Gayet, P., Benoit, J. P., and Saulnier, P. Nano-emulsions and nanocapsules by the PIT method: An investigation on the role of the temperature cycling on the emulsion phase inversion. *Int J Pharm.* 2007; 344 (1): 44–52.

Anton, N., and Vandamme, T. The universality of low-energy nano-emulsification. *Int J Pharm.* 2009; 377 (1–2): 142–147.

AquaNova. Available online at: http://www. aquanova.de. (Accessed on 20 July, 2017).

Bai, L., and McClements, D. J. Development of microfluidization methods for efficient production of concentrated nanoemulsions: Comparison of single-and dual-channel microfluidizers. *J Colloid Interface Sci.* 2016; 466: 206–212.

Balakumar, K., Raghavan, C. V., and Abdu, S. Self nanoemulsifying drug delivery system (SNEDDS) of rosuvastatin calcium: Design, formulation, bioavailability and pharmacokinetic evaluation. *Colloids Surf B Biointerfaces.* 2013; 112: 337–343.

Borrin, T. R., Georges, E. L., Moraes, I. C., and Pinho, S. C. Curcumin-loaded nanoemulsions produced by the emulsion inversion point (EIP) method: An evaluation of process parameters and physico-chemical stability. *J Food Eng.* 2016; 169: 1–9.

Bouchemal, K., Briançon, S., Perrier, E., and Fessi, H. Nano-emulsion formulation using spontaneous emulsification: Solvent, oil and surfactant optimisation. *Int J Pharm.* 2004; 280 (1): 241–251.

Bouwmeester, H., Dekkers, S., Noordam, M. Y., Hagens, W. I., Bulder, A. S., de Heer, C. *et al.* Review of health safety aspects of nanotechnologies in food production. *Regul Toxic Pharmac.* 2009; 53 (1): 52–62.

Capek, I. Degradation of kinetically-stable O/W emulsions. *Adv Coll Inter Sci.* 2004; 107 (2–3): 125–155.

Chaudhry, Q., Scotter, M., Blackburn, J., Ross, B., Boxall, A., Castle, L., Aitken, R., and Watkins, R. Applications and implications of nanotechnologies for the food sector. *Food Add Contam.* 2008; 25 (3): 241–258.

Chen, F., Fan, G. Q., Zhang, Z., Zhang, R., Deng, Z. Y., and McClements, D. J. Encapsulation of omega-3 fatty acids in nanoemulsions and microgels: Impact of delivery system type and protein addition on gastrointestinal fate. *Food Res Int.* 2017 (*In press*).

Chen, H., Khemtong, C., Yang, X., Chang, X., and Gao, J. Nanonization strategies for poorly water-soluble drugs. *Drug Dis Today.* 2011; 16 (7): 354–360.

Chen, L., Remondetto, G. E., and Subirade, M. Food protein based materials as nutraceutical delivery systems. *Trends Food Sci Tech.* 2006; 17 (5): 272–283.

Chiu, W. The preparation and physical properties of jojoba oil nano-emulsion. Doctoral Dissertation, Department of Applied Chemistry, China. 2006.

Chu, B. S., Ichikawa, S., Kanafusa, S., and Nakajima, M. Preparation of protein-stabilized β-carotene nanodispersions by emulsification–evaporation method. *J Amer Oil Chemists Soc.* 2007a; 84 (11): 1053–1062.

Chu, B. S., Ichikawa, S., Kanafusa, S., and Nakajima, M. Preparation and characterization of beta-carotene nanodispersions prepared by solvent displacement technique. *J Agri Food Chem.* 2007b; 55 (16): 6754–6760.

Chu, B. S., Rich, G. T., Ridout, M. J., Faulks, R. M., Wickham, M. S. J., and Wilde, P. J. Modulating pancreatic lipase activity with galactolipids: Effects of emulsion interfacial composition. *Langmuir.* 2009; 25 (16): 9352–9360.

Coles, R., McDowell, D., and Kirwan, M. J. *Food packaging technology.* Blackwell Publishing, CRC Press, Oxford; 2003; p. 346.

Constantinides, P. P., Chaubal, M. V., and Shorr, R. Advances in lipid nanodispersions for parenteral drug delivery and targeting. *Adv Drug Del Rev.* 2008; 60 (6): 757–767.

Dasgupta, N., Ranjan, S., Mundra, S., Ramalingam, C., and Kumar, A. Fabrication of food grade vitamin E nanoemulsion by low energy approach, characterization and its application. *Int J Food Prop.* 2016; 19 (3): 700–708.

Date, A. A., Desai, N., Dixit, R., and Nagarsenker, M. Self-nanoemulsifying drug delivery systems: Formulation insights, applications and advances. *Nanomedicine.* 2010; 5 (10): 1595–1616.

deOca-Ávalos, J. M., Candal, R. J., and Herrera, M. L. Colloidal properties of sodium caseinate-stabilized nanoemulsions prepared by a combination of a high-energy homogenization and evaporative ripening methods. *Food Res Int.* 2017 (*In press*).

des Rieux, A., Fievez, V., Garinot, M., Schneider, Y. J., and Préat, V. Nanoparticles as potential oral delivery systems of proteins and vaccines: A mechanistic approach. *J Control Rel.* 2006; 116 (1): 1–27.

Dickinson, E. *Introduction to food colloids.* Royal Society of Chemistry, Cambridge, UK; 1992.

Esquena, J., Ravi, G. S. R., and Solans, C. Highly concentrated w/o emulsions prepared by the PIT method as templates for solid foams. *Langmuir.* 2003; 19 (7): 2983–2988.

Feliu, N., Docter, D., Heine, M., del Pino, P., Ashraf, S., Kolosnjaj-Tabi, J., Macchiarini, P., Nielsen, P., Alloyeau, D., Gazeau, F., and Stauber, R. H. *In vivo* degeneration and the fate of inorganic nanoparticles. *ChemSoc Rev.* 2016; 45 (9): 2440–2457.

Galus, S., and Kadzińska, J. Food applications of emulsion-based edible films and coatings. *Trends Food Sci Technol.* 2015; 45 (2): 273–283.

García-Márquez, E., Higuera-Ciapara, I., and Espinosa-Andrews, H. Design of fish oil-in-water nanoemulsion by microfluidization. *Innov Food SciEmerg Technol.* 2017; 40: 87–91.

Gutiérrez, J. M., González, C., Maestro, A., Sole, I., Pey, C. M., and Nolla, J. Nano-emulsions: New applications and optimization of their preparation. *Curr Opin Coll Inter Sci.* 2008; 13 (4): 245–251.

Henry, J., Fryer, P., Frith, W., and Norton, I. Emulsification mechanism and storage instabilities of hydrocarbon-in-water sub-micron emulsions stabilized with tweens (20 and 80), brij 96v and sucrose monoesters. *J Coll Inter Sci*. 2009; 338 (1): 201–206.

Hoeller, S., Sperger, A., and Valenta, C. Lecithin based nanoemulsions: A comparative study of the influence of nonionic surfactants and the cationic phytosphingosine on physicochemical behavior and skin permeation. *Int J Pharm*. 2009; 370 (1–2): 181–186.

Hu, L., Mao, Z. W., and Gao, C. Y. Colloidal particles for cellular uptake and delivery. *J Mat Chem*. 2009; 19 (20): 3108–3115.

Imeson, A. *Food stabilizers, thickeners and gelling agents*. John Wiley & Sons, Chichester, UK; 2010.

Israelachvili, J. *Intermolecular and surface forces*, Second Edition. Academic Press, London; 1992.

Jafari, S., He, Y., and Bhandari, B. Optimization of nano-emulsions production by microfluidization. *Eur Food Res Tech*. 2007; 225 (5): 733–741.

Jiang, J., Oberd¨orster, G., and Biswas, P. Characterization of size, surface charge, and agglomeration state of nanoparticle dispersions for toxicological studies. *J Nano Res*. 2009; 11 (1): 77–89.

Katouzian, I., and Jafari, S. M. Nano-encapsulation as a promising approach for targeted delivery and controlled release of vitamins. *Trends Food Sci Technol*. 2016; 53: 34–48.

Kesisoglou, F., Panmai, S., and Wu, Y. H. Application of nanoparticles in oral delivery of immediate release formulations. *Curr Nanosci*. 2007; 3 (2): 183–190.

Komaiko, J. S., and McClements, D. J. Formation of food-grade nanoemulsions using low-energy preparation methods: A review of available methods. *Compr Rev Food Sci Food Saf*. 2016; 15 (2): 331–352.

Leong, T., Wooster, T., Kentish, S., and Ashokkumar, M. Minimising oil droplet size using ultrasonic emulsification. *Ultra Sono*. 2009; 16 (6): 721–727.

Littoz, F., and McClements, D. J. Bio-mimetic approach to improving emulsion stability: Cross-linking adsorbed beet pectin layers using laccase. *Food Hydrocoll*. 2008; 22 (7): 1203–1211.

Lundin, L., Golding, M., and Wooster, T. J. Understanding food structure and function in developing food for appetite control. *Nutr Dietetics*. 2008; 65: S79–S85.

Manchun, S., Dass, C. R., and Sriamornsak, P. Designing nanoemulsion templates for fabrication of dextrin nanoparticles via emulsion cross-linking technique. *Carbohydr Polym*. 2014; 101: 650–655.

Martins, J. T., Ramos, O. L., Pinheiro, A. C., Bourbon, A. I., Silva, H. D., Rivera, M. C., Cerqueira, M. A., Pastrana, L., Malcata, F. X., González-Fernández, A., and Vicente, A. A. Edible bio-based nanostructures: Delivery, absorption and potential toxicity. *Food Eng Rev*. 2015; 7 (4): 491–513.

McClements, D. J., Decker, E. A., and Weiss, J. Emulsion-based delivery systems for lipophilic bioactive components. *J Food Sci*. 2007a; 72 (8): R109–R124.

McClements, D. J., Decker, E. A., and Park, Y. Physicochemical and structural aspects of lipid digestion. *In:* DJ McClements, ed., *Understanding and controlling the microstructure of complex foods*, CRC Press, Boca Raton, FL, 2007b; pp. 483–503.

McClements, D. J. Edible nanoemulsions: Fabrication, properties, and functional performance. *Soft Matter*. 2011; 7 (6): 2297–2316.

McClements, D. J. *Food emulsions: Principles, practice, and techniques*. CRC Press, Boca Raton, FL; 2005.

Moghimi, R., Ghaderi, L., Rafati, H., Aliahmadi, A., and McClements, D. J. Superior antibacterial activity of nanoemulsion of *Thymus daenensis* essential oil against E. coli. *Food Chem*. 2016; 194: 410–415.

Möller, M., Eberle, U., Hermann, A., Moch, K., and Stratmann, B. *Nanotechnology in the food sector*. Zürich: TA-SWISS; 2009.

Morais, J., David Henrique dos Santos, O., Delicato, T., AzziniGonçalves, R., and Alves da Rocha Filho, P. Physicochemical characterization of canola oil/water nanoemulsions obtained by determination of required HLB number and emulsion phase inversion methods. *J DisperSci Tech*. 2006; 27 (1): 109–115.

Morales, D., Gutiérrez, J. M., García-Celma, M. J., and Solans, Y. C. A study of the relation between bicontinuous microemulsions and oil/water nano-emulsion formation. *Langmuir*. 2003; 19 (18): 7196–7200.

Nachay, K. Analyzing nanotechnology. *Food Technol*. 2007; 1: 34–36.

Neethirajan, S., and Jayas, D. S. Nanotechnology for the food and bioprocessing industries. *Food Biopr Tech*. 2011; 4 (1): 39–47.

Ntim, S. A., and Noonan, G. O. Nanotechnology in food packaging. Vol. 2, Nanotechnologies in Food. *In:* Q. Chaudhry, L. Castle, R. Watkins, *eds.*, Nanoscience & Nanotechnology Series, Royal Society of Chemistry, UK, 2017; 118–142.

NutraLease (2011a). Available online at: http://www.nanotechproject.org/cpi/search-products/?title=vitamin (Accessed on 26 July, 2018).

NutraLease (2011b). Available online at: http://www.nanotechproject.org/cpi/products/marie-louise-vital-nanoemulsion/ (Accessed on 26 July, 2018).

NutraLease (2011c). Available online at: http://www.nanotechproject.org/cpi/products/soothing-moisturizing-lotion-nanoemulsion-10-9/. (Accessed on 26 July, 2018).

O'Sullivan, J. J., Park, M., Beevers, J., Greenwood, R. W., and Norton, I. T. Applications of ultrasound for the functional modification of proteins and nanoemulsion formation: A review. *Food Hydrocoll.* 2017; 71: 299–310.

Otles, S., and Yalcin, B. Intelligent food packaging. *LogForum.* 2008; 4 (4): 3.

Özer, E. A., Özcan, M., and Didin, M. Nanotechnology in food and agriculture industry. In *Food processing: Strategies for quality assessment.* Springer New York; 2014; pp. 477–497.

Pavelková, A. Time temperature indicators as devices intelligent packaging. *Acta Universitatis Agriculturaeet Silviculturae Mendelianae Brunensis.* 2013; 61 (1): 245–251.

Pena, A. A., and Miller, C. A. Transient behavior of polydisperse emulsions undergoing mass transfer. *Indus EngiChem Res.* 2002; 41 (25): 6284–6296.

Peng, L. C., Liu, C. H., Kwan, C. C., and Huang, K. F. Optimization of water-in-oil nanoemulsicns by mixed surfactants. *Colloid Surf A Physico Engi Asp.* 2010; 370 (1–3): 136–142.

Pereira de Abreu, D. A., Cruz, J. M., and Losada, P. P. Active and intelligent packaging for the food industry. *Food Rev Inter.* 2012; 28 (2): 146–187.

Porter, C. J., Trevaskis, N. L., and Charman, W. N. Lipids and lipid-based formulations: Optimizing the oral delivery of lipophilic drugs. *Nat Rev Drug Dis.* 2007; 6 (3): 231–248.

Qian, C., Decker, E. A., Xiao, H., and McClements, D. J. Nanoemulsion delivery systems: Influence of carrier oil on β-carotene bioaccessibility. *Food Chem.* 2012; 135 (3): 1440–1447.

Raikos, V., and Ranawana, V. Designing emulsion droplets of foods and beverages to enhance delivery of lipophilic bioactive components—A review of recent advances. *Int J Food Sci Tech.* 2017; 52 (1): 68–80.

Rao, J. J., and Mcclements, D. J. Stabilization of phase inversion temperature nanoemulsions by surfactant displacement. *J Agri Food Chem.* 2010; 58 (11): 7059–7066.

Ravi, T. P. U., and Padma, T. Nanoemulsions for drug delivery through different routes. *Res Biotech.* 2011; 2 (3): 1–13.

Ruiz-Montañez, G., Ragazzo-Sanchez, J. A., Picart-Palmade, L., Calderón-Santoyo, M., and Chevalier-Lucia, D. Optimization of nanoemulsions processed by high-pressure homogenization to protect a bioactive extract of jackfruit (*Artocarpusheterophyllus* Lam). *Innov Food SciEmerg Technol.* 2017; 40: 35–41.

Sahay, G., Alakhova, D. Y., and Kabanov, A. V. Endocytosis of nanomedicines. *J Control Rel.* 2010a; 145: 182–195.

Sahay, G., Kim, J. O., Kabanov, A. V., and Bronich T. K. The exploitation of differential endocytic pathways in normal and tumor cells in the selective targeting of nanoparticulate chemotherapeutic agents. *Biomaterials.* 2010b; 31 (5): 923–933.

Sajjadi, S. Nanoemulsion formation by phase inversion emulsification: On the nature of inversion. *Langmuir.* 2006, 22 (13): 5597–5603.

Salminen, H., Gömmel, C., Leuenberger, B. H., and Weiss, J. Influence of encapsulated functional lipids on crystal structure and chemical stability in solid lipid nanoparticles: Towards bioactive-based design of delivery systems. *Food Chem.* 2016; 190: 928–937.

Salvia-Trujillo, L., Martín-Belloso, O., and McClements, D. J. Excipient nanoemulsions for improving oral bioavailability of bioactives. *Nanomaterials.* 2016; 6 (1): 17.

Sandra, S., Decker, E. A., and McClements, D. J. Effect of interfacial protein cross-linking on the *in vitro* digestibility of emulsified corn oil by pancreatic lipase. *J Agri Food Chem.* 2008; 56 (16): 7488–7494.

Sekhon, B. Food nanotechnology—An overview. *Nanotech Sci Appli.* 2010; 3: 1–15.

Shafiq-un-Nabi, S., Shakeel, F., Talegaonkar, S., Ali, J., Baboota, S., Ahuja, A., Khar, R. K., and Ali, M. Formulation development and optimization using nanoemulsion technique: A technical note. *AAPS PharmSciTech.* 2007; 8 (2): E12-7.

Shakeel, F., Baboota, S., Ahuja, A., Ali, J., Aqil, M., and Shafiq, S. Nanoemulsions as vehicles for transdermal delivery of aceclofenac. *AAPS PharmSciTech,* 2007; 8 (4): 191–199.

Shinoda, K., and Saito, H. The effect of temperature on the phase equilibria and the types of dispersions of the ternary system H20-C6H12-nonionic agent. *J Coll Inter Sci.* 1968; 26: 70.

Singh, H., Ye, A., and Horne, D. Structuring food emulsions in the gastrointestinal tract to modify lipid digestion. *Prog Lipid Res.* 2009; 48 (2): 92–100.

Singh, Y., Meher, J. G., Raval, K., Khan, F. A., Chaurasia, M., Jain, N. K. *et al.* Nanoemulsion: Concepts, development and applications in drug delivery. *J Control Rel.* 2017. (*In press*).

Solans, C., Morales, D., and Homs, M. Spontaneous emulsification. *Curr Opin Coll Inter Sci.* 2016; 22: 88–93.

Solè, I., Maestro, A., González, C., Solans, C., and Gutiérrez, J. M. Optimization of nano-emulsion preparation by low-energy methods in an ionic surfactant system. *Langmuir.* 2006a; 22 (20): 8326–8332.

Solè, I., Maestro, A., Pey, C. M., González, C., Solans, C., and Gutiérrez, J. M. Nano-emulsions preparation by low energy methods in an ionic surfactant system. *Colloid Surf A Physic Engi Asp.* 2006b; 288 (1): 138–143.

Spernath, A., and Aserin, A. Microemulsions as carriers for drugs and nutraceuticals. *AdvColl Inter Sci.* 2006; 128–130: 47–64.

Tadros, T., Izquierdo, R., Esquena, J., and Solans, C. Formation and stability of nano-emulsions. *AdvColl Inter Sci.* 2004; 108–109: 303–318.

Tan, T. B., Yussof, N. S., Abas, F., Mirhosseini, H., Nehdi, I. A., and Tan, C. P. Forming a lutein nanodispersion via solvent displacement method: The effects of processing parameters and emulsifiers with different stabilizing mechanisms. *Food Chem.* 2016; 194: 416–423.

Teo, A., Dimartino, S., Lee, S. J., Goh, K. K., Wen, J., Oey, I. *et al.* Interfacial structures of whey protein isolate (WPI) and lactoferrin on hydrophobic surfaces in a model system monitored by quartz crystal microbalance with dissipation (QCM-D) and their formation on nanoemulsions. *Food Hydrocoll.* 2016; 56: 150–160.

Thakur, R. K., Villette, C., Aubry, J. M., and Delaplace, G. Dynamic emulsification and catastrophic phase inversion of lecithin-based emulsions. *Colloid Surf A PhysicoEngi Asp.* 2008; 315 (1–3): 285–293.

Turner, R. Natural sweetener. US 15/291710; 2016.

vanAken, G. A. Modelling texture perception by soft epithelial surfaces. *Soft Matter.* 2010a. 6 (5): 826–834.

vanAken, G. A. Relating food emulsion structure and composition to the way it is processed in the gastrointestinal tract and physiological responses: What are the opportunities? *Food Biophysics.* 2010b; 5 (4): 258–283.

Wani, T. A., Shah, A. G., Wani, S. M., Wani, I. A., Masoodi, F. A., Nissar, N. *et al.* Suitability of different food grade materials for the encapsulation of some functional foods well reported for their advantages and susceptibility. *Crit Rev Food SciNutr.* 2016; 56 (15): 2431–2454.

Weiss, J., Canceliere, C., and McClements, D. J. Mass transport phenomenain oil-in-water emulsions containing surfactant micelles: Ostwald ripening. *Langmuir.* 2000; 16 (17): 6833–6838.

Weiss, J., Decker, E. A., McClements, D. J., Kristbergsson, K., Helgason, T., and Awad, T. Solid lipid nanoparticles as delivery systems for bioactive food components. 2nd International Symposium on Delivery of Functionality in Complex Food Systems, Amherst, MA. 2007.

Weiss, J., Decker, E. A., McClements, D. J., Kristbergsson, K., Helgason, T., and Awad, T. Solid lipid nanoparticles as delivery systems for bioactive food components. *Food Biophysics.* 2008 1; 3 (2): 146–154.

Witthayapanyanon, A., Acosta, E., Harwell, J., and Sabatini, D. Formulation of ultralow interfacial tension systems using extended surfactants. *J Surf Deter.* 2006; 9 (4): 331–339.

Wooster, T. J., Golding, M., and Sanguansri, P. Impact of oil type on nanoemulsion formation and Ostwald ripening stability. *Langmuir.* 2008; 24 (22): 12758–12765.

Xu, J., Mukherjee, D., and Chang, S. K. Physicochemical properties and storage stability of soybean protein nanoemulsions prepared by ultra-high pressure homogenization. *Food Chem.* 2017 (*In press*).

Xue, J., and Zhong, Q. Thyme oil nanoemulsionscoemulsified by sodium caseinate and lecithin. *J Agric Food Chem.* 2014; 62 (40): 9900–9907.

Yao, M., McClements, D. J., Zhao, F., Craig, R. W., and Xiao, H. Controlling the gastrointestinal fate of nutraceutical and pharmaceutical-enriched lipid nanoparticles: From mixed micelles to chylomicrons. *Nano Impact.* 2017; 5: 13–21.

Zhang, J., Yuan, H., Zhao, J., and Mei, N. Viscosity estimation and component identification for an oil-water emulsion with the inversion method. *Appl Therm Eng.* 2017; 111: 759–767.

Zhang, J., Bing, L., and Reineccius, G. A. Comparison of modified starch and Quillajasaponins in the formation and stabilization of flavor nanoemulsions. *Food Chem.* 2016; 192: 53–59.

9 Nanoprobiotics: Progress and Issues

Kamla Pathak and Nida Akhtar*

CONTENTS

9.1 INTRODUCTION

Nutraceutical has gained its origin from a blend of two words, *nutrition* and *pharmaceutical*. The terminology conceptually has gained its origin from the United Kingdom, Germany, and France. The meaning of nutraceuticals was transformed by Canada Health Ministry, which describes nutraceuticals as food products having physiological benefits to the host. The Britain Ministry of Agriculture, Fisheries and Food, describes functional food as food comprised of component(s) that provide physiological or specific medical benefits, rather than mere nutritional benefits. However, there exists little difference between the nutraceuticals and functional foods. When the food is developed by applying the scientific principles, with or without prior information on how/why it is being employed, it is known to be a functional food. The functional foods impart the body with optimum amounts of fats, vitamins, carbohydrates, proteins, etc. However, when such functional foods help in treating or preventing disease/disorder(s), except anemia, these are said to be nutraceuticals (Kalra, 2003).

According to Dr. Stephen L. De Felicel (1989), nutraceuticals constitute those products that have been obtained from foods and are purified before being developed as a medicine. These possess prominent health benefits and provide protection against various chronic diseases. Nutraceuticals have been categorized based on chemical constitution or source. On the basis of source, these have been categorized as products derived from minerals, animals, plants, or microbial sources. However, chemically, these are available as amino acids, minerals, vitamins, and fatty acids. Nutraceuticals have also been classified on the basis of action mechanism, food sources, chemical properties, etc. However, the food sources are natural and are classified as probiotics, prebiotics, dietary fibers, polyunsaturated fatty acids, antioxidants, and vitamins (Chauhan *et al.*, 2013).

* Corresponding author.

9.2 NANONIZATION

Despite the fact that they demonstrate several health benefits, there are certain limitations that restrict their applications for human use. The reduced stability, permeability, and bioavailability of the bioactive components in the gastrointestinal fluids lead to their partial absorption during the hepatic first-pass metabolism. These limitations result in minimal, negligible, or even nil biological action. Further, from the conventional approaches (e.g., capsules, pills, powders, tinctures, or suppositories) developed so far for the delivery of nutraceutical(s), limited portions of the administered dose undergo absorption and reach the actual site of action. The remaining portion either initiates adverse effects/toxicity or undergoes excretion. Thus, to avoid these limitations, methods based on encapsulation of nutraceuticals have been devised that comprise microencapsulation and nanoencapsulation. The latter is being investigated for efficient and safe transport of nutraceuticals. To resolve the issue of nutraceuticals absorption, the concept based on nanodelivery carrier system has gained emergence (Aswathy, 2015). In recent years, there has been tremendous developmental progress in nanonutraceuticals as well as in the transport of lipophilic nutraceuticals through nanotechnology.

Probiotics, a type of nutraceutical, are the live microbes that provide beneficial health benefits on a host cell when taken in adequate amounts. Probiotics have many benefits on the health of humans and play an important role in the process of digestion (Guhannath *et al.*, 2014). Nanoencapsulation technology implemented to formulate designer probiotic bacterial preparations can transport the probiotic bacteria encapsulated within them to the body parts, for example, the gastro-intestinal tract. These get to interact with specific receptors and show effective results. The designer nanoprobiotics may work as *de novo* vaccines by modifying the immune response. These were also found to be quite effective in supplementing various therapies like gastrointestinal infections and irritable bowel syndrome. Therefore, evaluating various aspects of probiotics could be interesting and highly challenging. Thus, this chapter is designed with the main aim to explore the use of nanonutraceuticals, specifically probiotics, for their significant health benefits and role in disease management.

9.3 PROBIOTICS

Probiotics are described as those microorganisms that are in their live state and can result in tremendous benefits to the health of host organism on administration. These are the mixtures of live bacterial species and are available in the form of foods such as cheese, yogurts, and yogurt-type fermented milk, puddings, and fruit based drinks (Guhannath *et al.*, 2014). The most commonly investigated probiotic microorganisms that have attained commercialization include *Lactobacillus*, *Lactococcus*, *Bacillus*, *Bifidobacterium*, *Pidiococcus*, *Enterococcus*, *Weisella*, *Leuconostocs*, and *Saccharomyces*. Table 9.1 highlights vital commercial preparations of probiotics. Probiotic microorganisms (specifically bacteria) are remarkably similar to the functionally active microorganisms present in the human gut (Sharif *et al.*, 2017). When administered, these acquire the ability to provide benefits to the consumers by retaining/improvising the microbial flora present in the intestine. Also, these probiotics have been found to be useful in treating and managing numerous diseases/disorders as depicted in Table 9.2. These are available mainly in the form of dietary supplements, foods, and as pharmaceutical preparations such as tablets, capsules, and powders. Probiotic formulations have gained prominent introduction in international market in the form of food and dietary supplements, functional foods, and natural health products. Various patents have also been reported

TABLE 9.1
Commercial Preparations of Probiotics

Product	Microrganism	Manufacturer
Florastor	*Saccharomyces cerevisiae* boulardii	Biocodex (Creswell, OR)
Align	*B. infantis* 35624	Procter & Gamble (Mason, OH)
Yakult	*L. casei* Shirota *B. breve* Yakult	Yakult (Torrence, CA)
DanActive fermented milk	*L. casei* DN-114 001 *B. animalis* DN-173 010	Dannon (Tarrytown, NY)
Activia yogurt	*L. johnsonii* Lj-1 (NCC533; *L. acidophilus* La-1)	Nestle (Lausanne, Switzerland)
Ingredient Good Belly	*L. plantarum* 299V	Probi AB (Lund, Sweden) NextFoods (Boulder, CO)
Nature Made Digestive Probiotic Health		Pharmavite
BioGaia Probiotic chewable tablets or drops	*L. reuteri* ATCC 55730 ("Protectis")	Biogaia (Stockholm, Sweden)
Culturelle	*L. rhamnosus* GG ("LGG")	Valio Dairy (Helsinki, Finland)
Sustenex, Digestive Advantage Ingredient for food use	*Bacillus coagulans* BC30	Schiff Nutrition International Ganeden Biotech Inc. (Cleveland, OH)

that specify their commercial feasibility (Table 9.3). An assortment of mechanisms of action by which probiotics confer potential health benefits are depicted in Figure 9.1.

Generally, most probiotics are Gram-positive, usually catalase-negative, rods having rounded ends, and found in the form of pairs (either short or long chains). *Lactobacillus* and *Bifidobacterium* constitute the most common form of probiotics (Guhannath *et al.*, 2014). They are non-motile, non-flagellated, non-spore forming and are salt-intolerant. Optimum temperature for growth of most probiotics was found to be 37°C, but few strains like *L. casei* preferably can grow at 303.15 K. *L. acidophilus* is microaerophilic in nature with anaerobic referencing and is capable to grow aerobically. *Bifidobacterium* species, on the other hand, are anaerobic forms, but few species are aero-tolerant. Most probiotic bacteria are observed to be fastidious in their nutritional requirements (Song *et al.*, 2012). With reference to fermentation, probiotics are either obligate homofermentative (e.g., *L. acidophilus*, *L. helvelicas*) or obligate heterofermentative (e.g., *L. brevis*, *L. reuteri*), or facultative heterofermentative (e.g., *L. casei*, *L. plantarum*). Moreover, probiotics result in the production of various beneficial compounds such as antimicrobials, hydrogen peroxide, lactic acid, and a variety of bacteriocins.

Despite the multiple functions exhibited by probiotic microorganisms, maintenance of the cell viability in probiotic-containing products is still a considerable challenge. First, during industrial processing, the microorganisms must undergo survival. Second, as these products are intended for oral use, the remaining viable cells undergo further stresses during the gastrointestinal passage due to changes in pH or contact with bile salts, which might disrupt the membrane of microorganisms. As a consequence, additional loss in cell viability is expected, before they even reach their target (Haffner *et al.*, 2016). In this aspect, encapsulation of probiotic bacteria provides beneficial effects to enhance the viability as well as the survival rates of probiotic bacteria.

TABLE 9.2
Applications of Probiotics in Disease Management

Probiotics	Disease Targeted	Inference	Reference
Bifidobacterium and *Lactobacillus*	Periodontitis	Improved the recognized clinical signs of chronic and aggressive periodontitis.	(Matsubara *et al.*, 2016)
Probiotic bacteria	Common infectious disease	Probiotics reduced the symptoms of Upper Respiratory Tract Infection (URTI).	(Hao *et al.*, 2011)
Lactobacilli species	Infectious diarrhea	Probiotics showed reduction in the duration of diarrhea.	(Allen *et al.*, 2010)
Lactobacillus, *Bifidobacterium*, *Streptococcus*, and various combination products	Irritable bowel syndrome (IBS)	Probiotics significantly reduced IBS symptoms.	(Moayyedi *et al.*, 2010)
Probiotic bacteria	Allergy/Atopic dermatitis	Interrupted the disease progression in atopic dermatitis/allergy.	(Shane *et al.*, 2010; Folster-Holst, 2010)
Probiotic strains (*Bifidobacterium*, *Lactobacillus*, *Saccharomyces*, and/or *S. thermophilus*)	Necrotizing enterocolitis (NEC)	Prevented NEC and showed reduction in the frequency and reduction in overall mortality.	(Deshpande *et al.*, 2010)
Lactobacillus and *Bifidobacterim*	Gastric ulcer	Suppression rate was improved.	(Prakash and Urbanska, 2008)
L. acidophilus and *B. animalis* subsp. *Lactis*	Collagenous colitis	Affected the disease course of collagenous colitis.	(Prakash and Urbanska, 2008)
S. boulardii and a mixture of *L. acidophilus* and *B. bifidum*	Traveler's diarrhea	Probiotics showed significant efficacy with no serious adverse effects.	(McFarland, 2007)
Lactobacilli species (*L. rhamnosus* GG), *L. acidophilus*	Acute diarrhea including antibiotic-associated diarrhea and travelers' diarrhea and other acute diarrhea	Reduction in the risk for diarrhea was observed.	(Sazawal *et al.*, 2006)
L. rhamnosus GG	Ulcerative colitis	Remarkably reduced the number of first episodes (primary prevention) of pouchitis in patients who received probiotics daily.	(Gosselink *et al.*, 2004)
Probiotic containing both *L. acidophilus* and *B. bifidum*	*C. difficile*–associated diarrhea	Suggested the use of probiotics in reducing the likelihood of successful colonization of the gut by *C. difficile*.	(Plummer *et al.*, 2004)
Lactobacillus species	Colonic cancer	Yielded consistent and beneficial effects in controlling cancer.	(Saikali *et al.*, 2004)

TABLE 9.3
Patent Reports Highlighting Nanoparticulate Delivery of Probiotics

Patent Publication Number/Publication Year/Filing Year	Applicant	Highlight	Reference
US 8968721 B2/2015	Advanced Bionutrition, Columbia, USA	Disclosed a solid glass matrix of monosaccharides, disaccharides, polysaccharides, or in association with polyols to deliver probiotics.	(Harel, 2015)
US 9044497 B2/2015	Advanced Bionutrition, Columbia, USA	Disclosed a solid glass matrix of monosaccharides, disaccharides, polysaccharides, or in association with polyols to deliver probiotics.	(Harel and Kohavi-Beck, 2015)
US 9072310 B2/2015	Advanced Bionutrition, Columbia, USA	Disclosed a probiotic delivery system.	(Harel and Bennett, 2015)
CA 2618655 C/2015	University of Massachusetts Lowell, Massachusetts, USA	Disclosed a nanoemulsion based preparation that enhanced bioavailability of incorporated pharmaceuticals and nutraceuticals.	(Nicolosi and Wilson, 2015)
EP 2734049 A2/2014	H. J. Heinz Company, Pittsburgh, Pennsylvania, USA	Demonstrated a composition and its preparation methods. The compositions comprised of foods fermented by probiotic organism. The composition may include a colloidal liposome, microparticle, hydrogel, nanoparticle, or a block copolymer micelle.	(Budelli *et al.*, 2014)
EP 2818056 A1/2014	Biosearch SA, Granada, Spain	Invention related to a probiotic bacteria like *Lactobacillus* and *Bifidobateria*, containing metals and or metallic nanoparticles, foodstuff and pharmaceutical composition containing bacteria.	(Dominguez *et al.*, 2014)
US 8642088 B2/2014	Wisconsin Alumni Research Foundation, Madison, State of Wisconsin, USA	Described chitosan-tannin composite material that might be a nanoparticle, a hydrogel film, etc. The invention also provides a method to deliver a bioactive agent including omega-6 fatty acids, fat soluble vitamins, antibiotics, probiotics, etc.	(Reed *et al.*, 2014)
CA 2840030 A1/2013	University of Munster, Germany	Depicted an invention containing nucleic acid molecule in a host cell. The host cell includes probiotic bacteria.	(Schmidt *et al.*, 2013)
CA 2864578 A1/2013	Bionanoplus, S.L., Navarra, Spain	Depicted the preparation and use of nanoparticles for incorporating probiotic bacteria.	(Hesham *et al.*, 2013)

(Continued)

TABLE 9.3 (CONTINUED)
Patent Reports Highlighting Nanoparticulate Delivery of Probiotics

Patent Publication Number/Publication Year/Filing Year	Applicant	Highlight	Reference
CA 2914782 A1, 2013	Phosphorex, Inc., Hopkinton, Massachusetts 01748, USA	Disclosed the nanoparticles containing indirubin, their method of preparation and uses. The nutraceutical may include fatty acids, vegetable extracts, and probiotics.	(Wu, 2013)
EP 2625966 A1/2013	Bionanoplus, S.L., Navarra, Spain	Depicted the preparation and used of nanoparticles for incorporating a probiotic bacteria.	(Hesham *et al.*, 2013)
CA 2844474 A1/2012	Degama Smart Ltd., Cambridge, UK	Invented heat-processed/processible health food products specifically liquid-based products containing probiotics.	(Penhasi, 2012)
US 20100022680/2010	Massachusetts Institute of Technology, Cambridge, Massachusetts, US; The Brigham and Women's Hospital, Inc., Boston, Massachusetts, USA	Invented as microfluidic systems and method of preparation on nanoparticles for drug transport and also certain nutraceuticals like plant or animal extracts, vegetable extracts, lutein, phosphatidylserine, lipoid acid,, flaxseeds, fish and marine animal oils (e.g., cod liver oil), and probiotics, etc.	(Karnik *et al.*, 2010)
EP 2099496 A2/2009	Massachusetts Institute of Technology, Cambridge, Massachusetts, USA; University of California-San Diego, La Jolla, California, USA	Highlighted compositions and methods for targeted delivery via nanoparticles containing one modulating entity. In some cases a nutraceutical agent including plant or animal extracts, such as fatty acids and/or omega-3 fatty acids and probiotics.	(Agarwal *et al.*, 2009)
CA 2688415 A1/2008	Anterios, Inc., New York, USA	Presented an invention related to nanoparticle compositions containing one or more nucleic acids with nutraceutical agents including fatty acids, probiotics, etc.	(Edelson *et al.*, 2008)
US 20070077305/2007	Le Tien C, Tran Tu H, Mircea-Alexandru Mateescu	Disclosed a biocompatible polymeric matrix that can be developed as microbeads, capsules, tablets, nanoparticles, etc. The bioactive agent may be selected from drug, an alkaloid, RNA, DNA, probiotics, etc.	(Le *et al.*, 2007)

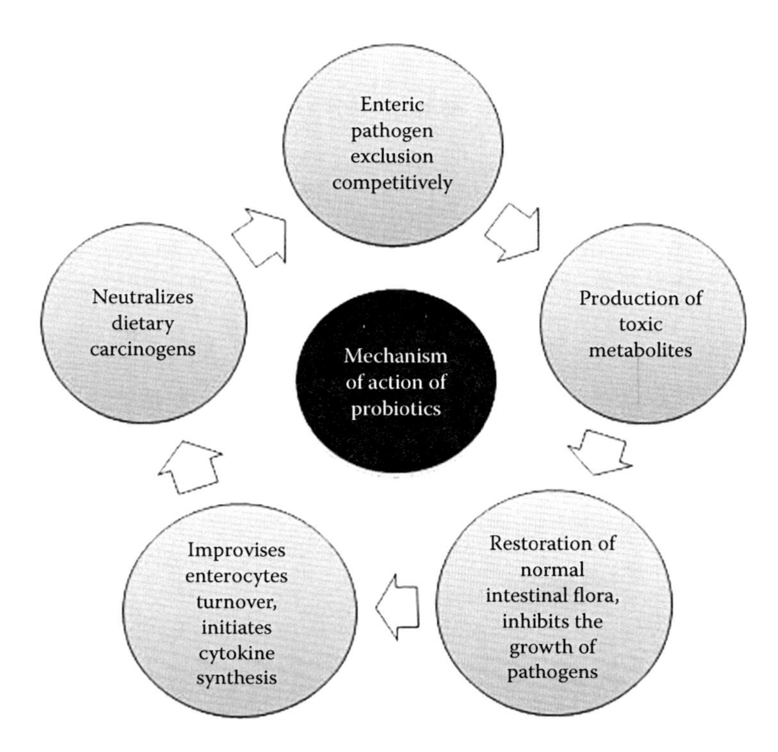

FIGURE 9.1 Plausible mechanisms underlying the activity of probiotics.

9.4 ENCAPSULATION OF PROBIOTICS

To improvise the probiotic bacterial viability, microencapsulation has emerged as an important methodology. It helps in delivering viable bacteria to the gastrointestinal tract of the host (Shori, 2017). This methodology provides a protective coating to the probiotic bacteria and separates it from surrounding environment, and also appears to provide protection of cells from mild heat treatment. Several studies have reported microencapsulation of probiotics by the use of vegetable gum and gelatin for protection against acid-sensitive *Lactobacillus* and *Bifidobacterium*. Song *et al.* (2012) showed better survival rate of alginate microencapsulated cells as compared to the free cells and were remarkably more effective in delivering viable cells to the colon than its non-encapsulated form. Microencapsulation technology has been used by various researchers to encapsulate probiotic bacteria as depicted in Table 9.4. However, to successfully encapsulate viable cells into foodstuff, it is of utmost importance to preserve the bacterial viability in all of the procedures along with the judicious selection of encapsulation materials (Haffner *et al.*, 2016).

More recently, nanoencapsulation is being explored for encapsulating probiotic bacteria. Nanotechnology, comprised of development of structures at nanometric scale, provides numerous opportunities for the establishment of innovative products. Application of nanotechnology in designing nanoparticles based on probiotic bacteria has gained tremendous potential worldwide in the field of nutraceuticals and is expected to grow rapidly in the future. Nanosized materials have a distinctive potential to enhance the bioavailability or functionality of nutrients as well as ingredients, and thus, reduce the concentration required in the food products. Nanoparticle-based

TABLE 9.4
A Cross-Section of Reports on Delivery of Encapsulated Probiotics via Multiparticulate Carriers

Probiotics	Delivery System	Inference	Reference
L. acidophilus (NCDC 014) and *L. casei* (NCDC 018)	Alginate microspheres	Provide significant protection to probiotic bacteria and effectively deliver the bacterial to colon.	(Mathews, 2017)
L. paracasei A13 and *L. salivarius subsp. salivarius* CET 4063	Microcapsules	Microencapsulation resulted in the production fermented milk with enhanced functionality and sensory features.	(Patrignani *et al.*, 2017)
L. plantarum	Gelatin coated whey protein concentrate microcapsules	Decreased initial viability with negative effect on the survival of bacteria was observed.	(Gomez-Mascaraque *et al.*, 2017)
Tuna oil (O)/*L. casei*	Microcapsules	Synergistic effect was observed between probiotic bacteria and oxidative stability of omega-3 oil.	(Eratte *et al.*, 2016)
L. rhamnosus GG probiotics	Alginate-gum microcapsules	Developed microcapsules showed improved viability of the probiotics upon their incorporation into soy milk.	(Cheow *et al.*, 2016)
L. salivarius NRRL B-30514	Emulsions	Depicted the feasibility of s/o/w emulsion to transport bacteria in foods in order to improve the viable nature at the time of storage, processing, and digestion.	(Zhang *et al.*, 2016)
L. plantarum 33	Microcapsules in olive paste	Extended survival of cells upon refrigeration.	(Alves *et al.*, 2015)
Pediococcus pentosaceus OZF	Microcapsules	Chitosan based microcapsules depicted protective effect on the viability of cells.	(Kiran *et al.*, 2015)
L. acidophilus	Ovalbumin bi-layered mini-tablet-in-tablet	Proved to be efficient in providing protected probiotic transport in colonic as well as simulated human intestinal environment.	(Govender *et al.*, 2015)
L. reuteri KUB-AC5	Microcapsules	Proved to be an effective intestine-targeted probiotic system.	(Rodklongtan *et al.*, 2014)
Metronidazole and probiotics	Polysaccharides based bigels	Developed bigels that could be employed effectively in the transport of probiotics and drug.	(Behera *et al.*, 2014)
Recombinant L. plantarum 25	Microcapsules	Promising delivery carrier for oral administration of probiotic expressing vaccine.	(Jiang *et al.*, 2014)
L. plantarum CG ATCC 53103/*L. reuteri* ATCC 55730/*L. acidophilus* DSM 20079	Pullulan/Starch edible films	Provided higher relative cell viability up to 70–80%.	(Kanmani and Lim, 2013)
L. fermentum NCIMB 5221	Carboxymethyl cellulose films	Carboxymethyl cellulose based oral thin films maintained probiotic viability and antioxidant activity.	(Saha *et al.*, 2013)
B. bifidum	Microcapsules	Microcapsules formulated using chitosan and alginate employing emulsion method showed improved protection of *B. bifidum* against adverse environment.	(Zhang *et al.*, 2013)

(Continued)

TABLE 9.4 (CONTINUED)
A Cross-Section of Reports on Delivery of Encapsulated Probiotics via Multiparticulate Carriers

Probiotics	Delivery System	Inference	Reference
L. plantarum CRL 1815	Microparticles	Specified the role of biological compatibility in microencapsulating material, as modification of gut microbiota occurred by them, and explored the use of bacterial exopolysaccharides as a new targeted-delivery coating material.	(Jimenez-Pranteda *et al.*, 2012)
Ginger extract and *L. acidophilus*	Floating beads	Encapsulation of *Lactobacillus acidophilus* and ginger extract floating beads together could help in exploiting their benefits as therapeutic curative agents.	(Singh and Kaur, 2012)
B. lactis and *L. acidophilus*	Solid lipid microparticles	Provide protection against simulated gastric/intestinal fluid.	(Pedroso *et al.*, 2012)
L. bulgaricus FTDC 1511	Soy protein isolate (SPI) gel carriers	Incorporation of agro wastes into SPI carriers resulted in remarkably lower release of *L. bulgaricus* FTDC 1511 in simulated gastric fluid and a higher release in simulated intestinal fluid with reference to the control.	(Yew *et al.*, 2011)
B. longum HA-135	β-Lactoglobulin tablets	Succinylated β-lg tablets provided the protection to acid-sensitive bacteria during transit in the upper gastro-intestinal tract.	(Poulin *et al.*, 2011)
L. rhamnosus	Mucoadhesive coacervate microparticles	Extended bacterial release provided by the system.	(Alli *et al.*, 2011)

emulsions are currently produced by food companies to modify the material based properties of foodstuffs, like in ice cream manufacturing, to enhance its texture uniformity (Paul and Dewangan, 2015). Effects of nanosizing on encapsulated probiotics are depicted in Figure 9.2.

9.4.1 Nanoencapsulated Probiotics

Nanoparticles, because of their specific features, like small size, shape, charge, large surface area, chemical properties, solubility, and degree of agglomeration, are able to cross cell boundaries or pass directly and reach to all the body organs. Effective encapsulation, solubilization, and delivery as biocompatible systems based on nanoscience and technology are bestowed with features such as reduced frequency of administration, magnified absorption in lower doses, and amended therapeutic index. Encapsulated ingredients exhibited increased product shelf-life. Application of nanotechnology also imparts benefits like developing a product with novel characteristics, improved security, processing, new packaging, and improved microbial food safety (Grumezescu, 2016). The nanoparticulate carriers developed using proteins, synthetic polymers, or polysaccharides, possess specific features to be employed for administration of nutraceuticals (Prudhviraj *et al.*, 2015). Enhanced focus has been applied to depict the potentiality of nanocarriers for the delivery of probiotics. Several nanocarriers such as nanoparticles, nanofibers, nanobeads, and nanolayers

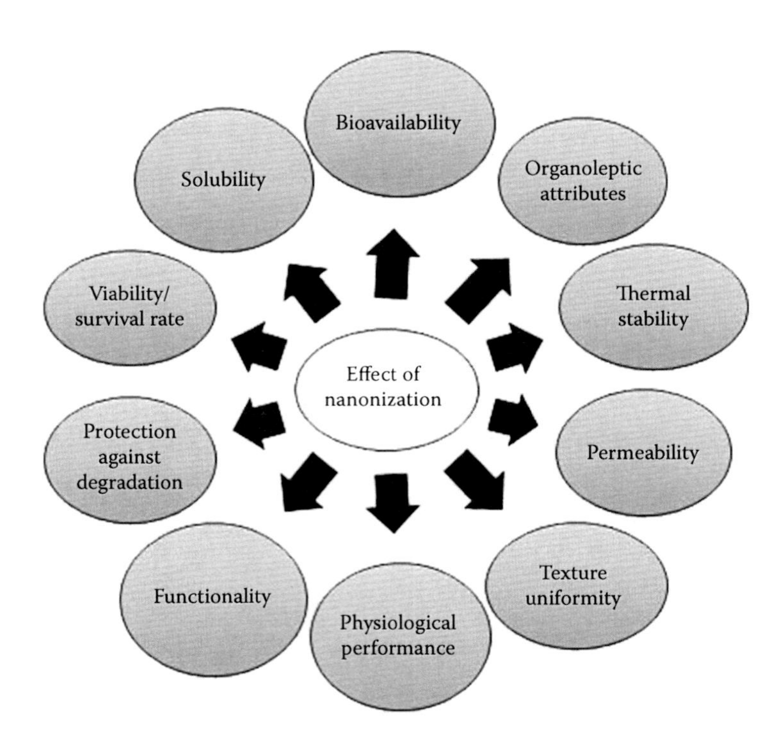

FIGURE 9.2 Effect of nanonization on properties of probiotics.

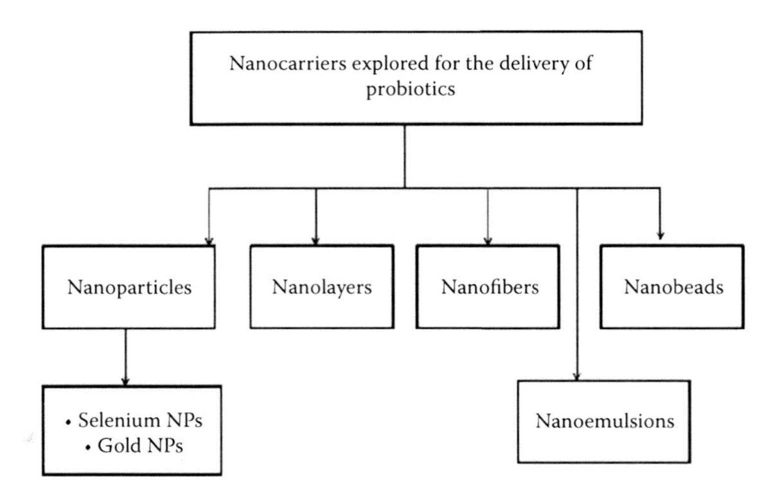

FIGURE 9.3 Various nanocarriers explored for encapsulation of probiotics.

based on layer-by-layer approach have been devised to encapsulate the probiotic bacteria. Figure 9.3 comprehensively depicts the development of various nanocarriers that have been researched to be formulated using probiotic bacteria or encapsulated within them. Encapsulation of several probiotic bacteria via these carriers has been discussed in the preceding text.

9.4.1.1 Nanoparticles

Nanoparticles in the size range of 10–1000 nm have been explored to encapsulate the nutraceuticals. More specifically, selenium (Se) and gold (Au) nanoparticles have been used to encapsulate probiotic bacteria as discussed in the preceding text. Yazdi *et al.* (2012) selected a strain of *Lactobacillus plantarum* and enriched these with selenium nanoparticles to be employed as an immunomodulating agent. Female inbred Balb/c mice ($n = 30$) were equally divided into two groups: a control and a test group. Each mouse was administered with a daily dose of 2.5×10^8 cfu × 1000/m^3 of *L. plantarum* orally enriched with selenium nanoparticles, for 2 weeks prior to tumor induction, followed by subcutaneous injection of 1×10^6 4T1 cells. The daily administration of nanoparticles (NPs) was then repeated for 3 cycles of 7 days on/3 days off, after tumor induction. Immunological factors like levels of cytokines, tumor growth factor, natural killer cells activity, and mouse survival were assessed. Pro-inflammatory cytokines IL-2, IFN-γ, and TNF-α production in spleen cell cultures was found to be enhanced in case of test mice administered with SeNP-enriched *L. plantarum*. The test mice also demonstrated remarkable enhancement in NK cell activity. Volumes of tumor in case of treated mice were reduced and their survival rate significantly enhanced in comparison to the mice under *L. plantarum* alone, treatment, or control mice. It was interpreted that the usage of SeNP-enriched *L. plantarum* may have caused induction in an immune response by elevating pro-inflammatory cytokines TNF-α, IFN-γ, and IL-2 levels, and increased the NK cell activity. Thus, the treatment may provide better cancer prognosis.

Kazmierczak (2015) described loading of NPs on to a biological "mailman," a therapeutically active strain of *Salmonella typhimurium*. This strain was designed to advantageously and specifically seek out, undergo penetration, and hinder the prostate cancer cells. These NPs were engineered to transport the therapeutic actives to the target site. In this investigation, first the authors had associated the *Salmonella* bacteria's surface with the sucrose-linked gold NPs. Second, the membrane of *Salmonella* was biotinylated to link the streptavidin-conjugated fluorophores. It was reported that, by this processing, biotin concentration was increased on the membrane surface, thus leading to an improvement in the tumor destruction.

Hu (2015) devised the NPs-coated bacteria to deliver DNA vaccine and to stimulate the immune system to destroy the cancer cells. In comparison to the uncoated bacteria, the coated bacteria were able to bypass several roadblocks that restricted the immune response and posed a threat to DNA vaccination. The NPs-coated bacteria demonstrated much better chance than uncoated bacteria in overcoming these hurdles and evoking significantly stronger immune response. The NPs were comprised of positively charged polymers, which self-assembled themselves over the negatively charged cell walls of the *Salmonella* due to electrostatic interactions. The coated bacteria could better tolerate the acidic stomach and intestines, partly because the coating acted as a buffer, resisting changes in pH, and partly due to the physical protection it offered. The coating also helped the bacteria escape the phagosomes, because the cationic polymers initiated a "proton sponge effect," soaking up protons and causing the phagosomes to swell and rupture due to osmosis. Due to these benefits, the NPs-coated bacteria were more likely to reach the lower gut to conquer the infection. The authors observed that 60% of mice that were vaccinated with NPs-coated *Salmonella* showed survival time of 35 days without tumor growth. Also, the tested animals showed negligible loss in body weight, thus depicting the low toxicity of the vaccine.

Feher (2012) invented a method for the preparation of probiotic NPs from natural sources. A physical preparation phase was developed by ultracentrifugation, disruption in bead mill, colloid mill, by using French press, cryofracturing, osmotic shock, microwave exposure, gamma ray exposure, or UV light exposure. A formulation preparation phase also comprised of powderized drying, desiccation to abolish hygroscopic nature of the dried lysate by the addition of glycogen or maltodextrin, or packaging and storage of prepared final products including powder, watery solutions, or lipid emulsion. The invention depicted a preparation containing probiotic NPs for medical, nutritional, or cosmetic purposes, for systemic or topical application, enteral or parenteral, oral, intranasal, gastric, or parenteral administration(s). The prepared probiotic NPs were suggested for the treatment of infective diseases, traumas, age-related diseases, autoimmune diseases, malignancies, inherited diseases, or connatal diseases, as well as functional diseases and disorders. On the basis of results, it was postulated that the preparation and administration of NPs from dead probiotics may guarantee better bioavailability and lesser adverse effects compared to live probiotics.

9.4.1.2 Nanofibers

Nanofibers present novel improvised products and carrier systems for supplementary foodstuffs. Developed electrospun nanofibers can be used as the delivery carriers for nutrients to provide protection at the time of processing and storage or in transferring the components to the desired site in the body. Lopez-Rubio (2009) utilized electrospinning for encapsulating bifidobacterial strain (*Bifidobacterium animalis*) to prepare nanofibers to improve the viability and stability of strains. Poly (vinyl alcohol) was used as an encapsulating agent. Encapsulated nanofibers showed an optimum size of 150 nm. Encapsulation of bacterial strain resulted in the reduction of crystallinity and melting point of the nanofibers; and enhancement in polymeric glass transition temperature. The viability tests conducted at room temperature depicted that encapsulated *B. animalis Bb12* in nanofibers showed enhanced viability up to 40 days (at room temperature) and for 130 days (under refrigerated conditions) in comparison to non-encapsulated bacteria.

Fung (2011) explored the feasibility of soluble dietary nanofibers in the nano-encapsulation of *Lactobacillus acidophilus* using 8% poly (vinyl alcohol) by electrospinning. *L. acidophilus* was then incorporated into the spinning solution to form nanofibers-encapsulated probiotic, measuring 229–703 nm, visible using fluorescence microscopy. Viability analysis depicted good bacterial survivability (78.6–90%) under electrospinning conditions and retained the viability at refrigeration conditions during the storage period of 21 days. FTIR spectra of nanofibers depicted the existence of hemicellulose in soluble dietary nanofibers and thermal behavior of nanofibers revealed thermal protection of probiotics.

Plantaricin 423, formed by *L. plantarum 423*, was encapsulated in nanofibers produced by electrospinning 18% w/v polyethylene oxide. The average diameter of nanofibers was found to be 288 nm. The activity of *Plantaricin 423* significantly decreased after exposure to electrospinning, as determined against *L. sakei DSM 20017* and *Enterococcus faecium HKLHS*, respectively. Nanofibers encapsulated *L. plantarum 423*, showed reduction in cells from $2.3 \times 10^{10} \times 10^3$ cfu/m^3 before electrospinning to $4.7 \times 10^8 \times 10^3$ cfu/m^3 after electrospinning. Interestingly, the cells entrapped in the nanofibers continued to produce *L. plantaricin 423*. This methodology might be employed to develop a delivery system for bacteriocins and for encapsulating probiotic lactic acid bacteria (Heunis *et al.*, 2010).

Nagy (2014) demonstrated the potential of electrospinning-based biodrug delivery and produced an electrospinning-based technology to formulate vaginal delivery systems. *L. acidophilus* encapsulated nanofibers were developed using polyvinyl alcohol and polyvinyl pyrrolidone. The viability test depicted that the nanofibers could prolong the stability of live bacteria at or below 280.15 K. Thus, electrospinning resulted in a stable solid formulation of probiotic bacteria and the shelf life could be further extended using stabilizer excipients or by using oxygen excluding packaging. The formed nanofibrous product can be applied as an inexpensive and easy-to-use dosage form that can be recommended for the treatment of bacterial vaginosis. The dosage form (vaginal douche) based on biohybrid nano-webs constitutes a novel way to cure bacterial vaginosis.

9.4.1.3 Nanolayers

Layer-by-layer (LbL) employed to form the assembly of layers on the surface of solids is a novel approach in the development of nanolayers or thin films to be used in drug, nutraceutical, as well as gene delivery and in biosensing. These nanolayers are comprised of at least three layers. Here, the first layer consists of a positively charged polyelectrolyte, the second layer a polymer and an outer layer comprising a functionalized polysaccharide or a polyether. The agent to be released is then incorporated into the layers. These layers have been found to be useful for release of one or more therapeutic agents.

Franz (2010) developed LbL nano self-assembled coated *Allochromatium vinosum* using several polyelectrolyte combinations. Effect of surface charge and development of physical barrier imparted new apprehensions in contact mechanisms between insoluble elemental sulfur and cell surface. Further, sulfide uptake by encapsulated cells was also observed. Growth experiments using coated cells revealed that surface charge neither influenced the uptake of sulfur nor the contact formation between the solid sulfur and cells. However, increasing layers slowed down or reduced the uptake of sulfide and elemental sulfur.

A year later, Priya (2011) explored the encapsulating potential of nanostructured polyelectrolyte on probiotic *L. acidophilus*. The effect LbL self-assembly of polyelectrolyte chitosan and carboxy methyl cellulose was studied to improvise the bacterial strain's survival in adverse conditions of gastrointestinal tract. Significantly enhanced survival of strain was observed in comparison to the non-encapsulated form of bacterial strain both in simulated gastric and intestinal fluid. Complete death of free cells was observed upon access to the media for about 120 × 60 s. However, cells coated with nanolayers of polyelectrolyte chitosan and carboxy methylcellulose showed 33 log% survival rate of cells under similar conditions. This enhancement in bacterial survival was attributed to the fact that polyelectrolyte nanolayers imparted impermeability to the enzymes like pepsin and proteolytic that resulted in proteolysis. Further, reduction in viability losses in case of nanoencapsulated bacteria was also observed, upon freeze drying and freezing.

9.4.1.4 NanoBEADS

Nanosized bacteria-enabled autonomous delivery systems (NanoBEADS) have been developed recently to encapsulate probiotic bacteria. Flagellated bacteria possess specific self-propulsion features, capable of moving through the highly viscous fluid and porous semisolid environments effectively. This inherent feature may be used for the development of bio-hybrid microrobotic systems showing applicability in directed transport and microassembly. Traore (2014) developed nanoscale bacteria-enabled autonomous delivery system, termed NanoBEADS. NanoBEADS comprise of flagellated *E. coli* bacterium loaded with an assembly of spherical polystyrene NPs. It was demonstrated by the authors that assembly of NanoBEADS was entirely based on the ratio of bacteria to nanoparticles. Further, the stochastic motion of nanoparticles was characterized as a function of size and quantity of NPs, and examined the influence of the nanoparticle load on the drag force. It was reported that the swimming speed of NanoBEADS was reduced by 65% of the free-swimming bacteria speed at the highest possible load. Thus, it was concluded that NanoBEADS might be employed as single agent or collaborative swarm to be used in several applications starting from drug delivery to whole cell biosensing.

9.4.1.5 Nanoemulsions

Nanostructured delivery systems, comprised of emulsions, constitute a system consisting of liquid phase dispersion in another liquid phase with droplet size less than 200 nm. Nanoemulsions (NEs) are formulated using coarse emulsions by reducing the emulsion droplet dimension with high energy techniques like high-pressure homogenization, microfluidization, and high-power ultrasound. NEs provide effective drug delivery and initiate the release and absorption of loaded bioactives and food agents. NEs-based systems have acquired both food and pharmaceutical potential, and their development has paralleled the evolution of several effective emulsification methodologies. Two nanoemulsion-based approaches have been devised based on nanoemulsion excipient systems as well as nanoemulsion delivery systems. These two approaches, termed as nanoemulsion delivery systems and nanoemulsion excipient systems, facilitated the development of new formulations to improvise the bioavailability of nutraceuticals. In case of approach based on nanoemulsion delivery system, hydrophobic bioactives are allowed to dissolve in lipid phase of oil-in-water NEs. On the other hand, NE excipient systems consist of the bioactives in conventional drug, food, or supplement. These approaches were designed to enhance the bioavailability of bioactives. Furthermore, to ensure the commercial viability of these systems, and to address their safety and efficacy issues, extensive research is required (Aboalnaja *et al.*, 2016). The feasibility of NEs in improvising the bioaccessibility of numerous nutraceuticals in mangoes was studied. The findings suggested the potential of excipient nanoemulsions in boosting the bioavailability of lipophilic bioactive agents in fruits and vegetables (Liu *et al.*, 2016).

9.5 ISSUES

One of the challenging translational applications of synthetic biology involves the engineering and integration of microbes. In the near future, functional foods and nutraceuticals might be customized depending on the metabolic needs of individuals and linked up to the genetic makeup of each person. Bioactive ingredients provide health-stimulating nutrients and protection through encapsulation of actives, from the deleterious environment (Onwulata, 2012). Further, encapsulation of probiotic bacteria as well as bioactive compounds within the probiotics is a matter of great concern in food industry as well as in academia, as it helps in providing protection and improving the survival rate. Various technologies have been investigated, studied, and evaluated to encapsulate bioactive compounds and probiotics. However, difficult processing conditions employed in these methods can remarkably minimize bacterial viability or destruct the structural integrity of target molecules. Electrospinning or electrospraying show promising results to work as an innovative vehicle for the delivery of food compound. Production of nanofibers via electrospinning has gained tremendous focus in food industry due to their potential to serve as a vehicle for both controlled and sustained release. The room temperature process route is in accordance with the food grade polymers/biopolymers and provides effective encapsulation. Accordingly, electrospun fibrous assemblies can be clearly developed to gain advancement in designing novel products as well as delivery systems for supplementary food substances. However, to optimize the production conditions and enhance the throughput, the mechanism of electrospinning should be clearly understood (Ghorani and Tucker, 2015).

Another issue of concern in probiotic therapy is the usage of inactivated, unviable form of bacteria present in many probiotic products. With an increase in such kinds of formulations, studies have been undertaken to observe the effectiveness of bacteria present in them as probiotics. While functionality was observed in nonviable bacteria to provide health benefits to some extent, no substitute can be found to be as effective as viable bacteria. These kinds of bacteria need to be explored more to determine their probiotic potential. With current research on probiotics, one of the prominent issues of concern is the lack of original research. It has been highlighted by Stevenson and Blaauw (2011) that approximately 26% among all probiotics-based publications are only review articles, indicating that probiotic research *per se* lacks the original data. With numerous unanswered and unresolved questions and issues related to safety and effectiveness of probiotics as well as the constant research to find more safe and effective probiotic bacteria, there is tremendous scope of further research work to be undertaken on probiotics, a field that remains quite untapped.

9.6 SAFETY AND EFFICACY

Issues that have gained attention in probiotic therapy include the effectiveness of several probiotic-based formulations in the market. Numerous formulations, though reported to deliver multiple bacterial strains, are not as effective as probiotics and do not comply with the probiotic requirements. Single formulation containing multiple strains has shown not to be much effective in supplementing probiotic bacteria and was found to be a major issue in probiotics-based therapy (Kopp-Hoolihan, 2001). Another issue of concern is the misconception about the bacteria, that all of them from the same class possess analogous probiotic effects. Further, formulations may possess probiotic bacteria within the product, but may inefficiently transfer the bacteria for the supplementation of intestinal flora; with others, these may deliver pathogenic bacteria capable of inducing antibiotic resistance and other pathological conditions in the patient.

Intestinal flora vary from race or ethnicity, person to person, and within the age groups. Children tend to acquire higher amounts of *Bifidobacterium* bacteria as compared to the elderly with several *in vitro* studies not having taken into consideration the host-specific factors that influence the efficacy of delivered probiotic bacteria. For instance, it is important to have the knowledge of the intestinal flora of the patient to ensure maximum health benefits. It is, however, recommended that formulations must be patient-specific (Makinen *et al.*, 2012). Most important is the consideration of the age group, which would provide maximum health benefits to the patient. However, lack of safety issues and industry standardization has plagued the probiotics usage with many having a negative opinion on probiotics instead of their health benefits (Kopp-Hoolihan, 2001). Recent regulations and legislations in various countries are, however, now requiring research and validation of the bacterial probiotic cultures that would probably lead to improvised safety and efficacy for all patients (Makinen *et al.*, 2012).

FDA regulations do not exercise control over premarket approval of food/supplementary products including probiotics (Govender *et al.*, 2014). Only a few products exhibit the effectiveness and state the health benefits of these products, with few accurately proving the effectiveness in human trials. Thus, it depends on the patients to read the labels properly, do extensive research, and restrain from any misconception. Furthermore, education of appointed healthcare professionals will also prove to be effective as they can provide required information to patients about the misinterpretations, safety, effectiveness, and risks of the formulations that have not yet been proved or tested (Douglas and Sanders, 2008).

9.7 REGULATIONS

Nanoformulations should comply with the regulatory requirements to ensure the efficacy, safety, quality testing, and marketing authorization procedures (Shane *et al.*, 2010). For decades, the FDA has been regulating these formulations to ensure their safety and efficacy. To regulate them, the FDA has created eight centers/offices: Center for Drug Evaluation and Research (CDER), Center for Biologics Evaluation and Research (CBER), Center for Devices and Radiological Health (CDRH), Center for Veterinary Medicine (CVM), Office of the Commissioner (OC), Office of Regulatory Affairs (ORA), Center for Food Safety and Applied Nutrition (CFSAN), and National Center for Toxicological Research (NCTR). The nanoformulations are regulated on product-by-product basis considering their premarket acceptance, market approval, and post-market review. The current requirements of CDER/FDA for safety testing of products are quite exhaustive. Extensive testing requirements, nevertheless, may eventually become essential, if research finds any toxicological risks that are unique and related to nanomaterials.

REFERENCES

Aboalnaja, K. O., Yaghmoor, S., Kumosani, T. A., and McClements, D. J. Utilization of nanoemulsions to enhance bioactivity of pharmaceuticals, supplements, and nutraceuticals: Nanoemulsion delivery systems and nanoemulsion excipient systems. *Exp Opin Drug Deliv.* 2016, 1: 1–10.

Agarwal, A., Bhatia, S. N., Derfus, A. M., Harris, T., Min, D. H., and Von, M. G. Delivery of nanoparticles and/or agents to cells. 2009; EP 2099496 A2.

Allen, S. J., Martinez, E. G., Gregorio, G. V., and Dans, L. F. Probiotics for treating acute infectious diarrhoea. *Cochrane Database Syst Rev.* 2010; 11: CD003048F.

Alli, S. M. Preparation and characterization of a coacervate extended-release microparticulate delivery system for *Lactobacillus rhamnosus*. *Int J Nanomed.* 2011; 6: 1699–1707.

Alves, M., Peres, C. M., Hernandez-Mendonza, A., Bronze, M. R., Peres, C., and Malcata, F. X. Olive paste as vehicle for delivery of potential probiotic *Lactobacillus plantarum* 33. *Food Res Int.* 2015; 75: 61–70.

Aswathy, R. G. (2015). Prospects of nano-nutraceuticals for better and healthier future. *World Nutraceutical Conference and Expo*, July 13–15, Philadelphia, USA.

Behera, B., Sagiri, S., Singh, V. K., Pal, K., and Anis, A. Mechanical properties and delivery of drug/probiotics from starch and non-starch based novel bigels: A comparative study. *Standardization News.* 2014; 66: 865–879.

Budelli, A., Fasano, F. R., Terzano, M., and Bramati, L. Probiotic compositions and methods. 2014; EP 2734049 A2.

Chauhan, B., Kumar, G., Kalam, N., and Ansari, S. H. Current concepts and prospects of herbal nutraceutical: A review. *J Adv Pharm Tech Res.* 2013; 4: 4–8.

Cheow, W. S., Kiew, T. Y., and Hadinoto, K. Effects of adding resistant and waxy starches on cell density and survival of encapsulated biofilm of *Lactobacillus rhamnosus GG* probiotics. *LWT-Food Sci Tech.* 2016; 69: 497–505.

Deshpande, G., Rao, S., Patole, S., and Bulsara M. Updated meta-analysis of probiotics for preventing necrotizing enterocolitis in preterm neonates. *Pediatrics.* 2010; 125: 921–930.

Dominguez, V., Manuel, J., and Rodriguez, G. Probiotic bacteria comprising metals, metal nanoparticles and uses thereof. 2014; EP 2818056 A1.

Douglas, L. C., and Sanders, M. E. Probiotics and prebiotics in dietetics practice. *J Am Diet Assoc.* 2008; 108: 510–521.

Edelson, J., Kotyla, T., and Zhnag, B. Nucleic acid nanoparticles and uses thereof. 2008; CA 2688415 A1.

Eratte, D., Wang, B., Dowling, K., Barrow, C. J., and Adhikari, B. Survival and fermentation activity of probiotic bacteria and oxidative stability of omega-3 oil in co-microcapsules during storage. *J Functional Foods.* 2016; 23: 485–496.

Feher, J. Methods for preparing probiotic nanoparticles. US 20120114776 A1; 2012.

Folster-Holst, R. Probiotics in the treatment and prevention of atopic dermatitis. *Ann Nutr Metab.* 2010; 57: 16–19.

Franz, B., Balkundi, S. S., Dahl, C., Lvov, Y. M., and Prange, A. Layer-by-layer nano-encapsulation of microbes: Controlled cell surface modification and investigation of substrate uptake in bacteria. *Macromol Biosci.* 2010; 11: 164–172.

Fung, W. Y., Yuen, K. H., and Liong, M. T. Agrowaste based nanofibres as a probiotic encapsulant: Fabrication and characterization. *J Agr Food Chem.* 2011; 59: 8140–8147.

Gomez-Mascaraque, L. G., Ambrosio-Martin, J., Perez-Masia, R., and Lopez-Rubio, A. Impact of acetic acid on survival of *L. Plantarum* upon microencapsulation by coaxial electrospraying. *J Healthcare Eng.* 2017; 17: 1–6.

Ghorani, B., and Tucker, N. Fundamentals of electrospinning as a novel delivery vehicle for bioactive compounds in food nanotechnology. *Food Hydrocoll.* 2015; 227–240.

Gosselink, M. P., Schouten, W. R., van Lieshout, L. M., Hop, W. C., Laman, J. D., and Ruseler-van Embden, J. G. Eradication of pathogenic bacteria and restoration of normal pouch flora: Comparison of metronidazole and ciprofloxacin in the treatment of pouchitis. *Dis Colon Rectum.* 2004; 47: 1519–1525.

Govender, M., Choonara, Y. E., Kumar, P., Du Toit, R. C., Van Vuuren, S., and Pillay, V. A review of the advancements in probiotic delivery: Conventional *vs.* non-conventional formulations for intestinal flora supplementation. *AAPS PharmSciTech.* 2014; 15: 29–43.

Govender, M., Choonara, Y. E., Kumar, P., Du Toit, R. C., Van Vuuren, S., and Pillay, V. A gastro-resistant ovalbumin bi-layered mini-tablet-in-tablet system for the delivery of *Lactobacillus acidophilus* probiotic to stimulated human intestinal and colon conditions. *J Pharm Pharmacol.* 2015; 67: 939–950.

Guhannath, S., Aaron, A. I., Raj, A. A. S., and Ranganathan, T. V. Recent innovations in nanotechnology in food processing and its various applications: A review. *Int J Pharm Sci Rev Res.* 2014; 29: 116–124.

Haffner, F. B., Diab, R., and Pasc, A. Encapsulation of probiotics: Insights into academic and industrial approaches. *AIMS Mat Sci.* 2016; 3: 114–136.

Hao, Q., Lu, Z., Dong, B. R., Huang, C. Q., and Wu, T. Probiotics for preventing acute upper respiratory tract infections. *Cochrane Database Syst Rev.* 2011; 9: CD006895.

Harel, M., and Bennett, A. Dry food product containing live probiotic. US 9072310 B2; 2015.

Harel, M., and Kohavi-Beck, K. Delivery vehicle for probiotic bacteria comprising a dry matrix of polysaccharides, saccharides and polyols in a glass form and methods of making same. US 9044497 B2; 2015.

Harel, M. Delivery vehicle for probiotic bacteria comprising a dry matrix of polysaccharides, saccharides and polyols in a glass form and methods of making same. US 8968721 B2; 2015.

Hesham, H. A. S., Azcarate, I. G., and Catalan, I. E. Nanoparticles comprising a vegetable hydrophobic protein and a water miscible non-volatile organic solvent and uses thereof. CA 2864578 A1; 2013.

Hesham, H. A. S., Azcarate, I. G., and Catalan, I. E. Nanoparticles comprising a vegetable hydrophobic protein and a water miscible non-volatile organic solvent and uses thereof. EP 2625966 A1; 2013.

Heunis, T. D. J., Botes, M., and Dicks, L. M. T. Encapsulation of *Lactobacillus plantarum* 423 and its bacteriocin in nanofibres. *Probiotics Antimicrob Proteins.* 2010; 2: 46–55.

Hu, Q., Wu, M., Fang, C., Cheng, C., Zhao, M., Fang, W. *et al.* Tang GS. Engineering nanoparticle-coated bacteria as oral DNA vaccines for cancer immunotherapy. *Nano Lett.* 2015; 15: 2732–2739.

Jiang, T., Singh, B., Maharjan, S., Li, H. S., Kang, S. K., Bok, J. D. *et al.* Oral delivery of probiotic expressing M cell homing peptide conjugated BmpB vaccine encapsulated into alginate/chitosan/alginate microcapsules. *Eur J Pharm Biopharm.* 2014; 88: 768–777.

Jimenez-Pranteda, M. L., Aguilera, M., Mccartney, A. L., Jimenez-Valera, M., Nader-Macias, M. E., Ramos-Cormenzana, A. *et al.* Investigation of the impact of feeding *Lactobacillus plantarum* CRL 1815 encapsulated in microbially derived polymers on the rat faecal microbiota. *J Appl Microbiol.* 2012; 113: 399–410.

Kalra, E. K. Nutraceutical-definition and introduction. *AAPS PharmSciTech.* 2003; 5: 2–3.

Kanmani, P., and Lim, S. T. Development and characterization of novel probiotic-residing pullulan/starch edible films. *Food Chem.* 2013; 141: 1041–1049.

Karnik, R., Gu, F. X., Basto, P., Cannizzaro, C., Khademhosseini, A., Langer, R. S., and Farokhzad, O. C. Microfluidic synthesis of organic nanoparticles. US 20100022680 A1; 2010.

Kazmierczak, R., Choe, E., Sinclair, J., and Eisenstark, A. Direct attachment of nanoparticle cargo to *Salmonella typhimurium* membranes designed for combination bacteriotherapy against tumors. Salmonella: Methods and Protocols. *Methods Mol Biol.* 2015; 1225: 151–163.

Kiran, F., Mokrani, M., and Osmanagaoglu O. Effect of encapsulation on viability of *Pediococcus pentosaceus* OZF during its passage through the gastrointestinal tract model. *Curr Microbiol.* 2015; 71: 95–105.

Kopp-Hoolihan, L. Prophylactic and therapeutic uses of probiotics: A review. *J Am Diet Assoc.* 2001; 101: 229–238.

Le, T., Tran, T., and Mateescu, M. A. Biocompatible polymeric matrix and preparation thereof. US 20070077305 A1; 2007.

Liu, X., Bi, J., Xiao, H., and McClements, D. J. Enhancement of nutraceutical bioavailability using excipient nanoemulsions: Role of lipid digestion products on bioaccessibility of carotenoids and phenolics from mangoes. *J Food Sci.* 2016; 81: N754–N761.

Lopez-Rubio, A., Sanchez, E., Sanz, Y., and Lagaron, J. M. Encapsulation of living *Bifidobacteria* in ultrathin PVOH electrospun fibers *Biomacromolecules.* 2009; 10: 2823–2829.

Makinen, K., Berger, B., Bel-Rhlid, R., and Ananta, E. Science and technology for the mastership of probiotic applications in food products. *J Biotechnol.* 2012; 162: 356–365.

Mathews, S. Microencapsulation of probiotics by calcium alginate and gelatin and evaluation of its survival in simulated human gastro-intestinal condition. *Int J Curr Microbiol Appl Sci.* 2017; 6: 2080–2087.

Matsubara, V. H., Bandara, H. M., Ishikawa, K. H., Mayer, M. P., and Samaranayake, L. P. The role of probiotic bacteria in managing periodontal disease: A systematic review. *Expert Rev Anti Infect Ther.* 2016; 3: 1–13.

McFarland, L. V. Meta-analysis of probiotics for the prevention of traveler's diarrhea. *Travel Med Infect Dis.* 2007; 5: 97–105.

Moayyedi, P., Ford, A. C., Talley, N. J., Cremonini, F., Foxx-Orenstein, A. E., Brandt, L. J., and Quigley, E. M. The efficacy of probiotics in the treatment of irritable bowel syndrome: A systematic review. *Gut.* 2010; 59: 325–332.

Nicolosi, R., and Wilson, T. Compositions and methods for making and using nanoemulsions. CA 2618655 C; 2015.

Nagy, Z. K., Wagner, I., Suhajda, A., Tobak, T., Harasztos, A. H., Vigh, T. *et al.* Nanofibrous solid dosage form of living bacteria prepared by electrospinning. *Express Polymer Lett.* 2014; 8: 352–361.

Onwulata, C. I. Encapsulation of new active ingredients. *Ann Rev Food Sci Tech.* 2012; 3: 183–202.

Patrignani, F., Siroli, L., Serrazanetti, D. I., Braschi, G., Betoret, E., and Reinheimer, J. A. Microencapsulation of functional strains by high pressure homogenization for a potential use in fermented milk. *Food Res Int.* 2017; 97: 250–257.

Paul, S. D., and Dewangan, D. Nanotechnology and neutraceuticals. *Int J Nanomater Nanotech Nanomed.* 2015; 2: 9–12.

Pedroso, D. L., Thomazini, M., Heinemann, R. J. B., and Favaro-Trindade, C. S. Protection of *Bifidobacterium lactis* and *Lactobacillus acidophilus* by microencapsulation using spray chilling. *Int Dairy J.* 2012; 26: 127–132.

Penhasi, A. Probiotic liquid food products. CA 2844474 A1; 2012.

Plummer, S., Weaver, M. A., Harris, J. C., Dee, P., and Hunter, J. *Clostridium difficile* pilot study: Effects of probiotic supplementation on the incidence of *C. difficile* diarrhoea. *Int Microbiol.* 2004; 7: 59–62.

Poulin, J. F., Caillard, R., and Subirade, M. β-Lactoglobulin tablets as a suitable vehicle for protection and intestinal delivery of probiotic bacteria. *Int J Pharm.* 2011; 405: 47–54.

Prakash, S., and Urbanska, A. M. Colon targeted delivery of live bacterial cell biotherapeutics including microencapsulated live bacterial cells. *Biologics.* 2008; 2: 355–378.

Priya, A. J., Vijayalakshmi, S. P., and Raichur, A. M. Enhanced survival of probiotic lactobacillus acidophilus by encapsulation with nanostructured polyelectrolyte layers through layer-by-layer approach. *J Agr Food Chem.* 2011; 59: 11838–11845.

Prudhviraj, G., Vaidya, Y., Singh, S. K., Yadav, A. K., Kaur, I. P., Gulati, M., and Gowthamarajan, K. Effect of co-administration of probiotics with polysaccharide based colon targeted delivery systems to optimize site specific drug release. *Eur J Pharm Biopharm.* 2015; 97: 164–172.

Reed, J. D., Krueger, C. G., and Madrigal-Carballo, S. Tannin-chitosan composites. US 8642088 B2; 2014.

Rodklongtan, A., La-Ongkham, O., Nitisinpirasert, S., and Chitprasert, P. Enhancement of *Lactobacillus reuteri* KUB-AC5 survival in broiler gastrointestinal tract by microencapsulation with alginate-chitosan semi-interpenetrating polymer networks. *J Appl Microbiol.* 2014; 117: 227–238.

Saha, S., Tomaro-Duchesneau, C., Daoud, J. T., Tabrizian, M., and Prakash, S. Novel probiotic dissolvable carboxymethyl cellulose films as oral health biotherapeutics: *In vitro* preparation and characterization. *Exp Opin Drug Deliv.* 2013; 10: 1471–1482.

Saikali, J., Picard, C., Freitas, M., and Holt, P. Fermented milks, probiotic cultures, and colon cancer. *Nutr Cancer.* 2004; 49: 14–24.

Sazawal, S., Hiremath, G., Dhingra, U., Malik, P., Deb, S., and Black, R. E. Efficacy of probiotics in prevention of acute diarrhea: A meta-analysis of masked, randomized, placebo-controlled trials. *Lancet Infect Dis.* 2006; 6: 374–382.

Schmidt, A., Vetmsn, K., and Cichon, C. Novel compounds for the treatment of inflammatory bowel disease. 2013; CA 2840030 A1.

Shane, A. L., Cabana, M. D., Vidry, S., Merenstein, D., Hummelen, R., Ellis, C. L. *et al.* Guide to designing, conducting, publishing and communicating results of clinical studies involving probiotic applications in human participants. *Gut Microbes.* 2010; 1: 243–253.

Sharif, A., Kheirkhah, D., Esfandabadi, P. S., Masoudi, S. B., Ajorpaz, N. M., and Sharif, M. R. Comparison of regular and probiotic yogurts in treatment of acute watery diarrhea in children. *J Prob Health.* 2017; 5: 164–169.

Shori, A. B. Microencapsulation improved probiotics survival during gastric transit. *Hayati J Biosci.* 2017; 17: 1–5.

Song, D., Ibrahim, S., and Hayek, S (2012). Recent application of probiotics in food and agricultural science. *In*: Everlon Rigobelo, eds, (ISBN 978-953-51-0776-7), InTech Publisher, pp. 1–34.

Stevenson, S., and Blaauw, R. Probiotics, with special emphasis on their role in the management of irritable bowel syndrome. *S Afr J Clin Nutr.* 2011; 24:63–73.

Traore, M. A., Damico, C. M., and Behkam, B. Biomanufacturing and self-propulsion dynamics of nanoscale bacteria-enabled autonomous delivery systems. *Appl Phys Lett.* 2014; 105: 112–138.

Wu, B. Nanoparticles of indirubin, derivatives thereof and methods of making and using same. CA 2914782 A1; 2013.

Yazdi, M. H., Mahdavi, M., Kheradmand, E., and Shahverdi, A. R. The preventive oral supplementation of a selenium nanoparticle-enriched probiotic increases the immune response and lifespan of 4T1 breast cancer bearing mice. *Arzneimittel-Forschung/Drug Res.* 2012; 62: 525–531.

Yew, S. E., Lim, T. J., Lew, L. C., Bhal, R., Mat-Easa, A., and Liong, M. T. Development of a probiotic delivery system from agrowastes, soy protein isolate, and microbial transglutaminase. *J Food Sci.* 2011; 76: 108–115.

Zhang, F.., Li, X. Y., Park, H. J., and Zhao, M. Effect of microencapsulation methods on the survival of freeze-dried *Bifidobacterium bifidum*. *J Microencapsulation.* 2013; 30: 511–518.

Zhang, Y., Lin, J., and Zhong, Q. S/O/W emulsions prepared with sugar beet pectin to enhance the viability of probiotic *Lactobacillus salivarius NRRL B-30514. Food Hydrocoll.* 2016; 56: 804–810.

10 Nano Resveratrol: A Promising Future Nanonutraceutical

*Chahinez Houacine and Kamalinder K. Singh**

CONTENTS

10.1 INTRODUCTION

10.1.1 History and Source of Resveratrol

Until 1940, resveratrol (RES) was isolated from the roots of white hellebore (*Veratrum grandiflorum*). Afterwards, in 1963, it was recognized as the active constituent of the dried roots of Japanese knotweed, *Polygonum cuspidatum* (PC), also known as Ko-jokon in Japan, which has been used in the Chinese and Japanese traditional medicine as a remedy for gonorrhea, vessels inflammation,

* Corresponding author.

dermatitis, heart ailments, and hyperlipidemia (Das and Maulik, 2006). PC has various positive medicinal uses, principally from resveratrol, and its derivatives. Furthermore, the PC-dried root powder has been employed as a remedy for atherosclerosis with positive influence on modulation of age-related diseases like neurodegeneration (Sarubbo *et al.*, 2017), together with other medical conditions including diabetes, hypertension, cough, and cancer (McCubrey *et al.*, 2017).

Japanese knotweed incorporated in considerable quantities in herbal preparations that have shown therapeutic accomplishments, for example, relief in menstrual discomfort and relaxing muscles and joints. Itadori herbal tea prepared using Japanese knotweed plant forms an essential part of the diet in Japan (Charters *et al.*, 2003). Various research studies on Japanese knotweed reached the same conclusion in the treatment of common ailments comprising of cardiovascular disease and as an anticancer agent coupled with no reports of severe adverse reaction even when large doses have been employed (Udenigwe *et al.*, 2008) has led to generating lot of interest in this plant and resveratrol the compound responsible for the pharmacological activity as nutraceutical. These treatments can eventually lead to an extended life expectancy (Fischer *et al.*, 2017).

Later, resveratrol was identified in a broad range of around 70 plant species, as shown in Figure 10.1, which encompasses purple grapes, cranberries, blueberries, peanuts, rhubarb, mulberries, pines, and groundnuts (Das and Maulik, 2006). Various food sources of resveratrol have been summarized in Table 10.1. As resveratrol is found in end-products of grapes, too, like in wine, grapes perhaps become the indispensable source of resveratrol for humans. It was primarily discovered in grapevines (*Vitis vinifera*) in 1976 (Langcake and Pryce, 1976), and later in wine in 1992 (Siemann and Creasy, 1992). The highest concentration (i.e., 50–100 μg per gram) of resveratrol occurs in the skin and seeds of grapes (Pervaiz, 2003).

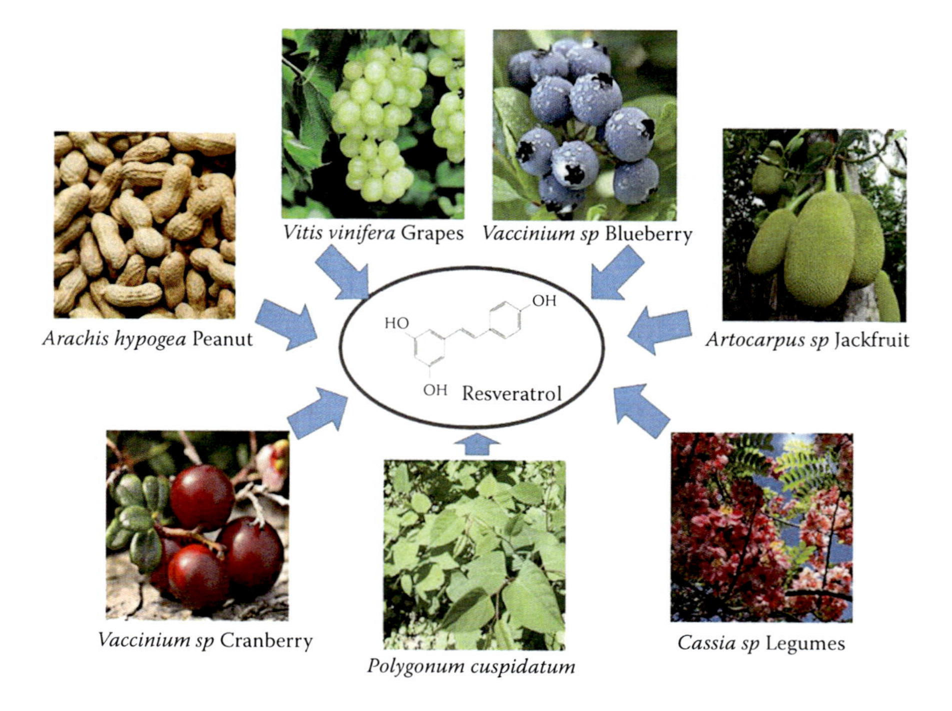

FIGURE 10.1 Sources of resveratrol from various plants.

TABLE 10.1
Content of Resveratrol in Different Plant Sources

Sources	Consumed Portion	Total Resveratrol (μg)
Grapes	100 g	150–780
Red wines	150 mL	80–2700
Grape juice	240 mL	0.12–0.26
Raw peanuts	28 g	0.6–50
Bilberry	100 g	~77
Cranberry	100 g	~90
Blueberry	100 g	86–170
Roasted peanuts	28 g	0.5–2.2
Peanut butter	32 g	4.7–24

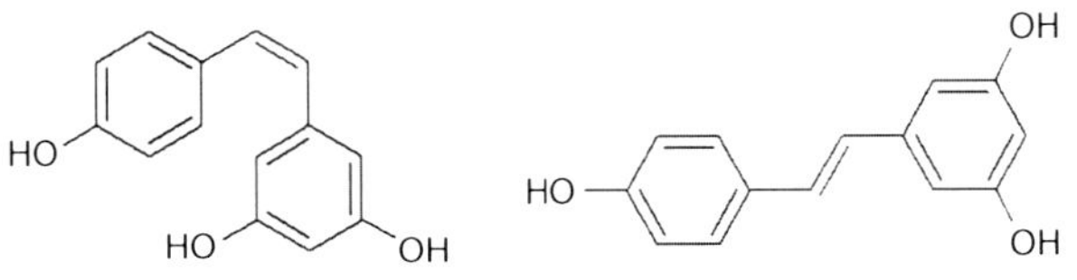

FIGURE 10.2 Isomers of resveratrol.

10.2 BIOSYNTHESIS OF RESVERATROL

Resveratrol (3, 5, 4'-trihydroxystilbene), a derivative of stilbene, belongs to a group of plant compounds called polyphenols. This polyphenolic stilbenoid is produced as a natural defense in response to damage to plants, bruise, or attack by microbes such as bacteria or fungi. It occurs in two isomeric configurations, cis- and trans-form, with the former occurring in abundance (Figure 10.2). The two isomers can occur either as free form or as bound form as glucoside, and the trans-form can isomerize to cis-form on exposure to light or UV radiation.

Synthesis of resveratrol in plants tends to involve an enzyme, stilbene synthase (STS), that gets stimulated in response to stress-inducing factors, like injury, UV radiation, and pathogen infestation (Fornara *et al.*, 2008). STS is involved in the catalysis of three condensation reactions among coumaroyl-coenzyme A (CoA) and three molecules of malonyl-CoA through cleavage of three carbon dioxide molecules (Figure 10.3). Furthermore, STS also plays a role in the catalysis of the terminal carboxyl group lost, which leads to the production of the C14 molecule, that is, resveratrol (Fornara *et al.*, 2008; Wang *et al.*, 2010).

10.3 RESVERATROL MECHANISM OF ACTION AND THERAPEUTIC POTENTIALS

Resveratrol has demonstrated numerous therapeutic effects, *in vitro* as well as *in vivo*, including chemopreventive, antioxidant, antiangiogenic, anti-obesity, anti-inflammatory, and neuroprotective effects, consequently generating wide interest in exploiting it as a plant bioactive to be employed in functional foods and pharmaceuticals.

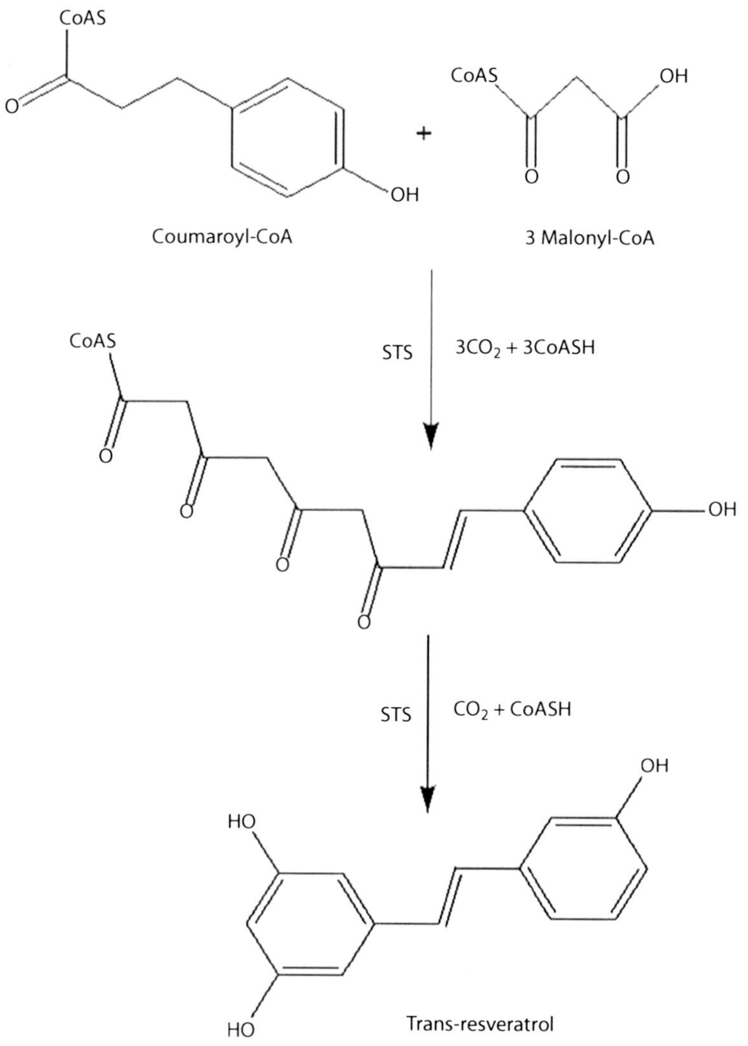

FIGURE 10.3 Resveratrol biosynthesis.

10.3.1 Cardiovascular Effects of Resveratrol

Mild-to-moderate wine consumption has been associated with reduced cerebrovascular, cardiovascular, and peripheral vascular risk. Intake of resveratrol found in wine has been demonstrated to account for its cardio-protection (Weiskirchen and Weiskirchen, 2016). Various reports have proposed different mechanisms to elucidate the cardiovascular effects of resveratrol. Some of these proposed mechanisms arise from the capacity of resveratrol to hinder eicosanoid synthesis from arachidonic acid, leading to decreased platelet aggregation, thus protecting from atherosclerosis (Olas *et al.*, 2001). Resveratrol has been reported to possess anticoagulant activity, thus enabling it to be effective in cardiovascular disorders, like cerebral and myocardial infarctions, by reducing embolism and thrombosis, leading to blocking of arteries. Besides, resveratrol inhibits platelet adhesion and coagulation and also protects fibrinogen from conformational changes promoted by epinephrine (Bonechi *et al.*, 2017). In another *in vitro* study, resveratrol was able to successfully inhibit the platelet aggregation in aspirin-resistant patients, recommending its use in treatment of these high-risk vascular patients (Gyorgyi *et al.*, 2006). Furthermore, resveratrol is believed to upregulate endothelial NO synthase in the vasculature, which shows anti-atherosclerotic and antithrombotic effect, and thus prevents vascular damage (Xia and Förstermann, 2017).

Few studies suggest that the presence of resveratrol may lead to upregulation of tumor suppressor gene p53 by this mechanism and protects the cardiovascular system against atherosclerosis, based on the ability of resveratrol to attenuate the secretion of catecholamine and reduce their production, thus reducing the risk of coronary heart and vascular diseases. This was also considered as a possible mechanism by which resveratrol offers significant cardio-protection (Neves *et al.*, 2012). Resveratrol may also be useful in the prevention of diabetes-associated atherosclerosis (Imamura *et al.*, 2017).

10.3.2 Antioxidant Effect of Resveratrol

The antioxidant properties of stilbenes and their analogues have also been widely investigated. Several studies have shown resveratrol to be an excellent scavenger of superoxide, hydroxyl, and other radicals, thereby protecting the cell membranes toward lipid peroxidation and avoiding DNA damage as triggered by the production of reactive oxygen species (ROS) (Leonard *et al.*, 2003). This protective effect is ascribed to the downregulation of Keap1 protein, thus stimulating the breakdown and formation of the nuclear factor (erythroid-derived 2)-like 2, that is in charge of the elimination of ROS via triggering antioxidant enzymes, such as glutathione peroxidase, catalase, glutathione reductase, superoxide dismutase, glutathione transferase, and oxidoreductases (Rubiolo *et al.*, 2008). It has also been suggested that resveratrol might improve the oxidative capacities of cancer cells through the CamKKB/AMPK pathway (Saunier *et al.*, 2017).

10.3.3 Obesity and Diabetes Prevention

Resveratrol is found to be useful in the therapeutic management of obesity and diabetes too. It has shown significant contribution in the prevention, management, and treatment of insulin resistance, type 2 diabetes, or dyslipidemia (Yonamine *et al.*, 2017).

Furthermore, resveratrol augments the lipolytic response to epinephrine and diminishes the ability of insulin to counterbalance lipolysis in adipose tissue (Szkudelska *et al.*, 2009). In diabetic rats, resveratrol was found to be capable of reducing hyperglycemia. Resveratrol affects pancreatic β-cells, which improves insulin secretion (Fiori *et al.*, 2013). Intracellular glucose transport induced by resveratrol is believed to involve many different mechanisms (Abbasi *et al.*, 2017). Translocation of glucose transporter (GLUT4) to the plasma membrane gets ameliorated with the intake of resveratrol, further allowing internalization of glucose in the cells. In diabetic rat's liver, resveratrol has exhibited an increase in glycogen synthase activity and a reduction in glycogen phosphorylase activity, with a resultant increase in liver glycogen storage (Palsamy and Subramanian, 2009).

10.3.4 Anti-Inflammatory Effects

Resveratrol has been shown to modulate numerous mediators of the inflammatory response. It diminished the production of nitric oxide and secretion of COX-2 dependent prostaglandin E2 and impaired the NF-κB transcription pathway in RAW 267.4 cells. It also diminished the mRNA levels of IL-1α, IL-1β, IL-6, TNF-α, CCL4/MIP-1β, and CCL5/RANTES in these calls. Additionally, resveratrol inhibited PGE2 production in THP-1 cells (Schwager *et al.*, 2017). Resveratrol has demonstrated an anti-carcinogenic effect by inhibiting the expression of various pro-inflammatory components (Palacz-Wrobel *et al.*, 2017).

When employed in therapeutic doses, resveratrol can also produce inflammation relief (Charters *et al.*, 2003). A variety of studies have shown that the topical application of PC extracts inhibits inflammation in mouse ears by controlling the key components of the immune response, that is, edema and neutrophil infiltration. Extract at the doses of 0.3, 1.25, and 2.5 milligrams was found to be as effective as indomethacin, a nonsteroidal anti-inflammatory drug, in decreasing the edema (Bralley *et al.*, 2008).

10.3.5 Anti-Aging Effects

Substantial investigations have been carried out to increase longevity and improve the lifespan. On exploration of various model organisms including fruit flies, yeast, and rodents, it is found that resveratrol triggers genes linked with longevity, especially a class of genes, known as sirtuins, which enhance the survival. It is the strongest activator of SIRT1, which acts in the same way as the pathways induced by caloric restriction, thus preventing chronic diseases and increasing the lifespan. Resveratrol's mechanisms of action are quite complex, with a plethora of effects in various tissues. The effects are frequently dependent on the physiologic status of the cells, for example, normal cells vs. specific cancer cell lines vs. cells with oxidative damage.

A study has shown that resveratrol could increase lifespan in obese mice, though its role in improving lifespan in healthy mammals has not yet been established. Resveratrol may not increase longevity, but it might extend the period of time in one's life before one develops a chronic disease, that is, health span, which is used to prevent or treat many chronic diseases related to aging in humans (Silk and Smoliga, 2014). Lately, resveratrol derivatives have been shown to extend the life span of *Caenorhabditis elegans*, making them promising candidates for investigation as anti-aging bioactives (Fischer *et al.*, 2017).

10.3.6 Anticancer Effect

Chemopreventive activity of resveratrol was first discovered in 1997 (Jang *et al.*, 1997). It is reported to exhibit antiproliferative properties against lymphoid and myeloid cancers, multiple myeloma, melanoma, squamous cell carcinoma, ovarian carcinoma, cervical carcinoma, and cancers of breast, prostate, stomach, colon, pancreas, thyroid, head, and neck. Its potential in colon cancer therapeutics has been shown through anticancer activity in HCT 116 cells, which is mediated by inhibiting PI3K/Akt signaling via upregulating BMP7 (Zheng *et al.*, 2017). The antitumor potential of resveratrol has been attributed to its ability to bind to Cu2+ and cancer-involved G-quadruplexes in human melanoma cells (Platella *et al.*, 2017).

Resveratrol causes apoptosis in several cancer cell lines by inhibiting the cyclooxygenase enzyme. High doses of resveratrol have shown to slow down breast cancer metastasis by inhibiting lamellipodia extension, thus proving resveratrol to be a potential chemopreventive agent of breast cancer. On the other hand, its low doses caused an increase in the metastasis and migration (Azios *et al.*, 2007). Resveratrol has shown to promote breast cancer cell breakdown by natural killer (NK) cells by upregulation of the protein, and mRNA expression of MICA and MICB (Pan *et al.*, 2017).

10.3.7 Neuroprotective Effects

Owing to the paucity of research on resveratrol effects directly on the "human" nervous system, its medical implications are entirely based on data obtained from various laboratory models. Cell line and animal studies proposed resveratrol to be effective in the prevention or slowing down of neurodegenerative diseases, including Parkinson's disease and Alzheimer's disease, where resveratrol was proved to improve mitochondrial activity via stimulation of a large number of metabolic sensors, leading eventually to the activation of PGC-1α. Resveratrol modulates the inflammatory response driven by IL-1β, downregulation of OPN and induced upregulation of iNOS, and thus provides the neuroprotective effect (Al Dera, 2017).

Resveratrol has also been successfully explored in various animal models for treating epilepsy and acute seizures. Moreover, it has been demonstrated to diminish the damage of brain cells induced by reoxygenation, consequently showing potential for treating stroke patients.

In rodents and nonhuman primates, a number of studies have proven that cognitive performance can be improved by resveratrol, though this improvement has not yet been consistently observed in humans (Silk and Smoliga, 2014). A recent report has disclosed resveratrol to be a potential antidepressant agent (de Oliveira *et al.*, 2017).

10.4 RESVERATROL AS NUTRACEUTICAL

Recently, resveratrol has received significant attention for its many beneficial effects on health. There is a body of evidence with various research studies demonstrating antioxidant, anti-inflammatory, anti-obesity, and anticancer effects of resveratrol in laboratory animals. With its beneficial effects contributing to improved health and extended life expectancy, resveratrol has emerged as a promising nutraceutical and functional food (Koga *et al.*, 2016; Al Dera *et al.*, 2017).

Resveratrol, in its free form or its plant sources like Japanese knotweed, have been incorporated in considerable quantities in herbal preparations with varied therapeutic indications, including relief from menstrual discomfort and relaxing muscles and joints (Charters *et al.*, 2003). Resveratrol has mostly been used in its free form, either as a solid (Cottart *et al.*, 2010) or as dissolved, diluted, or suspended in different vehicles (i.e., wine) (Soleas *et al.*, 2001), grape juice (Meng *et al.*, 2004), ethanolic extract (Walle *et al.*, 2004), ethanol + physiological saline (Asensi *et al.*, 2002), ethanol + corn or neobee oil (Yu *et al.*, 2002), propylene glycol + water (Vitrac *et al.*, 2003), and glycerol formal (Sale *et al.*, 2004), among others for various clinical studies. No severe adverse reaction has been reported with resveratrol even when large doses were employed (Udenigwe *et al.*, 2008).

Resveratrol is commercially available as various nutraceutical brands. Products consist of pure resveratrol capsules or/and multi-combination formulations with standardized concentration of resveratrol. A total of 14 brands of nutraceuticals containing resveratrol were evaluated to corroborate their actual resveratrol content and to check if their health-promoting properties were associated to manufacturing quality. Samples were analyzed for total trans-resveratrol, flavonoids, and polyphenol content and were compared vis-à-vis labeled content. Out of 14 brands, only five met the labeled values requirements, that is, 95–105% content, 3 were in the 8–64% range and 4 products were in 83–111% range. However, two of these samples were found to be below the detection limit (Rossi *et al.*, 2012).

On the other hand, the solid forms showed poor water solubility, and the liquid forms often made use of excipients (e.g., propylene glycol and ethanol), which are well-known to have a predictable effect. Studies have focused on the development of novel formulation(s) of resveratrol to improve its stability against any degradation, increase its solubility in an attempt to improve its bioavailability, along with controlled release characteristics (Amri *et al.*, 2012).

10.5 RESVERATROL DELIVERY CHALLENGES

Despite high bioactivity of resveratrol, its application as a nutraceutical in the food industry is presently limited due to its poor water solubility, low oral bioavailability, and high chemical instability. Its poor aqueous solubility is unfavorable for incorporation of high levels of resveratrol in aqueous based nutraceuticals. Also, it is prone to chemical degradation when exposed to high temperatures, pH changes, light, and certain enzymes. Owing to poor bioavailability of resveratrol, the concentrations of resveratrol at target tissues appear in far from sufficient levels to exhibit effectiveness in humans. The major reason for its low bioavailability is its rapid and extensive metabolism

(Walle, 2011). Resveratrol has a half-life of only 8–14 minutes owing to its metabolism into sulfate and glucuronide metabolites in liver and intestinal epithelial cells in human (Ponzo *et al.*, 2014).

In addition to the photo-stability issues, trans-resveratrol is prone to easy oxidation and erratic pharmacokinetics (Frozza *et al.*, 2010). Therefore, successful translational applications of resveratrol as a therapeutic or prophylactic agent pose a difficult challenge for the medical, pharmaceutical, and nutraceutical industries. Researchers have tried different methods to increase the solubility and bioavailability, and searching for analogs and explanation of new trans-resveratrol delivery systems by co-administration of metabolism inhibitors (Amiot *et al.*, 2013).

10.6 APPROACHES FOR OVERCOMING DELIVERY CHALLENGES OF RESVERATROL

Nanotechnology provides avenues for improving the properties of nutraceuticals and enables their efficient delivery (Aklakur *et al.*, 2016). Encapsulation of resveratrol into novel carrier system(s) can result in improving its water solubility, chemical stability, and bioavailability (Neves *et al.*, 2016). Several existing encapsulation technologies are able to overcome the challenges associated with resveratrol as a bioactive agent in foods. In particular, emulsion-based delivery systems have proved to be promising system(s) due to the fact that the hydrophobic core of the lipid droplets can easily encapsulate the lipophilic compound, thereby protecting the molecule from degradation during storage (Davidov-Pardo and McClements, 2015). However, it is important that factors for processing it as a nutraceutical should be borne in mind during development. This includes use of safe GRAS excipients, robust and commercially viable production process, favorable organoleptic properties in terms of taste and flavor, and reasonable storage shelf life.

10.6.1 MICROSPHERES

Microspheres are spherical microscopic particles with size ranging from 1 to 1,000 nm. However, defining these on the basis of size can sometimes be obscure, since spheres with a size of more than 1,000 nm may still be called microspheres. Such microparticles have wide-ranging possible applications (Figure 10.4a). Since resveratrol is an extremely light-sensitive compound, functionalized monodisperse porous polymeric microspheres of about 5 nm in diameter have been employed as material for preservation and stabilization of resveratrol. It was observed that cyano-functional

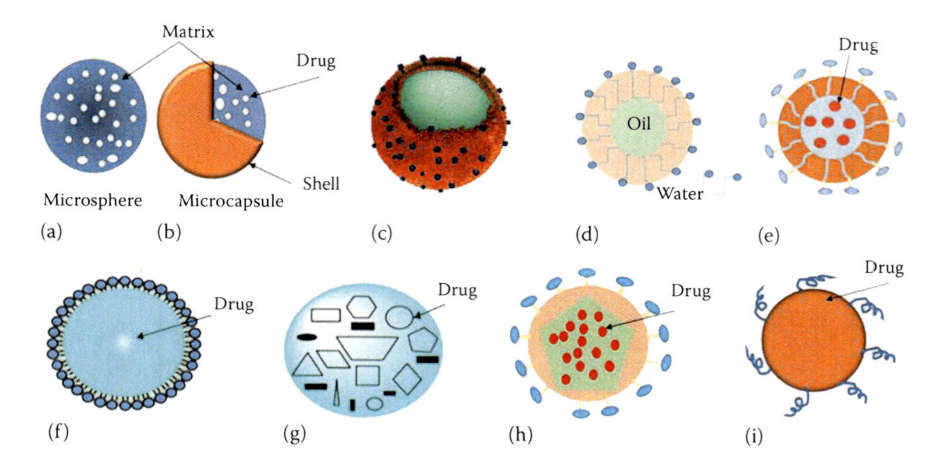

FIGURE 10.4 Resveratrol delivery systems: (a) microspheres, (b) microcapsules, (c) nanosponges, (d) nanoemulsions, (e) liposomes, (f) solid lipid nanoparticles, (g) nanostructured lipid carriers, (h) polymeric nanoparticles, and (i) nano-suspensions.

groups in the microspheres contributed greatly toward stability of resveratrol in the porous microparticles. The entrapped resveratrol maintained its antioxidant activity (93%), much better compared to pristine resveratrol, and slow release of entrapped resveratrol from porous microparticles sustained its bioactivity for more than 5 weeks (Nam *et al.*, 2005).

10.6.2 Microcapsules

Shi *et al.* (2008) established the yeast-encapsulated resveratrol using *Saccharomyces cerevisiae* as an encapsulating wall material. (Shi *et al.*, 2008). The formulation presented slower photo-degradation and better free radical scavenging properties than free resveratrol (Figure 10.4b). The *in vitro* release profile revealed 90% release of resveratrol from its microcapsules in simulated gastric fluid (pH 1.2) within 90 min. Thus, poor bioavailability of resveratrol, which is due to fast metabolism and elimination, could be improved with yeast cell encapsulation technology.

10.7 NANOSCALE DELIVERY OF RESVERATROL

The main reason for formulation of resveratrol nanoparticles is to increase the bioavailability of the bioactive, improve stability, and provide protection to it. The following sections describe different potential resveratrol formulations.

10.7.1 β-Cyclodextrin Nano-Sponges

Cyclodextrin-based nanosponges present a modern approach for highlighting the progression, which could be brought about by nanotechnology-based delivery systems (Chilajwar *et al.*, 2014). Nanosponges (Figure 10.4c) are the sponges with size of about a virus, which can be occupied with a wide variety of drugs. These minute sponges tend to circulate within the body until they meet the specific target site(s) to release the drug in a controlled and predictable manner. As they can cater to the site-directed drug delivery, they possess vibrant opportunities in drug delivery. Because of their high aqueous solubility, nanosponges have been found effective for bioactives with poor solubility (Subramanian *et al.*, 2012). Nanosponges offer high loading efficiency vis-à-vis other nanocarriers. Controlled release of the loaded bioactives and solubility enhancement of poor water-soluble substances are the major advantages of nanosponge delivery systems.

It is well-known that resveratrol is poorly soluble in aqueous media, complexation with β-cyclodextrin (β-CD), maltosyl-β-cyclodextrin (G2-β-CD), randomly methylated-β-cyclodextrin (RM-β-CD) and hydroxypropyl-β-CD (HP-β-CD) have all been investigated to enhance trans-resveratrol solubility. López-Nicolás and coworkers (2006) established that resveratrol formed a 1:1 complex with β-CD. Formation of inclusion complexes of trans-resveratrol with β-CD and HP-β-CD both improved the limited water solubility of resveratrol, though HP-β-CD displayed stronger inclusive capacity for resveratrol than β-CD. Thus, CDs could act as a useful carrier system for resveratrol and release the bioactive in a dosage-controlled manner (Lucas-Abellán *et al.*, 2007). Usage of trans-resveratrol-β-CD complexes could slow down prompt metabolism and elimination of trans-resveratrol, and thereby enhance its bioavailability (Das *et al.*, 2008).

Ansari *et al.* (2011) prepared resveratrol β-CD nano-sponges with particle size in the range of 400 to 500 nm by cross-linking β-CD, the formulated nanosponges afforded 30–40% encapsulation efficiency. It was found that encapsulating resveratrol in β-CD nanosponges enhanced its water solubility (33 to 48 fold) while improving its photo-stability and release profile (Ansari *et al.*, 2011).

10.7.2 Nanoemulsions

Nanoemulsions are submicron-sized colloidal particulate carriers of drugs and nutraceuticals (Figure 10.4d) with size varying from 10 to 1,000 nm. The main components of nanoemulsion include oily phase, emulsifier, and aqueous phase. The emulsification methods used to prepare nanoemulsions include high-energy stirring, ultrasonic emulsification, high-pressure homogenization, microfluidization, and membrane emulsification (Tadros *et al.*, 2004; Amiji and Tiwari, 2006). Nanoemulsion-based delivery systems can be efficiently used in the encapsulation of active compounds, improving their water solubility, protecting them from degradation, and preserving their antioxidant activity. As a delivery system, they improve the therapeutic efficacy of the encapsulated bioactive and diminish adverse effect and toxic reactions.

To enhance the oral bioavailability of resveratrol, different o/w nanoemulsion delivery systems have been explored on the basis of permeability tests on Caco-2 cells. Lecithin-based nanoemulsions are able to transport resveratrol through cell monolayers in relatively shorter times (1–6 h) than required for their metabolism (3–12 h), thus permitting better maintenance of the emulsion droplet integrity and offering better protection for resveratrol molecule. Flow cytometry studies have established that resveratrol was entrapped in the inner core of the nanoemulsions, offering protection against chemical degradation. Moreover, resveratrol incorporated into the nanoemulsions showed sustained release characteristics during *in vitro* release studies (Sessa *et al.*, 2014).

In another study, resveratrol nanoemulsions were formulated by the use of ternary-phase diagrams, isopropyl myristate as oil phase, poly (oxyethylene)-hydrogenated castor oil was selected as surfactant and ethanol as co-surfactant with resveratrol loading of 6.18 ± 0.11 mg. After three months of resveratrol nanoemulsion storage, the nanoemulsion was clear and transparent, without any phase separation or flocculation (Li *et al.*, 2009). Donsi *et al.* (2011) encapsulated resveratrol in peanut oil-based nanoemulsions, using different emulsifiers such as sugar esters and soy lecithin. Encapsulation of resveratrol in peanut oil-based nanoemulsions enhanced its stability by significantly reducing its chemical transformation from trans-resveratrol to cis-form (Donsì *et al.*, 2011). It was proved that the bioavailability of compounds encapsulated into emulsions is improved when emulsion droplets are of nanometric size. O/W nanoemulsions formulated by use of combination of lipophilic and hydrophilic emulsifiers have been reported to improve dispersibility of resveratrol in aqueous phase, along with protection of resveratrol from chemical degradation (Acosta, 2009).

10.7.3 Liposomes

Composed of spherical lipid carrier vesicles that differ in size from 20 nm to 20 µm, the liposomes consist of phospholipid bilayer enclosing an aqueous core (Figure 10.4e). These bilayers are either single or multi-concentric and have the capability to encapsulate both lipophilic and hydrophilic agents (Khan *et al.*, 2015). Addition of cholesterol to the phospholipid bilayer contributes to the stability of liposomes.

Resveratrol incorporated into liposomes showed improved efficacy on cell proliferation and enhanced photo-stability of resveratrol (Caddeo *et al.*, 2008). Before addition of resveratrol, small blank oligo-lamellar vesicles with particle size of 70–100 nm were formulated by sonication and extrusion with an entrapment efficiency of about 70% (Kristl *et al.*, 2009). These findings proved that inclusion of resveratrol into liposomes improved its biological activity and stability against UV-induced oxidative damage with sustained release properties to resveratrol.

Another study with liposomal resveratrol formulations has shown that resveratrol is integrated within the lipid bilayer, where the encapsulation process protects the drug against trans-cis isomerization, with 70% trans-form present after 16 min of UV light exposure compared to just 10% when resveratrol was exposed to UV light in its free form (Coimbra *et al.*, 2011). Photo-stability of resveratrol in liposomes has further been demonstrated by Detoni *et al.* (2012) with only 29.3% isomerisation to cis-resveratrol after 4h exposure to UV radiation. Soo *et al.* (2016) have co-encapsulated pristine resveratrol along with cyclodextrin-resveratrol inclusion complexes in the lipophilic and hydrophilic compartments of liposomes for its enhanced delivery. In another study Ethemoglu *et al.* (2017) have attempted to deliver resveratrol across the blood–brain barrier employing amphipathic liposomes for its effective delivery in penicillin-induced epileptic seizure model.

10.7.4 Solid Lipid Nanoparticles (SLNs) and Nanostructured Lipid Carriers (NLCs)

Solid lipid nanoparticles (SLN) have been introduced as novel nanoparticulate delivery systems, majorly consisting of solid lipids, as a surrogate to other nanocarriers systems. These offer stellar advantages of enhanced physical stability, protection to the incorporated sensitive bioactive from degradation, controlled release, and excellent tolerability (Figure 10.4f). However, the twin problems associated with SLNs are limited drug loading capacity and drug expulsion during storage. These problems can be overcome and/or minimized by newer generation nanostructured lipid carriers (NLCs) (Figure 10.4g), as these have imperfect crystal lattice structure that allows more drug incorporation in the lipid matrix. Although NLCs improved the drug loading capacity, it caused expulsion of drug at high oil concentration. Thus, to overcome this limitation, modified nano-lipid carriers (MNLCs) were introduced by modifying the lipid matrix so as to enhance the lipid solubility of the drug (Pople and Singh, 2011).

Solid lipid nanoparticles (SLNs) of resveratrol have been prepared through a melt-emulsification process with glyceryl behenate, poloxamer 188, and hydrogenated soybean lecithin, yielding particles with a size and loading capacity of 180 nm and 85%, respectively. Study findings showed that resveratrol was released from SLNs in a sustained manner. It was also observed that the use of SLNs as resveratrol carriers protected the keratinocytes against the cytotoxic effects induced by resveratrol alone (Teskac and Kristl, 2010). In another study, SLNs and NLCs were prepared employing a modified hot homogenization technique encapsulating resveratrol for its enhanced oral bioavailability. Morphologic microscopy studies showed spherical and uniform nanoparticles with a smooth surface and entrapment efficiency of ~70% for both SLNs and NLCs. The dynamic light scattering measurements gave a Z-average of 150–250 nm, polydispersity index of ~0.2, and a highly negative zeta potential of around –30 mV with statistically insignificant differences for two months, demonstrating good stability. Results ratified that resveratrol remained mostly associated with the lipid nanoparticles after their incubation in digestive fluids. Both SLNs and NLCs were found to be suitable carriers for oral administration, allowing protection to the incorporated resveratrol and conferring a controlled release (Neves *et al.*, 2013).

Resveratrol SLN (RES-SLN) prepared by the solvent emulsification-evaporation method had spherical shape with an average particle size of 96.7 nm, drug loading of 7.95 +/–0.21%, zeta potential –16.3 mV, and entrapment efficiency of (91.34 +/–0.18%). RES-SLN demonstrated high entrapment efficiency, drug loading and uniform particle size, sustained *in vitro* release, and enhanced anticancer effect (Zhang *et al.*, 2010).

RES-SLN have shown to ameliorate mitochondrial oxidative stress and mitigate cognitive decline observed in bilateral common carotid artery occlusion induced rat model and could thus provide for an alternate therapeutic strategy in vascular dementia and other age-related neurodegenerative disorders (Yadav *et al.*, 2017).

Oxyresveratrol (OXY) SLNs prepared by high-speed homogenization technique yielded mean particle size of 134.40 ± 0.57 nm. *In vitro* drug release study showed that the optimized formulation delayed the release profile for OXY with no initial burst release, compared to OXY suspension in the simulated gastrointestinal fluids. Thus, these SLNs could possibly be explored as another surrogate system for improving its oral bioavailability (Sangsen *et al.*, 2013).

10.7.5 Polymeric Nanoparticles

Polymeric nanoparticles (PNPs) are solid, colloidal polymeric particles in the size range of 10 to 1000 nm (Figure 10.4h). PNPs verily include any type of polymeric nanoparticles, nanospheres, and nanocapsules. Nanospheres are the matrix type of particles, wherein the entire mass is solid and molecules may be dispersed at the sphere surface or encapsulated within. Nanocapsules are a reservoir kind of vesicular system in which the entrapped substances are incorporated into a liquid core surrounded by a solid material shell (Rao and Geckeler, 2011).

Polymeric nanoparticles shield the encapsulating drugs from degradation and sustain their release with enhancement in intracellular penetration and improvement in bioavailability. Spherical resveratrol-loaded PCL-PLGA-PEG nanoparticles, with an average diameter of 150 nm, polydispersity index of 0.125, zeta potential of –25.7 mV, encapsulation efficiency range from 73% to 98%, and drug loading between 1.5% and 4%, have been reported by Sanna *et al.* (2013) with sustained release for 24 h. In another study, 7-folds increase in the absorption rate constant and 10-folds increase in the area under the curve have been reported with resveratrol-loaded PLGA nanoparticles vis-à-vis both pure drug and a marketed resveratrol product, indicating a significant enhancement in bioavailability using PNPs (Guo *et al.*, 2013).

Frozza *et al.* (2010) prepared resveratrol-loaded polymeric lipid core nanocapsules to improve its biodistribution and slow down its metabolism. Interfacial polymer deposition approach was employed to encapsulate trans-resveratrol nanocapsules, which resulted in 99.9% encapsulation efficiency. Nanocapsules had a size of about 240 nm, –14 mV of surface zeta potential, and 0.16 of polydispersity index. These physicochemical characteristics persisted after 3 months of storage at room temperature. *In vivo* distribution in healthy rats demonstrated significantly higher concentrations of trans-resveratrol in brain, kidney, and liver, in comparison to the free trans-resveratrol after intra-peritoneal and oral administration.

10.7.6 Nanosuspensions

Nanosuspensions are submicron colloidal dispersions of actives, stabilized with stabilizers such as surfactants, or mixture(s) of both, with the stabilizers causing an increase in the saturation solubility and speeding up the dissolution rate, thus allowing the increased surface to a volume ratio of the nanoparticles, especially for particle size below 1 µm (Figure 10.4i).

A study conducted by Hao *et al.* (2015) aimed to optimize and prepare resveratrol nanosuspensions by means of the anti-solvent precipitation method. Effects of process parameters on specified responses were determined applying Design of Experiment (DoE) approach using a three-factor three-level Box-Behnken design (BBD). Different mathematical polynomial models identified the influence of input variables on the chosen response variables and their interrelationship. The optimal formulation consisted of polyvinylpyrrolidone (PVP) K17 0.38%, drug 29.2 (mg/ml), and F188 3.63%. The morphology of nanosuspensions was found to be near-spherical shaped using scanning electron microscopy (SEM) observation. The X-ray powder diffraction (XRPD) and differential scanning calorimetry (DSC) analysis established that the nanoparticles were in the amorphous state. Furthermore, resveratrol nanosuspensions revealed augmentation in the saturation solubility and enhanced dissolution rate, owing to small particle size and amorphous nature of nanoparticles, compared to pristine resveratrol. In the intervening time, resveratrol nanosuspensions displayed equal antioxidant potency to that of pristine resveratrol, offering a rational strategy to widen the application range of this interesting bioactive (Hao *et al.*, 2015; Zeng *et al.*, 2017).

10.8 MISCELLANEOUS APPROACHES

Another approach of enhancing resveratrol bioavailability is making use of piperine as a bioenhancer to improve the pharmacokinetic parameters of resveratrol via inhibiting its glucuronidation, thereby slowing its elimination. A study conducted by Johnson *et al.* (2011) examined the enhancement of pharmacokinetic parameters of resveratrol along with piperine through inhibition of its glucuronidation. Mice were administered resveratrol (100 mg/kg) or resveratrol (100 mg/kg) + piperine (10 mg/kg) orally and the serum levels of resveratrol and resveratrol-3-O-β-D-glucuronide were examined at different times to study the effect of co-administration of piperine with resveratrol on the serum levels of free resveratrol, and it was found that the degree of exposure (i.e., AUC) to resveratrol was markedly enhanced (229%) and the maximum serum concentration (C_{max}) was increased enormously (1544%) with the addition of piperine, thus establishing that piperine significantly augments the bioavailability of resveratrol *in vivo* (Johnson *et al.*, 2011).

10.9 CONCLUSION AND FUTURE PERSPECTIVES

As resveratrol, a strong life prolonging bioactive agent, exhibits multiplicity of therapeutic actions, it is emerging as an important nutraceutical agent and an attractive molecule for the protection and treatment of several ailments. However, the poor biopharmaceutical and pharmacokinetic properties limit its futuristic potential to emerge as a powerful nutraceutical. Due to the fact that resveratrol undergoes extensive metabolism in the intestine and liver resulting in a low oral bioavailability (less than 1%) and short half-life, the perspectives now are to develop novel formulations that enable the drug to be carried out to the site of action, to overcome all the limitations governing resveratrol bioavailability and to be able to act effectively.

Development of nanoscale formulations to overcome some of the barriers to resveratrol's physicochemical properties has shown some success. A vast majority of the studies have intensively focused on developing controlled-release forms of resveratrol with improved resveratrol stability and solubility, further enhancing its *in vitro* release properties. Though a variety of nanocarriers are being investigated for incorporation of resveratrol, nano-encapsulation of resveratrol is still in its infancy. Several challenges such as long-term safety of nanoparticles, their interactions with biological systems, reproducibility and stability, lower loading capacity, and high costs still persist and need to be addressed. This calls for efforts to be undertaken to design nanoformulations of resveratrol with high loading capacity and colloidal stability, thus allowing maximum pharmacological activity of resveratrol. As a lot of these studies are still at preclinical stages, there is the need to perform extensive clinical studies to determine its safety. Another aspect that needs to be critically looked into is its manufacturing perspectives with focus on scale-up and technological transfers from bench to bedside before such nanonutraceuticals become a commercial reality.

REFERENCES

Abbasi, O. E., Goodarzi, M. T., Higgins, V., and Adeli, K. Role of resveratrol in the management of insulin resistance and related conditions: Mechanism of action. *Crit Rev Clin Lab Sci.* 2017; 54(4): 267–293.

Acosta, E. Bioavailability of nanoparticles in nutrient and nutraceutical delivery. *Curr Opin Colloid Interface Sci.* 2009; 14: 3–15.

Aklakur, M., Asharf Rather, M., and Kumar, N. Nanodelivery: An emerging avenue for nutraceuticals and drug delivery. *Crit Rev Food Sci Nutr.* 2016; 56(14): 2352–2361.

Amiji, M., and Tiwari, S. Nanoemulsion formulations for tumor-targeted delivery. *Nanotechnology for Cancer Therapy.* 2006; pp. 723–739.

Amiot, M. J., Romier, B., Dao, T. M., Fanciullino, R., Ciccolini, J., Burcelin, R. *et al.* Optimization of trans-resveratrol bioavailability for human therapy. *Biochimie.* 2013; 95(6): 1233–1238.

Amri, A., Chaumeil, J. C., Sfar, S., and Charrueau, C. Administration of resveratrol: What formulation solutions to bioavailability limitations. *J Control Rel.* 2012; 158(2): 182–193.

Al Dera, H. Neuroprotective effect of resveratrol against late cerebral ischemia reperfusion induced oxidative stress damage involves upregulation of osteopontin and inhibition of interleukin-lbeta. *J Physiol Pharmacol.* 2017; 68(1): 47–56.

Ansari, K. A., Vavia, P. R., Trotta, F., and Cavalli R. Cyclodextrin-based nanosponges for delivery of resveratrol: *In vitro* characterisation, stability, cytotoxicity and permeation study. *AAPS PharmSciTech.* 2011; 12(1): 279–286.

Asensi, M., Medina, I., Ortega, A., Carretero, J., Bano, M. C., Obrador, E. *et al.* Inhibition of cancer growth by resveratrol is related to its low bioavailability. *Free Radic Biol Med.* 2002; 33(3): 387–398.

Azios, N. G., Krishnamoorthy, L., Harris, M., Cubano, L. A., Cammer, M., and Dharmawardhane, S. F. Estrogen and resveratrol regulate Rac and Cdc42 signaling to the actin cytoskeleton of metastatic breast cancer cells. *Neoplasia.* 2007; 9(2): 147–158.

Bonechi, C., Lamponi, S., Donati, A., Tamasi, G., Consumi, M., Leone, G., Rossi, C., and Magnani, A. Effect of resveratrol on platelet aggregation by fibrinogen protection. *Biophys Chem.* 2017; 222: 41–48.

Bralley, E. E., Greenspan, P., Hargrove, J. L., Wicker, L., and Hartle, D. K. Topical anti-inflammatory activity of Polygonum cuspidatum extract in the TPA model of mouse ear inflammation. *J Inflamm*. 2008; 5(1): 1–7.

Caddeo, C., Teskac, K., Sinico, C., and Kristl, J. Effect of resveratrol incorporated in liposomes on proliferation and UV-B protection of cells. *Int J Pharma*. 2008; 363(1–2): 183–191.

Charters, A. J., Selander, S., Morris, R., and Blackner, L. Herbal composition and method for combating inflammation. US Patent 6541945. 2003.

Chilajwar, S. V., Pednekar, P. P., Jadhav, K. R., Gupta, G. J., and Kadam, V. J. Cyclodextrin-based nanosponges: A propitious platform for enhancing drug delivery. *Expert Opin Drug Deliv*. 2014; 11(1): 111–120.

Coimbra, M., Isacchi, B., van Bloois, L., Torano, J. S., Ket, A., and Wu, X. Improving solubility and chemical stability of natural compounds for medicinal use by incorporation into liposomes. *Int J Pharm*. 2011; 416(2): 433–442.

Cottart, C. H., Nivet-Antoine, V., Laguillier-Morizot, C., and Beaudeux, J. L. Resveratrol bioavailability and toxicity in humans. *Mol Nutr Food Res*. 2010; 54(1): 7–16.

Das, D. K., and Maulik, N. Resveratrol in cardioprotection: A therapeutic promise of alternative medicine. *Mol Interv*. 2006; 6(1): 36–47.

Das, S., Lin, H. S., Ho, P. C., and Ng, K. Y. The impact of aqueous solubility and dose on the pharmacokinetic profiles of resveratrol. *Pharm Res*. 2008; 25(11): 2593–2600.

Davidov-Pardo, G., and McClements, D. J. Nutraceutical delivery systems: Resveratrol encapsulation in grape seed oil nanoemulsions formed by spontaneous emulsification. *Food Chem*. 2015; 167: 205–212.

de Oliveira, M. R., Chenet, A. L., Duarte, A. R., Scaini, G., and Quevedo, J. Molecular mechanisms underlying the anti-depressant effects of resveratrol: A review. *Mol Neurobiol*. 2018; 55(6): 4543–4559.

Detoni, C. B., Souto, G. D., da Silva, A. L., Pohlmann, A. R., and Guterres, S. S. Photostability and skin penetration of different E-resveratrol-loaded supramolecular structures. *Photochem Photobiol*. 2012; 88(4): 913–921.

Donsì, F., Sessa, M., Mediouni, H., Mgaidi, A., and Ferrari, G. Encapsulation of bioactive compounds in nanoemulsion-based delivery systems. *Procedia Food Sci*. 2011; 1: 1666–1671.

Ethemoglu, M. S., Seker, F. B., Akkaya, H., Kilic, E., Aslan, I., Erdogan, C. S., and Yilmaz, B. Anticonvulsant activity of resveratrol-loaded liposomes *in vivo*. *Neuroscience*. 2017; 357: 12–19.

Fischer, N., Büchter, C., Koch, K., Albert, S., Csuk, R., and Wätjen, W. The resveratrol derivatives trans-3,5-dimethoxy-4-fluoro-4'-hydroxystilbene and trans-2,4',5-trihydroxystilbene decrease oxidative stress and prolong lifespan in Caenorhabditis elegans. *J Pharm Pharmacol*. 2017; 69(1): 73–81.

Fiori, J. L., Shin, Y. K., Kim, W., and Egan, J. M. Resveratrol prevents β-cell dedifferentiation in nonhuman primates given a high-fat/high-sugar diet. *Diabetes*. 2013; 62(10): 3500–3513.

Fornara, V., Onelli, E., Sparvoli, F., Rossoni, M., Aina, R., and Marino, G. Localization of stilbene synthase in Vitis vinifera L. during berry development. *Protoplasma*. 2008; 233(1–2): 83–93.

Frozza, R. L., Bernardi, A., Paese, K., Hoppe, J. B., da Silva, T., Battastini, A. M. *et al.* Characterization of trans-resveratrol-loaded lipid-core nanocapsules and tissue distribution studies in rats. *J Biomed Nanotechnol*. 2010; 6(6): 694–703.

Guo, W., Li, A., Jia, Z., Yuan, Y., Dai, H., and Li, H. Transferrin modified PEG-PLA-resveratrol conjugates: *In vitro* and *in vivo* studies for glioma. *Eur J Pharmacol*. 2013; 718(1–3): 41–47.

Gyorgyi, S., Anna, C., Kenneth, L., Zoltan, U., and Gabor, V. Resveratrol inhibits aggregation of platelets from high-risk cardiac patients with aspirin resistance. *J Cardiovasc Pharmacol*. 2006; 48(2): 1–5.

Hao, J., Gao, Y., Zhao, J., Zhang, J., Li, Q., Zhao, Z. *et al.* Preparation and optimization of resveratrol nanosuspensions by antisolvent precipitation using Box-Behnken design. *AAPS PharmSciTech*. 2015; 16(1): 118–128.

Imamura, H., Yamaguchi, T., Nagayama, D., Saiki, A., Shirai, K., and Tatsuno, I. Resveratrol ameliorates arterial stiffness assessed by cardio-ankle vascular index in patients with type 2 diabetes mellitus. *Int Heart J*. 2017; 58(4): 577–583.

Jang, M., Cai, L., Udeani, G. O., Slowing, K. V., Thomas, C. F., and Beecher, C. W. Cancer chemopreventive activity of resveratrol, a natural product derived from grapes. *Science*. 275 5297 (1997): 218–220.

Johnson, J. J., Nihal, M., Siddiqui, I. A., Scarlett, C. O., Bailey, H. H., and Mukhtar, H. Enhancing the bioavailability of resveratrol by combining it with piperine. *Mol Nutr Food Res*. 2011; 55(8): 1169–1176.

Khan, I., Yousaf, S., Subramanian, S., Korale, O., Alhnan, M. A., Ahmed, W., Taylor, K. M., and Elhissi, A. Proliposome powders prepared using a slurry method for the generation of beclometasone dipropionate liposomes. *Int J Pharm*. 2015; 496(2): 342–350.

Koga, C. C., Lee, S. Y., and Lee, Y. Consumer acceptance of bars and gummies with unencapsulated and encapsulated resveratrol. *J Food Sci.* 2016; 81(5): S1222-S1229.

Kristl, J., Teskac, K., Caddeo, C., Abramovic, Z., and Sentjurc, M. Improvements of cellular stress response on resveratrol in liposomes. *Eur J Pharm Biopharm.* 2009; 73(2): 253–259.

Langcake, P., and Pryce, R. J. Production of resveratrol by Vitis-vinifera and other members of vitaceae as a response to infection or injury. *Physiol Plant Pathol.* 1976; 9(1): 77–86.

Leonard, S. S., Xia, C., Jiang, B. H., Stinefelt, B., Klandorf, H., and Harris, G. K. Resveratrol scavenges reactive oxygen species and effects radical-induced cellular responses. *Biochem Biophys Res Commun.* 2003; 309(4): 1017–1026.

Li, D. C., Zhong, X. K., Zeng, Z. P., Jiang, J. G., Li, L., and Zhao, M. M. Application of targeted drug delivery system in Chinese medicine. *J Control Rel.* 2009; 138(2): 103–112.

Lopez-Nicolas, J. M., Nunez-Delicado, E., Perez-Lopez, A. J., Barrachina, A. C., and Cuadra-Crespo, P. Determination of stoichiometric coefficients and apparent formation constants for beta-cyclodextrin complexes of trans-resveratrol using reversed-phase liquid chromatography. *J Chromatogr A.* 2006; 1135(2): 158–165.

Lucas-Abellán, C., Fortea, I., López-Nicolás, J. M., and Núñez-Delicado, E. Cyclodextrins as resveratrol carrier system. *Food Chem.* 2007; 104(1): 39–44.

McCubrey, J. A., Lertpiriyapong, K., Steelman, L. S., Abrams, S. L., Cocco, L., and Ratti, S. Regulation of GSK-3 activity by curcumin, berberine and resveratrol: Potential effects on multiple diseases. *Adv Biol Regul.* 2017; 65: 77–88.

Meng, X., Maliakal, P., Lu, H., Lee, M. J., and Yang, C. S. Urinary and plasma levels of resveratrol and quercetin in humans, mice, and rats after ingestion of pure compounds and grape juice. *J Agric Food Chem.* 2004; 52(4): 935–942.

Nam, J., Ryu, J., Kim, J., Chang, I., and Suh, K. Stabilization of resveratrol immobilized in monodisperse cyano-functionalized porous polymeric microspheres. *Polymer.* 2005; 46(21): 8956–8963.

Neves, A. R., Lucio, M., Lima, J. L., and Reis, S. Resveratrol in medicinal chemistry: A critical review of its pharmacokinetics, drug-delivery, and membrane interactions. *Curr Med Chem.* 2012; 19(11): 1663–1681.

Neves, A. R., Lucio, M., Martins, S., Lima, J. L., and Reis, S. Novel resveratrol nanodelivery systems based on lipid nanoparticles to enhance its oral bioavailability. *Int J Nanomedicine.* 2013; 8: 177–187.

Neves, A. R., Martins, S., Segundo, M. A., and Reis, S. Nanoscale delivery of resveratrol towards enhancement of supplements and nutraceuticals. *Nutrients.* 2016; 8(3): 131.

Olas, B., Wachowicz, B., Szewczuk, J., Saluk-Juszczak, J., and Kaca, W. The effect of resveratrol on the platelet secretory process induced by endotoxin and thrombin. *Microbios.* 2001; 105(410): 7–13.

Palacz-Wrobel, M., Borkowska, P., Paul-Samojedny, M., Kowalczyk, M., Fila-Danilow, A., Suchanek-Raif, R., and Kowalski, J. Effect of apigenin, kaempferol and resveratrol on the gene expression and protein secretion of tumor necrosis factor alpha (TNF-α) and interleukin-10 (IL-10) in RAW-264.7 macrophages. *Biomed Pharmacother.* 2017; 93: 1205–1212.

Palsamy, P., and Subramanian, S. Modulatory effects of resveratrol on attenuating the key enzymes activities of carbohydrate metabolism in streptozotocin-nicotinamide-induced diabetic rats. *Chem Biol Interact.* 2009; 179(2–3): 356–362.

Pan, J., Shen, J., Si, W., Du, C., Chen, D., Xu, L., Yao, M., Fu, P., and Fan, W. Resveratrol promotes MICA/B expression and natural killer cell lysis of breast cancer cells by suppressing c-Myc/miR-17 pathway. *Oncotarget.* 2017; 8(39): 65743–65758.

Pervaiz, S. Resveratrol: From grapevines to mammalian biology. *The FASEB Journal.* 2003; 17(14): 1975–1985.

Platella, C., Guida, S., Bonmassar, L., Aquino, A., Bonmassar, E., Ravagnan, G., Montesarchio, D., Roviello, G. N., Musumeci, D., and Fuggetta, M. P. Antitumour activity of resveratrol on human melanoma cells: A possible mechanism related to its interaction with malignant cell telomerase. *Biochim Biophys Acta.* 2017: S0304–4165(17) 30246-5.

Ponzo, V., Soldati, L., and Bo, S. Resveratrol: A supplementation for men or for mice. *J Transl Med.* 2014; 12: 158.

Pople, P. V., and Singh, K. K. Development and evaluation of colloidal modified nanolipid carrier: Application to topical delivery of tacrolimus. *European J Pharm Biopharm.* 2011; 79(1): 82–94.

Rao, J. P., and Geckeler, K. E. Polymer nanoparticles: Preparation techniques and size-control parameters. *Prog Polym Sci.* 2011; 36(7): 887–913.

Rossi, D., Guerrini, A., Bruni, R., Brognara, E., Borgatti, M., and Gambari, R. trans-Resveratrol in nutraceuticals: Issues in retail quality and effectiveness. *Molecules.* 2012; 17(10): 12393–12405.

Rubiolo, J. A., Mithieux, G., and Vega, F. V. Resveratrol protects primary rat hepatocytes against oxidative stress damage: Activation of the Nrf2 transcription factor and augmented activities of antioxidant enzymes. *Eur J Pharmacol.* 2008; 591(1–3): 66–72.

Subramanian, S., Krishnamoorthy, K., and Rajappan, M. Nanosponges: A novel class of drug delivery system. *J Pharm Pharm Sci.* 2012; 15(1): 103–111.

Sale, S., Verschoyle, R. D., Boocock, D., Jones, D. J., Wilsher, N., and Ruparelia, K. C. Pharmacokinetics in mice and growth-inhibitory properties of the putative cancer chemopreventive agent resveratrol and the synthetic analogue trans 3,4,5,4'-tetramethoxystilbene. *Br J Cancer.* 2004; 90(3): 736–744.

Sangsen, Y., Likhitwitayawuid, K., Sritularak, B., Wiwattanawongsa, K., and Wiwattanapatapee, R. Novel solid lipid nanoparticles for oral delivery of oxyresveratrol: Effect of the formulation parameters on the physicochemical properties and *in vitro* release. *Int J Med Sci Eng.* 2013; 17(12): 1–8.

Sanna, V., Siddiqui, I. A., Sechi, M., and Mukhtar, H. Resveratrol-loaded nanoparticles based on poly(epsilon-caprolactone) and poly(D,L-lactic-co-glycolic acid)-poly(ethylene glycol) blend for prostate cancer treatment. *Mol Pharm.* 2013; 10(10): 3871–3881.

Sarubbo, F., Moranta, D., Asensio, V. J., Miralles, A., and Esteban, S. Effects of resveratrol and other polyphenols on the most common brain age-related diseases. *Curr Med Chem.* 2017; 24(38): 4245–4266.

Saunier, E., Antonio, S., Regazzetti, A., Auzeil, N., Laprévote, O., Shay, J. W., Coumoul, X., Barouki, R., Benelli, C., Huc, L., and Bortoli, S. Resveratrol reverses the Warburg effect by targeting the pyruvate dehydrogenase complex in colon cancer cells. *Sci Rep.* 2017; 7(1): 6945.

Schwager, J., Richard, N., Widmer, F., and Raederstorff, D. Resveratrol distinctively modulates the inflammatory profiles of immune and endothelial cells. *BMC Complement Altern Med.* 2017; 17(1): 309.

Sessa, M., Balestrieri, M. L., Ferrari, G., Servillo, L., Castaldo, D., and D'Onofrio, N. Bioavailability of encapsulated resveratrol into nanoemulsion-based delivery systems. *Food Chem.* 2014; 147: 42–50.

Shi, G., Rao, L., Yu, H., Xiang, H., Yang, H., and Ji, R. Stabilization and encapsulation of photosensitive resveratrol within yeast cell. *Int J Pharm.* 2008; 349(1–2): 83–93.

Siemann, E. H., and Creasy, L. L. Concentration of the phytoalexin resveratrol in wine. *Am J Enol Vitic.* 1992; 43(1): 49–52.

Silk, J. J., and Smoliga, J. M. Resveratrol: Nutraceutical believed to counteract the detrimental effects of high-fat diet. *Lipid Technol.* 2014; 26(1): 15–17.

Soleas, G. J., Angelini, M., Grass, L., Diamandis, E. P., and Goldberg, D. M. Absorption of trans-resveratrol in rats. *Methods Enzymol.* 2001; 335: 145–154.

Soo, E., Thakur, S., Qu, Z., Jambhrunkar, S., Parekh, H. S., and Popat, A. Enhancing delivery and cytotoxicity of resveratrol through a dual nanoencapsulation approach. *J Colloid Interface Sci.* 2016; 462: 368–374.

Szkudelska, K., Nogowski, L., and Szkudelski, T. Resveratrol, a naturally occurring diphenolic compound, affects lipogenesis, lipolysis and the antilipolytic action of insulin in isolated rat adipocytes. *J Steroid Biochem Mol Biol.* 2009; 113(1–2): 17–24.

Tadros, T., Izquierdo, P., Esquena, J., and Solans, C. Formation and stability of nano-emulsions. *Adv Colloid Interface Sci.* 2004; 108–109: 303–18.

Teskac, K., and Kristl, J. The evidence for solid lipid nanoparticles mediated cell uptake of resveratrol. *Int J Pharm.* 2010; 390(1): 61–69.

Udenigwe, C. C., Ramprasath, V. R., Aluko, R. E., and Jones, P. J. Potential of resveratrol in anticancer and anti-inflammatory therapy. *Nutr Rev.* 2008; 66(8): 445–454.

Vitrac, X., Desmouliere, A., Brouillaud, B., Krisa, S., Deffieux, G., and Barthe, N. Distribution of [14C]-trans-resveratrol, a cancer chemopreventive polyphenol, in mouse tissues after oral administration. *Life Sci.* 2003; 72(20): 2219–2233.

Walle, T., Hsieh, F., DeLegge, M. H., Oatis, J. E., Jr., and Walle, U. K. High absorption but very low bioavailability of oral resveratrol in humans. *Drug Metab Dispos.* 2004; 32(12): 1377–1382.

Walle, T. Bioavailability of resveratrol. *Ann N Y Acad Sci.* 2011; 1215: 9–15.

Wang, W., Tang, K., Yang, H.-R., Wen, P.-F., Zhang, P., and Wang, H.-L. Distribution of resveratrol and stilbene synthase in young grape plants (Vitis vinifera L. cv. Cabernet Sauvignon) and the effect of UV-C on its accumulation. *Plant Physiol Biochem.* 2010; 48(2–3): 142–152.

Weiskirchen, S., and Weiskirchen, R. Resveratrol: How much wine do you have to drink to stay healthy? *Adv Nutr.* 2016; 7(4): 706–718.

Xia, N., Förstermann, U., and Li, H. Effects of resveratrol on eNOS in the endothelium and the perivascular adipose tissue. *Ann N Y Acad Sci.* 2017; 1403(1): 132–141.

Yadav, A., Sunkaria, A., Singhal, N., and Sandhir, R. Resveratrol loaded solid lipid nanoparticles attenuate mitochondrial oxidative stress in vascular dementia by activating Nrf2/HO1 pathway. *Neurochem Int.* 2017; 112: 239–254.

Yonamine, C. Y., Pinheiro-Machado, E., Michalani, M. L., Alves-Wagner, A. B., Esteves, J. V., Freitas, H. S., and Machado, U. F. Resveratrol improves glycemic control in type 2 diabetic obese mice by regulating glucose transporter expression in skeletal muscle and liver. *Molecules*. 2017; 14: 22(7): E1180.

Yu, C., Shin, Y. G., Chow, A., Li, Y., Kosmeder, J. W., and Lee, Y. S. Human, rat, and mouse metabolism of resveratrol. *Pharm Res*. 2002; 19(12): 1907–1914.

Zhang, Q. H., Xiong, Q. P., Shi, Y. Y., and Zhang, D. Y. Study on preparation and characterization of resveratrol solid lipid nanoparticles and its anticancer effects *in vitro*. *Zhong Yao Cai*. 2010; 33(12): 1929–1932.

Zeng, Y. H., Zhou, L. Y., Chen, Q. Z., Li, Y., Shao, Y., Ren, W. Y., Liao, Y. P., Wang, H., Zhu, J. H., Huang, M., He, F., Wang, J., Wu, K., and He, B. C. Resveratrol inactivates PI3K/Akt signaling through upregulating BMP7 in human colon cancer cells. *Oncol Rep*. 2017; 38(1): 456–464.

11 Nanotherapeutics: Enabling Vitamin D3 as a Multifaceted Nutraceutical

*Krantisagar S. More, Vinod S. Ipar, Amit S. Lokhande, Anisha A. D'souza, and Padma V. Devarajan**

CONTENTS

* Corresponding author.

11.1 INTRODUCTION

Nutraceuticals are bioactive molecules derived from food sources to provide extra health benefits. They are hybridized moieties, which could serve as nutrients and pharmaceuticals (Kesarwani *et al.*, 2013; Gupta *et al.*, 2015), acting either through specific or nonspecific biological therapies (McClements *et al.*, 2015; Gleeson *et al.*, 2016). The sale of nutraceutical products has unprecedentedly spiked in the last 20 years due to enhanced tendency of self-medication and an increase in the aging population (Lockwood, 2007). The Nutrition Business Journal has reported a business of $140 billion in the global nutrition and supplements market in 2016.

Nutraceuticals include macronutrients and micronutrients. Macronutrients are proteins, carbohydrates, and fats, usually needed in large amounts to provide energy. Micronutrients include vitamins and minerals used to build and repair tissues and regulate body processes. The functions of vitamins and minerals are interconnected (Prentice, 2005; Fogelholm *et al.*, 2012). Vitamins, the vital amines, have diverse biochemical functions such as antioxidant (vitamin E and C), hormone like function (vitamin D), enzyme cofactor or precursor (vitamin B), cell and tissue growth regulator (vitamin A), and so on (Pond *et al.*, 1995). Most of the vitamins can be taken as external supplements. Vitamin D, however, is obtained by exposure to sunlight and plays a crucial role in health management (Mostafa and Hegazy, 2015). It is a steroid hormone with pleiotropic action on cells and tissues of the body (Buttriss, 2015). Despite being a sunbeam vitamin, more than a billion people worldwide are vitamin D deficient (Holick, 2008; Naeem, 2010). The major cause of vitamin D deficiency in humans is lack of exposure to sunlight and limited food sources containing vitamin D (Holick and Chen, 2008; Johnson, 2010; Kumar *et al.*, 2015). While rickets is associated with severe vitamin D deficiency in children, bone disorders including osteoporosis in adults are also linked to insufficient vitamin D concentration (Holick, 2006; Thacher and Clarke, 2011). As a nutraceutical, vitamin D, especially vitamin D3, is gaining importance due to myriad activities linked to this nutraceutical including its role in modulating autoimmune diseases, cardiovascular diseases, stroke, birth defects, periodontal diseases, and 17 different types of cancers (Calvo *et al.*, 2004). With a brief introduction to vitamin D and conventional vitamin D formulations, this chapter will focus on the multifaceted role of vitamin D in various pathologies. Vitamin D as a nanonutraceutical and its reported application in a number of these pathologies is also summarized.

11.2 VITAMIN D

Vitamin D was discovered in the mid-1600s, and deficiency of the vitamin is associated with rickets (Rajakumar and Holick, 2012; Deluca, 2014; O'Riordan and Bijvoet, 2014). In the mid-1800s, it was observed that urban children were more susceptible to rickets than rural children (Wagner *et al.*, 2008). The Nobel Prize in Chemistry in 1928 was awarded to Windaus and colleagues, who separated the precursors of vitamin D3, 7-dehydrocholesterol from animal skin, which when exposed to sunlight gets converted to vitamin D. Hence, even today, the vitamin is known as the sunbeam vitamin. Vitamin D is the general name assigned to a group of natural fat soluble sterol-like constituents. It includes vitamin D2 (i.e., ergocalciferol) derived from plant source and vitamin D3 (i.e., cholecalciferol) derived from animal source. It is responsible for improving the intestinal absorption of inorganic ions such as calcium, zinc, iron, magnesium, and phosphate (Hartiti *et al.*, 1995; Pointillart *et al.*, 1995; Marks *et al.*, 2006). Among the vitamin D family, vitamin D3 is the most potent form of vitamin D (Castor, 2016). The family of vitamin D consists of different, chemically distinct types, including D1, D2, D3, D4 and D5, as listed in Table 11.1.

Among these forms, only 2 forms of vitamin D, D2 and D3, are stable. Vitamin D2 is mainly derived from plant and fungal sources (Dadoniene *et al.*, 2016). In humans, vitamin D2 has only 25%–30% of the biological activity as compared to vitamin D3. Further, it was observed that vitamin D3 concentration was almost 10-fold higher compared to vitamin D2 when administered orally (Armas *et al.*, 2004). Hence, vitamin D3, which maintains mineral homeostasis and skeletal health, is considered more important than any other derivative of vitamin D. Several guidelines on intake of vitamin D have been published by national and international organizations that range from 400–60,000 IU (Spears and Weiss, 2014). Some studies have revealed that an oral intake of 250 μg (10,000 IU) of vitamin D3 is equivalent to one whole day of exposure to sunlight (Vieth, 2007; Whayne, 2011).

The requirement of vitamin D3 intake varies with age; children, older people, and pregnant and postmenopausal women require additional vitamin D supplementation (Harinarayan and Joshi, 2009; Agarwal *et al.*, 2013; Viswanath 2013). Other indications for vitamin D3 supplementation include individuals not exposed to adequate sunlight or obese individuals wherein the vitamin is sequestered in the fat tissues. The relationship of vitamin D3 levels in serum and bone health status is depicted in Table 11.2 (Institute of Medicine, 2010).

11.2.1 Source of Vitamin D3

Vitamin D3 has limited availability from dietary sources. The prime source of vitamin D3 in humans is natural production in the skin upon exposure to ultraviolet-B (UV-B) radiation (290–315 nm) (Heinig, 2003; Grant and Holick, 2005). The richest dietary sources of vitamin D3 are cod liver oil (400 IU/tsp), egg yolk (approximately 20 IU), oil-rich salmon (approximately 400 IU per 3.5oz.), sardines and mackerel, poultry, meat and meat products like liver and kidney (0.1–1.5 μg/100 g) (McDonnell *et al.*, 2014), milk (approximately 20 IU), and milk products, with smaller amounts found in mushroom (Urbain *et al.*, 2011; Biancuzzo *et al.*, 2013; Keegan *et al.*, 2013). An important source today is vitamin D fortified foods and supplements (Dhaussy, 2014; Gupta, 2014).

11.2.2 Pharmacokinetics and Bioavailability of Vitamin D3

11.2.2.1 Absorption and Distribution

Vitamin D3 is a lipid soluble vitamin. Dietary vitamin D is absorbed through the small intestine as chylomicron fraction and then transported to the lymph. In plasma, vitamin D3 circulates as 25-hydroxy vitamin D3 and binds to a α2-globulin protein (Vitamin D-binding Protein, DBP) and accumulates in the liver. Vitamin D3 is hydroxylated in the liver by enzyme hydroxylase in the mitochondria and forms 25-hydroxyvitamin D3 (also known as 25-hydroxycholecalciferol, calcidiol or calcifediol) (Holick, 2007; Milne and Delander, 2008).

TABLE 11.1
Types of Vitamin D

Vitamin D Types	Synonym	Structure
Vitamin D1	Ergocalciferol with lumisterol 1:1	–
Vitamin D2	Ergocalciferol	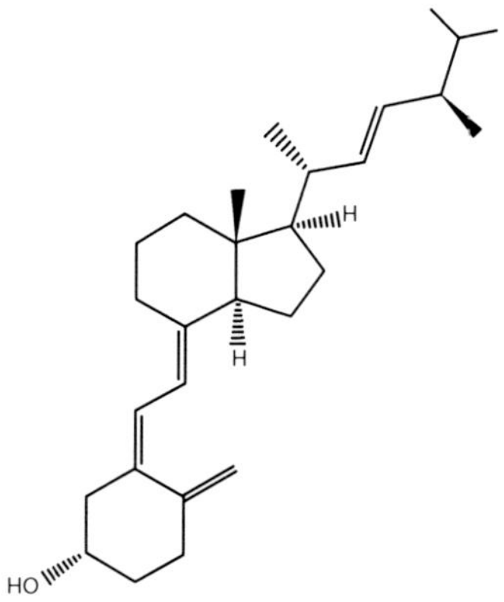
Vitamin D3	Cholecalciferol	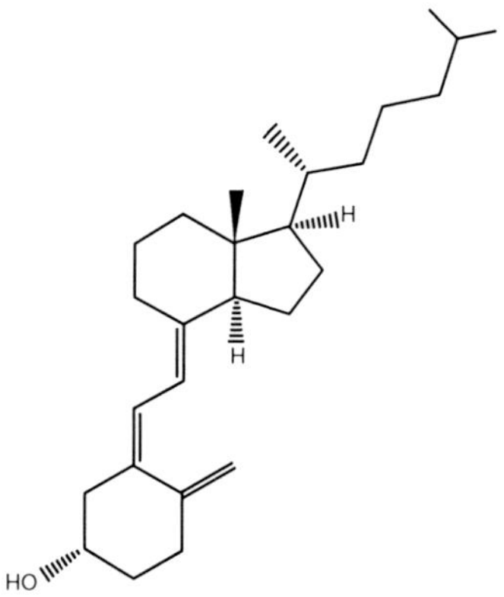
Vitamin D4	22-dihydroergocalciferol	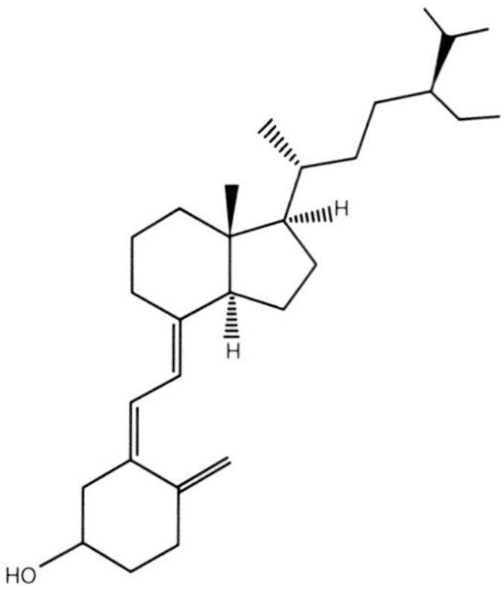

(Continued)

TABLE 11.1 (CONTINUED)
Types of Vitamin D

Vitamin D Types	Synonym	Structure
Vitamin D5	Sitocalciferol	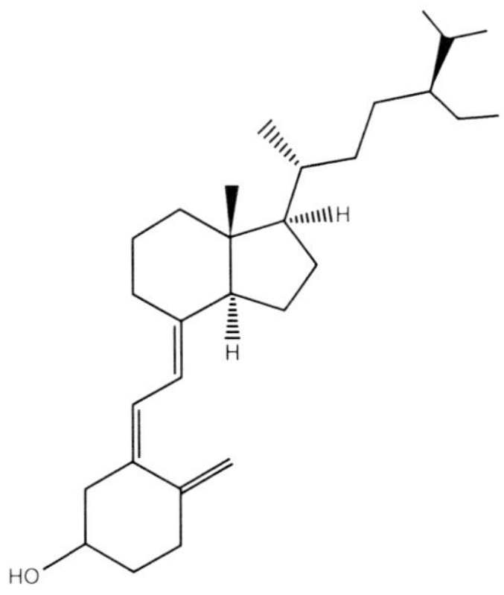

TABLE 11.2
Serum Vitamin D3 Concentrations and Bone Health Status

Serum Levels of Vitamin D3		
nmol/L*	ng/mL	Health Status
<30	<12	Associated with vitamin D3 deficiency, leading to rickets in infants and children and osteomalacia in adults
30 to <50	12 to <20	Generally considered inadequate for bone and overall health in healthy individuals
≥50	≥20	Generally considered adequate for bone and overall health in healthy individuals
>125	>50	Emerging evidence links potential adverse effects to such high levels, particularly >150 nmol/L (>60 ng/mL)

* 1 nmol/L = 0.4 ng/mL.

11.2.2.2 Metabolism and Excretion

The 25-hydroxyvitamin D3 is transported from the liver to kidney, where it is hydroxylated by the parathyroid hormone (PTH) to form 1, 25-dihydroxycholecalciferol or calcitriol, the active form of vitamin D3. The 25-hydroxycholecalciferol–24-hydroxylase (24-OH-ase) enzyme hydroxylates 25-hydroxycholecalciferol to form 24, 25-dihydroxycholecalciferol. Finally, vitamin D3 metabolizes to calcitroic acid, which is excreted in urine (Henry and Norman, 1984; Plum and DeLuca, 2010). Vitamin D and its metabolites can increase the calcium balance and ease mineral deposition in bone matrix without direct effects on bone cells, as vitamin D3 enhances bone resorption, simultaneously inhibiting bone mineralization (Eisman and Bouillon, 2014).

TABLE 11.3
Recommended Dietary Allowances (RDAs) for Vitamin D3

Age	Male	Female	Pregnancy	Lactation
0–12 months	400 IU(10 µg)	400 IU(10 µg)	–	–
1–13 years	600 IU(15 µg)	600 IU(15 µg)	–	–
14–18 years	600 IU(15 µg)	600 IU(15 µg)	600 IU(15 µg)	600 IU(15 µg)
19–50 years	600 IU(15 µg)	600 IU(15 µg)	600 IU(15 µg)	600 IU(15 µg)
51–70 years	600 IU(15 µg)	600 IU(15 µg)	–	–
>70 years	800 IU(20 µg)	800 IU(20 µg)	–	–

11.2.2.3 Bioavailability

Bioequivalence study of vitamin D2 and vitamin D3 in humans demonstrated higher levels of vitamin D3 in serum compared to vitamin D2 (Lehmann *et al.*, 2013). Oral bioavailability of vitamin D3 increased to 60–80%, when administered along with milk, long-chain fatty acids, triglycerides (Glossmann, 2011), and also as fortified food (Wagner *et al.*, 2008). Vitamin D3 concentration in serum was reduced in alcoholics and smokers and in subjects who were on a high fiber diet due to rapid elimination from the body (Brot *et al.*, 1999).

11.3 VITAMIN D DIETARY SUPPLEMENTS

The recent Dietary Reference Intake (DRI) values of vitamin D3 for adults are approximately half the amount of a teaspoon of cod-liver oil, the traditional source of vitamin D3 (Vieth, 1999). The upper levels of vitamin D3 intake are based on the highest safe amount an individual can take. The Dietary Reference Intake (DRI) for vitamin D3 is given in Table 11.3 (Institute of Medicine, 2010).

The maximum recommended dietary intake (RDI) of vitamin D3 is 800 IU/day (Ross *et al.*, 2011; Viswanath, 2013). Moreover, for therapeutic use, the intake of vitamin D3 goes up to 60,000 IU/week depending on the severity of the disease state (Harinarayan *et al.*, 2012; Cashman, 2015). Therefore, oral supplementation of vitamin D3 can be considered as a safe and practical strategy to achieve optimum dose levels of vitamin D3, thereby providing an alternative to sunlight.

11.4 CONVENTIONAL APPROACHES FOR VITAMIN D3 DELIVERY

Conventional delivery systems of vitamin D3 include tablets, capsules, oral liquid drops, and topical creams. Vitamin D3 supplements are available as liquid solutions for infants. Daily consumption of 10 mg vitamin D3 as a fish oil capsule showed increased levels of serum calcidiol (precursor of vitamin D3) concentration as compared to the same dose of vitamin D3 administered as a multivitamin tablet (Ashwell *et al.*, 2010; Jarrett *et al.*, 2014). Vitamin D3 is sensitive to environmental factors, namely temperature, oxygen, pressure, light, etc. (Luo *et al.*, 2012). Vitamin D3 is a BCS class II molecule. The highly lipophilic nature of vitamin D3 precludes its ease of dispersion into an aqueous solution. Appropriate formulation strategies are therefore essential to enhance stability, solubility, and bioavailability. Encapsulation of vitamin D3 can increase its stability during preparation and storage (Das *et al.*, 2010). Vitamin D3 tablets dispersed in sterile water, when administered through an enteral feeding tube in ventilated intensive care unit patients, at high doses of 250,000 IU and 500,000 IU, increased plasma concentrations to the required and safe range (Han *et al.*, 2016). Vitamin D3 oral solution at a dose of 500 IU/kg/day improved functional recovery such as locomotor activity, H-reflex depression, breathing, and improved spasticity when delivered during the acute phase after a spinal cord trauma (Gueye *et al.*, 2015). Intramuscular (IM) administration is also an effective route to increase vitamin D3 levels and balance its performance (Tellioglu *et al.*, 2012). IM injection of vitamin D3 (600,000 IU) in adults increased the serum vitamin D3 level to normal (>30 ng/mL) levels in all subjects within two months. However, it reached inadequate ranges within the next six months. The increase in vitamin D3 level was dependent on BMI with subjects having normal BMI demonstrating higher serum vitamin D3 than those with higher body mass index (BMI) (Purohit, 2015). Once-yearly IM injection of a mega dose of vitamin D3 (600,000 IU) (Diamond *et al.*, 2005) in an arachis oil depot formulation as an adjunct to oral supplementation proved to be safe and effective in improving serum vitamin D3 concentrations for a period of 12 months, but did not increase the bone mineral density (Einarsdóttir *et al.*, 2010). Ointment containing a vegetable oil like soya oil, coconut oil, and polyhydric alcohols with vitamin D3 is also available (Makino *et al.*, 1989). Vitamin D3 cream containing aromatic oils and aloe vera gel as permeation enhancers safely and effectively delivered vitamin D3 at dose of 5,000 IU/day through dermal route to achieve a minimum plasma concentration level of 30 ng/mL after 90 days of treatment (Sadat-Ali *et al.*, 2014).

Absorption of vitamin D3 is facilitated by bile acid and fat complex through action of chylomicrons, which enable dissolution and absorption of vitamin D3 as a fine dispersion (Lehmann and Meurer, 2010; Keegan *et al.*, 2013). Once absorbed, vitamin D3, being fat-soluble, is deposited mainly in adipose tissues and the skeletal muscle. However, the conventional delivery of vitamin D3 exhibits concerns of short half-life *in vivo* (Yin *et al.*, 2010) and rapid first-pass effect (Plum and DeLuca, 2010). To improve its short half-life, vitamin D3 may also be designed as sustained release formulations (Bishop *et al.*, 1998).

Vitamin D3-loaded polylactic acid (PLA) microspheres (42 μm) could provide sustained release and an effective strategy for the treatment of diabetes (Chung *et al.*, 2001; Li *et al.*, 2013). These particles could protect bone marrow stromal cell line (BMSCs) in a diabetic periodontitis-like environment and prevent alveolar bone loss in a rat model of diabetic periodontitis. Cellulose acetate microsphere of vitamin D3 increased the insulin output in T1DM patients and revealed anti-oxidizing effect to enhance pancreatic islet cell viability and function (Luca *et al.*, 2003). Targeted delivery of vitamin D3 to the colon using β-glucuronides revealed promising reduction of colitis lesions and symptoms (Goff *et al.*, 2012). A formulation for preventing or treating osteoporosis comprised a solid dispersion of vitamin D and a bisphosphonate (Evangelos Karavas *et al.*, 2015). This therapeutic approach provides attractive benefits of targeted delivery systems.

11.5 STABILIZATION OF VITAMIN D3

Solid dispersion, comprising of vitamin D3 or its derivatives with cyclodextrin, prepared for prevention or treatment of osteoporosis revealed enhanced stability (Woo *et al.*, 2007). Solid dispersions using lipophilic surfactant D-α-tocopheryl polyethylene glycol 1000 succinate (TPGS) and SiO_2 as solidifier could modify the release and enhance stability of vitamin D3 (Yuan *et al.*, 2013). The β-lactoglobulin (β-lg)/D3 complexes exhibited significantly improved stability at 4°C and when exposed to ultraviolet-C light. Enhanced bioavailability demonstrated in rats suggested application of this complex for the fortification of milk products and low-fat content food (Diarrassouba *et al.*, 2014). Vitamin D3 was successfully protected from oxidative and environmental stress by encapsulation in a matrix of carbohydrates and hydrogenated carbohydrates using a special extrusion technique (Petritz *et al.*, 2006). While a solid dispersion comprising of vitamin D3 or its derivatives with cyclodextrin, enhanced stability (Woo *et al.*, 2007) loss of vitamin D3 was observed when processed at 50°C (Delaurent *et al.*, 1998). Polylactide microspheres of vitamin D3 as a drug model increased the vitamin D3 stability (Pavanetto *et al.*, 1993). Microencapsulation of a combination of multiple functional lipophilic ingredients including tuna oil, vitamin A, vitamin D3, vitamin E, vitamin K2, coenzyme Q10, and curcumin prepared by complex coacervation improved oxidative stability index of vitamin D3 (Wang *et al.*, 2015).

Nanoencapsulation approaches to enhance stability are also reported. Nanocarriers of vitamin D3 encapsulated in high amylose starch conferred thermal degradation. Physicochemical stability of vitamin D3 was improved using casein micelles, chitosan-zein complexes, and chitosan-soy protein complexes (Haham *et al.*, 2012; Luo *et al.*, 2012; Luo *et al.*, 2013). Vitamin D3 along with flavors encapsulated in surfactant-based colloidal delivery systems exhibited stabilization (Ziani *et al.*, 2012). Nanoliposomal formulation of vitamin D3 with lecithin and cholesterol exhibited stability for one month at refrigeration in beverages (Mohammadi *et al.*, 2014). Self-assembled vitamin D3-hydrophobin complexes provided excellent protection to vitamin D3 against degradation and presents as a promising nanovehicle for food and clear beverage enrichment (Israeli-Lev and Livney, 2014). Other nanoparticles investigated are vitamin D3 complex with carboxymethyl chitosan (CMCS) and soy protein isolate (SPI) (Teng *et al.*, 2013), and vitamin D3 encapsulated within zein nanoparticles with carboxymethyl chitosan (CMCS) coatings (Luo *et al.*, 2012). Stability of vitamin D3 in the presence of oxygen also increased upon entrapment in whey protein isolate (WPI) nanoparticles. These NPs had application for clear beverages (Abbasi *et al.*, 2014). Physicochemical stability of vitamin D3 increased using casein micelles, chitosan-zein complexes, and chitosan soy protein complexes (Haham *et al.*, 2012; Luo *et al.*, 2012; Luo *et al.*, 2013).

Nanocomplexes of corn protein hydrolysate based vitamin D3 enhanced physicochemical stability, bioavailability, and bioactivity (Lin *et al.*, 2016). For nutritional purposes, these encapsulated vitamins were recommended for fortification of foods and as antioxidants (Wilson and Shah, 2007). Yuan *et al.* (2013) developed solid dispersions of vitamin D3 by using melt-mixing method with lipophilic surfactants D-α-tocopheryl polyethylene glycol 1000 succinate (TPGS) and SiO_2 as solidifier to modify the release and enhance the stability of vitamin D3. During release study of calcitriol lipid dispersions, no significant change was observed at various storage conditions (when stored at 25°C and 30°C), which revealed that the solid dispersion remained stable. Nanoemulsion of vitamin D3 revealed that long chain triglycerides-based nanoemulsions were most suitable for increasing the bioavailability of vitamin D3 (Ozturk *et al.*, 2015). Soybean β-conglycinin entrapped vitamin D3 nanoparticles revealed enhanced stability at pH 6.8 and pH 2.5, which reduces vitamin D losses during pasteurization. Improved stability to gastric digestion was also observed (Levinson *et al.*, 2014). Core-shell micelles of an amphiphilic chitosan derivative of N,N-dimethylhexadecyl carboxymethyl chitosan were reported as promising carriers for vitamin D3 to improve aqueous solubility, provide sustained release, and improve stability and bioavailability (Li *et al.*, 2014).

11.6 THERAPEUTIC INDICATIONS OF VITAMIN D3

Although initially recognized as a vitamin to prevent rickets in children and bone-related disorders in adults, the myriad roles that vitamin D3 could play in the therapy of various ailments are fast gaining recognition. The following section highlights the proposed role of vitamin D3 in some of the major therapeutic applications.

11.6.1 Bone Disorders

11.6.1.1 Rickets

Vitamin D deficiency and "Rickets" is well-established. Genetic abnormalities in vitamin D regulation can also cause rickets. Oral supplementation of vitamin D3 enables absorption through incorporation in chylomicrons into the lymphatic system and then into circulation by binding to DBP and lipoproteins into the liver, where it is released from the DBP. Subsequent sequential hydroxylation on C-25 by the vitamin D-25-hydroxylases mainly regulated by CYP enzymes (25-OHase; CYP27A1, CYP3A4, CYP2R1, CYP2J3) produces 25-hydroxyvitamin D [25(OH)D], which has a plasma half-life >2 weeks. 25(OH)D is again bound to DBP, and this complex binds to megalin on the plasma membrane of the renal tubule cell in the kidney and is transported into the cell.

There, it gets converted to the active form 1,25-dihydroxyvitamin D [1,25(OH)2D] in the mitochondria by the 25-hydroxyvitamin D-1α-hydroxylase [1-OHase; CYP27B1], which is responsible for maintaining calcium and phosphorus homeostasis by interacting with its vitamin D receptor (VDR) in the small intestinal cells (Holick, 2006). This 1,25(OH)2D-VDR structure forms complex with retinoic acid X receptor (RXR) in the nucleus. The 1,25(OH)2D-VDR-RXR complexes then bind to the vitamin D–responsive element (VDRE) for the transcription of TR PV 6 (a member of the transient receptor potential vanilloid family), which encodes a critical calcium transport channel. The increased expression causes enhanced transport of calcium into the bloodstream with the help of vitamin D–dependent calcium-binding protein calbindin. The 1,25(OH)2D also enhances phosphorus absorption in the small intestine. Thus, the major physiologic function of vitamin D is to maintain serum calcium and phosphorus levels within the normal physiologic range to support most metabolic functions, neuromuscular transmission, and bone mineralization as shown in Figure 11.1 (Holick, 2006).

11.6.1.2 Bone Loss and Osteoporosis

Vitamin D deficiency leads to inequity in calcium and phosphorous homeostasis with increased PTH secretion, resulting in increased bone loss and osteoporosis, as shown in Figure 11.1 (Sahota, 2000). The bone loss in patients with vitamin D deficiency is mainly due to secondary hyperparathyroidism and is for a large part irreversible (Tsai *et al.*, 1984; Lips and van Schoor, 2011).

11.6.1.3 Bone Related Inflammation

Rheumatoid arthritis (RA) is a progressive inflammatory disease characterized by inflammation of the synovium that leads to the destruction of joint bone and cartilage (Marques *et al.*, 2010). One factor that could regress RA and has been studied extensively in the past decade is vitamin D (Pelajo *et al.*, 2010). The immunomodulatory actions of vitamin D are largely dependent on the availability of precursor 25(OH)D for conversion to 1,25(OH)2D with the aid of dendritic cells (DCs) and macrophages expressing 1α-hydroxylase encoded by *CYP27B1* gene (Marques *et al.*, 2010). Notably, inflammatory cytokines such as TNF, IL-15 and IL-1β stimulate expression of *CYP27B1* gene in monocytes, DCs, and endothelial cells. VDR expression has since been described in many other cells of the immune system, supporting a wide array of potential actions for 1,25(OH)2D. These include innate antibacterial responses in monocytes and neutrophils, effects on antigen presentation by DCs, and modulation of T-cell and B-cell function. Multiple immune-cell and stromal-cell involvement in RA pathogenesis leads to greater inflammatory connections associated with joint destruction as the imbalance between osteoclast and osteoblast activities is increased. T cells, synovial macrophages, and fibroblast-like synoviocytes (FLSs) seem to be the major cellular players in RA, but B cells and DCs also have important roles, as shown in Figure 11.2. CD4+ T helper (TH) cells are the key players in immune responses. Within the inflamed synovium, both macrophages and DCs can function as "professional" APCs, but FLSs, B cells, and osteoclasts can also process and present antigen to T cells. APCs and FLSs direct T-cell differentiation through secreted cytokines into TH1 cells (which produce mainly IFNγ) and TH17 cells (which produce IL-17A and IL-17F, as well as TNF, IL-21, IL-22, and IL-26), which play pathogenetic roles in RA. IL-17 exerts inflammatory effects within the joint by releasing proinflammatory cytokines (IL-1β, IL-6, IL-8, and TNF-α), which can also promote osteoclast activity. The active 1,25(OH)2D can act on DCs to suppress TH1-inducing cytokines by promoting expression of anti-inflammatory IL-10. The 1,25(OH)2D has been shown to suppress TNF, IL-17, IL-1, and IL-6 production, by producing its anti-inflammatory and immunomodulatory effects (Figure 11.2) (Arnson *et al.*, 2007; Marques *et al.*, 2010; Jeffery *et al.*, 2016).

B cells express *CYP27B1*, suggesting that they can act as an alternative source of locally synthesized 1,25(OH)2D. Fibroblasts (FLSs) also play an active role in the pathogenesis of RA, releasing inflammatory cytokines, leukocyte-attracting chemokines, and leukocyte survival factors (Marques *et al.*, 2010; Hart *et al.*, 2011; Jeffery *et al.*, 2016).

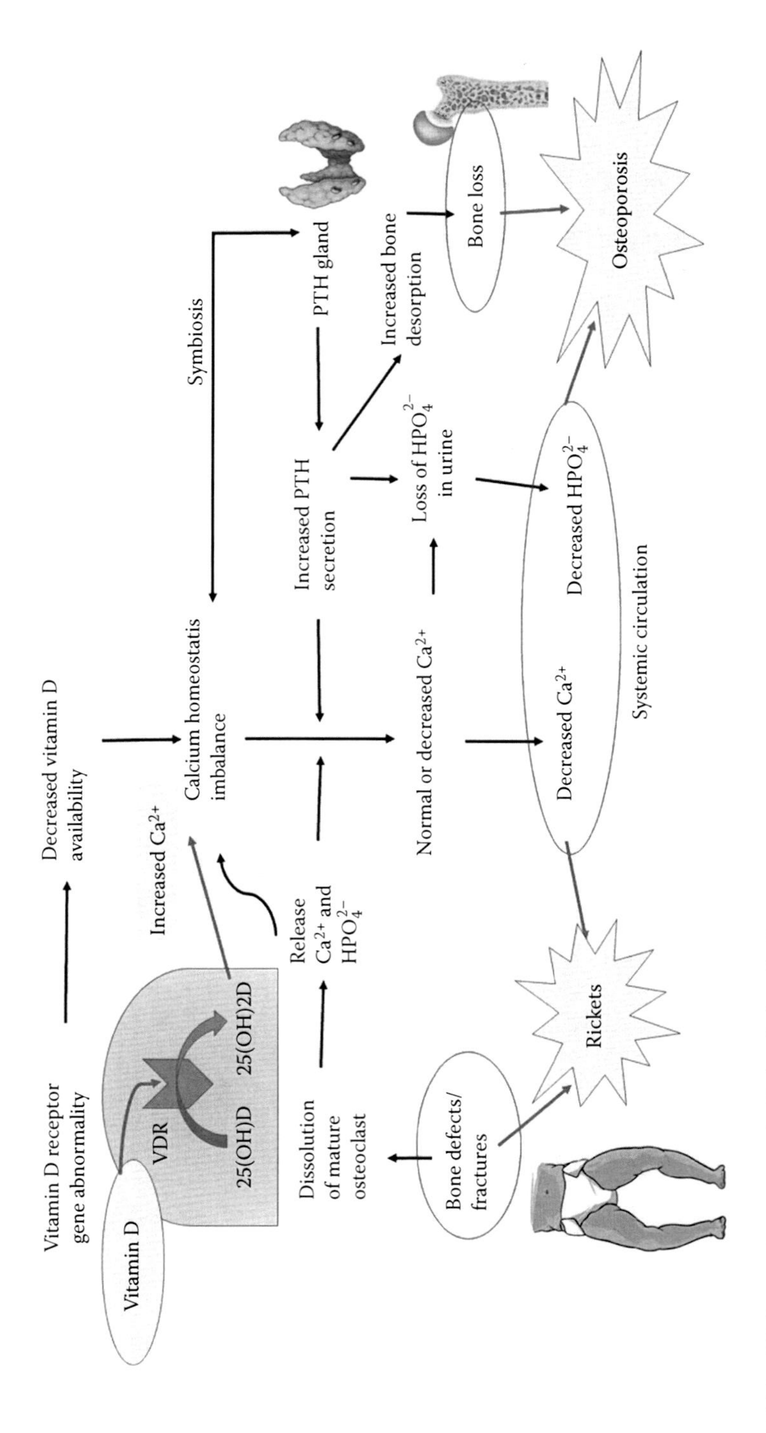

FIGURE 11.1 Rickets and osteoporosis: Role of vitamin D.

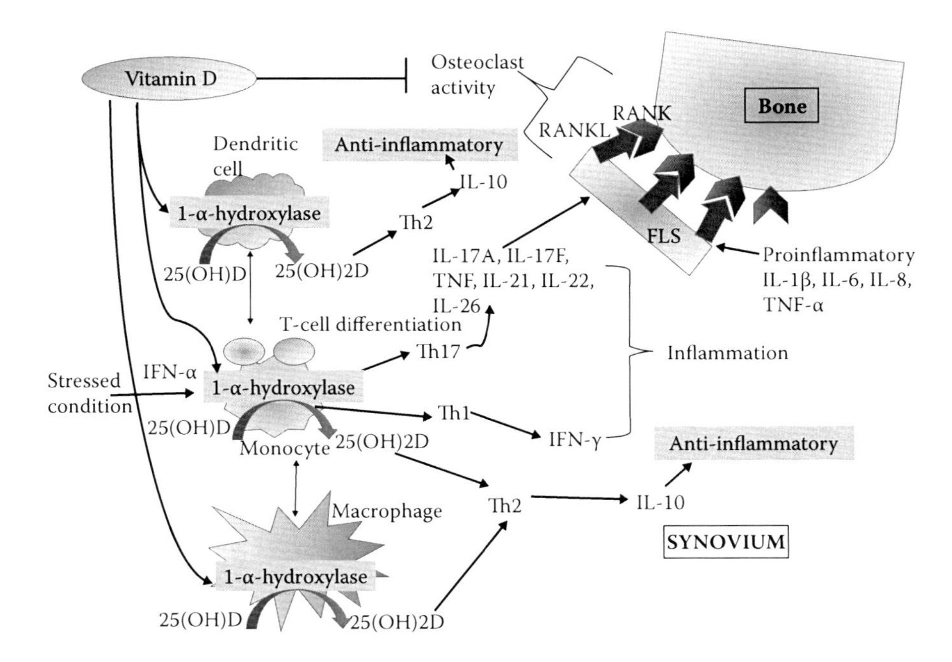

FIGURE 11.2 Immunomodulatory activity of vitamin D.

11.6.2 Psoriasis

Psoriasis is a common, chronic autoimmune inflammatory skin disorder, characterized by sharply demarcated, erythematous patches or plaques, with potential systemic complications (Soleymani *et al.*, 2015). In psoriatic skin, the erythematous scaling plaques are the result of keratinocyte hyperproliferation and abnormal differentiation. Vitamin D has been known to inhibit keratinocyte proliferation and exerts a selective proapoptotic effect as shown in Figure 11.3. Psoriasis is a disease

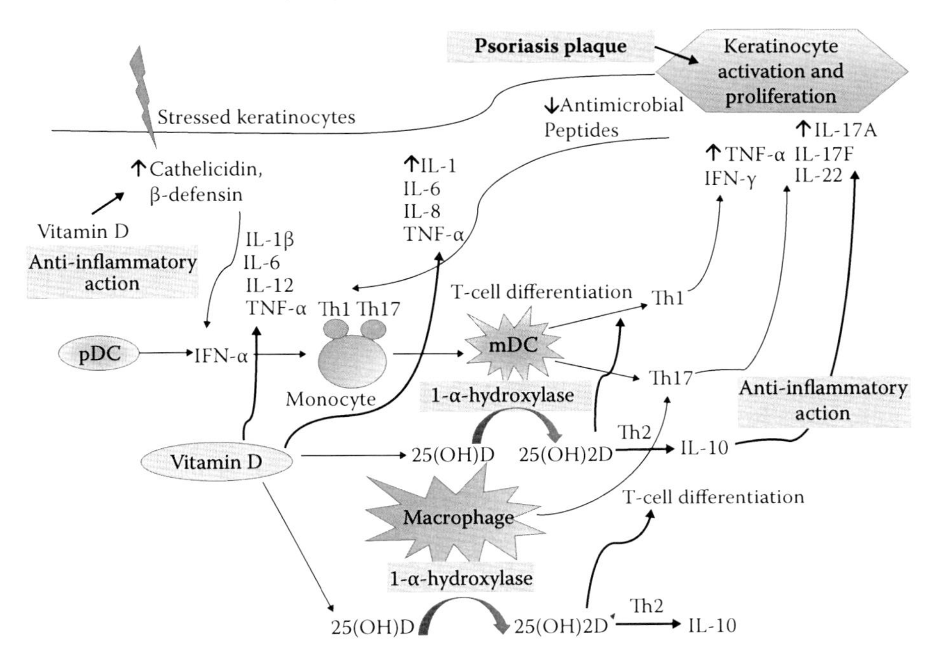

FIGURE 11.3 Psoriasis: Role of vitamin D.

based on T-cell dysregulation and cytokine dysregulation. During stressed conditions, proinflammatory cytokines (TNF-α, IFN-α, IL-2, and IL-8) induce T cell differentiation into Th1 and Th17 releasing inflammatory cytokines, which cause keratinocyte proliferation (Fu and Vender, 2011; Soleymani *et al.*, 2015). Vitamin D modulates immune cell expression by inhibiting the production of cytokines required for Th1 and Th17 differentiation and stimulate T cells to produce anti-inflammatory Th2 cytokines such as IL-10, thus reducing the production of inflammatory cytokines (IL-2, IL-8, IFN-γ, and TNF-α) in dendritic cells. Vitamin D has also been shown to stimulate the antimicrobial peptide cathelicidin (LL-37) in keratinocytes. Cathelicidin serves as a first-line defense mechanism in the innate immune system and has anti-inflammatory effects. Cathelicidin is upregulated during wound healing and tissue repair. A study demonstrated that topical application of calcitriol and calcipotriol on psoriatic lesions induced Th2 differentiation and inhibited IL-12/23 production, with upregulation of cathelicidin (Fu and Vender, 2011; Soleymani *et al.*, 2015).

11.6.3 Neurological Disorders

Central or peripheral neurological diseases are world health challenges. These diseases require long-term or lifelong treatments. It has been established that cerebral 1α-hydroxylase enables local transformation of cholecalciferol into the active form calcitriol. In addition, genetic polymorphism in the expression of vitamin D receptor (VDR) presents various degenerative diseases, including amyotrophic lateral sclerosis (ALS) and Parkinson's disease (PD). Studies have also demonstrated the immunomodulatory role of vitamin D within the nervous system and on various neurological functions, notably brain development, reduction of oxidative stress, neuroplasticity, neurotransmission, and neuroprotection, as shown in Figure 11.4. These different considerations have led to renewed interest in the usefulness of vitamin D for the management of neurological disorders where therapeutic options remain limited (Mpandzou *et al.*, 2016).

11.6.3.1 Alzheimer's Disease

Alzheimer's disease is a neurodegenerative condition characterized clinically by progressive cognitive decline and histologically by senile plaques and neurofibrillary tangles. The major component of senile plaques is the amyloid β protein (Aβ), whose accumulation is accompanied by increased inflammatory responses in the brain. Such inflammatory responses cause increased influx of Ca^{2+}

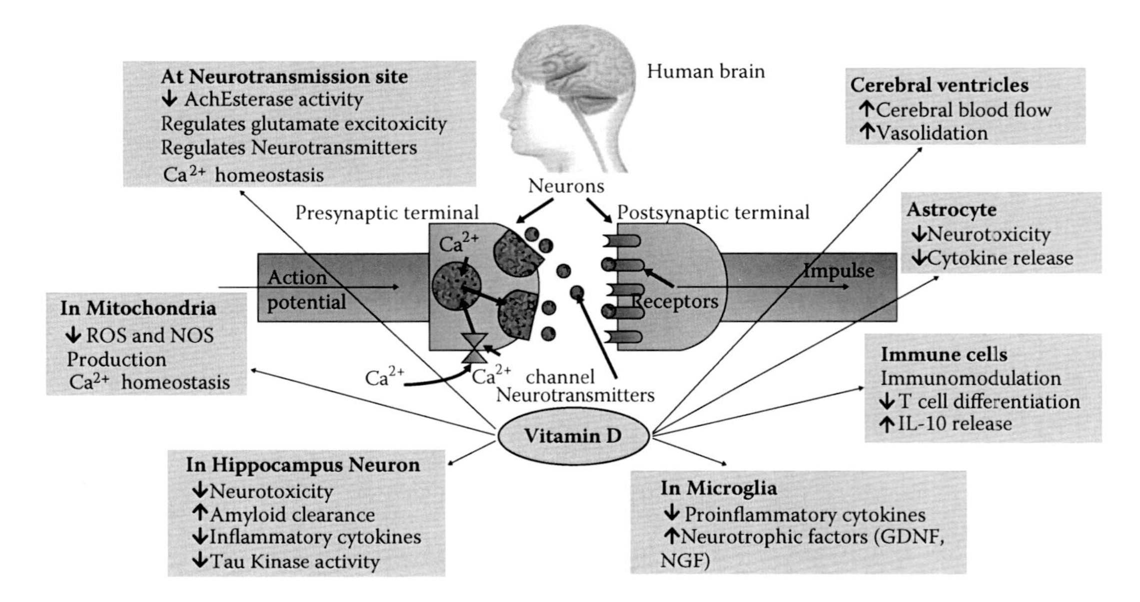

FIGURE 11.4 Neurological disorders: Role of vitamin D.

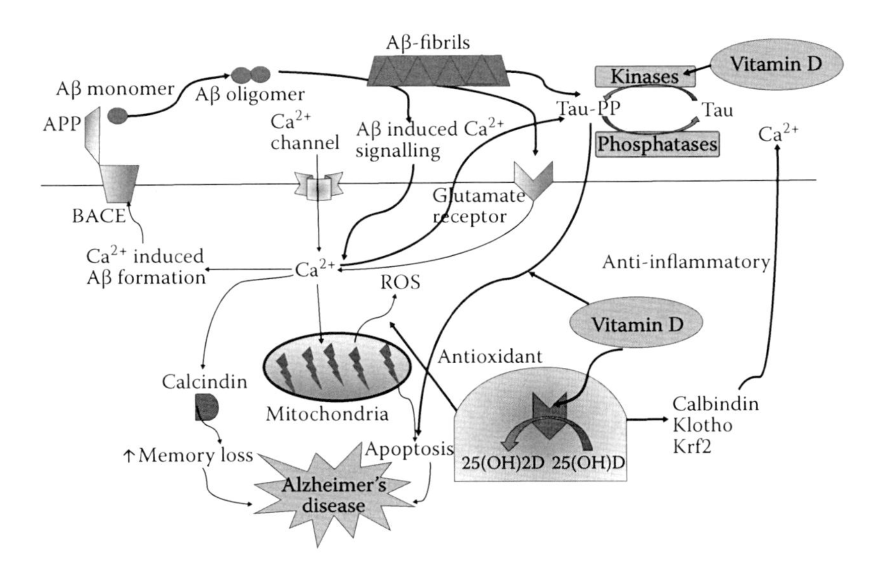

FIGURE 11.5 Alzheimer's disease: Role of vitamin D.

in neurons, which causes destruction of mitochondria by releasing reactive oxygen species (ROS). This Ca^{2+} then binds with calcium binding proteins (calcindin, calbindin), where calcindin binding causes memory loss whereas calbindin increases Ca^{2+} outflow. Increased Ca^{2+} concentration induces Aβ formation, which on binding with AβPP (Amyloid β precursor protein) causes Aβ oligomerization and fibrillation (Gezen-Ak *et al.*, 2014). This Aβ fibrillation enhances Tau Kinase expression, which causes phosphorylation of Tau protein and neuronal death. However, vitamin D is known to exhibit neuroprotection, where it increases calbindin expression, reduces Tau Kinase activity, reacts with ROS, and reduces glutamate excitotoxicity, as shown in Figure 11.5 (Banerjee *et al.*, 2015).

11.6.3.2 Parkinson's Disease

Parkinson's disease is characterized by substantial loss of dopaminergic neurons. These dopaminergic neurons are protected by glial cell derived neurotrophic factor (GDNF), which prevents loss of dopamine neurotransmitter and also increases Tyrosine to L-dopa conversion. It has been reported that vitamin D increases the neurotrophic factors such as GDNF in the substratum nigra or striatum, and also protects dopaminergic neurons from cell death when exposed to toxic agents. Thus, vitamin D has a therapeutic role in Parkinson's disease (Knekt *et al.*, 2010; Lương and Nguyễn, 2012; Mpandzou *et al.*, 2016).

11.6.3.3 Amyotrophic Lateral Sclerosis

Vitamin D3 deficiency exacerbates the pathophysiology of ALS, a disease that weakens muscles and affects physical function, by its action on markers of oxidative stress (i.e., antioxidant enzymes), inflammation, apoptosis, and neuron quantification. Vitamin D exerts anti-inflammatory effect on inflammation and inhibits oxidative stress by reducing TNF-α, IL-1b, and nitric oxide synthase (NOS). It prevents neuronal loss by increasing certain neuroprotector factors and regress striated muscle atrophy by increasing calcium absorption and synthesis of contractile proteins. It also decreases glutamate excitotoxicity by increasing calcium binding proteins parvalbumin, calbindin, and calretinin (Gianforcaro and Hamadeh, 2014; Mpandzou *et al.*, 2016).

11.6.3.4 Multiple Sclerosis (MS)

Vitamin D ameliorates MS through its immunomodulatory and anti-inflammatory actions, where active form of vitamin D is thought to play an important function. This active form 1,25(OH)2D exerts it activity through widespread availability of VDR and overexpressed target genes in the

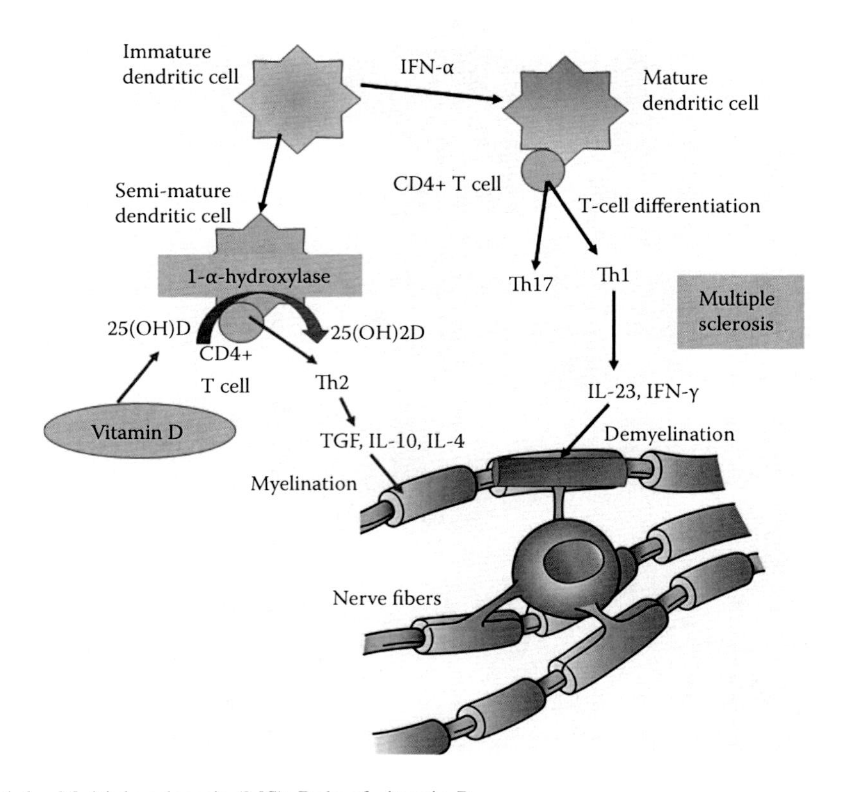

FIGURE 11.6 Multiple sclerosis (MS): Role of vitamin D.

immune cells. The immune cells (e.g., macrophages, dendritic cells, T and B lymphocytes) also have the ability to convert 25(OH)D to 1,25(OH)2D with the help of 1-α-hydroxylase (Figure 11.6) (Hayes *et al.*, 1997).

Activation of T lymphocytes is a critical step in the pathogenesis of MS. CD8+ T cells can attack neurons and oligodendrocytes and modulate oligoclonal expansion of lymphocytes in brain lesions and cerebrospinal fluid. Vitamin D increases the expression of some interleukins (IL-4 and IL-10) and transforming growth factor beta-1 (TGFβ-1), thus exhibiting immune regulation in MS. Vitamin D also inhibits the expression of certain pro-inflammatory cytokines, including IL-12 and IL-17, IFN-γ, and tumor necrosis factor (TNF)-α in peripheral blood mononuclear cells, thus controlling the infiltration of CD8+ T cells in neurons. This immunomodulatory effect of vitamin D may also help to protect the blood–brain barrier (BBB). One study found that 1,25(OH)2D prevents accumulation of inflammatory cells in the CNS and helps to maintain activated Th1/Th17 cells in the circulation, preventing them from crossing the BBB and destroying myelin sheath, as shown in Figure 11.6 (Harandi *et al.*, 2014; Saini *et al.*, 2015).

11.6.3.5 Huntington's Disease

Huntington's disease is a neurodegenerative disorder characterized by motor impairments, psychiatric problems, and dementia. It is caused by an elongated trinucleotide (CAG) repeat (36 repeats or more) on the short arm of chromosome p16 in the huntingtin gene. In this disease Ca^{2+} signalling is disturbed with less influx of Ca^{2+}. This causes transcription of the mutant Huntington gene within the nucleus, which directly leads to altered neurotransmission and muscle atrophy. Vitamin D deficiency is associated with type II muscle fiber atrophy as observed in Huntington's disease. Vitamin D acts through VDR receptors to produce antioxidant effects, thereby preventing gene mutation and transcription. It also stimulates neurotrophic growth factors, which help protect neurons and promote proper function (Chel *et al.*, 2013).

11.6.3.6 Epilepsy

Epilepsy is a chronic disorder, the hallmark of which is recurrent, unprovoked seizures. Many people with epilepsy experience more than one type of seizure and may have other symptoms of neurological problems as well. Vitamin D produces anti-epileptic action after binding of 1,25(OH)D to VDR and regulating the expression of several proteins expressed in the nervous system including neurotrophins such as neurotrophin-3, neurotrophin-4, and nerve growth factor and glial cell derived neurotrophic factor as well as parvalbumin a calcium binding protein and inhibiting the synthesis of the nitric oxid synthetase (Mpandzou *et al.*, 2016). Parvalbumin, known for its antiepileptic effects while inhibiting nitric oxide synthetase, is thought to convey general neuroprotective effects. Vitamin D, after binding to membrane-bound VDR, modulates the expression of GABA(A) receptor, thus controlling neuronal excitability (Hollo *et al.*, 2014; Pendo and DeGiorgio, 2016).

11.6.3.7 Fibromyalgia

Fibromyalgia syndrome is a common chronic musculoskeletal disorder, characterized by the presence of widespread pain and multiple tender points on physical examination. There has been accumulating evidence that chronic pain is associated with changes in brain anatomy, particularly related to a decrease in the gray matter in chronic patients. The central sensitization of nociceptive neurons in the dorsal horn is because of the activation of N-methyl-D-aspartic (NMDA) acid receptors and the resultant triggering of pain appears to be the cause of hypersensitivity to pain. The distribution of 1,25(OH)2D receptor (VDR), and la-hydroxylase (1α-OHase), activates anti-inflammatory events in the brain, which reduces hypersensitivity to pain and inflammation (Jesus *et al.*, 2013; Karras *et al.*, 2016; Mpandzou *et al.*, 2016).

11.6.3.8 Schizophrenia

Vitamin D stimulates the expression of several neurotrophic factors, nerve growth factor (NGF), glial cell line-derived neurotrophic factor (GDNF), which modulate the dopaminergic (DA) neuronal development, survival, and function and neurotrophin 3. It decreases the levels of neurotrophin 4. Vitamin D deficiency reduces the proline dehydrogenase gene expression (located on chromosome 22q11, a commonly deleted region that confers a genetic risk for schizophrenia), an enzyme that converts proline to glutamate. Phosphatidylinositide 3-kinase–protein kinase B pathway is involved in neurogenesis. It has been suggested that vitamin D may activate this pathway, thus its impaired signaling is observed with vitamin D deficiency (Mpandzou *et al.*, 2016; Samoes and Silveira, 2017).

11.6.3.9 Vascular Risk Factors and Cerebral Infarction

A high risk of cerebral infarction relative to vitamin D deficiency was associated independently with certain polymorphisms of the VDR and GC genes and fibroblast growth factor-23 (FGF-23) in patients with a left ventricular assistance device. Supplementation with vitamin D will help reduce the risk of ischemic stroke (Mpandzou *et al.*, 2016; Zittermann *et al.*, 2016).

11.6.4 Diabetes

The association between vitamin D and diabetes mellitus is explained by the discovery of receptors of vitamin D (VDR) and 1α-hydroxylase enzyme inside beta cells, calcium-linking protein vitamin D-dependent (DBP) in the pancreatic tissue, and an increase in the association between acquired and innate immunity. The 1α-hydroxylase enzyme, within beta cells, converts 25-hydroxy vitamin D into 1,25(OH)2D, which is connected to the VDRs, which will form the heterodimer VDR/RXR (retinoid x receptor). After translocation to the core of the cell, the complex is connected to VERE (vitamin D's element of response) in the promoter of the gene of the insulin and will activate the transcription of the insulin's gene, which will promote the cell's proliferation, differentiation, and the immunomodulation (Linhares *et al.*, 2016; Palomer *et al.*, 2008). While in Type 1 diabetes, a chronic inflammatory process occurs due to the presence of infiltration of the T (CD4+ and CD8+) cells, macrophages, lymphocytes B and NK cells in the beta cells of the pancreas. The macrophages and the dendritic cells secrete IL-12, which promotes the differentiation of the Th0 cells in Th1 stimulation of those to secrete IFN-γ and IL-2, which induces the migration of the T cytotoxic CD8+ cells, and causes the destruction of the beta cells. The usage of 1,25(OH)2D or analogues inhibit the secretion of cytokines pro-inflammatory Th1, Th9, and Th22 and promote the production of anti-inflammatory cytokines Th2, inhibits the maturation of the dendritic cells, and modulates the development of lymphocytes CD4, avoiding stress in the beta cells, and avoiding apoptosis in the beta cells. Vitamin D stimulates signal transduction and activation of the glucose 4 transporter (GLUT4) for better insulin response. The contribution of vitamin D to the metabolism of glucose also implicates modifications in the concentration extra and intracellular of calcium in the pancreatic beta cells. The secretion of insulin is a calcium-dependent process mediated by 1,25(OH)2D3 and by the PTH (Seshadri *et al.*, 2011).

11.6.5 Inflammatory Bowel Disease

Vitamin D3 has been shown to inhibit development of dextran sodium sulfate (DSS)-induced colitis in mice but can also cause hypercalcemia (Laverny *et al.*, 2010). Goff *et al.* reported that targeted delivery of vitamin D3 to the colon using β-glucuronides showed the highest reduction of colitis lesions and symptoms in mice with reduced the risk of hypercalcemia (Goff *et al.*, 2012).

Vitamin D and its receptors play important roles in maintaining gut integrity and protecting the intestine from pathogenic enteric bacterial infection. Vitamin D suppresses bacterial-induced NF-κB activity in the intestine. Vitamin D inhibits inflammatory cytokines. The presence of CYP27B1 enzymes would readily allow the conversion of 25(OH)D to 1,25(OH)2D and exhibit local immune regulatory effects (Mouli and Ananthakrishnan, 2014).

11.6.6 Obesity

Vitamin D3 is effective for inducing apoptosis in mature adipocytes via activation of the Ca^{2+}/calpain/caspase-dependent pathway, which indicates that targeting of Ca^{2+} signaling to induce apoptotic death in adipocytes may represent an effective strategy for prevention and treatment of obesity (Kong and Li, 2006; Sergeev, 2009).

11.6.7 Infectious Diseases

11.6.7.1 Tuberculosis

Tuberculosis (TB) is a respiratory illness that usually affects the lungs. The TB bacteria infect the body by attacking the macrophages. Vitamin D3 boosts our immune system by making macrophages stronger and prevents TB bacteria (Nnoaham and Clarke, 2008; Ralph *et al.*, 2013). Innate immune systems are the first line of defense against pathogen infections. Upon pathogen detection such as *Mycobacterium tuberculosis* (*M. tuberculosis*), toll-like receptors (TLR) on the macrophage membrane are activated to induce transcriptional up-regulation of the VDR and enhance CYP27B1 expression. This increases the synthesis of vitamin D3 and VDR, two essential components responsible for the VDR-dependent regulation of a variety of genes including the up-regulation of cathelicidin expression (Adams and Hewison, 2008).

A study on adult patients (n =129) with recently diagnosed sputum-positive pulmonary TB suggested that 79% of patients with deficiency of vitamin D3 levels were five-folds more prone to TB. Serum levels >20 ng/mL calcitriol provided initiated innate immune response to *M. tuberculosis* (Talat *et al.*, 2010). Vitamin D3 supplementation accelerated resolution of inflammatory responses associated with increased risk of mortality during treatment of pulmonary infections. Adjunctive vitamin D3 can thus be used for treatment of active TB. Clinical study was carried out on patients (n = 247) with vitamin D3 supplementation (2.5 mg for 0, 2, 6 weeks, i.e., 100,000 IU) and suggested that supplementation with vitamin D3 helps to reduce TB infection (Sutaria *et al.*, 2014). Vitamin D3 showed noticeable suppressive effect in α-chemokines rather than β-chemokine. This proposed that vitamin D3 may down-regulate the recruitment and stimulation of T-cells through α-chemokines at the site of infection and may act as a promotable anti-inflammatory agent (Selvaraj *et al.*, 2012). It was also observed that vitamin D3 down-modulates the pro-inflammatory cytokine profile, which stimulates an antimicrobial response, with a 16-fold increase in the antimicrobial peptide cathelicidin production with vitamin D3 treatment (Khoo *et al.*, 2011). A double-blind clinical trial of administering 0.25 mg vitamin D3 per day during the sixth week of TB treatment in 67 patients improved percentage sputum conversion and radiological features (Nursyam *et al.*, 2006). However, another double-blind randomized clinical trial of administration of 100000 IU vitamin D3 in 281 patients with TB revealed no influence on sputum conversion rate or TB severity score (Wejse *et al.*, 2009). Vitamin D3 supplementation in children either given as 800 IU of vitamin D3 every day for 6 months or a dummy pill prevented TB infection on tuberculin skin test (TST) (Ganmaa *et al.*, 2012). Solid lipid nanoparticles (SLN) of vitamin A and vitamin D3 were studied to control TB (Kumar *et al.*, 2014a). While here the focus is on TB, vitamin D, likewise, can also play a role in the treatment of other infectious diseases in a similar fashion, like pneumonia, HIV, etc.

11.6.8 Cardiovascular Diseases

Vitamin D deficiency is related to a broad spectrum of cardiovascular disease and its risk factors (Giovannucci *et al.*, 2008). Hypertension is associated with inappropriate stimulation of the renin-angiotensin-aldosterone system (RAAS). Vitamin D is known to regulate the expression of RAAS. Vitamin D inhibits renin gene expression by interfering with cyclic adenosine monophosphate. Activation of the vitamin D receptor suppresses the RAAS suggesting that this activation could exert protective effects on blood pressure, and on vascular and cardiac tissue. Vitamin D suppresses the pro-inflammatory state by downregulating nuclear factor-κB activity and decreasing IL-6, IL-12, interferon-γ, and TNF-α production. At the same time, it increases anti-inflammatory cytokines. Vitamin D deficiency therefore leads to a reduction in NF-κB activity, a decrease in IL-10 production, and an increase in IL-6, IL-12, interferon-γ, and TNF-α production, all of which lead to the CHF (Fanari *et al.*, 2015).

11.6.9 Cancer

As the active form of vitamin D has anti-inflammatory, antioxidant, and anti-proliferative properties, it has been tested against a variety of cancers, including

- Brain cancer (Hähnlein *et al.*, 2003; van Ginkel *et al.*, 2007)
- Breast cancer (Sequeira *et al.*, 2012; Haussler *et al.*, 2013)
- Lung cancer (Norton and O'Connell, 2012)
- Ovarian cancer (Zhang *et al.*, 2006)
- Pancreatic cancer (Kawa *et al.*, 1997)
- Prostate cancer (Chen and Holick, 2003)
- Skin cancer (Tang *et al.*, 2011; Jiang and Bikle, 2014)
- Several other different types of cancers

The exact mechanism of vitamin D3 in specific cancers is not yet clearly established. Nevertheless, we present below the various mechanisms, by which vitamin D could inhibit cancer progression via its actions on various cancer signaling pathways, besides others.

11.6.9.1 Vitamin-D-VDR Signaling

The VDR is a type II nuclear receptor that interacts with specific DNA-binding sites (vitamin D-response elements, VDREs) within VDRE genes. Vitamin D after binding triggers VDR-RXR complex, which further induces transcriptional changes. This leads to antioxidant and anti-inflammatory cytokine release. These properties tend to inhibit cancer progression (Ben-Shoshan *et al.*, 2007; Fleet *et al.*, 2012).

11.6.9.2 Wnt/β-Catenin Signalling

In the cytoplasm, β-catenin is found in association with APC (adenomatous polyposis coli). Activation of Wnt signalling leads to β-catenin accumulation and its subsequent release from APC. This free β-catenin then interferes with DNA in the nucleus and activates transcription of genes, which regulates proliferation. Mutations in the APC gene lead to cancer progression. 1,25(OH)2D can block β-catenin-mediated gene transcription by inducing binding of VDR to β-catenin (Ben-Shoshan *et al.*, 2007; Fleet *et al.*, 2012).

11.6.9.3 TGFβ2 (Transforming Growth Factor 2) Signalling

TGFβ2 is essential for the maintenance of tissue homoeostasis and is an anti-proliferative factor. The 1,25(OH)2D and its analogue induce expression of TGFβ1, thus exhibiting anti-proliferative effects. Vitamin D also suppresses negative regulators of TGF β, LTBP (latent TGF β-binding protein) 1 and LTBP2 (Yang *et al.*, 2001; Buschke *et al.*, 2011; Fleet *et al.*, 2012).

11.6.9.4 IGF (Insulin-Like Growth Factor) Signalling

The 1,25(OH)2D indirectly acts on the growth rate of cells by interfering with action of proliferative growth factors. In breast cancer cells, it was demonstrated that vitamin D inhibits IGF-stimulated cell growth by releasing IGFBP3 (IGF-binding protein 3) into the medium, which exhibits anti-proliferative activity (Yang *et al.*, 2001; Fleet *et al.*, 2012).

11.6.9.5 Apoptosis

1,25(OH)2D regulates pro- and anti-apoptotic proteins, which are known to control apoptosis. Vitamin D suppresses the expression of such apoptotic proteins (Pan *et al.*, 2010; Fleet *et al.*, 2012).

11.6.9.6 Autophagy

Autophagy is a cellular process of macromolecules and organelles degradation in lysosomes. The 1,25(OH)2D acts on caspase- and p53-independent pathway, which regulates the expression of anti-apoptotic protein, Bcl-2 (Fleet *et al.*, 2012).

11.6.9.7 Antioxidant Defense and DNA Repair

Oxidative stress causes DNA damage and loss of DNA repair mechanisms, which are attributed to cancer. And, 1,25(OH)2D induces the expression of several enzymes to increase the antioxidant defense mechanism (Swami *et al.*, 2003; Peehl *et al.*, 2004; Fleet *et al.*, 2012).

11.7 VITAMIN D AS A NANO-NUTRACEUTICAL

The role of vitamin D3 in various therapeutic indications exerts one commonality, that of uptake into cells, wherein vitamin D3 or cholecalciferol is converted to its active form, that is, to 1,25-dihydroxyvitamin D [1,25(OH)2D] inside the cell in the mitochondria by the enzyme, 25-hydroxyvitamin D-1α-hydroxylase [1-OHase; CYP27B1]. This enzyme is ubiquitous in the body, hence suggesting that manifold therapeutic applications of vitamin D3 may be relied on. The disposition of vitamin D3 in the body also suggests the great role of nanopharmaceutical carriers for exploiting various activities. Intracellular delivery may rely on the VDR, wherever applicable. Alternatively, various receptor-mediated strategies or even passive targeting to RES cells can be a practical and considered approach. The following section discusses applications of Vitamin D3 as a nano-nutraceuticals.

11.7.1 Polymeric Nanoparticles

Nanoparticles from hydrophobically modified alginate oleoyl alginate ester (OAE) and vitamin D3 were prepared by self-aggregation. A dermal nanoformulation of vitamin D3 based on a modified chitosan polymer with fatty acids, amino acids, alginate collagen, PLGA, or in combination was reported to treat psoriasis and for antiaging effects (Mousa, 2013; Mousa, 2015). Vitamin D3 apolipoprotein E-based nanoparticles (94 nm), taken up via receptor-mediated transcytosis by glioma tumor cell line (C6), provoked 89% of G1 arrest, 53% of S arrest, and 48% of apoptosis and reduced colony-formation (Maleklou *et al.*, 2016). Polylactic acid (PLA)-nanoencapsulated calcitriol significantly inhibited breast cancer cell growth compared to free calcitriol. PLA NPs also enhanced the intracellular delivery of the vitamin in breast cancer cells (Almouazen *et al.*, 2012). The nanoparticles also showed an antiproliferative activity for more than 10 days against breast cancer cell line, MCF7, which may offer a new and potentially effective strategy for vitamin D3–based chemotherapy (Almouazen *et al.*, 2013). Poly lactic-co-glycolic acid (PLGA)-encapsulated calcitriol nanoparticles revealed enhanced inhibitory effect on pancreatic cancer cell growth at 91% and inhibitory effect on the lung cancer cell line, A549 (70%) because of efficient cell internalization by an endocytosis mechanism compared to plain calcitriol (Ramalho *et al.*, 2015). Thermosensitive and magnetic nanoparticles, composed of poly (*N*-isopropylacrylamide) coated silica/magnetic (PNIPAM/SiO_2/Fe_3O_4) nanoparticles, inhibited viability of liver cancer cell (HepG2) up to 85% (Lien *et al.*, 2013)

Combination of vitamin D3 doxorubicin nanoparticles and vitamin D3 epirubicin nanoparticles increased the cytotoxicity in C6 glioma cells (Maleklou *et al.*, 2016). Vitamin D3 nanoparticlesn conjugated to cytotoxic drugs paclitaxel and doxorubicin and phosphatidyl-inositol-3-kinase (PI3K) inhibitor (PI103) via succinic acid linker, demonstrated *in vitro* cytotoxicity in HeLa cervical cancer cell lines by internalization through lysosomal compartments (Patil *et al.*, 2013). PI103 (PI3K/Akt/mTOR signaling pathway inhibitor) and cisplatin or doxorubicin or proflavine (dual drug) loaded with vitamin D3 phosphatidylcholine-based lipidic nanoparticles less than 200 nm in size were designed to target drug resistance in human hepatocellular carcinoma. Intravenous injection of NPs into the bloodstream accumulated drug into tumor tissue through the enhanced permeability and retention (EPR) effect by passive targeting. The NPs increased cell death of Hep3B cell line (human hepatocellular carcinoma) by inducing apoptosis through DNA damage at 24 h compared to monotherapy (Palvai *et al.*, 2014).

11.7.2 Lipidic Nanoparticles

11.7.2.1 Solid Lipid Nanoparticles

Paucar (2016) prepared solid lipid nanoparticles by spray chilling method using vegetable fat and beeswax as lipids. The lipids increased vitamin D3 stability and bioavailability. Solid lipid nanoparticles of vitamin A and vitamin D3 were studied to control TB (Kumar *et al.*, 2014a; Kumar *et al.*, 2014b).

11.7.2.2 Nanostructured Lipid Carriers

Nanostructured lipid carriers of vitamin D3 enhanced drug permeation and exhibited limited irritation in the dorsal skin of female nude mice, and hence found useful for delivery of antipsoriatic drugs (Lin *et al.*, 2010). However, calcipotriol with methotrexate-loaded nanostructured lipid carriers systems showed reduced skin permeation of calcipotriol, but not of methotrexate.

11.7.3 Liposomes

Liposomes for topical application of vitamin D3 and its derivatives, particularly calcitriol, were less greasy than the usual paraffin ointments, and at the same time, provided good skin penetration ability (Naeff *et al.*, 1998). Merz (1994) investigated incorporation of four different vitamin D-derivatives (lα, 25-Dihydroxycholecalciferol, 25-Hydroxycholecalciferol, Pro- 25$(OH)_2$,-D_3, Calcipotriol) in liposomal membranes to optimize their use for psoriasis treatment. However, it is also reported that under certain conditions, vitamin D3 derivatives show a hypercalcemic activity that can be a serious limitation in their use for a dermatological application like psoriasis

(Soleymani *et al.*, 2015). Nevertheless, one report suggests that liposomal vitamin D3 preparations can achieve a given antipsoriatic effect with reduced concentration of the active substance, thereby reducing the risk of skin irritation and of hypercalcemia (Dattagupta *et al.*, 1998). The antiparakeratotic potency of liposomal vitamin D3 in a concentration of 2 µg/g was twice compared to commercial preparation containing a higher concentration of 4 µg/g in a mouse tail test (Korbel *et al.*, 2001). Vitamin D_3-3-bromoacetate derivative (AMPI-109) was encapsulated into phospholipid nanosomes using the SuperFluids™ critical fluid nanosome (CFN) process. The nanosomes reduced tumor size up to 39% (Castor, 2016).

11.7.4 Microemulsion and Self-Emulsifying Drug Delivery Systems (SEDDS)

Microemulsions are nanosystems reported to enhance bioavailability. Dahan and Hoffman (2006) studied the bioavailability of vitamin D3 (1 mg/kg) as self-emulsifying systems and microemulsions administered orally and by intravenous injection. The study showed higher bioavailability with long chain fatty acid vis-à-vis the medium and small chain triglyceride.

11.7.5 Quantum Dots

Quantum dots (QDs) are small semiconductor nanoparticles that are fluorescent under UV light. They are engineered light-emitting nanoscale delivery vehicles (Derfus *et al.*, 2004). QDs, conjugated to vitamin D3, increased delivery of calcitriol to inflammatory breast cancer in mice. The QDs-vitamin D3 conjugated to Mucin1 (MUC1) antibody facilitated imaging of drug-tumor interactions *in vivo*, and monitor delivery of vitamin D3 to the tumor and metastasized sites (Bonor *et al.*, 2012; Bonor *et al.*, 2013).

11.7.6 Nanofiber-Based Mats

Electrospinning is a technique that uses electric force alone to drive the spinning process and to produce polymer fibers from solutions or melts (Bellan and Craighead, 2006; Aluigi *et al.*, 2007). Fast-dissolving electrospun nanofibrous mats of poly (vinyl-pyrollidone) containing vitamin D3 were prepared using electrospinning technology. Vitamin D3 changed into an amorphous form and was uniformly distributed in PVP fibrous matrix. These nanofibrous mats provided rapid dissolution of vitamin D3 compared to plain vitamin D3 powder (Li *et al.*, 2013). Various patents based on Vitamin D nanonutraceuticals are summarized in Table 11.4.

11.8 TOXICITY OF VITAMIN D3

Vitamin D3 toxicity is caused by excessively high doses, which can result in serum levels of 750 nmol/L and greater (Jones, 2008). Hypercalcemia and hyperphosphatemia are responsible for producing most of the symptoms of vitamin D3 toxicity (Vieth, 2007; Alshahrani and Aljohani, 2013). Early symptoms of vitamin D3 toxicity include gastrointestinal disorders such as anorexia, diarrhea, constipation, nausea, and vomiting, headache, weakness, dry mouth, somnolence, muscle pain, metallic taste, and bone pain. Symptoms and signs of hypercalciuria, nephrocalcinosis, ectopic calcification, nephrocalcinosis, hypertension, and cardiac arrhythmias and endocrine disorders such as primary hyperparathyroidism are also indicators of vitamin D toxicity (Ketha *et al.*, 2015; Schwalfenberg, 2007).

11.9 LEGISLATION AND REGULATIONS

US FDA has set regulations under Code of Federal Regulations (CFR) Title 21 with guidelines under food for human consumption. Vitamin D2 and vitamin D3 crystals should meet the

TABLE 11.4
Patents on Vitamin D Nanonutraceuticals

Sr. No.	System	Ailments	Composition	Patent No.
1	Conjugates	Cancer, osteoarthritis	Vitamin D moiety associated with target molecule moiety and suitable pharmaceutically carrier	US6929797 B2
2	Conjugates	Insulin dependent diabetes mellitus	Insulin peptides, a stably attached scaffold, non-hormonal vitamin D, analog or metabolite	WO2016065052A1
3	Emulsions	Oral inflammation	Vitamin D, esters of 1α,25-dihydroxy vitamin D3, esters of 1,25-dihydroxy vitamin D3, polyethylene glycol	US20130095154
4	Emulsions	Supplement for animal and poultry	Vitamin A palmitate, vitamin E, vitamin D2, vitamin D3 or a mixture of two or more vitamins, polyoxyethylene ether of castor oil, glycerine	US3384545 A
5	Emulsions	Treatment of skin and mucous membranes	Cholecalciferol, Xylitol, glycerol, polyethylene glycol, stabilizers, acids, bases, buffers, antioxidants, emulsifiers, suspending	US20160331764 A1
6	Emulsions	Alopecia	Vitamin D3, propylene glycol, anhydrous absolute ethanol, ethoxydiglycol or transcutol	WO2011019617 A2
7	Emulsions	Nutritional supplement	25-hydroxyvitamin D3, encapsulation agent, coconut oil or palm kernel oil	WO2003059358 A1
8	Emulsions	Neurodegenerative disease	Glatiramer acetate, vasoactive intestinal peptide, vitamin D, granulocyte macrophage colony stimulating factor and transforming growth factor beta	US20140004148
9	Emulsions	Osteoporosis, cancers	Mono-, di- and triglycerides, phospholipids, sugars, bisphosphonates, lidocaine	WO2011063952 A1
10	Emulsions	Osteoporosis, renal osteodystrophy, Paget's disease	Vitamin D3 analogue, pharmaceutically acceptable excipients, diluents or carriers	US20080032955
11	Emulsions, Stents	Arterial Restenosis	Calcitriol, myglyol 812, vitamin E TPGS, BHA, BHT and therapeutic agents	US20070142339A1
12	Liposomes	Psoriasis	Vitamin D or derivative calcitriol, lecithin, cholesterol, polyethylene glycol derivatives, alcohol	US5834016A
13	Liposomes	HIV infection	Novel ammonium or sulfonium ion, cationic lipids, vitamin D3, pH sensitive amphiphilic lipids, sterols, phosphatidylglycerol, lysolipids, glycolipids	US5851548A
14	Liposomes	Arthritis	Derivative of a sterol or an organic acid derivative of cholesterol, vitamin D, phytosterol or steroidal hormone	WO1985004578A1

(Continued)

TABLE 11.4 (CONTINUED)
Patents on Vitamin D Nanonutraceuticals

Sr. No.	System	Ailments	Composition	Patent No.
15	Liposomes	Cancer, vitamin D deficiency, autoimmune disease, hypertension, osteoporosis, psoriasis and infectious disease	Vitamin D3 or vitamin D3 analog, dimyristoyl-phosphatidylcholine, cholesterol, vitamin D3-3-epoxide, vitamin D3-3-bromoacetate	US20160008377 A1
16	Micro-implants	Cancer and neurodegenerative disorders	Vitamin D3, biocompatible polymers polyanhydride copolymer of 1,3-bis(p-carboxyphenoxy) propane, sebacic acid	US20020076442
17	Nanoparticles	Viral infection, cancer, hepatitis B, hepatitis C, diabetes	Chitosan polymer with vitamin D derivative	US20130149385 A1
18	Nanoparticles	Osteoporosis	Strontium fortified calcium ions, strontium ions and its combinations, anions, vitamin D, growth factors and its combinations	WO2009000158 A1
19	Nanoparticles	Osteoporosis	Lepidium sativum or other Lepidium extracts, calcium, vitamin D, and antioxidants	US8563053 B2
20	Nanoemulsion	Alzheimer's, cardiovascular, and diabetes	Vitamin D3, rapamycin or Sirolimus, cycloastragenol and withanolides	US20130338039
21	Nanodispersion	Nutraceutical and cosmeceutical	Vitamin D3, sucrose and glycerol viscosifying agent, dispersing agent, sweetening agent, preservative, chelating agent, coloring agent, flavoring agent, and an antioxidant	WO2015155703 A2
22	Solid lipid nanoparticles	Psoriasis, acne, ichthyosis, rosacea, actinic keratosis	Calcipotriol or calcipotriol monohydrate, medium chain triglycerides caprylic, long chain triglycerides castor oil and hydrophilic surfactant	WO2012127037 A3
23	Solid lipid nanoparticles	Tuberculosis, macular degeneration, diabetic retinopathy, cancers, hyperpigmentation, acne, osteoporosis	Vitamin D3, retinoic acid, tri-glycerides, emulsifiers or surfactants polysorbate 80 and soya lecithin	US20140348938 A1
24	Solid lipid nanoparticles	Treatment of skin, scalp, hair, and nails	Vitamin D3 and its derivatives, liquid lipids vegetable oils, solid lipids fatty acids, polymers, cationic polymer	WO2011116963 A2

specifications of the Food Chemicals Codex, 3d Ed. (1981), pages 344 and 345. In accordance with Section 184.1(b) (2), the ingredients used in food as the sole source of added vitamin D3 should be within the specific limitations for their functional use, as listed in Table 11.5.

Vitamin D3 may be used in infant formula in accordance with section 412 (g) of the Federal Food, Drug and Cosmetic Act or with regulations promulgated under section 412 (a) (2) of the act. Vitamin D3 may be used in margarine, in accordance with 166.110 of this chapter (USFDA).

TABLE 11.5
Recommended Limits of Vitamin D3 in Foods

Category of Food	Maximum Levels of Vitamin D3 in Food (as served)
Breakfast cereals	350 (IU/100g)
Grain products and pastas	90 (IU/100g)
Milk	42 (IU/100g)
Milk products	89 (IU/100g)

11.10 CONCLUSION

A dietary supplement to prevent rickets in children, vitamin D, has travelled a long way to find a place among drugs and foods as a nutraceutical. The multifaceted applications propose expansion of the role of vitamin D in various therapies. More importantly, with the emergence of nanomedicine as a strong hold in therapeutics, vitamin D in nanostructured delivery systems could expand the horizons of its application, taking it to the status of a wonder nutraceutical.

REFERENCES

Abbasi, A., Emam-Djomeh, Z., Mousavi, M. A., and Davoodi, D. Stability of vitamin D(3) encapsulated in nanoparticles of whey protein isolate. *Food Chem.* 2014; 143: 379–383.

Adams, J. S., and Hewison, M. Unexpected actions of vitamin D: New perspectives on the regulation of innate and adaptive immunity. *Nat Clin Pract Endocrinol Metab.* 2008; 4: 80–90.

Agarwal, N., Mithal, A., Dhingra, V., Kaur, P., Godbole, M. M., and Shukla, M. Effect of two different doses of oral cholecalciferol supplementation on serum 25-hydroxy-vitamin D levels in healthy Indian postmenopausal women: A randomized controlled trial. *Indian J Endocrinol Metab.* 2013; 17: 883.

Almouazen, E., Bourgeois, S., Boussaid, A., Valot, P., Malleval, C., Fessi, H. *et al.* Development of a nanoparticle-based system for the delivery of retinoic acid into macrophages. *Int J Pharm.* 2012; 430: 207–215.

Almouazen, E., Bourgeois, S., Jordheim, L. P., Fessi, H., and Briancon, S. Nano-encapsulation of vitamin D3 active metabolites for application in chemotherapy: Formulation study and *in vitro* evaluation. *Pharm Res.* 2013; 30: 1137–1146.

Alshahrani, F., and Aljohani, N. Vitamin D: Deficiency, sufficiency and toxicity. *Nutrients.* 2013; 5: 3605–3616.

Aluigi, A., Varesano, A., Montarsolo, A., Vineis, C., Ferrero, F., Mazzuchetti, G. *et al.* Electrospinning of keratin/poly (ethylene oxide) blend nanofibers. *J Appl Polym Sci.* 2007; 104: 863–870.

Armas, L. A., Hollis, B. W., and Heaney, R. P. Vitamin D2 is much less effective than vitamin D3 in humans. *J Clin Endocrinol Metab.* 2004; 89: 5387–5391.

Arnson, Y., Amital, H., and Shoenfeld, Y. Vitamin D and autoimmunity: New aetiological and therapeutic considerations. *Ann Rheum Dis.* 2007; 66: 1137–1142.

Ashwell, M., Stone, E. M., Stolte, H., Cashman, K. D., Macdonald, H., Lanham-New, S. *et al.* UK Food Standards Agency Workshop Report: An investigation of the relative contributions of diet and sunlight to vitamin D status. *Br J Nutr.* 2010; 104: 603–611.

Banerjee, A., Khemka, V. K., Ganguly, A., Roy, D., Ganguly, U., and Chakrabarti, S. Vitamin D and Alzheimer's disease: Neurocognition to therapeutics. *Int J Alzheimers Dis.* 2015; 2015: 192747.

Bellan, L. M., and Craighead, H. Control of an electrospinning jet using electric focusing and jet-steering fields. *J Vac Sci Technol.* 2006; 24: 3179–3183.

Ben-Shoshan, M., Amir, S., Dang, D. T., Dang, L. H., Weisman, Y., and Mabjeesh, N. J. 1α, 25-dihydroxyvitamin D3 (Calcitriol) inhibits hypoxia-inducible factor-1/vascular endothelial growth factor pathway in human cancer cells. *Mol Cancer Ther.* 2007; 6: 1433–1439.

Biancuzzo, R. M., Clarke, N., Reitz, R. E., Travison, T. G., and Holick, M. F. Serum concentrations of 1,25-dihydroxyvitamin D2 and 1,25-dihydroxyvitamin D3 in response to vitamin D2 and vitamin D3 supplementation. *J Clin Endocrinol Metab.* 2013; 98: 973–979.

Bishop, C. W., Knutson, J. C., and Valliere, C. R. Method of treating prostatic diseases using delayed and/or sustained release vitamin D formulations. Google Patents. 1998.

Bonor, J., Schaefer, R., and Nohe, A. Using calcitriol conjugated quantum dots to target inflammatory breast cancer tumors and metastasis *in vivo*. *Biophys J.* 2013; 104: 574a.

Bonor, J. C., Schaefer, R. J., Menegazzo, N., Booksh, K., and Nohe, A. G. Design of 1,25 dihydroxyvitamin D3 coupled quantum dots, a novel imaging tool. *J Nanosci Nanotechnol.* 2012; 12: 2185–2191.

Brot, C., Jorgensen, N. R., and Sorensen, O. H. The influence of smoking on vitamin D status and calcium metabolism. *Eur J Clin Nutr.* 1999; 53: 920–926.

Buschke, S., Stark, H. J., Cerezo, A., Pratzel-Wunder, S., Boehnke, K., Kollar, J. et al. A decisive function of transforming growth factor-beta/Smad signaling in tissue morphogenesis and differentiation of human HaCaT keratinocytes. *Mol Biol Cell.* 2011; 22: 782–794.

Buttriss, J. L. Vitamin D: Sunshine vs. diet vs. pills. *Nutr Bull.* 2015; 40: 279–285.

Calvo, M. S., Whiting, S. J., and Barton, C. N. Vitamin D fortification in the United States and Canada: Current status and data needs. *Am J Clin Nutr.* 2004; 80: 1710S–1716S.

Cashman, K. D. Vitamin D: Dietary requirements and food fortification as a means of helping achieve adequate vitamin D status. *J Steroid Biochem Mol Biol.* 2015; 148: 19–26.

Castor, T. P. Formulations and compositions of vitamin D analogs for treating and preventing cancer and other diseases. Google Patents. 2016.

Chel, V. G., Ooms, M. E., van der Bent, J., Veldkamp, F., Roos, R. A., Achterberg, W. P. *et al.* High prevalence of vitamin D deficiency and insufficiency in patients with manifest Huntington disease: An explorative study. *Dermatoendocrinol.* 2013; 5: 348–351.

Chen, T. C., and Holick, M. F. Vitamin D and prostate cancer prevention and treatment. *Trends Endocrinol Metab.* 2003; 14: 423–430.

Chung, T. W., Huang, Y. Y., and Liu, Y. Z. Effects of the rate of solvent evaporation on the characteristics of drug loaded PLLA and PDLLA microspheres. *Int J Pharm.* 2001; 212: 161–169.

Dadoniene, J., Cypiene, A., Rinkuniene, E., Badariene, J., Burca, J., Sakaite, I. *et al.* Vitamin D and functional arterial parameters in postmenopausal women with metabolic syndrome. *Adv Med Sci.* 2016; 61: 224–230.

Dahan, A., and Hoffman, A. Use of a dynamic *in vitro* lipolysis model to rationalize oral formulation development for poor water soluble drugs: Correlation with *in vivo* data and the relationship to intra-enterocyte processes in rats. *Pharm Res.* 2006; 23: 2165–2174.

Das, R. K., Kasoju, N., and Bora, U. Encapsulation of curcumin in alginate-chitosan-pluronic composite nanoparticles for delivery to cancer cells. *Nanomedicine.* 2010; 6: 153–160.

Dattagupta, N., Das, A. R., Sridhar, C. N., and Patel, J. R. Liposomes containing cationic lipids and vitamin D. Google Patents. 1998.

Delaurent, C., Siouffi, A., and Pepe, G. Cyclodextrin inclusion complexes with vitamin D3: Investigations of the solid complex characterization. *Chem Analityczna.* 1998; 43: 601–616.

Deluca, H. F. History of the discovery of vitamin D and its active metabolites. *Bonekey Rep.* 2014; 3: 479.

Derfus, A. M., Chan, W. C., and Bhatia, S. N. Probing the cytotoxicity of semiconductor quantum dots. *Nano Lett.* 2004; 4: 11–18.

Dhaussy, A. Vitamin D recommendations, fortification in France, and communication. *OCL.* 2014; 21: D305.

Diamond, T. H., Ho, K. W., Rohl, P. G., and Meerkin, M. Annual intramuscular injection of a megadose of cholecalciferol for treatment of vitamin D deficiency: Efficacy and safety data. *Med J Aust.* 2005; 183: 10–12.

Diarrassouba, F., Garrait, G., Remondetto, G., Alvarez, P., Beyssac, E., and Subirade, M. Increased stability and protease resistance of the beta-lactoglobulin/vitamin D3 complex. *Food Chem.* 2014; 145: 646–652.

Einarsdóttir, K., Preen, D. B., Clay, T. D., Kiely, L., Holman, C. A. J., and Cohen, L. D. Effect of a single "megadose" intramuscular vitamin D (600,000 IU) injection on vitamin D concentrations and bone mineral density following biliopancreatic diversion surgery. *Obes Surg.* 2010; 20: 732–737.

Eisman, J. A., and Bouillon, R. Vitamin D: Direct effects of vitamin D metabolites on bone: Lessons from genetically modified mice. *Bonekey Rep.* 2014; 3: 499.

Fanari, Z., Hammami, S., Hammami, M. B., Hammami, S., and Abdellatif, A. Vitamin D deficiency plays an important role in cardiac disease and affects patient outcome: Still a myth or a fact that needs exploration? *J Saudi Heart Assoc.* 2015; 27: 264–271.

Fleet, J. C., DeSmet, M., Johnson, R., and Li, Y. Vitamin D and cancer: A review of molecular mechanisms. *Biochem J.* 2012; 441: 61–76.

Fogelholm, M., Anderssen, S., Gunnarsdottir, I., and Lahti-Koski, M. Dietary macronutrients and food consumption as determinants of long-term weight change in adult populations: A systematic literature review. *Adv Food Nutr Res.* 2012; 56.

Fu, L. W., and Vender, R. Systemic role for vitamin D in the treatment of psoriasis and metabolic syndrome. *Dermatol Res Pract.* 2011; 2011: 276079.

Ganmaa, D., Giovannucci, E., Bloom, B. R., Fawzi, W., Burr, W., Batbaatar, D. *et al.* Vitamin D, tuberculin skin test conversion, and latent tuberculosis in Mongolian school-age children: A randomized, double-blind, placebo-controlled feasibility trial. *Am J Clin Nutr.* 2012; 96: 391–396.

Gezen-Ak, D., Yılmazer, S., and Dursun, E. Why vitamin D in Alzheimer's disease? The hypothesis. *J Alzheimers Dis.* 2014; 40: 257–269.

Gianforcaro, A., and Hamadeh, M. J. Vitamin D as a potential therapy in amyotrophic lateral sclerosis. *CNS Neurosci Ther.* 2014; 20: 101–111.

Giovannucci, E., Liu, Y., Hollis, B. W., and Rimm, E. B. 25-hydroxyvitamin D and risk of myocardial infarction in men: A prospective study. *Arch Intern Med.* 2008; 168: 1174–1180.

Gleeson, J. P., Ryan, S. M., and Brayden, D. J. Oral delivery strategies for nutraceuticals: Delivery vehicles and absorption enhancers. *Trends Food Sci Technol.* 2016; 53: 90–101.

Glossmann, H. Pharmacology of vitamin D. *Osteologie.* 2011; 20: 299–303.

Goff, J. P., Koszewski, N. J., Haynes, J. S., and Horst, R. L. Targeted delivery of vitamin D to the colon using β-glucuronides of vitamin D: Therapeutic effects in a murine model of inflammatory bowel disease. *Am J Physiol Gastrointest Liver.* 2012; 302: G460–G469.

Grant, W. B., and Holick, M. F. Benefits and requirements of vitamin D for optimal health: A review. *Altern Med Rev.* 2005; 10: 94–111.

Gueye, Y., Marqueste, T., Maurel, F., Khrestchatisky, M., Decherchi, P., and Feron, F. Cholecalciferol (vitamin D3) improves functional recovery when delivered during the acute phase after a spinal cord trauma. *J Steroid Biochem Mol Biol.* 2015; 154: 23–31.

Gupta, A. Fortification of foods with vitamin D in India. *Nutrients.* 2014; 6: 3601–3623.

Gupta, S., Parvez, N., and Sharma, P. K. Nutraceuticals as functional foods. *J Nutr Ther.* 2015; 4: 64–72.

Haham, M., Ish-Shalom, S., Nodelman, M., Duek, I., Segal, E., Kustanovich, M. *et al.* Stability and bioavailability of vitamin D nanoencapsulated in casein micelles. *Food Funct.* 2012; 3: 737–744.

Hähnlein, W., Hansen, M. M., Olesen, J. E., and Tobiasen, A. G. Stable emulsions and dry powders of mixtures of fat-soluble vitamins, their preparations and use. Google Patents. 2003.

Han, J. E., Jones, J. L., Tangpricha, V., Brown, M. A., Brown, L. A., Hao, L. *et al.* High dose vitamin D administration in ventilated intensive care unit patients: A pilot double blind randomized controlled trial. *J Clin Transl Endocrinol.* 2016; 4: 59–65.

Harandi, A. A., Harandi, A. A., Pakdaman, H., and Sahraian, M. A. Vitamin D and multiple sclerosis. *Iran J Neurol.* 2014; 13: 1–6.

Harinarayan, C., and Joshi, S. R. Vitamin D status in India—Its implications and remedial measures. *J Assoc Physicians India.* 2009; 57: 40-8.

Harinarayan, C. V., Appicatlaa, L., Nalini, B. A., and Joshi, S. 60,000 IU vitamin D weekly with daily calcium was safe but not enough in the winter–2012. *Endocrinol Metabol Syndrome* 2012; S: 4. 2012.

Hart, P. H., Gorman, S., and Finlay-Jones, J. J. Modulation of the immune system by UV radiation: More than just the effects of vitamin D? *Nat Rev Immunol.* 2011; 11: 584.

Hartiti, S., Barrionuevo, M., López-Aliaga, I., Lisbona, F., Pallares, I., Alferez, M. *et al.* Effects of intestinal resection, cholecalciferol and ascorbic acid on iron metabolism in rats. *Br J Nutr.* 1995; 73: 871–880.

Haussler, M. R., Whitfield, G. K., Kaneko, I., Haussler, C. A., Hsieh, D., Hsieh, J. C. *et al.* Molecular mechanisms of vitamin D action. *Calcif Tissue Int.* 2013; 92: 77–98.

Hayes, C. E., Cantorna, M. T., and DeLuca, H. F. Vitamin D and multiple sclerosis. *Proc Soc Exp Biol Med.* 1997; 216: 21–27.

Heinig, M. J. Vitamin D and the breastfed infant: Controversies and concerns. *J Hum Lact.* 2003; 19: 247–249.

Henry, H. L., and Norman, A. W. Vitamin D: Metabolism and biological actions. *Annu Rev Nutr.* 1984; 4: 493–520.

Holick, M. F. Resurrection of vitamin D deficiency and rickets. *J Clin Invest.* 2006; 116: 2062–2072.

Holick, M. F. The vitamin D deficiency pandemic and consequences for nonskeletal health: Mechanisms of action. *Mol Aspects Med.* 2008; 29: 361–368.

Holick, M. F. Vitamin D deficiency. *N Engl J Med.* 2007; 357: 266–281.

Holick, M. F., and Chen, T. C. Vitamin D deficiency: A worldwide problem with health consequences. *Am J Clin Nutr.* 2008; 87: 1080S–1086S.

Hollo, A., Clemens, Z., and Lakatos, P. Epilepsy and vitamin D. *Int J Neurosci.* 2014; 124: 387–393.

Institute of Medicine FaNB. *Dietary reference intakes for calcium and vitamin D.* Washington, DC National Academy Press. 2010.

Israeli-Lev, G., and Livney, Y. D. Self-assembly of hydrophobin and its co-assembly with hydrophobic nutraceuticals in aqueous solutions: Towards application as delivery systems. *Food Hydrocoll.* 2014; 35: 28–35.

Jarrett, F., Ducasa, G. M., Buller, D. B., and Berwick, M. The effect of oral supplementation of vitamin D3 on serum levels of vitamin D: A review. *Epidemiology: Open Access*. 2014; 2014.

Jeffery, L. E., Raza, K., and Hewison, M. Vitamin D in rheumatoid arthritis-towards clinical application. *Nat Rev Rheumatol*. 2016; 12: 201–210.

Jesus, C. A., Feder, D., and Peres, M. F. The role of vitamin D in pathophysiology and treatment of fibromyalgia. *Curr Pain Headache Rep*. 2013; 17: 355.

Jiang, Y. J., and Bikle, D. D. LncRNA: A new player in 1α, 25 (OH) 2 vitamin D3/VDR protection against skin cancer formation. *Exp Dermatol*. 2014; 23: 147–150.

Johnson, L. R. Vitamin D insufficiency due to insufficient exposure to sunlight and related pathology. *Student Pulse*. 2010; 2.

Jones, G. Pharmacokinetics of vitamin D toxicity. *Am J Clin Nutr*. 2008; 88: 582S–586S.

Karras, S., Rapti, E., Matsoukas, S., and Kotsa, K. Vitamin D in fibromyalgia: A causative or confounding biological interplay? *Nutrients*. 2016; 8: 343.

Kawa, S., Nikaido, T., Aoki, Y., Zhai, Y., Kumagai, T., Furihata, K. *et al*. Vitamin D analogues up-regulate p21 and p27 during growth inhibition of pancreatic cancer cell lines. *Br J Cancer*. 1997; 76: 884–889.

Keegan, R. J., Lu, Z., Bogusz, J. M., Williams, J. E., and Holick, M. F. Photobiology of vitamin D in mushrooms and its bioavailability in humans. *Dermatoendocrinol*. 2013; 5: 165–176.

Kesarwani, K., Gupta, R., and Mukerjee, A. Bioavailability enhancers of herbal origin: An overview. *Asian Pac J Trop Biomed*. 2013; 3: 253–266.

Ketha, H., Wadams, H., Lteif, A., and Singh, R. J. Iatrogenic vitamin D toxicity in an infant—A case report and review of literature. *J Steroid Biochem Mol Biol*. 2015; 148: 14–18.

Khoo, A. L., Chai, L. Y., Koenen, H. J., Oosting, M., Steinmeyer, A., Zuegel, U. *et al*. Vitamin D(3) down-regulates proinflammatory cytokine response to Mycobacterium tuberculosis through pattern recognition receptors while inducing protective cathelicidin production. *Cytokine*. 2011; 55: 294–300.

Knekt, P., Kilkkinen, A., Rissanen, H., Marniemi, J., Saaksjarvi, K., and Heliovaara, M. Serum vitamin D and the risk of Parkinson disease. *Arch Neurol*. 2010; 67: 808–811.

Kong, J., and Li, Y. C. Molecular mechanism of 1,25-dihydroxyvitamin D3 inhibition of adipogenesis in 3T3-L1 cells. *Am J Physiol Endocrinol Metab*. 2006; 290: E916–E924.

Korbel, J. N., Sebok, B., Kerenyi, M., and Mahrle, G. Enhancement of the antiparakeratotic potency of calcitriol and tacalcitol in liposomal preparations in the mouse tail test. *Skin Pharmacol Appl Skin Physiol*. 2001; 14: 291–295.

Kumar, M., Kakkar, V., Mishra, A. K., Chuttani, K., and Kaur, I. P. Intranasal delivery of streptomycin sulfate (STRS) loaded solid lipid nanoparticles to brain and blood. *Int J Pharm*. 2014a; 461: 223–233.

Kumar, M., Sharma, G., Singla, D., Singh, S., Sahwney, S., Chauhan, A. S. *et al*. Development of a validated UPLC method for simultaneous estimation of both free and entrapped (in solid lipid nanoparticles) all-trans retinoic acid and cholecalciferol (vitamin D3) and its pharmacokinetic applicability in rats. *J Pharm Biomed Anal*. 2014b; 91: 73–80.

Kumar, V., Kumar, A., Tuli, I. P., and Rai, A. K. Therapeutic significance of vitamin D in allergic rhinitis. *Astrocyte*. 2015; 2: 8.

Laverny, G., Penna, G., Vetrano, S., Correale, C., Nebuloni, M., Danese, S. *et al*. Efficacy of a potent and safe vitamin D receptor agonist for the treatment of inflammatory bowel disease. *Immunol Lett*. 2010; 131: 49–58.

Lehmann, B., and Meurer, M. Vitamin D metabolism. *Dermatol Ther*. 2010; 23: 2–12.

Lehmann, U., Hirche, F., Stangl, G. I., Hinz, K., Westphal, S., and Dierkes, J. Bioavailability of vitamin D2 and D3 in healthy volunteers, a randomized placebo-controlled trial. *J Clin Endocrinol Metab*. 2013; 98: 4339–4345.

Levinson, Y., Israeli-Lev, G., and Livney, Y. D. Soybean β-conglycinin nanoparticles for delivery of hydrophobic nutraceuticals. *Food Biophysics*. 2014; 9: 332–340.

Li, H., Wang, Q., Xiao, Y., Bao, C., and Li, W. 25-Hydroxyvitamin D3-loaded PLA microspheres: *In vitro* characterization and application in diabetic periodontitis models. *AAPS PharmSciTech*. 2013; 14: 880–889.

Li, W., Peng, H., Ning, F., Yao, L., Luo, M., Zhao, Q. *et al*. Amphiphilic chitosan derivative-based core-shell micelles: Synthesis, characterisation and properties for sustained release of Vitamin D3. *Food Chem*. 2014; 152: 307–315.

Li, X., Lin, L., Zhu, Y., Liu, W., Yu, T., and Ge, M. Preparation of ultrafine fast-dissolving cholecalciferol-loaded poly(vinyl pyrrolidone) fiber mats via electrospinning. *Polym Compos*. 2013; 34: 282–287.

Lien, Y.-H., Wu, J.-H., Liao, J.-W., and Wu, T.-M. *In vitro* evaluation of the thermosensitive and magnetic nanoparticles for the controlled drug delivery of vitamin D3. *Macromol Res*. 2013; 21: 511–518.

Lin, Y., Wang, Y.-H., Yang, X.-Q., Guo, J., and Wang, J.-M. Corn protein hydrolysate as a novel nano-vehicle: Enhanced physicochemical stability and *in vitro* bioaccessibility of vitamin D3. *LWT-Food Sci Technol.* 2016; 72: 510–517.

Lin, Y. K., Huang, Z. R., Zhuo, R. Z., and Fang, J. Y. Combination of calcipotriol and methotrexate in nano-structured lipid carriers for topical delivery. *Int J Nanomedicine.* 2010; 5: 117–128.

Linhares, C., de Sá, L. B. P. C., Rocha, D. R. T. W., and Arbex, A. K. Vitamin D and diabetes mellitus: A review. *Open J Endocr Metab Dis.* 2016; 6: 1.

Lips, P., van Schoor, N. M. The effect of vitamin D on bone and osteoporosis. *Best Pract Res Clin Endocrinol Metab.* 2011; 25: 585–591.

Lockwood, B. *Nutraceuticals.* Pharmaceutical Press: London, 2007.

Luca, G., Basta, G., Calafiore, R., Rossi, C., Giovagnoli, S., Esposito, E. *et al.* Multifunctional microcapsules for pancreatic islet cell entrapment: Design, preparation and *in vitro* characterization. *Biomaterials.* 2003; 24: 3101–3114.

Luo, Y., Teng, Z., and Wang Q. Development of zein nanoparticles coated with carboxymethyl chitosan for encapsulation and controlled release of vitamin D3. *J Agric Food Chem.* 2012; 60: 836–843.

Luo, Y., Wang, T. T., Teng, Z., Chen, P., Sun, J., and Wang, Q. Encapsulation of indole-3-carbinol and 3,3'-diindolylmethane in zein/carboxymethyl chitosan nanoparticles with controlled release property and improved stability. *Food Chem.* 2013; 139: 224–230.

Lương, K., and Nguyễn, L. Role of vitamin d in Parkinson's disease. *ISRN Neurology.* 2012.

Makino, Y., Suzuki, Y., and Aoyagi, T. Method for treating psoriasis by externally administering to a patient a pharmaceutical composition containing active-type vitamin D. Google Patents. 1989.

Maleklou, N., Allameh, A., and Kazemi, B. Preparation, characterization and *in vitro*-targeted delivery of novel Apolipoprotein E-based nanoparticles to C6 glioma with controlled size and loading efficiency. *J Drug Target.* 2016; 24: 348–358.

Maleklou, N., Allameh, A., and Kazemi, B. Targeted delivery of vitamin D3-loaded nanoparticles to C6 glioma cell line increased resistance to doxorubicin, epirubicin, and docetaxel *in vitro. In Vitro Cell Dev Biol Anim.* 2016: 1–12.

Marks, J., Srai, S. K., Biber, J., Murer, H., Unwin, R. J., and Debnam, E. S. Intestinal phosphate absorption and the effect of vitamin D: A comparison of rats with mice. *Exp Physio.* 2006; 91: 531–537.

Marques, C. D., Dantas, A. T., Fragoso, T. S., Duarte, A. L. The importance of vitamin D levels in autoimmune diseases. *Rev Bras Reumatol.* 2010; 50: 67–80.

McClements, D. J., Zou, L., Zhang, R., Salvia-Trujillo, L., Kumosani, T., and Xiao, H. Enhancing nutraceutical performance using excipient foods: Designing food structures and compositions to increase bioavailability. *Compr Rev Food Sci Food Saf.* 2015; 14: 824–847.

McDonnell, S. L., French, C. B., and Heaney, R. P. Quantifying the food sources of basal vitamin d input. *J Steroid Biochem Mol Biol.* 2014; 144 Pt A: 149–151.

Merz, K., and Sternberg, B. Incorporation of vitamin D3-derivatives in liposomes of different lipid types. *J Drug Target.* 1994; 2: 411–417.

Milne, G. W., and Delander, M. *Vitamin D handbook: Structures, synonyms, and properties.* John Wiley & Sons, 2008.

Mohammadi, M., Ghanbarzadeh, B., and Hamishehkar, H. Formulation of nanoliposomal vitamin d3 for potential application in beverage fortification. *Adv Pharm Bull.* 2014; 4: 569–575.

Mostafa, W. Z., and Hegazy, R. A. Vitamin D and the skin: Focus on a complex relationship: A review. *J Adv Res.* 2015; 6: 793–804.

Mouli, V. P., and Ananthakrishnan, A. N. Review article: Vitamin D and inflammatory bowel diseases. *Aliment Pharmacol Ther.* 2014; 39: 125–136.

Mousa, S. A. Nanoformulation of vitamin D derivatives and/or vitamin D metabolites. Google Patents. 2015.

Mousa, S. A. Nanoformulation of vitamin D derivatives and/or vitamin D metabolites. Google Patents. 2013.

Mpandzou, G., Haddou, E. A. B., Regragui, W., Benomar, A., and Yahyaoui, M. Vitamin D deficiency and its role in neurological conditions: A review. *Revue Neurologique.* 2016; 172: 109–122.

Naeem, Z. Vitamin d deficiency: An ignored epidemic. *Int J Health Sci (Qassim).* 2010; 4: V–VI.

Naeff, R., Delmenico, S., Spycher, R., Corbo, M., and Flother, F. Liposome-based topical vitamin D formulation. Google Patents. 1998.

Nnoaham, K. E., and Clarke, A. Low serum vitamin D levels and tuberculosis: A systematic review and meta-analysis. *Int J Epidemiol.* 2008; 37: 113–119.

Norton, R., and O'Connell, M. A. Vitamin D: Potential in the prevention and treatment of lung cancer. *Anticancer Res.* 2012; 32: 211–221.

Nursyam, E. W., Amin, Z., and Rumende, C. M. The effect of vitamin D as supplementary treatment in patients with moderately advanced pulmonary tuberculous lesion. *Hemoglobin.* 2006; 1500: 1500.

O'Riordan, J. L., and Bijvoet, O. L. Rickets before the discovery of vitamin D. *Bonekey Rep.* 2014; 3: 478.

Ozturk, B., Argin, S., Ozilgen, M., and McClements, D. J. Nanoemulsion delivery systems for oil-soluble vitamins: Influence of carrier oil type on lipid digestion and vitamin D3 bioaccessibility. *Food Chem.* 2015; 187: 499–506.

Palomer, X., Gonzalez-Clemente, J. M., Blanco-Vaca, F., and Mauricio, D. Role of vitamin D in the pathogenesis of type 2 diabetes mellitus. *Diabetes Obes Metab.* 2008; 10: 185–197.

Palvai, S., Nagraj, J., Mapara, N., Chowdhury, R., and Basu, S. Dual drug loaded vitamin D3 nanoparticle to target drug resistance in cancer. *RSC Advances.* 2014; 4: 57271–57281.

Pan, L., Matloob, A. F., Du, J., Pan, H., Dong, Z., Zhao, J. *et al.* Vitamin D stimulates apoptosis in gastric cancer cells in synergy with trichostatin A/sodium butyrate-induced and 5-aza-2′-deoxycytidine-induced PTEN upregulation. *The FEBS Journal.* 2010; 277: 989–999.

Patil, S., Gawali, S., Patil, S., and Basu, S. Synthesis, characterization and *in vitro* evaluation of novel vitamin D3 nanoparticles as a versatile platform for drug delivery in cancer therapy. *J Mater Chem B.* 2013; 1: 5742–5750.

Paucar, O. C., Tulini, F. L., Thomazini, M., Balieiro, J. C. C., Pallone, E. M. J. A., and Favaro-Trindade, C. S. Production by spray chilling and characterization of solid lipid microparticles loaded with vitamin D3. *Food Bioprod Process.* 2016; 100: 344–350.

Pavanetto, F., Genta, I., Giunchedi, P., and Conti, B. Evaluation of spray drying as a method for polylactide and polylactide-co-glycolide microsphere preparation. *J Microencapsul.* 1993; 10: 487–497.

Peehl, D. M., Shinghal, R., Nonn, L., Seto, E., Krishnan, A. V., Brooks, J. D. *et al.* Molecular activity of 1,25-dihydroxyvitamin D3 in primary cultures of human prostatic epithelial cells revealed by cDNA microarray analysis. *J Steroid Biochem Mol Biol.* 2004; 92: 131–141.

Pelajo, C. F., Lopez-Benitez, J. M., and Miller, L. C. Vitamin D and autoimmune rheumatologic disorders. *Autoimmun Rev.* 2010; 9: 507–510.

Pendo, K., and DeGiorgio, C. M. Vitamin D3 for the treatment of epilepsy: Basic mechanisms, animal models, and clinical trials. *Front Neurol.* 2016; 7: 218.

Petritz, E., Tritthart, T., and Wintersteiger, R. Determination of phylloquinone and cholecalciferol encapsulated in granulates formed by melt extrusion. *J Biochem Biophys Methods.* 2006; 69: 101–112.

Plum, L. A., and DeLuca, H. F. Vitamin D, disease and therapeutic opportunities. *Nat Rev Drug Discov.* 2010; 9: 941–955.

Pointillart, A., Denis, I., and Colin, C. Effects of dietary vitamin D on magnesium absorption and bone mineral contents in pigs on normal magnesium intakes. *Magnes Res.* 1995; 8: 19–26.

Pond, W. G., Church, D. C., and Pond, K. R. *Basic animal nutrition and feeding.* John Wiley and Sons, 1995.

Prentice, A. M. Macronutrients as sources of food energy. *Public Health Nutr.* 2005; 8: 932–939.

Purohit, V. Effect of single injection of vitamin d (cholecalciferol, 6 lac iu) in adults: Does body mass index determine dosage frequency and rise in serum 25 (oh) D3 level? *Asian J Pharm Clin Res.* 2015; 8: 192–194.

Rajakumar, K., and Holick, M. F. Vitamin D deficiency: Historical perspectives. *Vitamin D and the lung.* Springer, 2012; pp. 3–23.

Ralph, A. P., Lucas, R. M., and Norval, M. Vitamin D and solar ultraviolet radiation in the risk and treatment of tuberculosis. *Lancet Infect Dis.* 2013; 13: 77–88.

Ramalho, M. J., Loureiro, J. A., Gomes, B., Frasco, M. F., Coelho, M. A., and Pereira, M. C. PLGA nanoparticles as a platform for vitamin D-based cancer therapy. *Beilstein J Nanotechnol.* 2015; 6: 1306–1318.

Ross, A. C., Manson, J. E., Abrams, S. A., Aloia, J. F., Brannon, P. M., Clinton, S. K. *et al.* The 2011 report on dietary reference intakes for calcium and vitamin D from the Institute of Medicine: What clinicians need to know. *J Clin Endocrinol Metab.* 2011; 96: 53–58.

Sadat-Ali, M., Bubshait, D. A., Al-Turki, H. A., Al-Dakheel, D. A., and Al-Olayani, W. S. Topical delivery of vitamin d3: A randomized controlled pilot study. *Int J Biomed Sci.* 2014; 10: 21.

Sahota, O. Osteoporosis and the role of vitamin D and calcium-vitamin D deficiency, vitamin D insufficiency and vitamin D sufficiency. *Age Ageing.* 2000; 29: 301–304.

Saini, V., Nadeem, M., Kolb, C., Gangloff, S., Zivadinov, R., Ramanathan, M. *et al.* Vitamin D: Role in pathogenesis of multiple sclerosis. *Multiple Sclerosis: A Mechanistic View.* 2015: 127.

Samoes, B., and Silveira, C. The role of vitamin D in the pathophysiology of schizophrenia. *Neuropsychiatry.* 2017; 7.

Schwalfenberg, G. Not enough vitamin D health consequences for Canadians. *Can Fam Physician.* 2007; 53: 841–854.

Selvaraj, P., Harishankar, M., Singh, B., Banurekha, V. V., and Jawahar, M. S. Effect of vitamin D3 on chemokine expression in pulmonary tuberculosis. *Cytokine*. 2012; 60: 212–219.

Sequeira, V. B., Rybchyn, M. S., Tongkao-on, W., Gordon-Thomson, C., Malloy, P. J., Nemere, I. *et al.* The role of the vitamin D receptor and ERp57 in photoprotection by 1α, 25-dihydroxyvitamin D3. *Mol Endocrinol*. 2012; 26: 574–582.

Sergeev, I. N. 1, 25-Dihydroxyvitamin D 3 induces Ca 2+-mediated apoptosis in adipocytes via activation of calpain and caspase-12. *Biochem Biophys Res Commun*. 2009; 384: 18–21.

Seshadri, K. G., Tamilselvan, B., and Rajendran, A. Role of vitamin D in diabetes. *Int J Endocrinol Metab Disord*. 2011; 1: 47–56.

Soleymani, T., Hung, T., and Soung, J. The role of vitamin D in psoriasis: A review. *Int J Dermatol*. 2015; 54: 383–392.

Spears, J. W., and Weiss, W. P. Invited review: Mineral and vitamin nutrition in ruminants 1. *The Prof Animal Scientist*. 2014; 30: 180–191.

Sutaria, N., Liu, C. T., and Chen, T. C. Vitamin D status, receptor gene polymorphisms, and supplementation on tuberculosis: A systematic review of case-control studies and randomized controlled trials. *J Clin Transl Endocrinol*. 2014; 1: 151–160.

Swami, S., Raghavachari, N., Muller, U. R., Bao, Y. P., and Feldman, D. Vitamin D growth inhibition of breast cancer cells: Gene expression patterns assessed by cDNA microarray. *Breast Cancer Res Treat*. 2003; 80: 49–62.

Talat, N., Perry, S., Parsonnet, J., Dawood, G., and Hussain, R. Vitamin D deficiency and tuberculosis progression. *Emerg Infect Dis*. 2010; 16: 853–855.

Tang, J. Y., Fu, T., Leblanc, E., Manson, J. E., Feldman, D., Linos, E. *et al.* Calcium plus vitamin D supplementation and the risk of nonmelanoma and melanoma skin cancer: Post hoc analyses of the women's health initiative randomized controlled trial. *J Clin Oncol*. 2011; 29: 3078–3084.

Tellioglu, A., Basaran, S., Guzel, R., and Seydaoglu, G. Efficacy and safety of high dose intramuscular or oral cholecalciferol in vitamin D deficient/insufficient elderly. *Maturitas*. 2012; 72: 332–338.

Teng, Z., Luo, Y., and Wang, Q. Carboxymethyl chitosan-soy protein complex nanoparticles for the encapsulation and controlled release of vitamin D(3). *Food Chem*. 2013; 141: 524–532.

Thacher, T. D., and Clarke, B. L. Vitamin D insufficiency. *Mayo Clin Proc*. 2011; 86: 50–60.

Tsai, K. S., Heath, H., III, Kumar, R., and Riggs, B. L. Impaired vitamin D metabolism with aging in women. Possible role in pathogenesis of senile osteoporosis. *J Clin Invest*. 1984; 73: 1668–1672.

Urbain, P., Singler, F., Ihorst, G., Biesalski, H. K., and Bertz, H. Bioavailability of vitamin D(2) from UV-B-irradiated button mushrooms in healthy adults deficient in serum 25-hydroxyvitamin D: A randomized controlled trial. *Eur J Clin Nutr*. 2011; 65: 965–971.

USFDA. CFR—Code of Federal Regulations Title 21. Available online at: https://www.accessdata.fda.gov (Accessed on 1 June 2017).

van Ginkel, P. R., Yang, W., Marcet, M. M., Chow, C. C., Kulkarni, A. D., Darjatmoko, S. *et al.* 1 alpha-Hydroxyvitamin D2 inhibits growth of human neuroblastoma. *J Neurooncol*. 2007; 85: 255–262.

Vieth, R. Vitamin D toxicity, policy, and science. *J Bone Miner Res*. 2007; 22 Suppl 2: V64–68.

Vieth, R. Vitamin D supplementation, 25-hydroxyvitamin D concentrations, and safety. *Am J Clin Nutr*. 1999; 69: 842–856.

Viswanath, D. Dietary reference intakes for calcium and vitamin D. *J Sci Innov Res*. 2013; 2: 710–715.

Wagner, C. L., Taylor, S. N., and Hollis, B. W. Does vitamin D make the world go 'round'? *Breastfeed Med*. 2008; 3: 239–250.

Wagner, D., Sidhom, G., Whiting, S. J., Rousseau, D., and Vieth, R. The bioavailability of vitamin D from fortified cheeses and supplements is equivalent in adults. *J Nutr*. 2008; 138: 1365–1371.

Wang, B., Vongsvivut, J., Adhikari, B., and Barrow, C. J. Microencapsulation of tuna oil fortified with the multiple lipophilic ingredients vitamins A, D3, E, K2, curcumin and coenzyme Q10. *J Funct Foods*. 2015; 19: 893–901.

Wejse, C., Gomes, V. F., Rabna, P., Gustafson, P., Aaby, P., Lisse, I. M. *et al.* Vitamin D as supplementary treatment for tuberculosis: A double-blind, randomized, placebo-controlled trial. *Am J Respir Crit Care Med*. 2009; 179: 843–850.

Whayne, T. F. Vitamin D: Popular cardiovascular supplement but benefit must be evaluated. *Int J Angiol*. 2011; 20: 063–072.

Wilson, N., and Shah, N. Microencapsulation of vitamins. *ASEAN Food J*. 2007; 14: 1–14.

Woo, J. S., Yi, H. G., and Jin, J. N. Complex formulation for preventing or treating osteoporosis which comprises solid dispersion of vitamin d or its derivative and bisphosphonate. Google Patents. 2007.

Yang, L., Yang, J., Venkateswarlu, S., Ko, T., and Brattain, M. G. Autocrine TGFbeta signaling mediates vitamin D3 analog-induced growth inhibition in breast cells. *J Cell Physiol.* 2001; 188: 383–393.

Yin, L., Grandi, N., Raum, E., Haug, U., Arndt, V., and Brenner, H. Meta-analysis: Serum vitamin D and breast cancer risk. *Eur J Cancer.* 2010; 46: 2196–2205.

Yuan, T., Qin, L., Wang, Z., Nie, J., Guo, Z., Li, G. *et al.* Solid lipid dispersion of calcitriol with enhanced dissolution and stability. *Asian J Pharm Sci.* 2013; 8: 39–47.

Zhang, X., Nicosia, S. V., and Bai, W. Vitamin D receptor is a novel drug target for ovarian cancer treatment. *Curr Cancer Drug Targets.* 2006; 6: 229–244.

Ziani, K., Fang, Y., and McClements, D. J. Encapsulation of functional lipophilic components in surfactant-based colloidal delivery systems: Vitamin E, vitamin D, and lemon oil. *Food Chem.* 2012; 134: 1106–1112.

Zittermann, A., Morshuis, M., Kuhn, J., Pilz, S., Ernst, J. B., Oezpeker, C. *et al.* Vitamin D metabolites and fibroblast growth factor-23 in patients with left ventricular assist device implants: Association with stroke and mortality risk. *Eur J Nutr.* 2016; 55: 305–313.

12 Nanotechnology and Nature's Miracle Compound: Curcumin

*Candace Minhthu Day, Ankit Parikh, Yunmei Song, and Sanjay Garg**

CONTENTS

12.1 INTRODUCTION

Natural plant products such as food and medicines are an integral part of human history. Many of these products have been successfully explored for development of drugs following modern scientific approaches. Turmeric (*Curcuma longa*) is well-known as the "golden spice" with a history of more than 4,000 years of usage (Prasad and Aggarwal, 2011; Gupta *et al.*, 2013).

* Corresponding author.

CUR, an active ingredient isolated from turmeric, has also received serious attention in recent years for its health-related applications, as indicated by its presence in more than 3,000 publications and nearly 2,000 patents in the last three decades (Anderson, 2000; Priyadarsini, 2014). CUR is the natural antioxidant compound, extracted from the rhizome of turmeric (Trujillo *et al.*, 2013). In term of therapeutic value, it is the most important active ingredient present in turmeric, unlike others such as the less active demothoxycurcumin or bisdemethoxycurcumin (Douglass and Clouatre, 2015; Panahi *et al.*, 2015). CUR is one of the most widely used herbal medicines, particularly in the Asian countries (Mishra and Palanivelu, 2008). Specifically, it has been used quite diversely as a supplement, spice, or dye in food and cosmetic industries across many countries (Gupta *et al.*, 2012; Prasad *et al.*, 2014).

12.2 CHEMISTRY OF CUR

CUR is chemically, 1, 7-bis-(4-hydroxy-3-methoxyphenyl)-hepta-1, 6-diene-3, 5-dione $C_{21}H_{20}O_6$ (Priyadarsini, 2014). With a molecular mass of 368.39 g/mol, it is known to be a lipophilic compound. Structurally, it comprises of two methoxy phenols, attached to α and β-diketone (Bose *et al.*, 2015; Ghalandarlaki *et al.*, 2014). These phenolic and diketone functional groups possess the antioxidant property, and therefore, are responsible for the antioxidant activity of CUR (Jitoe-Masuda *et al.*, 2013; Priyadarsini, 2014). CUR's action as a Michael acceptor is due to keto-enol tautomerization (Gupta, 2011). Its solubility in polar solvents such as methanol, ethanol and ethyl acetate is better than in aqueous solutions (Lee *et al.*, 2013).

12.3 CUR AS A THERAPEUTIC AGENT

CUR is one of the most bioactive molecules ever discovered from nature. Indeed, through its interaction with various intracellular and extracellular targets, CUR can affect different biological processes, including oxidative stress, inflammation, homoeostasis, survival, and proliferation of cells (Zhou *et al.*, 2011). The cells with irregular physiology presenting in abnormal states could also be targeted by CUR (Sa and Das, 2008). CUR's lipophilic nature favors its permeability across physiological barriers, including the blood–brain barrier (BBB) (Orlando *et al.*, 2012).

The pharmacological effects of CUR against various diseases such as inflammation, neurological disorders, cancer, infectious diseases, and lifestyle-associated diseases have been frequently reported (Gupta *et al.*, 2013). Heretofore, numerous *in vitro*, animal, and clinical studies have recommended its efficacy against a diverse range of diseases (Figure 12.1).

12.4 SAFETY OF CUR

Listed as generally recognized as safe (GRAS) by the U.S. FDA, it has now been marketed as a supplement all across the world (Basnet and Skalko-Basnet, 2011). The safety of CUR was also confirmed by the National Toxicology Program (NTP), performed in rat and mouse studies for the duration of up to 2 years, with dose-escalating studies from 50 to 2600 mg/kg (Aggarwal and Harikumar, 2009; Velusami *et al.*, 2013). A daily oral dose of 12 g was considered as safe and well-tolerated, as noticed from phase I and II clinical trials (Rahimi, 2015). Its safety has also been assessed in patients with oral leucoplakia, advanced pancreatic cancer, intestinal metaplasia (IM) of the gastric mucosa, multiple myeloma, Bowen disease of the skin relating to arsenic, rheumatoid arthritis, peptic ulcers, psoriasis, Dejerine-Sottas disease, uterine-cervical dysplasia, Alzheimer's disease, type 2 diabetic nephropathy, bladder cancer, and lupus nephritis (Gupta *et al.*, 2013).

12.5 PROBLEMS ASSOCIATED WITH DRUG DEVELOPMENT

To accomplish the desired therapeutic outcomes, it is necessary to achieve the required level of a drug at its target site (Bae and Park, 2011). Drug solubility and stability in biological fluids and

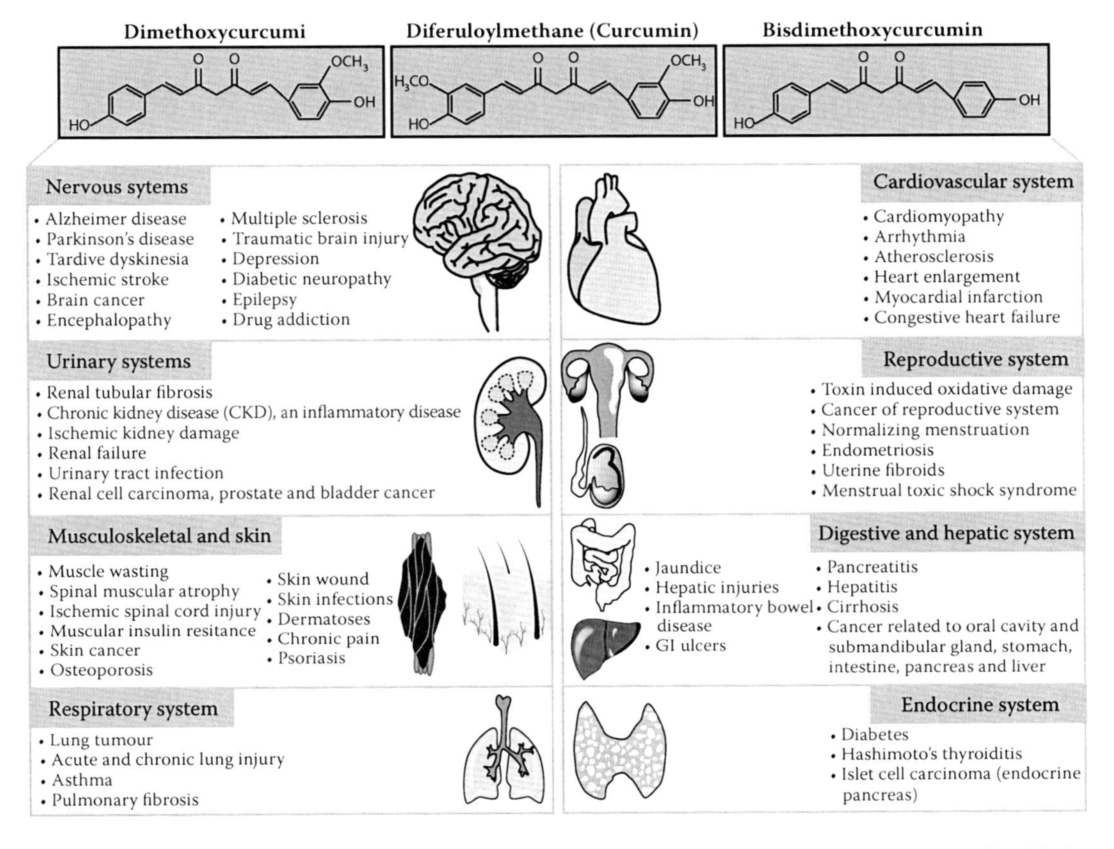

FIGURE 12.1 Therapeutic effects of curcuminoids (diferuloylmethane, demethoxycurcumin, bisdemethyoxycurcumin) on various diseases associated to different body systems.

physiological condition, metabolism, tissue distribution, and permeability are the major factors affecting achievement of drug levels needed to bring about therapeutic effects (Vaage *et al.*, 1994; Seo *et al.*, 2012; Bischoff *et al.*, 2014).

Regardless of proven efficacy and safety in multiple diseased conditions, through *in vitro* cell lines, animal and human trials, poor BA of CUR is considered as the major obstacle in its clinical development as a drug (Anand *et al.*, 2007). Moreover, the problems associated with the drug delivery of CUR include poor solubility, poor alkaline stability, photo-degradation, and extensive metabolism. These factors all account for CUR's poor BA (Figure 12.2) (Tonnesen *et al.*, 2002; Schneider *et al.*, 2015).

12.5.1 Solubility

CUR is a lipophilic compound, practically insoluble in water, with its solubility in aqueous solution (at pH 5.0) as only 11 ng/mL (Hu *et al.*, 2015). The lipophilic characteristic of CUR (log P of 3.29), though, facilitates its permeability, yet rapid degradation at physiological conditions and extensive metabolism hamper its permeability (Grynkiewicz and Slifirski, 2012). According to the BCS class, CUR belongs principally to a class II or class IV agent (Wahlang *et al.*, 2011).

12.5.2 Stability

The poor stability of CUR in the aqueous formulation, specifically at neutral to alkaline pH, could be considered as the second major issue (Thangavel *et al.*, 2015). The phenolate ion formed in water

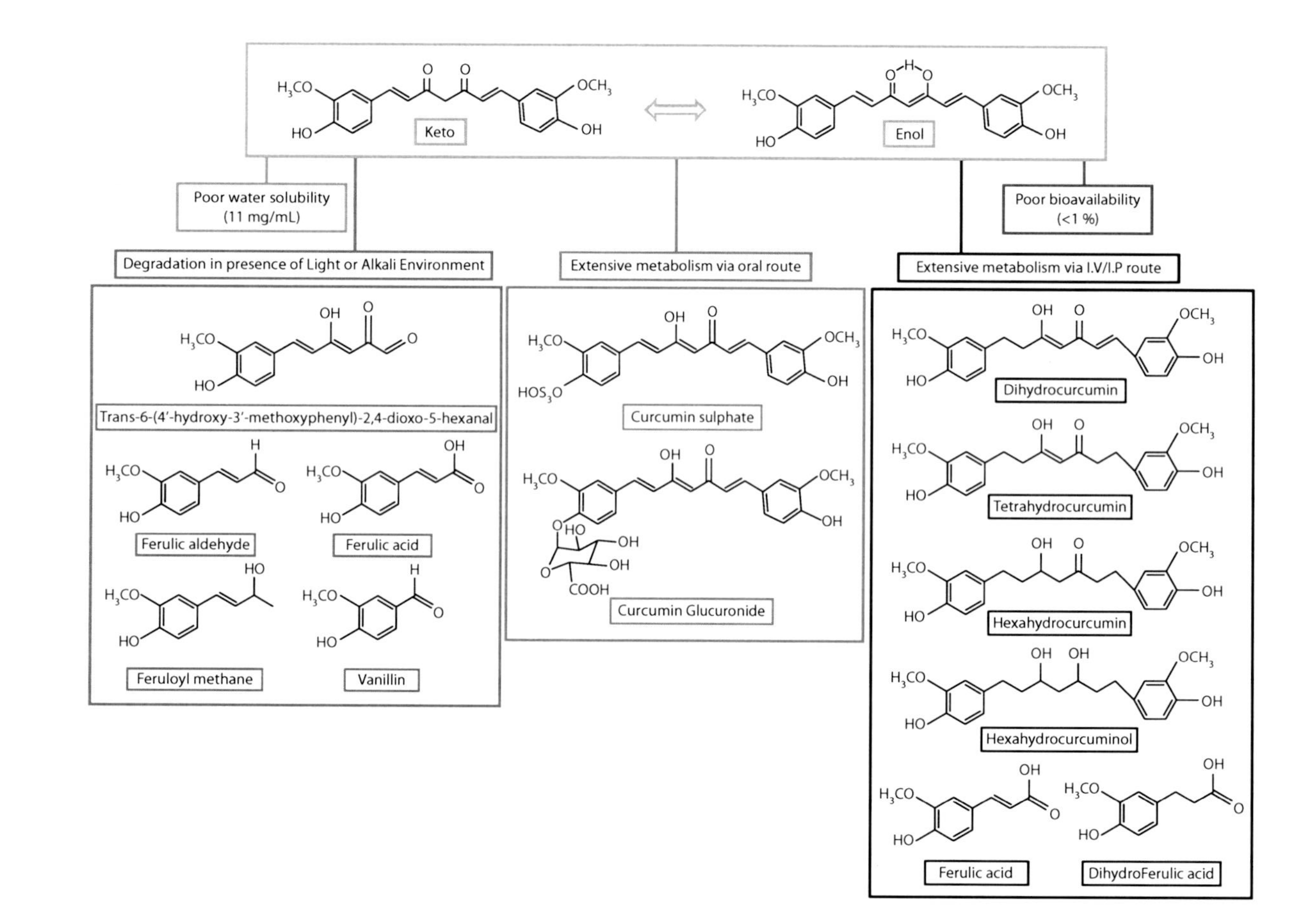

FIGURE 12.2 Problems associated with the use of curcumin, including poor solubility, poor stability in physiological milieu, alkaline condition or in presence of light, extensive metabolism and poor bioavailability. Curcumin has β-diketone chain, and displays a keto-enol tautomerism. Trans-6-(40-hydroxy-30-methoxyphenyl)-2, 4-dioxo-5-hexenal, ferulic aldehyde, ferulic acid, feruloyl methane and vanillin are the major degradants of curcumin.

has shown low stability at pH 7 and above, resulting in the formation of degradants (Kumavat *et al.*, 2013). In neutral to basic environments, CUR undergoes rapid degradation, such as possessing over 90% degradation in 30 minutes at pH 7.2 (Wang *et al.*, 1997). CUR is relatively stable at acidic pH, showing less than 20% degradation within 1 hour, as its dissociation shifts to neutral forms and the conjugated diene structures (Lee *et al.*, 2013). CUR also shows poor stability in cell culture medium, though on addition of human blood, fetal calf serum, or antioxidants such as ascorbic acid, the degradation of CUR decreases (Palanikumar, 2009). In addition to this, CUR is also considered as a light-sensitive drug. It is relatively quite stable at high temperature but suffers significant degradation in the presence of light, either in solution or at solid state (Zebi *et al.*, 2010). A CUR sample prepared in transparent glassware showed 5% less CUR content, compared to formulations prepared in amber-colored glassware (Wang *et al.*, 1997).

12.5.3 Metabolism

The CUR metabolism depends on the route of administration, as it goes for phase I metabolism based on reduction mechanism, when administrated via parenteral routes. For phase II metabolism, glucuronide and sulphates are involved both in humans and rodents (Shen and Ji, 2012). The major metabolites with their chemical structures are shown in Figure 12.2. Over 99% of CUR presented in its metabolite forms revelled from plasma hydrolysis study (Pan, 1999). Liver is responsible for the extensive metabolism of CUR (Prasad *et al.*, 2014). In fact, UDP-glucuronosyltransferase (UGT) is considered as the prominent hepatic enzyme for CUR metabolism. Evidently, when CUR is co-administrated with the specific inhibitors for this enzyme, significant improvement in its BA was observed (Grill *et al.*, 2014). CUR or its metabolite were not detected in plasma or urine in a trial conducted on patients with colorectal cancer, during a period of 29 days with daily dosing of 440–2200 mg CUR, given orally (Sharma *et al.*, 2001). In another study of 4-month duration, a concentration of 10 nM CUR and its metabolite was observed after 0.45–3.6 g/day. However, no conclusion could be available on the comparative activity of CUR and its metabolite (Anand *et al.*, 2007).

12.5.4 Bioavailability

In Sprague-Dawley rats, insignificant plasma levels of CUR were reported after administrating an oral dosing of 1 g/kg, suggesting its poor BA (Lao *et al.*, 2006). After 50 minutes, a plasma level of 15 ng/mL was detected following the same dose and route of administration. At 500 mg/kg dose, only 1% oral BA was obtained from rats' plasma (Prasad *et al.*, 2014). Besides this, a C_{max} value of 6.6 μg/mL was achieved following an intravenous injection of 2 mg/kg dose (He *et al.* 2015). Some studies also reported a high amount of CUR in serum compared to plasma (Gupta *et al.*, 2013). For instance, serum availability of CUR of 1.35 ± 0.23 μg/mL was observed in rats 0.83 hours after 2 g/kg oral intake of CUR (Anand *et al.*, 2007). Moreover, a 0.06 μg/mL serum level of CUR was achieved via oral route at 500 mg/kg dose, compared to that of 0.36 μg/mL following an intravenous administration of 10 mg/kg dose (Farooqui, 2016). The BA observations suggest the superiority of a parenteral formulation over oral ones for crude CUR.

In humans, a CUR plasma amount of 0.41–1.75 μM was determined after an hour, following an oral dosing of 4–8 g (Gopal, 2015). Likewise, an amount of 11.1 nM was also reported after 1 hour, post 3.6 g oral administration (Sharma *et al.*, 2004). With 10 g and 12 g oral administrations, 50.5 ng/mL and 51.2 ng/mL CUR serum levels were detected, respectively (Lao *et al.*, 2006).

12.6 NANOTECHNOLOGY AS EXPLORED FOR CURCUMIN

Nanotechnology, often referred to as "technology of the future," involves studying and controlling matters on atomic, molecular, or macromolecular scales, usually below 100 nm (de Villiers *et al.*, 2009). In recent years, nanotechnology has greatly fascinated scientists around the world for its

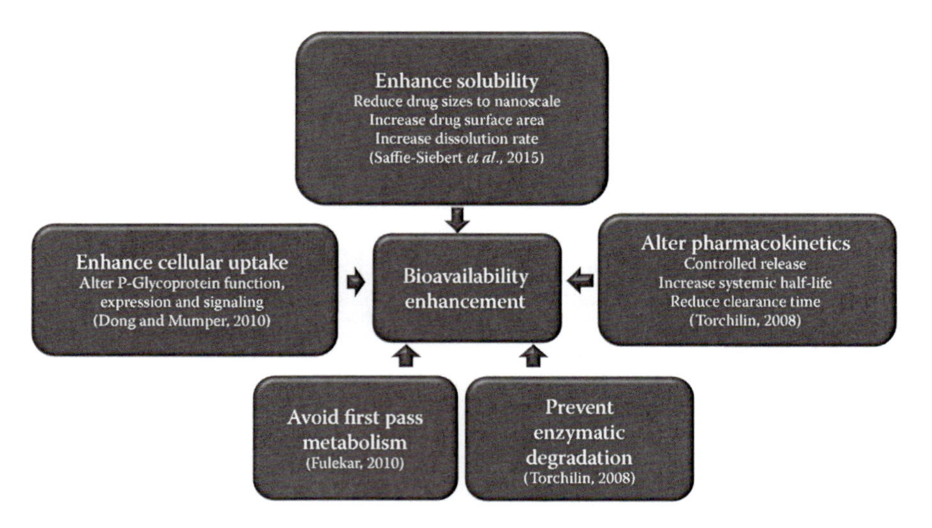

FIGURE 12.3 Key role of nanotechnology in pharmaceutical development.

widespread potential applications in pharmaceutical industry (Figure 12.3), particularly in drug delivery, despite its initial popularity and utility in mechanical engineering, magnetic, or chemical fields (Fulekar, 2010).

Nanotechnology has been widely recognized for its crucial role in pharmaceutical development. The incredibly rising trend in pharmaceutical nanotechnology applications has led to the synthesis of new "nanomaterials" or "nanocarriers," defined as "having one or more dimensions of the order of 100 nm or less structurally" (CIEL, 2014), opening a window to new therapeutic agents with greater efficacy and stability.

In an effort to overcome the obstacles in CUR delivery, numerous strategies have been proposed, including the use of adjuvants, use and synthesis of CUR derivatives, and analogues. However, nanotechnological approaches have emerged as the most prominent answer.

Various studies on nano-vesicles for CUR delivery have been explored by scientists around the world, resulting in different CUR nanostructures with enhanced solubility, stability, and pharmacokinetic profile (Mohanty *et al.*, 2012). All nano-enabled delivery systems devised to carry CUR could be mainly classified as liposomes, niosomes, nanoparticles, microemulsions, nanocrystals, micelles, cyclodextrin (CD) inclusions, and dendrimers. The percentage distribution of studied nanoformulation is pictographically portrayed in Figure 12.4.

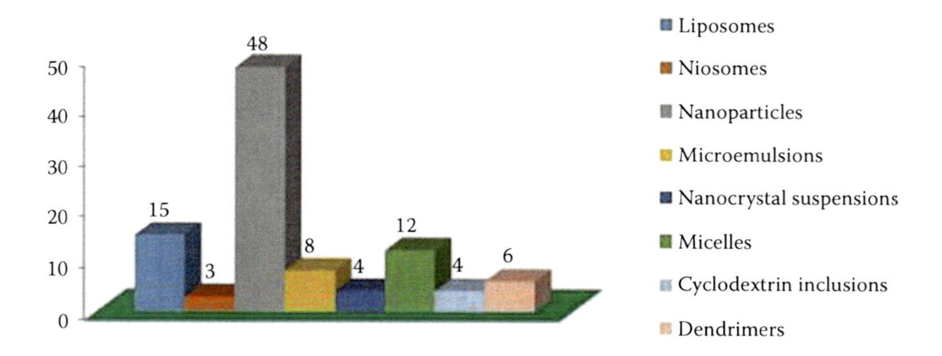

FIGURE 12.4 Percentage distribution among various nanocarriers of curcumin (CUR).

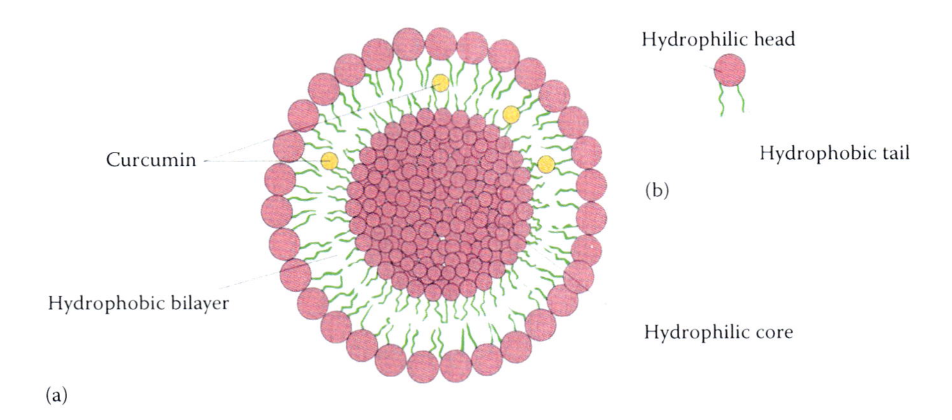

FIGURE 12.5 (a) Curcumin-loaded liposomes, (b) phospholipid.

12.6.1 Liposomes

Liposomes are small spherical self-assembling vesicles, consisting of a closed bilayer phospholipid systems (Figure 12.5). If there is more than one bilayer presented within liposomal structures, they are classified as "multilamellar vesicles" (Emeje *et al.*, 2012). Liposomal carriers were the first nanostructures approved by the U.S. Food and Drug Administration (FDA) in 1995, owing to their ability to encapsulate both hydrophilic drugs (within their aqueous core) and lipophilic drugs (inside their bi-lipid layer) (Torchilin, 2005; Immordino *et al.*, 2006; Barenholz, 2012). Therefore, liposomes are considered effective in protecting therapeutic compounds from degradation. In addition to this, liposomal carriers also have the ability to prolong systemic circulation half-life of a drug following its administration, once they are coated with polyethylene glycol (PEG) (Emeje *et al.*, 2012). CUR can be encapsulated into liposomes to improve its systemic BA and stability. The reported liposomes-based formulations of CUR are compiled as Table 12.1.

12.6.2 Niosomes

Niosomes were first introduced in the 1970s as alternative colloidal vesicular delivery systems to liposomes (Azmin *et al.*, 1985). As these vesicles employ non-ionic surface-active agents, they were given the name as niosomes. Albeit, structurally, niosomes also consist of a bilayer, yet unlike liposomes, niosomal bilayers are made up of non-ionic surfactants rather than phospholipids, giving them micellar structures upon interacting with aqueous solutions (Makeshwar and Wasankar, 2013) (Figure 12.6).

Niosomes are considered to be desirable carriers for anti-infection and anticancer medications, as they enhance the targeted delivery of these therapeutic agents, reducing cytotoxicity and maximising oral BA of insoluble hydrophobic drugs (Mehanny *et al.*, 2016a). Specifically, for poorly absorbed medications, liposomes are also believed to aid their permeability through the skin (Nauman *et al.*, 2015).

Niosomes are a better delivery system than the other surfactants forming micelles, in terms of protecting CUR from degradation. In fact, the rigid and confined niosomal microenvironment, owing to the presence of cholesterol molecules, elevate the steady state fluorescence intensity and fluorescence lifetime of CUR to a greater extent, compared to conventional micelles. The reported niosomes-based formulations are shown in Table 12.2.

TABLE 12.1
Liposomal Nanostructures of Curcumin and Improvements Compared to Naive Curcumin

System/Composition	Techniques/ Characteristics	Improvements	References
pH-dependant liposomal CUR	• pH-driven method • Thin-film hydration • Ethanol injection	• High bioaccessibility • Stability during storage • Easy scalability • No need of organic solvents	(Cheng *et al.*, 2017)
Cholesterol, Phospholipid; CUR and Tween 80	• Thin-film dispersion and dynamic high-pressure microfluidization (DHPM) • EE = 57.1 ± 1.1% • PS: 68.1 ± 1.5 nm • PDI: 0.246 • ZP: −3.16 mV	• Improved CUR stability in basic conditions • Good stability against metal ions such as Fe^{2+}, Fe^{3+}, Al^{3+}, Cu^{2+}, K^{+}, Li^{+} • Very good stability on storage at 4°C & 25°C (90 days)	(Chen *et al.*, 2015)
Multilamellar vesicles of CUR and lecithin (from rapeseed, soya and salmon)	• High-pressure homogenization method • EE: 63.2 ± 0.7% (rapeseed) 65.0 ± 1.1% (soya) 67.3 ± 1.1% (salmon) • Fine droplets, largest droplets diameters > 200 nm	• High EE (%) • Improved cellular effects of CUR	(Hasan *et al.*, 2014)
Conjugation of: 1,2-dipalmitoyl-sn-glycero-3-phosphothioethanol (DPSH), CUR and N,N-diisopropylethylamine (DIPEA)	• Michael addition (of CUR) reaction • PS: 63–200 nm • PDI: 0.26 • ZP: −6.44 to −10.5 mV	• In Alzheimer's disease mouse model, down-regulated amyloid peptide secretion • Prevented Aβ-induced toxicity • Monodisperse and stable • Cellular protective effects against the Aβ cytotoxicity	(Lazar *et al.*, 2013)
CUR; Silica-coated flexible liposomes	• Thin film hydration • EE (%): 93.28 ± 2.51 • PS: 157 nm	• BA of CUR increased 3.3-folds	(Li *et al.*, 2012)
Lipo Polyethylene glycol polyethylenimine (PEG-PEI) and CUR	• Thin film hydration • EE (%): 45 ± 0.2 • PS: 258–269 nm	• Better cytotoxic effects against CUR-sensitive and CUR-resistant cells (20-folds more active) • Led to cell cycle arrests, resulting in cytotoxic activity • Rapid CUR accumulation into cells • Tumor growth in mice models was suppressed (60–90%)	(Lin *et al.*, 2012)
4-[3,5-bis(2-chlorobenzylidene-4-oxo-piperidine-1-yl)-4-oxo-2-butenoic acid (CLEFMA); Hydroxypropyl-β-CD (HPβCD) and CUR	• Solvent evaporation and high-pressure homogenization • PS: 310.3 ± 4.7 nm	• Non-invasive imaging of CUR bio-distribution • More potent antitumor effects • Harmless against normal lung fibroblasts • Reduced mean volume of tumors • Yielded 94% tumor inhibition	(Agashe *et al.*, 2011)

Abbreviations: BA: Bioavailability; CUR: Curcumin; EE: Entrapment efficiency; PDI: Polydispersity index; PS: Particle size; ZP: Zeta potential.

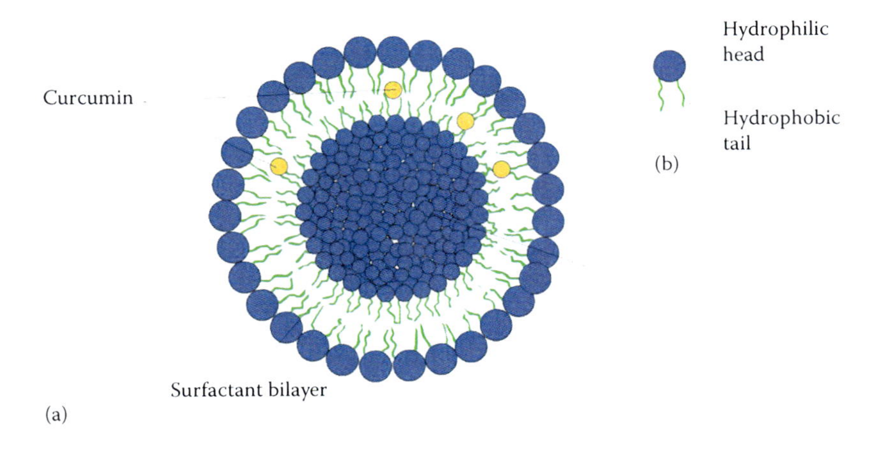

FIGURE 12.6 (a) Curcumin-loaded niosome, (b) surfactant.

TABLE 12.2
Niosomal Nanostructures of Curcumin and the Improvements

System/Composition	Techniques/ Characteristics	Improvements	References
Doxorubicin (DOX) and CUR-loaded polyethylene glycosylated niosomes (PEGNIO)	• Thin-film hydration	• Increased cytotoxicity of against U87 cells	(Ag Seleci *et al.*, 2017)
Tween 20 and CUR	• Hand shaking method	• Decreased the degradation of CUR by half, increased CUR stability • Greater fluorescence intensity and lifetime of CUR	(Sarthak Mandal *et al.*, 2013)
Sorbitan Monooleate, Cholesterol, Solulan C-24 and CUR	• EE: 83% • PS: 12.25 ± 5.00 nm	The number of curcuminoids niosomes traversed the shed snake skin increased from undetectable to 4%	(Rungphanichkul *et al.*, 2011)
CUR, Span 80, and Cholesterol	• Reserve Phase Evaporation (RPE) technique • EE: 82.3–86.8% • PS: 3.84–4.66 μm	• In refrigerated condition: 99.5% stability • At room and elevated temperature: 93% stability • Exhibited prolonged *in vitro* drug release of 61.8% (24 h)	(Kumar et al., 2011)

Abbreviations: CUR: Curcumin; EE: Entrapment efficiency; PS: Particle size.

12.6.3 Nanoparticles

Nanoparticles range in size between 10 to 1000 nm, synthesized from synthetic or natural polymers, and are used as a substitute to liposomes as drug carriers (Mukherjee *et al.*, 2009). Over the years, successful encapsulation of drugs inside nanoparticles has attracted tremendous attention due to resultant controlled release profile of drug contents and their stability brought about by these nanometric-sized carriers (De Jong and Borm, 2008). Moreover, nanoparticles are considered as ideal carriers, suited to optimize drug delivery, and reduce toxicity and unwanted effects (Mukherjee *et al.*, 2009).

Nanoparticles could be further classified into four main subtypes, which are polymeric nanoparticles, solid lipid nanoparticles (SLNs), metallic nanoparticles (MNs), and mesoporous nanoparticles. CUR is usually formulated as nanoparticles in an attempt to overcome its poor aqueous solubility. Moreover, CUR nanoparticles have been shown to have better antimicrobial and wound healing properties.

12.6.3.1 Polymeric Nanoparticles (PNPs)

Polymers are large molecules made up of repeated subunits, playing an important role in our everyday life. Polymers are classified into natural and synthetic subtypes. Water-based natural polymers including silk, wool, cellulose, and proteins are derived and extracted in nature. Synthetic polymers such as polyethylene, polyester, and nylon are mostly derived from petroleum oil and are developed by humans (Dan Mogoşanu *et al.*, 2016).

PNPs (Figure 12.7) are another coherent delivery system for insoluble hydrophobic agents, such as CUR, having the particles size within the range of 1–100 nm (Mehanny *et al.*, 2016). These nanoparticles have the ability to protect therapeutic agents from degradation, and modifying their pharmacokinetic parameters such as absorption, metabolism, clearance, and elimination, to achieve the drug controlled release patterns (Mehanny *et al.*, 2016). Therefore, PNPs are believed to be desirable for different routes of administration, ranging from oral, nasal, to the parenteral pathways (Sun *et al.*, 2013; Kumari *et al.*, 2016; Shakeri and Sahebkar, 2016).

The most common polymers used to encapsulate therapeutic agents are poly D, L-lactic-co-glycolide (PLGA), starch, and chitosan (Emeje *et al.*, 2012; Mehanny *et al.*, 2016). These polymers are highly acknowledged for their safety profile, biocompatibility, biodegradability, and cost-effectiveness (Emeje *et al.*, 2012; Li *et al.*, 2016; Mehanny *et al.*, 2016).

Another interesting type of PNP is adopted from protein structures, known as peptide/protein carriers. Proteins are natural molecules that occur in all living cells, synthesized from a long chain of amino acids. They are also well-known for their unique amphiphilicity and are considered as ideal materials for the preparation of nanoparticles (Lohcharoenkal *et al.*, 2014). The formulations in which CUR encapsulated in PNPs are shown in Table 12.3.

12.6.3.2 Solid Lipid Nanoparticles (SLNs)

Another efficient technique deployed to enhance the efficacy and stability of pharmaceutical agents is encapsulating it within SLNs (Figure 12.8). SLNs usually have mean diameters in the range of 50–1000 nm, consisting of solid lipidic cores, inside which active compounds can be dispersed, and are coated with a phospholipid monolayer on the outside (Emeje *et al.*, 2012).

SLNs offer several advantages over other nanostructures, as these can improve the solubility of active ingredients, provide higher drug loading capacity, and offer protection against light, heat, and lipid content degradation (Gasco, 2007; Souto and Müller, 2010; Carmelo *et al.*, 2015). Owing

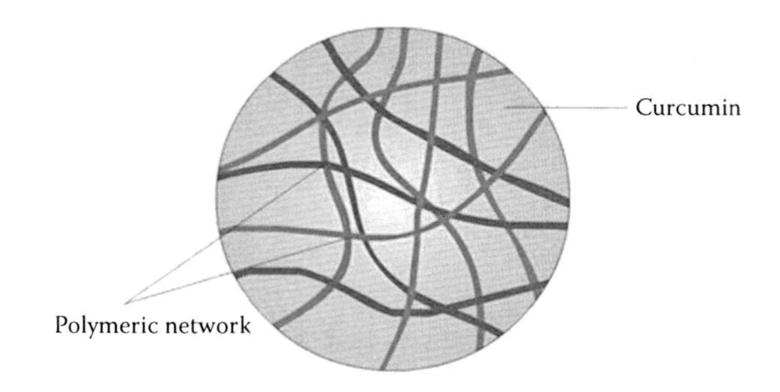

FIGURE 12.7 Curcumin loaded-polymeric nanoparticle.

TABLE 12.3
Curcumin-Loaded Polymeric Nanoparticles and the Improvements Thereof

System/Composition	Techniques/ Characteristics	Improvements	References
Glycyrrhetinic acid (GA)-modified CUR hydrogel	• Surface modification	• Sustainedly released of CUR • Better anticancer efficacy • Higher cellular uptake	(Chen *et al.*, 2017)
Starch (3%) nanoparticles, CUR and Tween 80	• Facile solution mixing • PS: 100 nm • DL: 3%	• At least 715-fold increase in solubility • Increased stability against UV irradiation • Simple procedure • Low price	(Li *et al.*, 2016)
CUR and Eudragit EPO	• Molecular complexation • High drug loading: 280 μg/mL	• Solubility increased to > 2 mg/mL • Improved CUR BA by 20-folds • Increased peak plasma concentration by 6-folds • Did not precipitate at pH 7 (intestine)	(Kumar *et al.*, 2016)
CUR, Poly (D, L-lactic-co-glycolic acid (PLGA), Dicyclohexylcarbodiimide (DCC) and 4-(Dimethylamino) pyridine (DMAP)	• Conjugation CUR–PLGA formed via ester link	• Inhibited the proliferation of cancer cell lines • Improved inhibiting effects of clonogenic and mutagenic ability • Increased cellular uptake of CUR • Achieved a better-sustained release	(Waghela *et al.*, 2015)
CUR, Gum Arabic, Tween 20, CS and Acetic acid (1%)	• Emulsification solvent diffusion technique • High-speed homogenization • EE: 95% • PS: 136 nm	• Higher stability against gastric juice and intestinal enzymes • Enhanced CUR colonic delivery • Greater cellular uptake of CUR • Greater anticancer properties	(Udompornmongkol and Chiang, 2015)
Human serum albumin (HSA), CUR and Mercaptoethanol (β-ME)	• Novel technique, facile route • PS: 130 nm	• Solubility increased more than 500-folds • Higher accumulation in the cytoplasm of tumor cells	(Guangming *et al.*, 2015)
PLGA, Eudragit S 100 and CUR	• Emulsification solvent diffusion method (SESD) • PS: 100 nm • ZP: –40 mV	• Increased CUR permeation • Reduced neutrophil and tumor necrosis factor (TNF)-a secretion • Colonic structure in Dextran sulfate sodium (DSS) induced colitis models were maintained	(Beloqui *et al.*, 2014)
CUR, Phytosome (PS) and CS	• DL: 2.67 ± 0.23% • PS: 68.59 ± 10.16 nm (CUR-PS) and 23.21 ± 6.72 nm (CUR-PS-CS)	• CUR was protected from plasma degradation • Promoted CUR half-life (82.66% longer) • Obtained a sustained release profile up to 12 h • Increased BA indicated via an AUC of 7.49-fold greater than native CUR	(Zhang *et al.*, 2013)

(Continued)

TABLE 12.3 (CONTINUED)
Curcumin-Loaded Polymeric Nanoparticles and the Improvements Thereof

System/Composition	Techniques/ Characteristics	Improvements	References
Tween 80, CS, Acetic acid, and CUR	• Nano-encapsulation • EE > 90% • PS: 20–25 nm	• Greater antioxidant and chelating potentials • Neuroprotective effects improved • Novel therapeutic regime for preventing arsenic toxicity. • Increased stability of CUR	(Yadav *et al.*, 2012)
CUR and PLGA	• Solvent evaporation and Sonication then emulsification • EE: 91.96% • DL: 5.75% • PS: < 200 nm	• 640-folds higher solubility • Absolute BA increased from 4.73 to 26.5% • Relative BA: 563% • Prolonged plasma half-life and lower clearance • Improved release profile in gastrointestinal fluids • Inhibited P-glycoprotein • Increased residence inside the intestine	(Xie *et al.*, 2011)
Human serum albumin (HSA) NPs (CCM-HSA-NPs)	• Albumin-bound technology • PS: 130–150 nm	• Water solubility of CUR increased 300-fold • Amounts of CUR penetrated into tumors were about 14-folds higher • Vascular endothelial cell binding increased 5.5-fold • Transportation across a vascular endothelial cell increased 7.7-folds • Greater therapeutic effect (50% or 66% tumor growth inhibition) compared to CCM (18% inhibition) in tumor xenograft HCT116 models • Nontoxic	(Kim *et al.*, 2011)

Abbreviations: BA: Bioavailability; CUR: Curcumin; DL: Drug Loading; EE: Entrapment efficiency; NPs: Nanoparticles; PDI: Polydispersity index; PS: Particle size; ZP: Zeta potential.

to their unique size-dependent properties, SLNs have great potential for controlled and site-specific drug delivery (Mukherjee *et al.*, 2009). The superior performance of CUR-SLNs was reported compared to naive CUR, as depicted in Table 12.4.

12.6.3.3 Mesoporous Silica Nanoparticles (MSNs)

MSNs have gained significant recognition worldwide, especially since the introduction of the well-ordered mesoporous silica in 1992 (Liu *et al.*, 2014). MSN surfaces contain a high density of negatively charged silanol groups, which could be further modified with various organic molecules, allowing active targeting by specific ligands such as peptides, folic acid, or positively charged PEG;

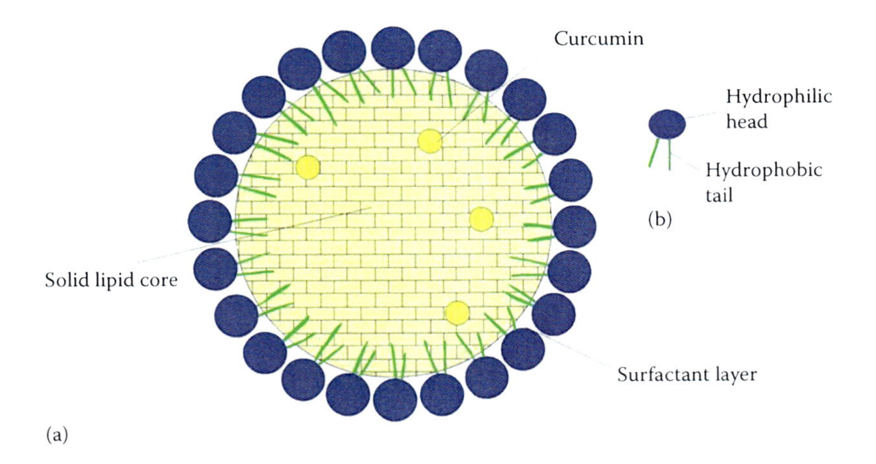

FIGURE 12.8 (a) Curcumin-loaded solid lipid nanoparticle, (b) surfactant.

and avoiding opsonization, which eventually results in brisk clearance of the nanoparticles (Behrens and Grier, 2008). Due to their unique honeycomb-like structures, together with large pore volumes (> 0.9 cm^3/g), miniscule pore diameters (ranging from 2 and 30 nm), and vast surface areas (> 900 m^2/g), MSNs are able to provide relatively large loading capacity of bioactive molecules (Slowing *et al.*, 2008).

TABLE 12.4
Curcumin Solid Lipid Nanoparticles (SLNs) and Their Achievements Thereof

System/Composition	Techniques/ Characteristics	Improvements	Reference(s)
CUR-loaded SLNs	• Emulsification • Low-temperature solidification	• 2.27-folds higher toxicity against non-small cell lung cancer A549 cells • Enhanced phototoxic effects	(Jiang *et al.*, 2017)
Compritol, CUR, Pluronic F68, lecithin and lipase	• High shear homogenization and Ultrasonication • EE = 80% • DL = 1.60% • PS: < 300 nm	• Stable at 4°C • Increased solubility indicated through drug loading, surfactant percentage and sonication time • Increased CUR permeability • A prolonged release profile indicated that CUR is soluble in solid lipid matrix	(Righeschi *et al.*, 2016)
CUR, glyceryl monostearate, soya lecithin, Brij 78 and D-α-tocopheryl polyethylene glycol succinate (TPGS)	• Emulsification and low-temperature solidification • EE: 91.09 ± 1.23% • PS: 135.3 ± 1.5 nm	• Relative BA of CUR reached 942.53% • Increased permeability • AUC increased 12.27-folds • T_{max} and half-life were both delayed, compared to untreated CUR ($p<0.01$), suggesting higher C_{max}	(Ji *et al.*, 2016)
Microcapsules of CUR with SLN core and a mesoporous silica shell	• PS: 2 μm • Mesoporous silica shell thickness: 0.3 μm	• pH-dependent release profile • Good cell tolerance • Useful for gut cells targeting	(Kim *et al.*, 2016)

(*Continued*)

TABLE 12.4 (CONTINUED)
Curcumin Solid Lipid Nanoparticles (SLNs) and Their Achievements Thereof

System/Composition	Techniques/ Characteristics	Improvements	Reference(s)
CUR and lipopolysaccharide stimulated RAW 264.7 cultured murine macrophages	Transient transfection experiments using an NF-κB reporter construct	• Improved solubility compared to raw CUR • Potential effects against infections • Decreased nitric oxide (NO), prostaglandin-E_2 (PGE_2), interleukin-6 (IL-6) • Significantly inhibited the transcriptional activities of NF-κB	(Nahar *et al.*, 2015)
Lyophilized mucoadhesive sponge and CUR	• High shear homogenization, Hot-melt extrusion method • EE: 88.07 ± 0.5% • PS: 289 ± 49.1 nm	• Sustained release patterns of CUR obtained • Increased buccal adhesion and residence time	(Hazzah *et al.*, 2015)
CUR, Tween 80 and soy lethicin	• Solvent evaporation • EE: 81.92 ± 2.91% • PS: 134.6 ± 15.4 nm	• Prolonged plasma residence time • Reduced CUR clearance • Novel approach to deliver CUR into the inflamed joints	(Arora *et al.*, 2015)
N-trimethyl chitosan surface-modified, CUR SLNs, palmitic acid, cholesterol	Homogenization	• Remained stable for days of sunlight and visible light exposure • More stable at room temperature and refrigerator condition • Improved oral BA indicated via: • Increased C_{max} compared to untreated CUR • Longer haft-life (12.26 ± 4.77 h) • Achieved a controlled release in simulated intestinal fluid • Higher penetration and distribution of CUR in brain tissues	(Ramalingam and Ko, 2015)
Dynasan® 114®, Sefsol-218®, CUR	• Hot high-pressure homogenization, High-shear dispersion • EE: 90% • PS: 152.8 ± 4.7 nm	• BA increased by 1.25-folds • Prolonged inhibition of MCF-7 cancer cell lines • Improved pharmacokinetic profile in rats	(Sun *et al.*, 2013)
Stearic acid, lecithin, CUR, Myrj53	• Homogenization • EE: 62–75% • PS: 20–80 nm	• Displaced cytotoxicity effects toward NCL-H460 and A549 cell lines • AUC increased by 26.4-folds • Better distribution of CUR	(Wang *et al.*, 2013)
Cur-loaded SLNS using three different sources of fatty acid: myristic, palmitic and behenic acid; chitosan hydrochloride for bioadhesion, polyvinyl alcohol (PVA) 9000, PVA L508	• Coacervation technique • EE: 28–81% • PS: < 300 nm	• Stable for 12 weeks • Fatty acid CUR-loaded formulations produced bioadhesive, positively changed nanoparticles; indicating water soluble CS-Cur complex	(Chirio *et al.*, 2011)

Abbreviations: BA: Bioavailability; CUR: Curcumin; DL: Drug Loading; EE: Entrapment efficiency; PDI: Polydispersity index; PS: Particle size; ZP: Zeta potential.

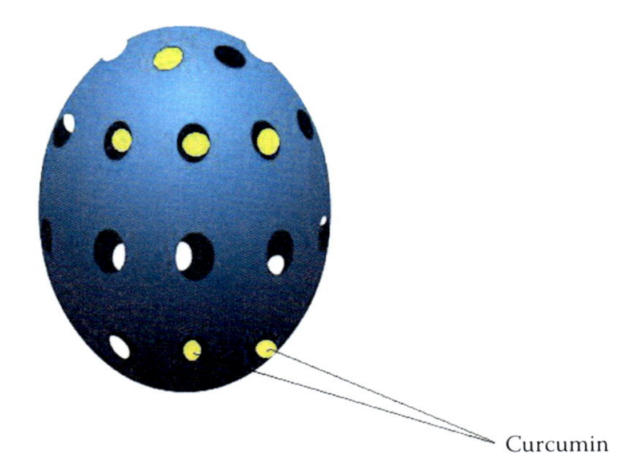

FIGURE 12.9 Curcumin-loaded mesoporous silica nanoparticle.

On the other hand, use of silica as nanobiomaterial is captivating, as it is chemically inert, thermally stable, harmless, and also inexpensive (Iskandar *et al.*, 2001). Because of the strong Si-O bond, silica-based MSNs demonstrate superior BA, as they can be rapidly dissolved in the bloodstream and absorbed at physiological pH, and then excreted, with nil to negligible toxicity (Bharti *et al.*, 2015). The most commonly synthesised types of MSNs are MCM-41 and SBA-15s. The synthesis procedure of these carriers involves the use of soft/hard "templates" or structure directing agents (SDA) (Trinh *et al.*, 2015). The soft templating method involves the use of micelles (i.e., soft materials) while hard-templating methods use silica, polymers, or carbon spheres as SDA (Zhang *et al.*, 2015). The reported CUR-MSNs (Figure 12.9) formulations are enumerated in Table 12.5.

12.6.3.4 Metallic Nanoparticles (MNs)

Since the investigation of Faraday colloidal gold in the 18th century, the optical properties of metallic nanoparticles (MNs) have been attracting interest in physical chemistry (Kelly *et al.*, 2003). Use of MNs has also heavily fascinated scientists because of their enormous potential in nanotechnology, biomedical sciences, and engineering (Mody *et al.*, 2010). Once functionalized with different chemicals, MNs have shown the ability to form conjugates with antibodies, ligands, and especially pharmaceutical agents (Mody *et al.*, 2010). Therefore, they exhibit a wide spectrum of potential applications, in drug targeting, gene delivery, and more importantly, in diagnostic imaging (Ahmad *et al.*, 2010).

CUR is able to form complexes with metal ions owing to its excellent chelating agent, α, β-unsaturated β-diketo moiety (Priyadarsini, 2014). CUR-metal complexes have the ability to alter the physicochemical properties of naive CUR to enhance its biological application. For instance, CUR-metal complexes are observed to possess greater anticancer activity than naive CUR itself (Priyadarsini, 2014). CUR-MNs are usually stabilized with polymers to overcome degradation (Priyadarsini, 2014) (Figure 12.10). The reported MNs-based formulations of CUR are mentioned in Table 12.6.

12.6.4 Microemulsions/Nanoemulsion

Emulsions are systems of two immiscible liquids formed by dispersing one in another (Tadros and Vincent, 1983). When the oil droplets are dispersed in a watery medium, the system is distinguished as oil-in-water (o/w). In contrast, when water droplets are dispersed in an oily medium, water-in-oil (w/o) emulsions are formed (Tadros, 2013).

The main difference between conventional emulsions and micro/nanoemulsions is the size and shape of dispersed particles (Jaiswal *et al.*, 2015). Micro/nanoemulsions are self-emulsifying isotropic nanostructured systems that will produce micro/nano-sized droplets, commonly oil droplets in

TABLE 12.5
Curcumin-Loaded Mesoporous Nanoparticles (MSNs) and the Improvements Thereof

System/Composition	Techniques/ Characteristics	Improvements	References
DOX and CUR loaded (GFLGHHHRRGDS) peptide-conjugated-MSNs	• Soft-templating process • Alkyne modification • Peptide conjugation	• Improve cytotoxicity of DOX + CUR against MCF-7/ADR drug resistant cell lines	(Sun *et al.*, 2017)
CUR-loaded silica materials (KIT-6,MSU-2 and MCM-41)	• Amine functionalization	• Higher cellular uptake of CUR • Exhibited the anticancer potentials of CUR • Inhibition of cancer cell viability • Down regulated of poly ADP ribose polymerase (PARP), leading to the activation of apoptosis (of cancer cells)	(Kotcherlakota *et al.*, 2016)
Bi-functionalized silica nanostars and magnetic nanostars	• Soft-templating process (using CTAT), Rotary evaporation method • EE: 76.51 ± 6.53% • PS: 249.09–299.07 nm	• Increased biocompatibility • Decreased toxic effects in normal cells • Increased cytotoxic effect against cancer cell lines • Magnetic response from magnetic-core nanostars	(Huang *et al.*, 2016)
CUR-MSNs coated with tannic acid-Fe(III) Templates: CTAB, P123	• EE: 100% • PS: 225–234 nm	• pH and glutathione-responsive • Improved cytotoxic effect against MRC5 cancer cell lines	(Kim *et al.*, 2015)
Guanidine functionalized PEGylated I3ad MSNs KIT-6 Templates: P123 tri block copolymers	• Hydrogen transfer nucleophilic addition reaction • PS: 60–70 nm	• High drug loading capacity • Sustained drug release profile • Long-term anticancer effects • pH-responsive controlled characteristics	(Mamani *et al.*, 2014)
CUR mesoporous nanospheres	• Rotavaps technique (to synthesise MCM-41 material) • EE: 85% • DL: 17% • PS: 190 nm • PDI: 0.31	• Sustained release patterns obtained • Exhibited cytotoxicity effects against skin cancer cell lines • Inhibition of PcG protein and caspade-3 • Increased solubility of CUR by 71%	(Jambhrunkar *et al.*, 2014)

Abbreviations: DL-Drug Loading; EE-Entrapment efficiency; PDI- Polydispersity index; PS-Particle size.

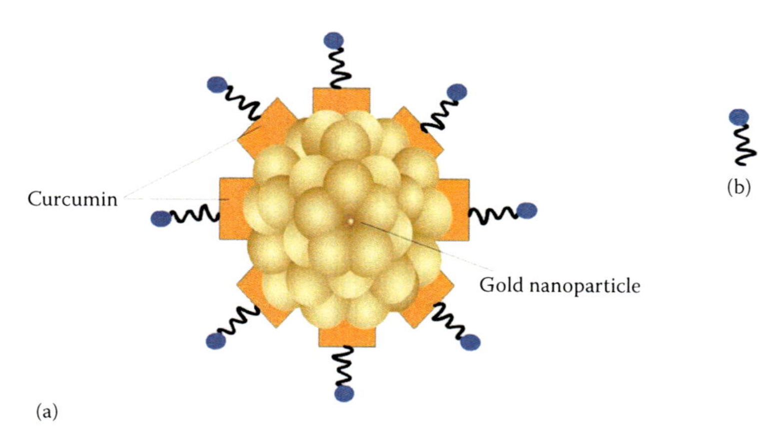

FIGURE 12.10 (a) Polymer-stabilizing curcumin-gold nanoparticle conjugation, (b) polymeric molecule.

TABLE 12.6
Curcumin-Loaded Metallic Nanoparticles (MNs) and the Improvements Thereof

System/ Composition	Techniques/ Characteristics	Improvements	References
CUR coated GNPs	• PS: 21.7–27.8 nm • ZP: > – 25 mV	• High stability (3 months) • Non toxic	(Shaabani *et al.*, 2017)
Alginate stabilised GNPs, CUR and Methotrexate	• Dual drug conjugation • "Green solvent" water simple reaction	Improved cytotoxicity against MCF-7 and C6 cancer cells	(Dey *et al.*, 2016)
Silver ion (Ag^+), CUR, Glycerol and PVP	• Green synthesis route using CUR to prepare mono-disperse CUR conjugated Ag NPs • Spherical shape • PS: 10–50 nm	• Exhibit antibacterial activity • Determination of DNA/RNA with a linear regression coefficient >0.997	(El Khoury *et al.*, 2015)
AgNPs hydrogel, CUR and Sodium Borohydride	• *In situ* method • Sun-shape (~ 5 nm) • Smaller nanoparticles (~ 1 nm)	• Good water absorption • Exhibited antimicrobial properties • Increased BA	(Guzman-Villanueva *et al.*, 2013)
CUR conjugation with PVP functionalized GNPs	• Reduction reaction of Chloroauric acid by tri-sodium citrate (reducing agent) • DL: 26%	• Highly stable in aqueous mediums at room temperature • Showed 100% cell viability	(Gangwar *et al.*, 2012)
CS and CUR encapsulating GNPs	• Solvent evaporation method • PS: 200–250 nm	• Obtained a controlled release profile • Uniformed distribution of CUR	(Satish *et al.*, 2012)

Abbreviations: AgNPs: Silver Nanoparticles; BA: Bioavailability; CS: Chitosan; CUR: Curcumin; DL: Drug Loading; EE: Entrapment efficiency; GNPs: Gold Nanoparticles; NPs: Nanoparticles; PS: Particle size; PVP: Polyvinylpyrolidone; ZP: Zeta potential.

the aqueous medium (Anton and Vandamme, 2011; Yallapu *et al.*, 2012). Incorporation of emulsifying agents into emulsions is essential, as they aid in stabilizing the structures of dispersed nanodroplets (Semiconductor International, 2009).

Microemulsions further improve on the benefits of conventional emulsions, opening the doors for therapeutic agents to be designed with better efficacy, better safety profiles, and dosing regimens. The emulsion droplets are extremely small in size, ranging from 50 to 200 nm, compared to conventional emulsions (1 to 100 µm) (Wang *et al.*, 2008). As a result, micro/nanoemulsions (Figure 12.11) have enormous potential applications toward enhancing the solubility of hydrophobic drugs, such as CUR, and improving the BA and stability of therapeutic agents (Saffie-Siebert *et al.*, 2005; Anton and Vandamme, 2011; Sutradhar and Amin, 2013; Jaiswal *et al.*, 2015) (Table 12.7).

12.6.5 Nanocrystal Suspensions

Nanocrystals or nanocrystal suspensions are nano-sized carrier-free systems, containing pure drug crystals and minimal quantities of stabilizing surfactants and/or polymer (Keck and Müller, 2006). Nanosuspensions are capable of increasing the solubility and dissolution rate of BCS class II and IV drugs. Therefore, they are excellent candidates for delivering such pharmaceutical compounds to

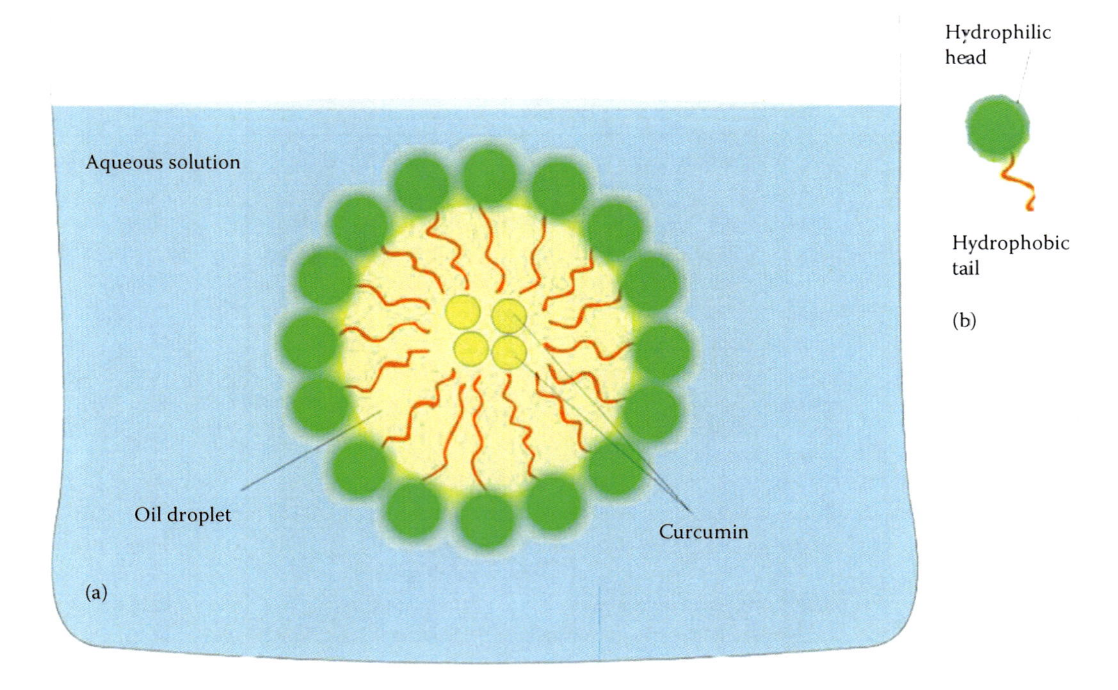

FIGURE 12.11 (a) Curcumin nanoemulsion system, (b) emulsifying agent.

the site of action (Rabinow, 2004; Patel and Agrawal, 2011). Furthermore, nano-crystalized drugs have greater dissolution rate because of their larger specific surface areas (Yallapu and Chauhan, 2012). As nanocrystals exhibit low level of irritability, they could be formulated as better-tolerated injections (Xiong *et al.*, 2008). CUR nanosuspensions (Figure 12.12) are promising drug delivery systems of CUR, as they offer the solution to its poor solubility and BA, thus improving its efficacy (Table 12.8). In addition, the preparation process of nanosuspensions is relatively simple and applicable to most drugs with poor aqueous solubility.

12.6.6 Micellar Systems

Micelles are formed through self-assembly amphiphilic substances containing hydrophobic and hydrophilic parts (Lindman and Wennerström, 1980). In an aqueous environment, when the concentration of the amphiphilic substances, such as polymers and surfactants, exceeds the "critical micelle concentration" (CMC) or the concentration above which micellar formation occurs, they arrange themselves in core–shell nanostructures consisting of a hydrophobic core and hydrophilic shell (Ana Domínguez *et al.*, 1997).

In pharmaceutical development, micelles have been regarded as useful vesicles for the delivery of lipophilic drugs (Xu *et al.*, 2013). In particular, polymeric micelles have been greatly attracting attention in diagnosis and imaging, due to their high stability, good biocompatibility profile, and drug targeting, owing to high abundance of such targeted moieties onto their surfaces (Xing *et al.*, 2014). Numerous biocompatible and biodegradable block copolymers such as polyethylene glycol-phosphatidyl ethanolamine (PEG-PE) have been intensively deployed for the synthesis of polymer micellar drug carriers (Xing *et al.*, 2014). Micellar systems are well-known as powerful tools to enhance the anticancer properties of CUR (Figure 12.13), owing to their ability of passive targeting solid tumors (Mehanny *et al.*, 2016b) (Table 12.9).

TABLE 12.7
Curcumin Microemulsions and Their Achievements Thereof

System/Composition	Techniques/ Characteristics	Improvements	References
CUR, Tween 80 microemulsion	Nanometric size	• Stable microemulsion • CUR reactive oxygen nature is maintained • Efficient activation of Keap1-Nrf2-EpRE pathway	(Greenwald *et al.*, 2017)
Magnetic PLGA micro-sphere, CUR and Glucose	• Used T-shaped microchannel reactor, based on composite emulsion with glucose • EE: 77.9% • DL: 3.2% • PS: 16–207 nm	• Fast magnetic response and mobility • Maximum magnetic entropy • Obtained slow-release patterns	(Hu *et al.*, 2016)
CUR, α-glucosyl hesperidin (hesperidin-G), and PVP K-30	Solvent evaporation method	• Solubility of CUR increased 2600-folds • Solution remained stable even after 24 h • Formation of an amorphous solid dispersion; protecting CUR from degradation and increasing its solubility	(Kadota *et al.*, 2015)
Tocopherol acetate and CUR	• Sonication and homogenization • PS: 246 ± 8 nm • Drug content: 1.25 mg/ml	• 24 h sustained release pattern • Eligible to be used for IV • Plasma AUC increased 8.8-folds • Better tissue distribution • Suppression of LPS-induced injuries	(Shukla *et al.*, 2014)
CUR dispersed in cellulose matrices: Hydroxypropylmethylcellulose acetate succinate (HPMCAS) or Carboxymethylcellulose acetate butyrate or Cellulose acetate adipate propionate (CAAdP)	Spay-dried solid dispersion	• Existed at amorphous state; indicating increased solubility • Higher stability against crystallization and degradation	(Li *et al.*, 2013)
CUR and Soy protein isolate (SPI)	Complexation	• Solubility increased 812-folds • >80% stability (12 h) when dissolved in water, stimulated gastric and intestinal fluid • Enhanced antioxidant activities	(Tapal and Tiku, 2012)
Dibenzoylmethane (DBM), β-diketone, CUR analog, Tween 20, Glycerol, and Monooleate	Oil-in-water nanoemulsion	• Increased oral BA of CUR, indicated by the increase in plasma exposure • Longer half-life and higher AUC • Down regulation of Sulfo transferase Family 1B Member 1 (SULT1B1) by 2.5 to 2.3-folds	(Lin *et al.*, 2011)

Abbreviations: CUR: Curcumin; DL: Drug Loading; EE: Entrapment efficiency; PS: Particle size.

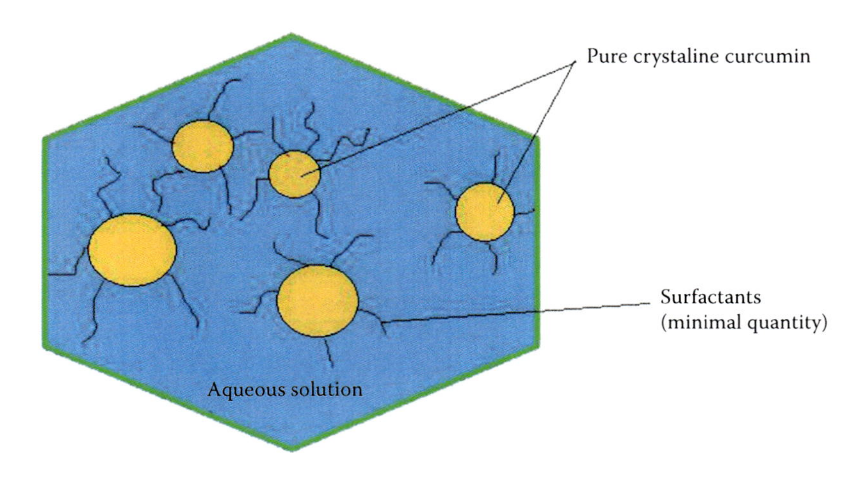

FIGURE 12.12 Curcumin nanosuspension system.

TABLE 12.8
Curcumin Nanosuspensions and their Achievements Thereof

System/Composition	Techniques/ Characteristics	Improvements	References
mPEG-DSPE stabilized CUR loaded nanosuspension, soybean lecithin	• Precipitation-ultrasonication • PS: 186.33 nm • ZP: −19.00 mV	• Enhanced cytotoxicity • Enhanced antitumor efficacy (from 40.03 to 70.34%) • Enhanced biodistribution	(Hong *et al.*, 2017)
CUR, and β-lactoglobulin (β-lg)	• Anti-solvent precipitation • PS: 150–175 nm	• Solubility increased 35-folds • Nanosuspensions prepared at pH 3.4 showed stability >90% after 30 day	(Aditya *et al.*, 2015)
CUR and Didecanoate	• Wet ball milling • PS: 500 nm	• Higher plasma concentration and brain distribution for up to 15 days	(Wei *et al.*, 2013)
CUR and TPGS	• High-pressure homogenization • PS: 210.2 nm	• AUC in plasma was 3.8-folds greater • Mean residence time was 11.2-folds longer • Increased solubility to 9.45 ± 0.33 mg/l from 11 ng/ml (35-folds)	(Gao *et al.*, 2010)
CUR-loaded nanocrystals	• Wet-milling, Lyophilization • PS: 196 nm	• BA increased 16-folds	(Onoue *et al.*, 2010)

Abbreviations: BA: Bioavailability; CUR: Curcumin; PS: Particle size; ZP: Zeta potential.

12.6.7 Cyclodextrin (CD) Inclusion Complexes

CD belongs to acyclic oligosaccharide family, consisting of a hydrophilic outer part and an inner hydrophobic matrix. Therefore, CD can form inclusion complexes and/or host numerous hydrophobic compounds (Mangolim *et al.*, 2014). Among all the CDs types; β-CD, which has seven 7-membered sugar units linked by α-(1,4)-glycosidic bonds, is the most often used, as it is able to host guest molecules with molecular weights up to 800 g/mol (Szente and Szejtli, 2004; Gomes *et al.*, 2014).

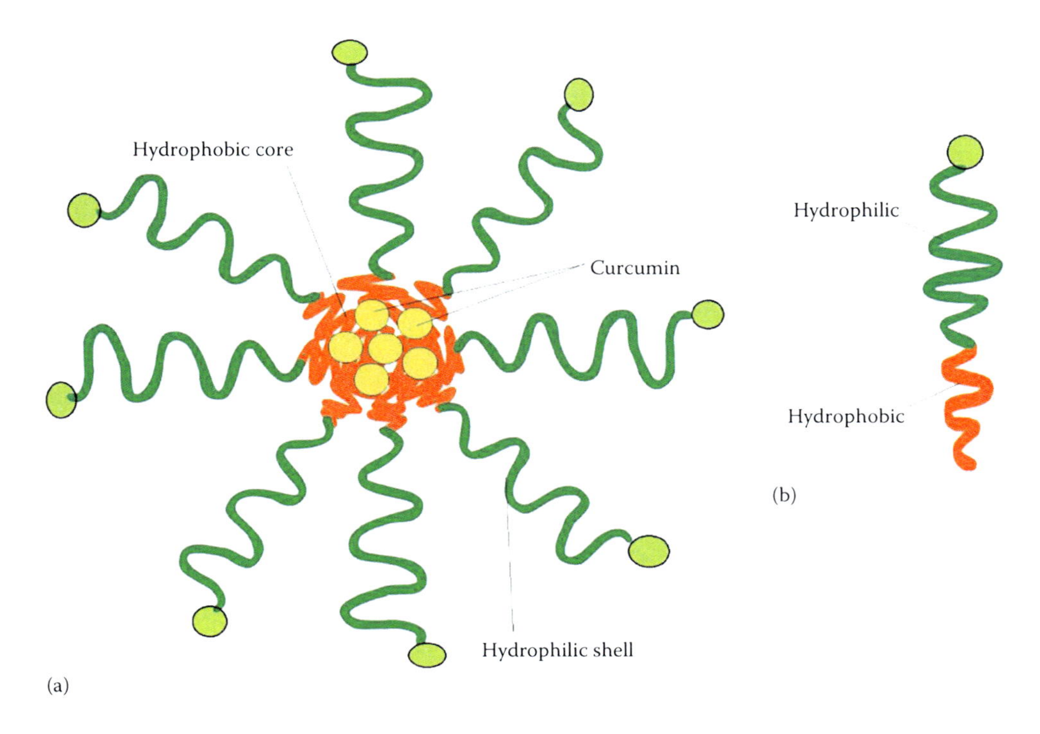

FIGURE 12.13 (a) Curcumin-loaded polymeric micelle, (b) amphiphilic copolymers.

TABLE 12.9
Micellar CUR Systems and the Improvements Thereof

System/Composition	Techniques/ Characteristics	Improvements	References
CUR loaded Polymeric Micelles	• Solvent evaporation	• Blood stable micellar system • Higher cellular uptake • Slower elimination • Delayed the growth of multi drug resistant erythroleukemia K562/ADR cells	(Gong *et al.*, 2017)
Bisdemethoxy CUR, Pluronic F68 and Pluronic F127	• Thin-film hydration method • PS: 25.27 ± 2.28 nm • EE: 89.1 ± 3.5%	• Biphasic release patterns with an initial burst release followed by a sustained effect • Enhanced solubility of CUR • Low CMC value • Displaced *in-vivo* longevity • Exhibited cytotoxicity in HepG-2 model	(Mehanny *et al.*, 2016b)
CUR and PEG-PLA	• Diameter: 1.1 ± 2.9 nm • ZP: -10.9 ± 0.7 mV • PS: 63.5 ± 7.1 nm	• Lower CMC indicated increased stability of CUR • pH-dependent release behavior • No cytotoxicity upon drug release	(Li *et al.*, 2016)

(Continued)

TABLE 12.9 (CONTINUED)
Micellar CUR Systems and the Improvements Thereof

System/Composition	Techniques/ Characteristics	Improvements	References
Micellar systems of CUR with Cationic and nonionic surfactants: Cationic: Tri (dodecyldimethylammonioacetoxy) tris (2-aminoethyl) amine trichloride (DDAD) Nonionic: Brik 96, Tyloxapol, and Tween 80	Sonication	High stability: only 20–25% of micellar CUR degraded after 10 h at pH 13	(Kumar *et al.*, 2016a)
Mixed surfactants micelles, CUR Double chained: Didodecyldimethyl-ammonium bromide (DiDDAB), Dimethylditetradecyl-ammonium bromide (DMDTAB), Dihexadecyldimethyl-ammonium bromide (DiCTAB) Dioctadecyldimethy-ammonium bromide (DODAB), Single chained Dodecylethyldimethyl-ammonium bromide (DDAB) 1:1 Molar ratio	• Self-assembling equimolar of double and single chained ionic surfactants	• Only 25 to 40% CUR degraded (after 10 h) at pH 13 • Increased stability • Interacted with DNA complex without adverse effects	(Kumar *et al.*, 2016b)
PCL-PEG-PCL and CUR	• Copolymers self-assembled into micelles • EE: 83 ± 1.29% • Spherical shapes • PS: 70.34 nm	• Proliferation inhibition • Apoptosis induction • Antitumor stimulation	(Manjili *et al.*, 2016)
CUR and Sophorolipid complex	• Co-processing CUR with biosurfactants to form complexes • PS: 6–7 nm	• Increased stability of CUR • Increase cellular uptake of CUR • The complex appeared transparent yellow while raw CUR looked turbid, indicating improved solubility	(Singh *et al.*, 2014)

(Continued)

TABLE 12.9 (CONTINUED)
Micellar CUR Systems and the Improvements Thereof

System/Composition	Techniques/ Characteristics	Improvements	References
Alginate-CUR micelles	• Diameter: 459 ± 0.32 nm • Negative zeta potential	• Increased the aqueous solubility of CUR • Increased stability of CUR in water at pH 7.4 • Excellent cytotoxic activity	(Dey and Sreenivasan, 2014)
Biodegradable CUR polymeric micelles	• One-step solid dispersion method • PS: 28.2 ± 1.8 nm • PDI: 0.136 ± 0.05 • EE: 98.9 ± 0.7% • DL: 14.8 ± 0.1%	• Potent anticancer effects against breast tumor animal models • Longer survival time in animal subjects bearing tumors • Sustained release patterns	(Liu *et al.*, 2013)
CUR encapsulated monomethyl poly (ε – caprolactone) MPEG-PCL	• One-step solid dispersion method • PS: 21.7–36.2 nm • EE: 97.7–99.1%	• Sustained release patterns (58% in 14 days) • Improved anti-angiogenic activity • More effective in tumor growth inhibition • Prolonged survival in IL/2 tumor models • Higher plasma concentration • Longer retention time	(Gong *et al.*, 2013)
CUR, Methoxy poly(ethylene glycol) and Zein	• PEGlyation • EE: 95 ± 4% • PS: 95–125 nm	• Aqueous solubility increased 1000 to 2000-folds • Stability increased 6 folds • More potent than free CUR (3-folds reduction in IC50 value) • Sustained release profile in 24 h • Higher cellular uptake and cytotoxic effects against drug-resistant NCI/ ADR-RES cancer cell lines	(Podaralla *et al.*, 2012)
Octenyl Succinic Anhydride modified ε-polylysine (M-EPL) and CUR	• Solvent evaporation, dialysis and high-speed homogenization • DL: 5.3 ± 1.9%	• Increased water solubility • Higher CUR encapsulation • Better stability at pH 7.4 • Elevated cellular antioxidant activity • Enhanced delivery of poorly water-soluble drug or phytochemicals and improved their bioactivities	(Yu *et al.*, 2011)

Abbreviations: CUR: Curcumin; DL: Drug Loading; EE: Entrapment efficiency; PS: Particle size.

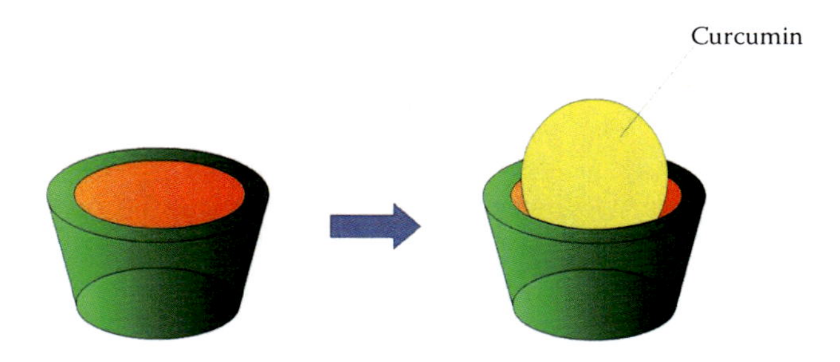

FIGURE 12.14 (a) β-cyclodextrin, (b) curcumin-β-cyclodextrin complex.

TABLE 12.10
Curcumin–CD Inclusion Systems and the Improvements Thereof

System/Composition	Techniques/ Characteristics	Improvements	References
Core shell nanofibers: CUR-CD (core), PLA (shell)	• Electrospinning • Yellow, different shapes	• Enhanced aqueous solubility	(Aytac and Uyar, 2017)
CUR complexation with β-CD	• Co-precipitation, freeze-drying, solvent evaporation and complexation • EE: 74%	• 18% higher rate of sunlight stability • Enhanced solubility by 31-folds • Stable to pH variations • Color retention of 99% • Intensified the color of ice-creams • Very good sensorial acceptance	(Mangolim *et al.*, 2014)
CUR and Sulfobutylether-β-CD (SBE–CD)	• Freeze-drying, kneading, co-evaporation and lyophilization	• Increases water solubility of CUR by 183-folds (from 0.56 to 102.78 μg/mL) • Positive impacts on CUR anticancer and antioxidant properties	(Cutrignelli *et al.*, 2014)
β-CD and CUR nanoparticle	• Solvent evaporation and homogenization • EE increased as the percentage of CUR in the complex nanoparticles raised up to 20%. • PS: 156 nm	• Increased CUR dissolution rate 10-folds • Enhance CUR permeability across skin model tissues up to 1.8 folds • Enhanced the solubility and penetration of CUR	(Rachmawati *et al.*, 2013)
B-CD and CUR	• Lyophilization • DL: 5–10% • PS: 250 nm	• Sustained release in PBS (pH = 7.4) • 70% drug retained • Possibly useful tools in CUR delivery and anti-prostate-cancer properties	(Yallapu *et al.*, 2010)

Abbreviations: CUR: Curcumin; DL: Drug Loading; EE: Entrapment efficiency; PS: Particle size.

CDs are nontoxic; therefore, they have been exploited for various pharmaceutical applications, including being used as complexing agents to increase the aqueous solubility of hydrophobic drugs; reducing their bitterness; and enhancing their BA and stability (Szente and Szejtli, 2004; Tiwari *et al.*, 2010). More than 30 marketed pharmaceutical products containing CD inclusion complexes are available up to now. The reported CUR-CD (Figure 12.14) based formulations are shown in Table 12.10.

12.6.8 Dendrimers

Dendrimers are macromolecular compounds with highly branched globular nanopolymeric 3D structures (Kesharwani *et al.*, 2014). These structures, usually consisting of symmetric cores, inner shells, and outer shells, were first discovered in 1978 by Fritz Vögtle (Vögtle *et al.*, 1978). Dendrimers have gathered significant recognition for their wide spectrum of potential applications in photodynamic therapy, diagnosis, and especially in drug delivery. Indeed, dendrimers have been intensively exploited as nanocarriers for numerous antimicrobial, anticancer, and anti-inflammatory drugs (Abbasi *et al.*, 2014; Noriega-Luna *et al.*, 2014).

Structurally, dendrimers have distinct and diverse branching patterns; their interior hydrophobic and hydrophilic cavities allow them to encapsulate various drug molecules (Noriega-Luna *et al.*, 2014). Poorly soluble drugs can be hosted inside the interior hydrophobic cavities of dendrimers, while their hydrophilic shells can interact with surrounding water (Tripathy and Das, 2013). Moreover, dendrimer-based systems offer protection of drugs against enzymatic degradation, generating favorable half-lives, drug distribution, and elimination patterns, improving drug BA (Fanun, 2016). Dendrimers are also considered as promising vectors for gene delivery, owing to their high ratio of surface groups to molecular volume (Madaan *et al.*, 2014). CUR-dendrimer complexes (Figure 12.15) have also been heavily studied and formulated by scientists around the world, in an attempt to overcome CUR drawback and enhance its potential biological applications (Table 12.11).

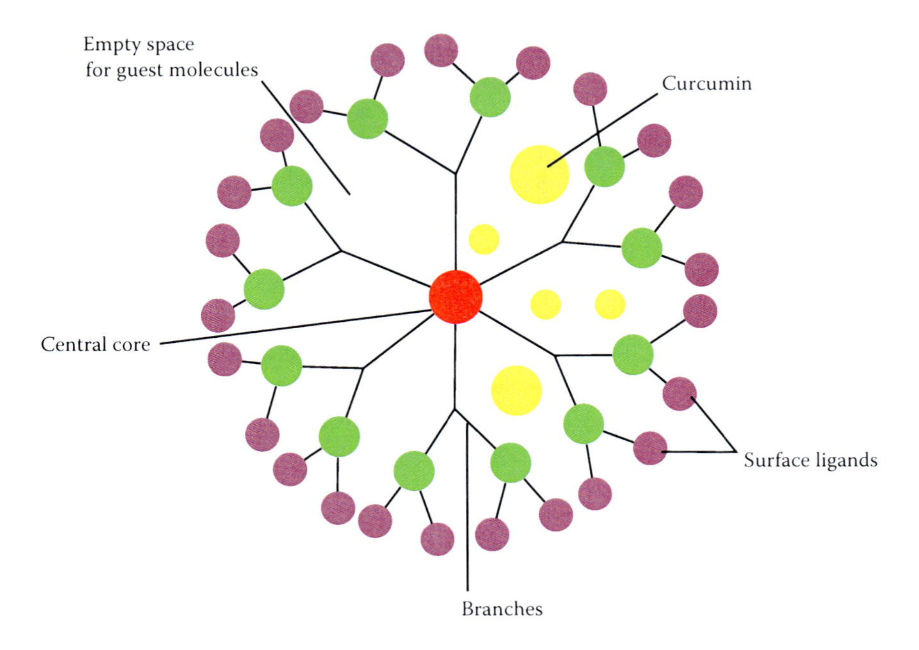

FIGURE 12.15 Dendrimer-curcumin conjugate.

TABLE 12.11
Dendrimer-Curcumin Conjugates and the Improvements Thereof

System/Composition	Techniques/ Characteristics	Improvements	References
CUR loaded polyamidoamine dendrimer (G0.5)	• Spherical nanoparticles • PS: 150 nm • EE: 62% • DL: 32%	• Solubility increased 150–450 folds • Prolonged and sustained CUR release patterns	(Falconieri *et al.*, 2017)
Non-toxic dendrosomal CUR	• Spherical in shape • PS: 142 nm	• Chemically and physically stable • Non-toxic, harmless to non-neoplastic cells • Potent anticancer properties	(Tahmasebi *et al.*, 2014)
Dendrimer-CUR conjugate	–	• More effective in cytotoxicity induction against Caspase-3 activation • Anticancer effects on breast cancer bearing animal subjects • Highly water soluble • Could be an effective anticancer drug for human	(Shawon *et al.*, 2013)
Acetyl-modified generation 5(G5) poly (amidoamine) (PAMAM) and CUR	Solvent evaporation and dialysis	• Solubility increased about 200-folds, increased BA significantly • Sustained release profile • Greater anticancer effects • Simpler procedure	(Wang *et al.*, 2013)
CUR-loaded dendrosomal NPs	NPs prepared by dissolving CUR in different dendrosomal liquids	• Processed a much higher BA than untreated CUR • Selective anticancer effects • Might have improved solubility • Prevented A431 and WEHI-164 cancer cells proliferation in a time- and dose-dependent manner	(Babaei *et al.*, 2012)

Abbreviations: CUR: Curcumin; DL: Drug Loading; EE: Entrapment efficiency; NPs: Nanoparticles; PS: Particle size.

12.7 CURCUMIN NANOFORMULATION PATENTS

There are more than 1,700 patents about CUR in the worldwide patent database (Espacenet, 2016), but nearly 50 patents on nanotechnology-applied CUR are found and nanotechnology-based delivery systems for CUR are shown in Table 12.12 (Office, 2016).

12.8 MARKETED NANO-ENABLED CURCUMIN PRODUCTS

There are more than a dozen nanoformulations of CUR available in the global market, as enumerated in Table 12.13.

TABLE 12.12
Patents in Nanotechnology-Based Delivery Systems for Curcumin

Delivery System(s)	Formulation and Characteristics	Improvement	Application	References
Liposomes	A preparation with the combination of tetrahydrocurcumin, ethoxydiglycol, turmerone, lecithin, EO-PO block-copolymer, preservative and water using thin-film, ultrasonic and/or high-pressure method.	2-folds increase in tetrahydrocurcumin concentration in the deep layers of the skin	Cosmetic transdermal formulation	(He *et al.*, 2016)
Amphiphilic peptide nanoparticles	Amphiphilic peptides that can self-associate at nonacidic pH to form positively charged, spherically micellar nanoparticles. CUR was incorporated in the core of the nanoparticles, while RGD peptides are conjugated on the surface.	Selectively targeted mammalian cells, bacterial cells and cancer cells	Drug carriers and antibacterial agents	(Chang *et al.*, 2017)
Nanoparticles	The matrix of the wound bed with a hydrophilic cotton fabric, zwitterionic low molecular weight chitosan, biosynthesized silver nanoparticles, CUR nanoparticles and other optional materials.	Enhancement of wound healing properties with synergistic action of all the components in the matrix	Wound-healing matrix	(Korneeva *et al.*, 2015)
Sophorolipid complex	• CUR's nano-encapsulation in acidic sophorolipid using sonication. • PS: 30 nm.	Improvement of solubility, stability and BA of CUR to enhance its therapeutic activity	Treatment of breast cancer	(Balaji *et al.*, 2015)
Nanomicelles	• An eye drops preparation with curcuminoid, chitosan, a surfactant, a co-surfactant and water using grind. • PS: 10–50 nm.	Improvement of stability and BA of CUR	A formulation for eye drops	(Singh *et al.*, 2016)
A dry liposome, and/or a dry nano-/ micro-emulsion	A preparation of dried CUR liposomes or nano/micro emulsion with a carrier.	• Improvement of stability of CUR, and facilitates and targets the entry into deep portions of lung tissues	A nanopowder inhalant for the treatment of acute lung injury	(Chen *et al.*, 2015)
Polymeric nanoparticles	• CUR is loaded on the matrix of nano compound that is formed by polyelectrolyte complexation reaction with the biodegradable chitosan and the Arabia adhesive. • PS: 250–360 nm.	Improvement of solubility, stability and BA of CUR	A functional health food	(Jin *et al.*, 2015)
Folate receptor-mediated CUR self-assembled polymer nanomicelles	CUR, polylactic acid-polyethylene glycol (mPEG-PLA) and the biodegradable Folate-PEG-PLA film are generated by a solvent evaporation technique before mixing with an aqueous phase.	• Improvement of water solubility and BA of CUR • Targeted delivery to tumor cells • Effectiveness enhancement and reduction of side effects	Treatment of cancer	(Xia *et al.*, 2015)

(*Continued*)

TABLE 12.12 (CONTINUED)
Patents in Nanotechnology-Based Delivery Systems for Curcumin

Delivery System(s)	Formulation and Characteristics	Improvement	Application	References
A high-load CUR soybean protein nanoparticles	Solvent Displacement/Precipitation method with soybean protein dispersed aqueous solution and a CUR ethanol solution.	Improvement of solubility, stability and BA of CUR	A functional food	(Zhai *et al.*, 2015)
Nanoemulsion	The nano-emulsion is prepared with the aqueous phase mixture and the oil phase mixture containing CUR using High-pressure homogenization (HPH) technique.	Improvement of solubility and BA of CUR	A functional food	(Tang and Chen, 2015)
Nano-CUR	• Combination of nano CUR and isoniazid. • The pure CUR is converted into nano CUR by a proprietary procedure. • PS: 58–197 nm.	More than 2-fold improvement in BA of CUR	Treatment of tuberculosis	(Jeong and Park, 2014)
Lipid nanoparticles	• A formulation with lipid and CUR using melt emulsification, high-pressure homogenization, solvent diffusion and/or a film dispersion and ultrasonic method. • PS: 232.7 nm.	Improvement of solubility, BA and metabolism of CUR	CUR oral formulation	(Kar *et al.*, 2014)
A gene transfer material of functionalized nano-CUR	• A formulation with zinc chloride and CUR by an approach of ultrasound in ethanol, washing with different solvents and drying at room temperature. • PS: 100 nm.	• A targeted delivery to human breast cancer cells (40%) • Effectiveness enhancement and reduction of side effects • Low cost and good biocompatibility	Treatment of breast cancer	(Zhang *et al.*, 2014)
A CUR lipid nanoparticle suspension	• The preparation is obtained by a reaction in a passage reactor with CUR and amphoteric degradable high-molecular polymer as the lipid phase solution and the aqueous phase solution containing surfactant before removing the organic solvent. • PS: 100 nm.	Improvement of solubility, stability and BA of CUR	Treatment of infection and cancer	(Wang *et al.*, 2014)
Nano CUR	CO_2 is pumped into a CUR organic solvent in a high-pressure reaction kettle with ultrasonic-assisted supercritical anti-solvent technique.	• Improvement of Solubility, stability and BA of CUR • Short treatment time of preparation with high purity and low residues of organic solvent	Treatment of inflammation, infection and cancer	(Guo *et al.*, 2014)

(*Continued*)

TABLE 12.12 (CONTINUED)
Patents in Nanotechnology-Based Delivery Systems for Curcumin

Delivery System(s)	Formulation and Characteristics	Improvement	Application	References
Lipid nanoparticles of tetrahydrocurcumin	A formulation produced by solidifying a water-in-oil emulsion containing tetrahydrocurcumin and surfactants with melt emulsification technique.	Improvement of stability and BA of tetrahydrocurcumin	Cosmetic formulation	(Xin *et al.*, 2014)
A multi-layer drug sustained-release nano fiber membrane	The inner layer containing Polylactic acid (PLA), CUR and cellulose acetate and the middle layer of preparation are formed via an electrostatic spinning process. The outer layer is generated by depositing PLA on the top of the middle layer.	• A film with a good biological compatibility, uniform thickness and smooth surface • sustained release and extend residence time of CUR • BA of CUR improvement	A film for anti-bacterial and anticoagulants	(Xia and Xia, 2014)
Nanoparticles	Taurocholic acid, heparin and CUR are conjugated through amido bond to form nano-granules. CUR is coated with taurocholic acid in the nano-granule.	• Improvement of solubility, absorption and BA of CUR • Enhancement of anti-thrombosis and anti-angiogenesis effect with Heparin in the formulation	An oral CUR formulation for inflammation and cancer treatment	(Wang *et al.*, 2014)
Nanoparticles with upconversion rare earth material	A preparation with CUR deposition on the nano-particles formed by an aqueous rare earth salt solution in a high-pressure vessel with solvent displacement/precipitation method.	Targeted drug delivery, higher tissue penetration, and high anti-tumor activity	A novel photosensitiser for photodynamic therapy	(Yang *et al.*, 2013)
Nanosuspension	• An injectable nano-suspension is prepared by suspending CUR dicaprate nano granules in the oil after wet milling followed by freeze drying technique. • PS: 500 nm.	• Stability and BA of CUR improvement • A good injectability and low cost	Treatment of depression and tumor	(Ye and Chen, 2013)
Nanocapsule	• Nano-capsules are obtained by dispersing water, biodegradable polymer and CUR organic solution and removing the solvent with solvent displacement/precipitation method. • PS: 50–500 nm	Increase the penetration of the CUR through the skin	Cosmetic composition	(Liao *et al.*, 2013)
CUR and albumin nano particles	The nano solid powder is obtained by mixing an aqueous phase containing human serum albumin and CUR ethanol solution and removing solvent with solvent displacement/Precipitation followed by Lyophilization method.	• Improvement of solubility, stability, dissolution and BA of CUR • Provide sustained release and targeted delivery	Treatment of cancer	(Yi *et al.*, 2013)

(*Continued*)

TABLE 12.12 (CONTINUED)
Patents in Nanotechnology-Based Delivery Systems for Curcumin

Delivery System(s)	Formulation and Characteristics	Improvement	Application	References
A CUR-polysaccharide conjugate	• A polysaccharide macromolecule is formed by grafting amino acid at one terminal of a diamine compound to polysaccharide through amidation before CUR is introduced into the frame of polysaccharide via a Schiff base reaction. • PS: 10–1000 nm.	• Solubility, stability and BA of CUR improvements • Efficacy enhancement and side effect reduction • A simple process and a low cost	Treatment of cancer	(Liu *et al.*, 2013)
SLN	SLN is prepared by the organic phase containing CUR and lipid and the surfactant water solution with melt emulsification followed by lyophilization method.	• 30-fold increase in CUR's BA • Targeted drug delivery • Efficacy enhancement and side effect reduction	Treatment of asthma	(Zhou *et al.*, 2013)
Nanospheres	A combination of paclitaxel, concordant remedies (e.g., CUR) and drug carriers (e.g., polycaprolactone-polyethylene glycol (mPEG-PCL)) produces nano-microspheres via an approach of solvent displacement/precipitation method.	Efficacy enhancement by the synergistic effect of paclitaxel and the concordant remedies in the formulation	Treatment of cancer	(Wang *et al.*, 2013)
Nanoparticle or nanobiopolymer	A biodegradable nontoxic nano-particles delivery system with various carriers, e.g., Polycefin, Poly (β-L-malic acid), mesoporous silica Nano rods or carbon nanotubes with iron oxide nanoparticle.	Targeted delivery to a brain or a retina	Imaging techniques to image amyloid beta plaques in the brain and or retina and treatment of vision and neurological disorders	(Li *et al.*, 2012)
Nanoemulsion	• A nanoemulsion formulation with a combination of CUR, co-solvent, stabilising agent, isotonicity agent, thickening agent, a buffering agent, electrolyte and surfactants. • PS: 5–500 nm.	Improvement in BA of CUR	Ophthalmic/nasal administration for the treatment of allergic conditions	(Verdooner, 2012)
Nano micelle preparation	• A film is generated by removing the organic solvent in CUR and the amphiphilic block copolymer solution with solvent evaporation and then dispersing the film with an aqueous phase and ultra-sonicating for a CUR micelle solution. • PS: <100 nm.	• Solubility, stability and BA of CUR improvements • Formulation with low toxicity, high safety, long *in vivo* circulation time and simple storage condition	Treatment of cancer	(Ramchand *et al.*, 2011)

(*Continued*)

TABLE 12.12 (CONTINUED)
Patents in Nanotechnology-Based Delivery Systems for Curcumin

Delivery System(s)	Formulation and Characteristics	Improvement	Application	References
Gold-magnetic nanoparticles	Multifunctional gold-magnetic nanoparticles are produced by coupling CUR and mercapto polyethylene glycol PEG-SH on magnetic nanoparticles, obtained by synthesis of a Fe_3O_4-SiO_2 nuclear shell structure, amine surface modification, protonation, and electrostatic adherence of gold nanoparticles.	• Targeted delivery to cancer site and tumor cells • Good biocompatibility and multi functionalization	Treatment of cancer	(Zhai *et al.*, 2011)
Nano-emulsion	• A nano-emulsion injection with a combination of CUR, phospholipids and ester polymers via a self-emulsifying process. • PS: 10–200 nm.	Improvement in solubility and stability of CUR	Antitumor preparation	(Chen *et al.*, 2010)
A nano-crystal formulation	• Nano-crystalline formulation with a combination of CUR and surfactants via high-pressure homogenization followed by lyophilisation. • PS: 30–500 nm.	Solubility and BA of CUR improvements • 300-fold improvement of solubility • 2–6 times increase gastrointestinal absorption	An oral preparation	(Liu *et al.*, 2009)
A nano-crystal formulation	Nano-crystal CUR is obtained by drying the nano-suspension which is prepared by solvent displacement or precipitation method?	Solubility and stability of CUR improvements	Aqueous dispersions or dry powders	(Zhai *et al.*, 2009)
A sustained release nanoparticle system	• A sustained release nanoparticle system is prepared with a combination of CUR nanoparticles and polylactic acid or polylactic acid-hydroxyacetic acid copolymer by emulsion-solvent diffusion technique. • PS: 150–500 nm.	• Solubility and stability of CUR improvements • Sustained-release CUR in more than 2 weeks • Extension of CUR retention and action time	CUR oral preparation for cancer treatment	(Auweter *et al.*, 2001)

Abbreviations: BA: Bioavailability; CUR: Curcumin; PS: Particle size.

TABLE 12.13
Current Curcumin Nanoformulations in the Market

Delivery System	Name	Claim	Company	Country	References
Nano-micelle	MetaCurcumin	277-folds improvement in absorption	RevGenetics	USA	(RevGenetics, 2016)
Nano-micelle	NovaSol® Curcumin	185-folds improvement in absorption	Molecular Health Technologies	USA	(LLC, 2014)
Liposome	Curcusome	–	NanoLife Tonics	USA	(Tonics, 2014)
Liposome	Optimal Liposomal Curcumin	–	Seeking Health	USA	(Health, 2014)
Liposome	Liposomal Turmeric Complex	–	BioBalance	New Zealand	(Zealand, 2016)
Liposome	Liposomal Structured Curcumin/ Resveratrol	–	Empirical Labs	USA	(Labs, 2016)
Liposome	Liposomal Curcumin 16 fluid ounces	–	Healthy Items	USA	(Items, 2007)
Phytosomes	Meriva®	31.5-folds improvement in absorption	Indena	Italy	(Indena, 2014)
Nanoparticles	Theracurmin	27-folds improvement in BA	Theravalues	Japan	(Corporation, 2015)
Colloidal emulsion nanoparticles	Nano Curcumin Elixir	10-folds improvement in BA	Asia Botanicals Sdn. Bhd	Malaysia	(Asia, 2016)
SLNs	Longvida®	65-folds improvement in BA	Verdure Sciences	USA	(Sciences, 2016)
Nanoemulsion	Full Spectrum Curcumin 60 Liquid Extract Softgels	185-folds improvement in BA	Solgar	USA	(Inc, 2016)
Nanocrystals	CurcuWIN	46-folds improvement in BA	Omniactive Health Technologies Ltd	USA	(Technologies, 2016)
Gamma-CD	Cavacurmin®	40-folds improvement in BA	Wacker Chemie AG	Germany	(Wacker Chemie AG, 2016)

Abbreviation: SLNs: Solid Lipid Nanoparticles.

12.9 CONCLUSION AND FUTURE PERSPECTIVES

CUR is undeniably an amazing natural compound, offering versatile biological applications, including anticancer, anti-inflammatory, and other disease-curative effects. However, its poor water solubility owing to its hydrophobic nature restricts its oral BA and remains as the major impediment in the development of CUR. It is strongly believed that nanotechnology has played a significant role in the development of CUR as a pharmaceutical product. Nano-enabled CUR products are formulated with superior traits such as improved aqueous solubility, enhanced BA, and improved stability. Despite some limitations, including difficulty in the preparatory process, cost, and resources, more and more nano-CUR formulations are devised every day, offering great hopes for use of this miracle bioactive compound.

REFERENCES

Abbasi, E., Aval, S. F., Akbarzadeh, A., Milani, M., Nasrabadi, H. T., Joo, S. W. *et al.* Dendrimers: Synthesis, applications, and properties. *Nanoscale Res Lett.* 2014; 9: 247.

Aditya, N. P., Yang, H., Kim, S., and Ko, S. Fabrication of amorphous curcumin nanosuspensions using β-lactoglobulin to enhance solubility, stability, and bioavailability. *Colloids Surf B.* 2015; 127: 114–121.

Ag Seleci, D., Seleci, M., Stahl, and Scheper, T. Tumor homing and penetrating peptide-conjugated niosomes as multi-drug carriers for tumor-targeted drug delivery. *RSC Advances.* 2017; 7: 33378–33384.

Agashe, H., Sahoo, K., Lagisetty, P., and Awasthi, V. Cyclodextrin-mediated entrapment of curcuminoid 4-[3,5-bis(2-chlorobenzylidene-4-oxo-piperidine-1-yl)-4-oxo-2-butenoic acid] or CLEFMA in liposomes for treatment of xenograft lung tumor in rats. *Colloids Surf B.* 2011; 84: 329–337.

Ahmad, M. Z., Akhter, S., Jain, G. K., Rahman, M., Pathan, S. A., Ahmad, F. J. *et al.* Metallic nanoparticles: Technology overview & drug delivery applications in oncology. *Expert Opin Drug Deliv.* 2010; 7: 927–942.

Anand, P., Kunnumakkara, A. B., Newman, R. A., and Aggarwal, B. B. Bioavailability of curcumin: Problems and promises. *Mol Pharm.* 2007; 4: 807–818.

Anon. https://www.semiconductors.org/.../2009_international_technology_roadmap. (Accessed on January 2018.)

Anderson, A. M., and Mohan, R. S. Isolation of curcumin from turmeric. *J Chem Educ.* 2000; 77: 359.

Anton, N., and Vandamme, T. F. Nano-emulsions and micro-emulsions: Clarifications of the critical differences. *Pharm Res.* 2011; 28: 978–985.

Arora, R., Kuhad, A., Kaur, I. P., and Chopra, K. Curcumin loaded solid lipid nanoparticles ameliorate adjuvant-induced arthritis in rats. *Eur J Pain.* 2015; 19: 940–952.

Asia, N. C. Nano curcumin elixir, nano curcumin Asia, Malasiya. Available online at www.nanocurcumin.asia. (Accessed on 26 July 2017.)

Auweter, H., Bohn, H., Hasselwander, O., Rieger, J., Schroeder, H., and Wegner, C. Formulations of curcumin. EP1103266 (A2); 2001.

Aytac, Z., and Uyar, T. Core-shell nanofibers of curcumin/cyclodextrin inclusion complex and polylactic acid: Enhanced water solubility and slow release of curcumin. *Int J Pharm.* 2017; 518: 177–184.

Azmin, M. N., Florence, A. T., Handjani-Vila, R. M., Stuart, J. F. B., Vanlerberghe, G., and Whittaker, J. S. The effect of non-ionic surfactant vesicle (niosome) entrapment on the absorption and distribution of methotrexate in mice. *J Pharm Pharmacol.* 1985; 37: 237–242.

Babaei, E., Sadeghizadeh, M., Hassan, Z. M., Feizi, M. A. H., Najafi, F., and Hashemi, S. M. Dendrosomal curcumin significantly suppresses cancer cell proliferation *in vitro* and *in vivo*. *Int Immunopharmacol.* 2012; 12: 226–234.

Bae, Y. H., and Park, K. Targeted drug delivery to tumors: Myths, reality and possibility. *J Control Release.* 2011; 153: 198–205.

Balaji, Avvari, B., Walia, Amita, Talwar, and Tulika. Multi-functional natural healing matrix wound bed for wound healing bandage/dressing. IN879DE2013 (A); 2015.

Barenholz, Y. Doxil®—The first FDA-approved nano-drug: Lessons learned. *J Control Release.* 2012; 160: 117–134.

Basnet, P., and Skalko-Basnet, N. Curcumin: An anti-inflammatory molecule from a curry spice on the path to cancer treatment. *Molecules.* 2011; 16: 4567–4598.

Behrens, S. H., and Grier, D. G. 2008. The charge of glass and silica surfaces. *Chem Phys.* 115, 6716–6721.

Beloqui, A., Coco, R., Memvanga, P. B., Ucakar, B., des Rieux, A., and Preat, V. pH-sensitive nanoparticles for colonic delivery of curcumin in inflammatory bowel disease. *Int J Pharm.* 2014; 473: 203–212.

Bharti, C., Nagaich, U., Pa, A. K., and Gulati, N. Mesoporous silica nanoparticles in target drug delivery system: A review. *Int J Pharm Investig.* 2015; 5: 124–133.

BioBalance New Zealand. Liposomal Turmeric Complex. BioBalance New Zealand. New Zealand. Available online at: www.biobalance.co.nz/product/liposomal-turmeric-complex/. (Accessed on 26 July 2017.)

Bischoff, S. C., Barbara, G., Buurman, W., Ockhuizen, T., Schulzke, J. D., Serino, M. *et al.* Intestinal permeability—A new target for disease prevention and therapy. *BMC Gastroenterol.* 2014; 14: 189.

Bose, S., Panda, A. K., Mukherjee, S., and Sa, G. Curcumin and tumor immune-editing: Resurrecting the immune system. *Cell Div.* 2015; 10: 6.

Center for International Environment Law (CIEL). Nanomaterials definition fact sheet. 2014. Available online at: http://www.ciel.org/reports/nanomaterials-definition-fact-sheet-nov-2014/. (Accessed 26 July 2017.)

Chang, R., Sun, L., Webster, T. J., and Mi, J. Amphiphilic peptide nanoparticles for use as hydrophobic drug carriers and antibacterial agents. US20170202783; 2017.

Chen, G., Li, J., Cai, Y., Zhan, J., Ga, J., Song, M., Yan, Z. *et al.* A glycyrrhetinic acid-modified curcumin supramolecular hydrogel for liver tumor targeting therapy. *Sci Rep.* 2017; 7: 44210.

Chen, S. W., Chen, S. M., Chen, J. W., and Zeng, S. Q. Curcumin nano micelle eye drops as well as preparation method and application. CN104644550 (A); 2015.

Chen, X., Zou, L.-Q., Niu, J., Liu, W., Peng, S.-F., and Liu, C.-M. The stability, sustained release and cellular antioxidant activity of curcumin nanoliposomes. *Molecules.* 2015; 20: 14293.

Chen, W., Xu, C. L., Xu, L. G., Xu, N. F., and Zhu, Y. Y. Construction and application of gold-magnetic nanoparticle-based medicament carrying platform. CN101716348; 2010.

Cheng, C., Peng, S., Li, Z., Zou, L., Liu, W., and Liu C. Improved bioavailability of curcumin in liposomes prepared using a pH-driven, organic solvent-free, easily scalable process. *RSC Advances.* 2017; 7: 25978–25986.

Chirio, D., Gallarate, M., Peira, E., Battaglia, L., Serpe, L., and Trotta, M. Formulation of curcumin-loaded solid lipid nanoparticles produced by fatty acids coacervation technique. *J Microencapsul.* 2011; 28: 537–548.

Corporation T. Theracurmin®—Our unique form of curcumin with markedly improved absorptivity. Theravalues Corporation, Japan. Available online at: http://theravalues.com/english/theracurmin-highly-bioavailable-curcumin/ (Accessed 26 July 2017.)

Cutrignelli, A., Lopedota, A., Denor, N., Iacobazzi, R. M., Fanizza, E., Laquintana, V. *et al.* A new complex of curcumin with sulfobutylether-beta-cyclodextrin: Characterization studies and *in vitro* evaluation of cytotoxic and antioxidant activity on HepG-2 cells. *J Pharm Sci.* 2014; 103: 3932–3940.

Dan, G., Mihai Grumezescu, A., Everard Bejenaru, L., and Bejenaru, C. Natural and synthetic polymers for drug delivery and targeting. *Nanobiomaterials in drug delivery.* William Andrew Publishing. 2016; pp. 229–284.

De Villiers, M., Aramwit, P., and G. Kwon (eds.). (2009). *Nanotechnology in drug delivery.* New York, Springer. X: 694, pp. 1–2. (Accessed on 19 May 2018.)

Dey, S., and Sreenivasan, K. Conjugation of curcumin onto alginate enhances aqueous solubility and stability of curcumin. *Carbohydr Polym.* 2014; 99: 499–507.

Dey, S., Sherly, M. C. D., Rekha, M., and Sreenivasan, K. Alginate stabilized gold nanoparticle as multidrug carrier: Evaluation of cellular interactions and hemolytic potential. *Carbohydr Polym.* 2016; 136: 71–80.

Douglass, B. J., and Clouatre, D. L. Beyond yellow curry: Assessing commercial curcumin absorption technologies. *J Am Coll Nutr.* 2015; 34: 347–358.

El Khoury, E., Abiad, M., Kassaify, Z. G., and Patra, D. Green synthesis of curcumin conjugated nanosilver for the applications in nucleic acid sensing and anti-bacterial activity. *Colloids Surf B.* 2015; 127: 274–280.

Emeje, M. O., Obidike, I. C., Akpabio, E. I., and Ofoefule, S. I. Nanotechnology in drug delivery. *Intech.* 2012; pp. 69–106.

Empirical Labs. Liposomal curcumin/resveratrol. Empirical Labs USA. Available online at: www.empirical-labs.com/vitamins-supplements/liposomal-curcumin-resveratrol. (Accessed on 26 July 2018.)

Falconieri, M., Adamo, M., Monasterol, C., Bergonzi, M., Coronnello, M., and Bilia, A. R. New dendrimer-based nanoparticles enhance curcumin solubility. *Planta Med.* 2017; 83: 420–425.

Fanun, M. *Colloids in drug delivery,* CRC Press; 2016; pp. 13–17.

Farooqui, A. *Therapeutic potentials of curcumin for Alzheimer disease.* Springer International Publishing. Switzerland; 2016; pp. 113–149.

Fulekar, M. H. *Nanotechnology: Importance and applications,* I. K. International Pvt Ltd.; 2010; pp. 10–13.

Gangwar, R. K., Dhumale, V. A., Kumari, D., Nakate, U. T., Gosavi, S. W., Sharma, R. B. *et al.* Datar S conjugation of curcumin with PVP capped gold nanoparticles for improving bioavailability. *Mater Sci Eng C Mater Biol Appl.* 2012; 32: 2659–2663.

Gao, Y., Li, Z., Sun, M., Li, H., Guo, C., Cui, J. *et al.* Preparation, characterization, pharmacokinetics, and tissue distribution of curcumin nanosuspension with TPGS as stabilizer. *Drug Dev Ind Pharm.* 2010; 36: 1225–1234.

Ghalandarlaki, N., Alizadeh, A. M., and Ashkani-Esfahani, S. Nanotechnology-applied curcumin for different diseases therapy. *Biomed Res Int.* 2014; 394264.

Gomes, L. M. M., Petito, N., Costa, V. G., Falcão, D. Q., de Lima Araújo, K. G. Inclusion complexes of red bell pepper pigments with β-cyclodextrin: Preparation, characterisation and application as natural colorant in yogurt. *Food Chem.* 2014; 148: 428–436.

Gong, C., Deng, S., Wu, Q., Xiang, M., Wei, X., Li, L. *et al.* Improving antiangiogenesis and anti-tumor activity of curcumin by biodegradable polymeric micelles. *Biomaterials.* 2013; 34: 1413–1432.

Gong, F., Chen, D., Teng, X., Ge, J., Ning, X., She, Y., Wang, S. *et al.* Curcumin-loaded blood-stable polymeric micelles for enhancing therapeutic effect on erythroleukemia. *Mol Pharm.* 2017; 14(8): 2585–2594.

Greenwald, B., Zlotkin, M., Soroka, Y., and Kohen, R. Curcumin protects skin against UVB-induced cytotoxicity via the Keap1-Nrf2 pathway: The use of a microemulsion delivery system. *Oxid Med Cell Longev.* 2017; 2017: 1–17.

Grill, A. E., Koniar, B., and Panyam, J. Co-delivery of natural metabolic inhibitors in a self-microemulsifying drug delivery system for improved oral bioavailability of curcumin. *Drug Deliv Transl Res.* 2014; 4: 344–352.

Grynkiewicz, G., and Slifirski, P. Curcumin and curcuminoids in quest for medicinal status. *Acta Biochim Pol.* 2012; 59: 201–212.

Guangming, G., Qinqin, P., Kaika, W., Rongchun, W., Yong, S., Ying, L. *et al.* Curcumin-incorporated albumin nanoparticles and its tumor image. *Nanotechnology.* 2015; 26: 045603.

Guo, F. Y., Yan, Q. Y., Guo, D. G., Hong, W. Y., and Yang, G. S. Method for preparing curcumin lipid nanoparticle suspension or nano-particles. CN103768012 (A), 2014.

Gupta, S. C., Kim, J. H., Patchva, S., Webb, L. J., Priyadarsini, I. K., and Aggarwala, B. B. Multitargeting by curcumin as revealed by molecular interaction studies. *Nat Prod Rep.* 2011; 28: 1937–1955.

Gupta, S. C., Patchva, S., and Aggarwal, B. B. Therapeutic roles of curcumin: Lessons learned from clinical trials. *AAPS J.* 2013; 15: 195–218.

Gupta, S. C., Patchva, S., Koh, W., and Aggarwal, B. B. Discovery of curcumin, a component of golden spice, and its miraculous biological activities. *Clin Exp Pharmacol Physiol.* 2012; 39: 283–299.

Gupta, S. C., Sun, B., Kim, J. H., Prasad, S., Li, S., and Aggarwal, B. B. Multitargeting by turmeric, the golden spice: From kitchen to clinic. *Mol Nutr Food Res.* 2013; 57: 1510–1528.

Guzman-Villanueva, D., El-Sherbiny, I. M., Herrera-Ruiz, D., and Smyth, H. D. C. Design and *in vitro* evaluation of a new nano-microparticulate system for enhanced aqueous-phase solubility of curcumin. *Biomed Res Int.* 2013; 724–763.

Hasan, M., Belhaj, N., Benachour, H., Barberi-Heyob, M., Kahn, C. J. F., Jabbar, E. *et al.* Liposome encapsulation of curcumin: Physico-chemical characterizations and effects on MCF7 cancer cell proliferation. *Int J Pharm.* 2014; 461: 519–528.

Hazzah, H. A., Farid, R. M., Nasra, M. M. A., El-Massik, M. A., and Abdallah, O. Y. Lyophilized sponges loaded with curcumin solid lipid nanoparticles for buccal delivery: Development and characterization. *Int J Pharm.* 2015; 492: 248–257.

He, Y., Yue, Y., Zheng, X., Zhang, K., Chen, S., and Du, Z. Curcumin, inflammation, and chronic diseases: How are they linked? *Molecules.* 2015; 20: 9183–9213.

Health, S. Optimal liposomal curcumin - 6 oz. Seeking Health. USA. Available online at: www.seekinghealth.com/optimal-liposomal-curcumin-6-oz.html. (Accessed on 26 July 2017.)

Hong, J., Liu, Y., Xiao, Y., Yang, X., Su, W., Zhang, M., Wang, X. *et al.* High drug payload curcumin nanosuspensions stabilized by mPEG-DSPE and SPC: *In vitro* and *in vivo* evaluation. *Drug Deliv.* 2017; 24: 109–120.

Hu, L., Shi, Y., Li, J. H., Gao, N., Ji, J., Niu, F. *et al.* Enhancement of oral bioavailability of curcumin by a novel solid dispersion system. *AAPS PharmSciTech.* 2015; 16: 1327–1334.

Hu, L. L., Huang, M., Wang, J. N., Zhong, Y., and Luo, Y. Preparation of magnetic poly(lactic-co-glycolic acid) microspheres with a controllable particle size based on a composite emulsion and their release properties for curcumin loading. *J Appl Polym Sci.* 2016; 133: 8.

Huang, P. L., Zeng, B. Z., Mai, Z. X., Deng, J. T., Fang, Y. P., Huang, W. H. *et al.* Novel drug delivery nanosystems based on out-inside bifunctionalized mesoporous silica yolk-shell magnetic nanostars used as nanocarriers for curcumin. *J Mater Chem B.* 2016; 4: 46–56.

Immordino, M. L., Dosio, F., and Cattel, L. Stealth liposomes: Review of the basic science, rationale, and clinical applications, existing and potential. *Int J Nanomed.* 2006; 1: 297–315.

Iskandar, F., Lenggoro, I. W., Kim, T. O., Naka, N., Shimada, M., Okuyama, K. *et al.* Fabrication and characterization of sio2 particles generated by spray method for standards aerosol. *J Chem Eng Jpn.* 2001; 34: 1285–1292.

Items, H. Liposomal curcumin 16 fluid ounces, Healthy Items. USA. Available online at: www.healthyitems.com/curcumin-p/607.htm. (Accessed on 26 July 2017.)

Jaiswal, M., Dudhe, Sharma, P. K. Nanoemulsion: An advanced mode of drug delivery system. *Biotech.* 2015; 5: 123–127.

Jambhrunkar, S., Karmakar, S., Popat, A., Yu, M., and Yu, C. Mesoporous silica nanoparticles enhance the cytotoxicity of curcumin. *RSC Advances.* 2014; 4: 709–712.

Jeong, H. J., and Park, H. J. Method for preparation of nanoemulsion comprising curcumin. KR20140115427 (A); 2014.

Ji, H., Tang, J., Li, M., Ren, J., Zheng, N., and Wu, L. Curcumin-loaded solid lipid nanoparticles with Brij78 and TPGS improved *in vivo* oral bioavailability and *in situ* intestinal absorption of curcumin. *Drug Deliv.* 2016; 23: 459–470.

Jiang, S., Zhu, R., He, X., Wang, J., Wang, M., Qian, Y., and Wang, S. Enhanced photocytotoxicity of curcumin delivered by solid lipid nanoparticles. *Int J Nanomed.* 2017; 12: 167–178.

Jin, Y. G., Li, M., Du, L. N., and Zhang, J. Application of dry curcumin nano-powder inhalant in treatment of acute lung injury. CN104415016 (A); 2015.

Jitoe-Masuda, A., Fujimoto, A., and Masuda, T. Curcumin: From chemistry to chemistry-based functions. *Curr Pharm Des.* 2013; 19: 2084–2092.

Kar, S. K., Das, G., and Suar, M. A pharmaceutical combination for treating tuberculosis. WO2014170820 (A2); 2014.

Kadota, K., Otsu, S., Fujimori, M., Sato, H., and Tozuka, Y. Soluble hydrolysis-resistant composite formulation of curcumin containing α-glucosyl hesperidin and polyvinylpyrrolidone. *Adv Powder Technol.* 2015; 27: 442–447.

Keck, C. M., and Müller, R. H. Drug nanocrystals of poorly soluble drugs produced by high pressure homogenization. *Eur J Pharm Biopharm.* 2006; 62: 3–16.

Kelly, K. L., Coronado, E., Zhao, L. L., and Schatz, G. C. The optical properties of metal nanoparticles: The influence of size, shape, and dielectric environment. *J Phys Chem B.* 2003; 107: 668–677.

Kesharwani, P., Jain, K., and Jain, N. K. Dendrimer as nanocarrier for drug delivery. *Prog Polym Sci.* 2014; 39: 268–307.

Kim, S., Diab, R., Joubert, O., Canilho, N., and Pasc, A. Core–shell microcapsules of solid lipid nanoparticles and mesoporous silica for enhanced oral delivery of curcumin. *Colloids Surf B.* 2016; 140: 161–168.

Kim, S., Philippot, S., Fontanay, S., Duval, R. E., Lamouroux, E., Canilho, N. *et al.* pH- and glutathione-responsive release of curcumin from mesoporous silica nanoparticles coated using tannic acid-Fe(iii) complex. *RSC Advances.* 2015; 5: 90550–90558.

Kim, T. H., Jiang, H. H., Youn, Y. S., Park, C. W., Tak, K. K., Lee, S. *et al.* Preparation and characterization of water-soluble albumin-bound curcumin nanoparticles with improved antitumor activity. *Int J Pharm.* 2011; 403: 285–291.

Korneeva, R. V. E., Kazanskij, A. L. V., Kljushnik, T. J. P., and Plakatina, T. J. P. Liposomal nano-means based on products derived from turmeric rhizomes. RU2571270 (C1); 2015.

Kotcherlakota, R., Barui, A. K., Prashar, S., Fajardo, M., Briones, D., Rodriguez-Dieguez, A. *et al.* Curcumin loaded mesoporous silica: An effective drug delivery system for cancer treatment. *Biomater Sci.* 2016; 4: 448–459.

Kumar, S., Kesharwani, S. S., Mathur, H., Tyagi, M., and Bhat, G. J. Molecular complexation of curcumin with pH sensitive cationic copolymer enhances the aqueous solubility, stability and bioavailability of curcumin. *Eur J Pharm Sci.* 2016; 82: 86–96.

Kumar, A., Kaur, G., Kansal, S. K., Chaudhary, G. R., and Mehta, S. K. (Cationic + nonionic) mixed surfactant aggregates for solubilisation of curcumin. *J Chem Thermodyn.* 2016a; 93: 115–122.

Kumar, A., Kaur, G., Kansal, S. K., Chaudhary, G. R., and Mehta, S. K. Enhanced solubilization of curcumin in mixed surfactant vesicles. *Food Chem.* 2016b; 199: 660–666.

Kumar, K., and Rai, A. K. Development and evaluation of proniosome-encapsulated curcumin for transdermal administration. *Trop J Pharm Res.* 2011; 10: 697–703.

Kumavat, S. D., Chaudhari, Y. S., Borole, P., Mishra, P., Shenghani, K., and Duvvuri, P. Degradation studies of curcumin. *Int J Pharm Sci Rev Res.* 2013; 3: 50–55.

Lao, C. D., Ruffin, M. T., Normolle, D., Heath, D. D., Murray, S. I., Bailey, J. M. *et al.* Dose escalation of a curcuminoid formulation. *BMC Complement Altern Med.* 2006; 6: 10.

Lazar, A. N., Mourtas, S., Youssef, I., Parizot, C., Dauphin, A., Delatour, B. *et al.* Curcumin-conjugated nanoliposomes with high affinity for Aβ deposits: Possible applications to Alzheimer disease. *Nanomedicine.* 2013; 9: 712–721.

Lee, W. H., Loo, C. Y., Bebawy, M., Luk, F., Mason, R. S., Rohanizadeh, R. *et al.* Curcumin and its derivatives: Their application in neuropharmacology and neuroscience in the 21st century. *Curr Neuropharmacol.* 2013; 11: 338–378.

Li, B., Konecke, S., Wegiel, L. A., Taylor, L. S., and Edgar, K. J. Both solubility and chemical stability of curcumin are enhanced by solid dispersion in cellulose derivative matrices. *Carbohydr Polym.* 2013; 98: 1108–1116.

Li, C., Zhang, Y., Su, T., Feng, L., Long, Y., and Chen, Z. Silica-coated flexible liposomes as a nanohybrid delivery system for enhanced oral bioavailability of curcumin. *Int J Nanomed.* 2012; 7: 5995–6002.

Li, J., Shin, G. H., Lee, I. W., Chen, X., and Park, H. J. Soluble starch formulated nanocomposite increases water solubility and stability of curcumin. *Food Hydrocoll.* 2016; 56: 41–49.

Li, M., Gao, M., Fu, Y., Chen, C., Meng, X., Fan, A. *et al.* Acetal-linked polymeric prodrug micelles for enhanced curcumin delivery. *Colloids Surf.* 2016; 140: 11–18.

Li, X. L., Liu, B. R., Xu, H., and Sun, Y. B. Anti-tumor double-drug nano drug carrying microsphere and preparation method; CN102641246 (A); 2012.

Liao, Y. H., Wei, X. L., Han, Y. R., Liu, C. Y., Quan, L. H., Yao, G. Y. *et al.* Curcumin dicaprate suspension injection and preparation method and application. CN103120643 (A); 2013.

Lin, W., Hong, J. L., Shen, G., Wu, R. T., Wu, Y., Huang, M. T. *et al.* Pharmacokinetics of dietary cancer chemopreventive compound dibenzoylmethane in the rats and impacts of nanoemulsion and genetic knockout of nrf2 on its disposition. *Biopharm Drug Dispos.* 2011; 32: 65–75.

Lin, Y. L., Liu, Y. K., Tsai, N. M., Hsieh, J. H., Chen, C. H., Lin, C. M. *et al.* A Lipo-PEG-PEI complex for encapsulating curcumin that enhances its antitumor effects on curcumin-sensitive and curcumin-resistance cells. *Nanomedicine.* 2012; 8: 318–327.

Lindman, B., and Wennerström, H. *Micelles.* Springer Berlin Heidelberg. Berlin. Heidelberg. 1980; pp. 1–83.

Liu, L., Sun, L., Wu, Q., Guo, W., Li, L., Chen, Y. *et al.* Curcumin loaded polymeric micelles inhibit breast tumor growth and spontaneous pulmonary metastasis. *Int J Pharm.* 2013; 443: 175–182.

Liu, K., Lang, Y. W., and Xu, H. High encapsulation rate curcumin albumin nano pharmaceutical composition. CN103054810 (A); 2013.

Liu, T., Liu, L., Liu, J., Liu, S., and Qiao, SZ. Fe3O4 encapsulated mesoporous silica nanospheres with tunable size and large void pore. *Front Chem Sci Eng.* 2014; 8: 114–122.

Liu, X. C., Li, Z. G., Sha, Y., and Chen, Y. H. Curcumin nano-lipid injection liquid, preparation method and application. CN101548946 (A); 2009.

Lohcharoenkal, W., Wang, L., Chen, Y. C., and Rojanasakul, Y. Protein nanoparticles as drug delivery carriers for cancer therapy. *Biomed Res Int.* 2014; 2014.

Madaan, K., Kumar, S., Poonia, N., Lather, V., and Pandita, D. Dendrimers in drug delivery and targeting: Drug-dendrimer interactions and toxicity issues. *J Pharm Bioallied Sci.* 2014; 6: 139–150.

Makeshwar, K. B., and Wasankar, S. R. Niosome: A novel drug delivery system. *Asian J Pharm Res.* 2013; 3: 16–20.

Mamani, L., Nikzad, S., Kheiri-manjili, H., al-Musawi, S., Saeedi, M., Askarlou, S. *et al.* Curcumin-loaded guanidine functionalized PEGylated I3ad mesoporous silica nanoparticles KIT-6: Practical strategy for the breast cancer therapy. *Eur J Med Chem.* 2014; 83: 646–654.

Mandal, S., Banerjee, C., Ghosh, S., Kuchlyan, J., and Sarkar, N. Modulation of the photophysical properties of curcumin in nonionic surfactant (tween-20) forming micelles and niosomes: A comparative study of different microenvironments. *J Phys Chem B.* 2013; 117: 6957–6968.

Mangolim, C. S., Moriwaki, C., Nogueira, A. C., Sato, F., Baesso, M. L., Neto, A. M. *et al.* Curcumin–β-cyclodextrin inclusion complex: Stability, solubility, characterization by FT-IR, FT-Raman, X-ray diffraction and photoacoustic spectroscopy, and food application. *Food Chem.* 2014; 153: 361–370.

Manjili, H. K., Sharafi, A., Danafar, H., Hosseini, M., Ramazani, A., and Ghasemi, M. H. Poly (caprolactone)-poly(ethylene glycol)-poly(caprolactone) (PCL-PEG-PCL) nanoparticles: A valuable and efficient system for *in vitro* and *in vivo* delivery of curcumin. *RSC Advances.* 2016; 6: 14403–14415.

Mehanny, M., Hathout, R. M., Geneidi, A. S., and Mansour, S. Bisdemethoxycurcumin loaded polymeric mixed micelles as potential anti-cancer remedy: Preparation, optimization and cytotoxic evaluation in a HepG-2 cell model. *J Mol Liq.* 2016a; 214: 162–170.

Mehanny, M., Hathout, R. M., Geneidi, A. S., and Mansour, S. Exploring the use of nanocarrier systems to deliver the magical molecule: Curcumin and its derivatives. *J Control Release.* 2016b; 225: 1–30.

Mishra, S., and Palanivelu, K. The effect of curcumin (turmeric) on Alzheimer's disease: An overview. *Ann Indian Acad Neurol.* 2008; 11: 13–19.

Mody, V. V., Siwale, R., Singh, A., and Mody, H. R. Introduction to metallic nanoparticles. *J Pharm Bioallied Sci.* 2010; 2: 282–289.

Mohanty, C., Das, M., and Sahoo, S. K. Emerging role of nanocarriers to increase the solubility and bioavailability of curcumin. *Expert Opin Drug Deliv.* 2012; 9: 1347–1364.

Molecular Health Technologies LLC MHT. NovaSOL® curcumin. Molecular Health Technologies LLC. USA. Available online at: http://novasolcurcumin.com/ (Accessed on 26 June 2017.)

Mukherjee, S., Ray, S., and Thakur, R. S. Solid lipid nanoparticles: A modern formulation approach in drug delivery system. *Indian J Pharm Sci.* 2009; 71: 349–358.

Nahar, P. P., Slitt, A. L., and Seeram, N. P. Anti-inflammatory effects of novel standardized solid lipid curcumin formulations. *J Med Food.* 2015; 18: 786–792.

NanoLife Tonics. What is curcumin, NanoLife Tonics. USA. Available online at: http://nanolifetonics.com/curcusome/what-is-curcumin. (Accessed on 26 July 2017.)

Nauman, R. K., Mohd, S. H., Asif, N., Nurulaini, H., and Tin, W. W. Nanocarriers and their actions to improve skin permeability and transdermal drug delivery. *Curr Pharm Des.* 2015; 21: 2848–2866.

Noriega-Luna, B., Godínez, L. A., Rodríguez, F. J., Rodríguez, A., Zaldívar-Lelo de Larrea, G., Sosa-Ferreyra, C. F. *et al.* Applications of dendrimers in drug delivery agents, diagnosis, therapy, and detection. *J Nanomater.* 2014; 2014: 1–19.

Office EP. Available online at: http://worldwide.espacenet.com. (Accessed on 26 July 2017.)

OmniActive Health Technologies. CurcuWIN, OmniActive Health Technologies. USA. Available online at: http://omniactives.com/curcuwin. (Accessed on 26 July 2017.)

Onoue, S., Takahashi, H., Kawabata, Y., Seto, Y., Hatanaka, J., Timmermann, B. *et al.* Formulation design and photochemical studies on nanocrystal solid dispersion of curcumin with improved oral bioavailability. *J Pharm Sci.* 2010; 99: 1871–1881.

Orlando, R. A., Gonzales, A. M., Royer, R. E., Deck, L. M., and Vander Jagt, D. L. A chemical analog of curcumin as an improved inhibitor of amyloid Abeta oligomerization. *PLoS One.* 2012; 7: e31869.

Palanikumar, L. Curcumin: A putative chemopreventive agent. *J Life Sci.* 2009; 3: 47–53.

Pan, M. H., Huang, T. M., and Lin, J. K. Biotransformation of curcumin through reduction and glucuronidation in mice. *Drug Metab Dispos.* 1999; 27: 486–494.

Panahi, Y., Ghanei, M., Bashiri, S., Hajihashemi, A., and Sahebkar, A. Short-term curcuminoid supplementation for chronic pulmonary complications due to sulfur mustard intoxication: Positive results of a randomized double-blind placebo-controlled trial. *Drug Res (Stuttg).* 2015; 65: 567–573.

Podaralla, S., Averineni, R., Alqahtani, M., and Perumal, O. Synthesis of novel biodegradable methoxy poly(ethylene glycol)–zein micelles for effective delivery of curcumin. *Mol Pharm.* 2012; 9: 2778–2786.

Prasad, S., and Aggarwal, B. B. Turmeric, the golden spice: From traditional medicine to modern medicine. *Herbal medicine: Biomolecular and clinical aspects.* 2nd edition. Boca Raton (FL): CRC Press/Taylor & Francis; 2011; pp. 263–282.

Prasad, S., Gupta, S. C., Tyagi, A. K., and Aggarwal, B. B. Curcumin, a component of golden spice: From bedside to bench and back. *Biotechnol Adv.* 2014; 32: 1053–1064.

Prasad, S., Tyagi, A. K., and Aggarwal, B. B. Recent developments in delivery, bioavailability, absorption and metabolism of curcumin: The golden pigment from golden spice. *Cancer Res Treat.* 2014; 46: 2–18.

Priyadarsini, K. I. The chemistry of curcumin: From extraction to therapeutic agent. *Molecules.* 2014; 19: 20091–20112.

Rachmawati, H., Edityaningrum, C. A., and Mauludin, R. Molecular inclusion complex of curcumin–β-cyclodextrin nanoparticle to enhance curcumin skin permeability from hydrophilic matrix gel. *AAPS PharmSciTech.* 2013; 14: 1303–1312.

Rahimi, H. R., Mohammadpour, A. H., Abnous, K., Ghayour-Mobarhan, M., Ramezanzadeh, E., Mousavi, F. *et al.* Curcumin: Reintroduced therapeutic agent from traditional medicine for alcoholic liver disease. *APJMT.* 2015; 4: 25–30.

Ramalingam, P., and Ko, Y. T. Enhanced oral delivery of curcumin from N-trimethyl chitosan surface-modified solid lipid nanoparticles: Pharmacokinetic and brain distribution evaluations. *Pharm Res.* 2015; 32: 389–402.

Ramchand, N. C., Anitha, K. N., Parthasarathy, K., Ganga, R. G., Rama, R. G., Bhupathiraju, K. *et al.* Curcuminoids and its metabolites for the application in allergic ocular/nasal conditions. CN102361552 (A); 2012. AU2011232745 (A1); 2011.

RevGenetics. Metacurcumin: Super curcumin. USA. Available online at: www.revgenetics.com/store/c-1-metacurcumin-super-curcumin.aspx. (Accessed on 26 July 2017.)

Righeschi, C., Bergonzi, M. C., Isacchi, B., Bazzicalupi, C., Gratteri, P., and Bilia, A. R. Enhanced curcumin permeability by SLN formulation: The PAMPA approach. *LWT–Food Sci Technol Res.* 2016; 66: 475–483.

Rungphanichkul, N., Nimmannit, U., and Muangsiri, W. Preparation of curcuminoid niosomes for enhancement of skin permeation. *Pharmazie.* 2011; 66: 570–575.

Sa, G., and Das, T. Anti-cancer effects of curcumin: Cycle of life and death. *Cell Div.* 2008; 3: 14.

Saffie-Siebert, R., Ogden, J., and Parry-Billings, M. Nanotechnology approaches to solving the problems of poorly water-soluble drugs. *DDW.* 2005; 71–76.

Schneider, C., Gordon, O. N., Edwards, R. L., and Luis, P. B. Degradation of curcumin: From mechanism to biological implications. *J Agric Food Chem.* 2015; 63: 7606–7614.

Seo, S. W., Han, H. K., Chun, M. K., and Choi, H. K. Preparation and pharmacokinetic evaluation of curcumin solid dispersion using Solutol(R) HS15 as a carrier. *Int J Pharm.* 2012; 424: 18–25.

Shaabani, E., Amini, S. M., Kharrazi, S., and Tajerian, R. Curcumin coated gold nanoparticles: Synthesis, characterization, cytotoxicity, antioxidant activity and its comparison with citrate coated gold nanoparticles. *Nanomedicine.* 2017; 4: 115–125.
Sharma, R. A., Euden, S. A., Platton, S. L., Cooke, D. N., Shafayat, A., Hewitt, H. R. *et al.* Phase I clinical trial of oral curcumin: Biomarkers of systemic activity and compliance. *Clin Cancer Res.* 2004; 10: 6847–6854.
Sharma, R. A., McLelland, H. R., Hill, K. A., Ireson, C. R., Euden, S. A., Manson, M. M. *et al.* Pharmacodynamic and pharmacokinetic study of oral Curcuma extract in patients with colorectal cancer. *Clin Cancer Res.* 2001; 7: 1894–1900.
Shawon, D., Darin, S., Sukanta, D., Chong, S., Saadyah, A., Krishnaswami, R. *et al.* Dendrimer-curcumin conjugate: A water soluble and effective cytotoxic agent against breast cancer cell lines. *Anticancer Agents Med Chem.* 2013; 13: 1531–1539.
Shen, L., and Ji, H. F. The pharmacology of curcumin: Is it the degradation products? *Trends Mol Med.* 2012; 18: 138–144.
Satish, K., Gnanaprakash, D., Mayilvaganan, K., Arunraj, C., and Mohankumar, S. Chitosan-gold nanoparticles as delivery systems for curcumin. *Int J Pharm Sci Res* 2012; 3: 4533–4539.
Shukla, P., Dwivedi, P., Gupta, P. K., and Mishra, P. R. Optimization of novel tocopheryl acetate nanoemulsions for parenteral delivery of curcumin for therapeutic intervention of sepsis. *Expert Opin Drug Deliv.* 2014; 11: 1697–1712.
Singh, P. K., Prabhune, A. A., and Ogale, S. B. Curcumin sophorolipid complex. WO2016013026 (A1); 2016.
Singh, P. K., Wani, K., Kaul-Ghanekar, R., Prabhune, A., and Ogale, S. From micron to nano-curcumin by sophorolipid co-processing: Highly enhanced bioavailability, fluorescence, and anti-cancer efficacy. *RSC Advances.* 2014; 4: 60334–60341.
Slowing, I. I., Vivero-Escoto, J. L., Wu, C. W., and Lin, V. S. Y. Mesoporous silica nanoparticles as controlled release drug delivery and gene transfection carriers. *Adv Drug Deliv Rev.* 2008; 60: 1278–1288.
Solgar Inc. Full Spectrum Curcumin. Solgar Inc. USA. Available online at: www.solgar.com/SolgarProducts /Full-Spectrum-Curcumin.htm. (Accessed on 26 July 2017.)
Sun, J., Bi, C., Chan, H. M., Sun, S., Zhang, Q., and Zheng, Y. Curcumin-loaded solid lipid nanoparticles have prolonged *in vitro* antitumor activity, cellular uptake and improved *in vivo* bioavailability. *Colloids Surf B.* 2013; 111: 367–375.
Sun, X., Luo, Y., Huang, L., Yu, B. Y., and Tian, J. A peptide-decorated and curcumin-loaded mesoporous silica nanomedicine for effectively overcoming multidrug resistance in cancer cells. *RSC Advances.* 2017; 7: 16401–16409.
Sutradhar, K. B., and Amin, M. L. Nanoemulsions: Increasing possibilities in drug delivery. *Eur J Nanomed.* 2013; 5: 2.
Szente, L., and Szejtli, J. Cyclodextrins as food ingredients. *Trends Food Sci Technol.* 2004; 15: 137–142.
Tadros, T. F. Emulsion formation, stability, and rheology. *Emulsion formation and stability.* Wiley-VCH Verlag GmbH & Co. KGaA; 2013; pp. 1–75.
Tadros, T. F., and Vincent, B. *Encyclopedia of emulsion technology.* Marcel Dekker: New York. 1983; pp. 385–424.
Tahmasebi, M., Isacch, B., Sadeghizadeh, M., Marr, F., Bilia, A. R., Mowla, S. J. *et al.* Dendrosomal curcumin nanoformulation downregulates pluripotency genes via miR-145 activation in U87MG glioblastoma cells. *Int J Nanomedicine.* 2014; 9: 403–417.
Tang, C. H., and Chen, F. P. Preparation method of high-load curcumin soybean protein nano product. CN104256048 (A); 2015.
Tapal, A., and Tiku, P. K. Complexation of curcumin with soy protein isolate and its implications on solubility and stability of curcumin. *Food Chem.* 2012; 130: 960–965.
Thangavel, S., Yoshitomi, T., Sakharkar, M. K., and Nagasaki, Y. Redox nanoparticles inhibit curcumin oxidative degradation and enhance its therapeutic effect on prostate cancer. *J Control Release.* 2015; 209: 110–119.
Tiwari, G., Tiwari, R., and Rai, A. K. Cyclodextrins in delivery systems: Applications. *J Pharm Bioallied Sci.* 2010; 2: 72–79.
Tonnesen, H. H., Masson, M., and Loftsson, T. Studies of curcumin and curcuminoids. XXVII. Cyclodextrin complexation: Solubility, chemical and photochemical stability. *Int J Pharm.* 2002; 244: 127–135.
Torchilin, V. P. Recent advances with liposomes as pharmaceutical carriers. *Nat Rev Drug Discov.* 2005; 4: 145–160.
Trinh, T. T., Tran, K. Q., Zhang, X. Q., van Santen, R. A., and Meijer, E. J. The role of a structure directing agent tetramethylammonium template in the initial steps of silicate oligomerization in aqueous solution. *Phys Chem Chem Phys.* 2015; 17: 21810–21818.

Tripathy, S., and Das, M. K. Dendrimers and their applications as novel drug delivery carriers. *J App Pharm Sci*. 2013; 3: 142–149.
Trujillo, J., Chirino, Y. I., Molina-Jijon, E., Anderica-Romero, A. C., Tapi, E., Pedraza-Chaverri, J. Reno. Protective effect of the antioxidant curcumin: Recent findings. *Redox Biol*. 2013; 1: 448–456.
Udompornmongkol, P., and Chiang, B. H. Curcumin-loaded polymeric nanoparticles for enhanced anti-colorectal cancer applications. *J Biomater Appl*. 2015; 30: 537–546.
Vaage, J., Barbera-Guillem, E., Abra, R., Huang, A., and Working, P. Tissue distribution and therapeutic effect of intravenous free or encapsulated liposomal doxorubicin on human prostate carcinoma xenografts. *Cancer*. 1994; 73: 1478–1484.
Velusami, C. C., Boddapati, S. R., Hongasandra Srinivasa, S., Richard, E. J., Joseph, J. A., Balasubramanian, M. *et al*. Safety evaluation of turmeric polysaccharide extract: Assessment of mutagenicity and acute oral toxicity. *Biomed Res Int*. 2013; 2013: 158348.
Verdooner, S. Chemical compositions to detect and treat amyloid in a patient's brain and retina. US2012207681 (A1); 2012.
Verdure Sciences. Longvida® optimized curcumin, Verdure Sciences, USA. Available online at: www.longvida.com. (Accessed on 26 July 2017.)
Vögtle, F., Buhleier, D., and Wehner, W. "Cascade"- and "Nonskid-chain-like" syntheses of molecular cavity topologies. *Synthesis*. 1978; 2: 155–158.
Wacker Chemie AG. Highly bioavailable curcumin, Wacker Chemie AG, Germany; 2016. Available online at: www.wacker.com/cms/en/industries/food/curcumin.jsp. (Accessed on 26 July 2017.)
Waghela, B. N., Sharm, A., Dhumale, S., Pandey, S. M., and Pathak, C. Curcumin conjugated with PLGA potentiates sustainability, anti-proliferative activity and apoptosis in human colon carcinoma cells. *PLoS ONE*. 2015; 10: e0117526.
Wahlan, B., Pawar, Y. B., and Bansal, A. K. Identification of permeability-related hurdles in oral delivery of curcumin using the Caco-2 cell model. *Eur J Pharm Biopharm*. 2011; 77: 275–282.
Wang, H. B., Qin, J. W., Wan, Y. Q., Gao, W. D., and Fu, J. J. Multi-layer drug sustain-release nano fiber membrane and preparation method, CN103599090 (A); 2014.
Wang, L., Xu, X., Zhang, Y., Zhang, Y., Zhu, Y., Shi, J. *et al*. Encapsulation of curcumin within poly(amidoamine) dendrimers for delivery to cancer cells. *J Mater Sci Mater Med*. 2013; 24: 2137–2144.
Wang, L. J. Controlled release microparticles of Nano medication of curcumin, and preparation method. CN1957926 (A); 2007.
Wang, P., Zhang, L., Peng, H., Li, Y., Xiong, J., and Xu, Z. The formulation and delivery of curcumin with solid lipid nanoparticles for the treatment of on non-small cell lung cancer both *in vitro* and *in vivo*. *Mater Sci Eng C Mater Biol Appl*. 2013; 33: 4802–4808.
Wang, S. L., Zhu, R. R., and Wang, W.R. Application of curcumin solid lipid nano-particle serving as medicament for treating asthma. CN102949344 (A); 2013.
Wang, S. M., Zhang, D. Y., Liu, Y., Xia, H. M., Zhou, H. Y., Li, Z. L. *et al*. Gene transfer material of functionalized nano curcumin and preparation method. CN103849650 (A); 2014.
Wang, X., Jiang, Y., Wang, Y. W., Huang, M. T., Ho, C. T., and Huang, Q. Enhancing anti-inflammation activity of curcumin through O/W nanoemulsions. *Food Chem*. 2008; 108: 419–424.
Wang, Y. J., Pan, M. H., Cheng, A. L., Lin, L. I., Ho, Y. S., Hsieh, C. Y. *et al*. Stability of curcumin in buffer solutions and characterization of its degradation products. *J Pharm Biomed Anal*. 1997; 15: 1867–1876.
Wei, X. L., Han, Y. R., Quan, L. H., Liu, C. Y., and Liao, Y. H. Oily nanosuspension for long-acting intramuscular delivery of curcumin didecanoate prodrug: Preparation, characterization and *in vivo* evaluation. *Eur J Pharm Sci*. 2013; 49: 286–293.
Xia, Q., and Xia, Y. C. Nano carrier loaded with tetrahydrocurcumin and preparation method. CN103655214 (A); 2014.
Xia, S. Q., Xie, J. H., Zhang, X. M., Tan, C., Jia, C. S., and Zhong, F. Curcumin nano compound and preparation method. CN104273522 (A); 2015.
Xie, X., Tao, Q., Zou, Y., Zhang, F., Guo, M., Wang, Y. *et al*. PLGA nanoparticles improve the oral bioavailability of curcumin in rats: characterizations and mechanisms. *J Agric Food Chem*. 2011; 59: 9280–9289.
Xin, N., Fatimasabily, Jia, J. F., Wang, W. C., and Zhao, Y. P. Method for preparing nano curcumin particles by utilizing ultrasonic-assisted supercritical anti-solvent. CN103705468 (A); 2014.
Xin, Y., Zhao, J., Conti, P. S., and Chen, K. Radiolabeled nanoparticles for multimodality tumor imaging. *Theranostics*. 2014; 4: 290–306.
Xiong, R. L., Lu, W. G., Li, J., Wan, P. Q., Xu, R., and Chen, T. T. Preparation and characterization of intravenously injectable nimodipine nanosuspension. *Int J Pharm*. 2008; 350: 338–343.

Xu, W., Ling, P., and Zhang, T. Polymeric micelles: A promising drug delivery system to enhance bioavailability of poorly water-soluble drugs. *J Drug Deliv.* 2013; 2013: 340315.

Yadav, A., Lomash, V., Samim, M., and Flora, S. J. Curcumin encapsulated in chitosan nanoparticles: A novel strategy for the treatment of arsenic toxicity. *Chem Biol Interact.* 2012; 199: 49–61.

Yallapu, M. M., Jaggi, M., and Chauhan, S. C. Poly (β-cyclodextrin)/Curcumin Self-Assembly: A novel approach to improve curcumin delivery and its therapeutic efficacy in prostate cancer cells. *Macromol Biosci.* 2010; 10: 1141–1151.

Yallapu, M. M., Jaggi, M., and Chauhan, S. C. Curcumin nanoformulations: A future nanomedicine for cancer. *Drug Discov Today.* 2012; 17: 71–80.

Yang, Y. L., Hu, Y. Q., Zhi, F., Jia, X. F., Shao, N. Y., and Zhou, Z. G. Oral curcumin nano-granule and preparation method. CN103446057 (A); 2014.

Ye, Y., and Chen, X. L. Nanocomposite of curcumin and upconversion rare earth, preparation method and application. CN103223169 (A); 2013.

Yi, J. G., Kim, D. M., Kim, D. H., and Choi, J. E. Cosmetic composition comprising nano capsule containing extract of camellia sinensis, curcuma longa, magnolia obovata and aralia, continentalis and manufacturing method. KR20130023850 (A); 2013.

Yu, H., Li, J., Shi, K., and Huang, Q. Structure of modified ε-polylysine micelles and their application in improving cellular antioxidant activity of curcuminoids. *Food Funct.* 2011; 2: 373–380.

Zebi, B., Mouloungui, Z., and Noirot, V. Stabilization of curcumin by complexation with divalent cations in glycerol/water system. *Bioinorg Chem Appl.* 2010; 292760.

Zhai, G. G., Lou, H. X., Gao, Y., Li, H. L., Liu, A. C., and Cao, F. L. Curcumin nano crystallization preparation and preparation method. CN101361713 (A); 2009.

Zhai, G. X., Zhao, L. Y., and Cao, F. L. Curcumin nano micelle preparation and preparation method. CN102274163 (A); 2011.

Zhai, G. X., Yang, C. F., and Yang, X. Y. Folate receptor-mediated curcumin self-assembled polymer nanomicelle and preparation method. CN104257629 (A); 2015.

Zhang, H., Xu, H., Wu, M., Zhong, Y., Wang, D., and Jiao, Z. A soft–hard template approach towards hollow mesoporous silica nanoparticles with rough surfaces for controlled drug delivery and protein adsorption. *J Mater Chem B.* 2015; 3: 6480–6489.

Zhang, J., Tang, Q., Xu, X., and Li, N. Development and evaluation of a novel phytosome-loaded chitosan microsphere system for curcumin delivery. *Int J Pharm.* 2013; 448: 168–174.

Zhang, J. Q., Yan, Z. J., Sun, L. L., Wan, K., Li, W. Y., and Hu, X. Y. Lipid carrier of curcumin in nano structure and preparation method of lipid carrier. CN103989659 (A); 2014.

Zhou, H., Beevers, C. S., and Huang, S. The targets of curcumin. *Curr Drug Targets.* 2011; 12: 332–347.

Zhou, J. P., Yao, J., Ni, J., and Yang, H. Curcumin-polysaccharide conjugate as well as preparation method and application. CN102988999 (A); 2013.

13 Nanoencapsulation of Iron for Nutraceuticals

*Naveen Shivanna, Hemanth Kumar Kandikattu, Rakesh Kumar Sharma, Teenu Sharma, and Farhath Khanum**

CONTENTS

* Corresponding author.

13.1 INTRODUCTION

13.1.1 Iron in Biological Systems

Iron is one of the most common and abundant elements in the inner and outer layer of earth. Iron plays important roles in our biological system and is found in hemoglobin, muscles, bone marrow, tissues, ferritin, hemosiderin, blood proteins, and enzymes. Iron forms complexes with molecular oxygen in two oxygen transport proteins, hemoglobin and myoglobin. Iron is present mainly at the active site of many redox enzymes that are responsible for cellular respiration, that is, oxidation and reduction reactions in animals and plants. Vital sources of iron in the diet are meat, nuts, vegetables, eggs, mushrooms, beans, and also a few fortified products such as wheat flour (Hunt *et al.*, 2009, Hurrell and Egli, 2010; Murray-Kolbe, 2010; Aggett, 2012).

13.1.2 Deficiency of Iron

Iron deficiency is a common nutritional deficiencies encountered across the globe. According to a World Health Organization (WHO) report, more than 2 billion people are suffering from iron deficiency, or anemia (Severance and Hamza, 2009). Whenever there is iron loss, it should be compensated through dietary intake, and if this does not happen, deficiency of iron will result and, over a period of time, turn to anemia. If the case is left untreated, it will result in insufficient number of RBCs, less hemoglobin with low oxygen supply, which would further lead to fatigue and tiredness (Zimmermann and Hurrell, 2007). Most commonly, children, premenopausal women, and people with malnutrition are most susceptible to iron-deficiency anemia (IDA). IDA further causes problems such as arrhythmia, pregnancy complications, and delayed growth milestones in infants and children. Iron deficiency, with hemoglobin less than 8 g/dl, leads to anemia of inflammation (AI), characterized by a reduction in red cell size as observed in inflammatory disorders and malignancy (Nemeth and Ganz, 2014).

13.1.3 Iron Excess

Excess of iron, or overload of iron, is termed as *hemochromatosis*, wherein iron accumulates in the body. The most important factor is heredity, a genetic disorder with a genetic defect in HLA-H gene region on chromosome 6, leading to low level of hepcidin, a key regulatory enzyme for entry of iron into the circulatory system, resulting eventually in excessive iron and hemochromatosis (Serra *et al.*, 2009). Repeated blood transfusion results in a condition called transfusional iron overload. Excess in the availability of iron to bind iron transport protein transferrin leads to iron toxicity. Excessive levels of free iron in the blood react with peroxides, leading to the generation of highly reactive free radicals that can damage macromolecules such as proteins, lipids, DNA, and other cellular components through Fenton reaction (Eaton and Qian, 2002). Iron toxicity is also observed in aging disorders such as atherosclerosis, Alzheimer's disease, and Parkinson's disease (Altamura and Muckenthaler, 2009).

13.1.4 Iron Metabolism

Total body content of iron is around 3–5 g. Out of this, 66% will be as hemoglobin, 4% as myoglobin, 29% will be stored in ferritin and hemosiderin, and the rest remains bound to enzymes such as cytochromes, peroxidases, and others involved in the citric acid cycle. Iron is also found in plasma, along with transferrin (0.1%). Almost every cell contains iron. Haem is the most predominant iron-containing substance, which is also a constituent of proteins/enzymes: haemoprotein-Hb, myoglobin, cytochromes, catalase, xanthenes oxidase, and tryptophan (Hurrell and Egli, 2010).

Completely balanced diet should contain good amounts of iron, sufficient enough for all the body's functions. Nearly 10% of dietary iron is absorbed each day, which suffices to balance the losses that occur daily from desquamation of epithelia. Higher iron requirement is observed in growing children and women, because there will be more loss of iron during menstruation in women, in hemorrhages and also during pregnancy, the latter state being marked with increased absorption efficiency of dietary iron to 20% for meeting the body requirements.

13.1.5 Iron Absorption

Iron is one of the most important micronutrients required for daily normal functioning, and overcoming its poor bioavailability will remain a challenge to pharmaceutical scientists. Iron bioavailability is predicted from the concentration of ferrous dialyzable iron in relation to controls, as this is one of the methods employed to predict bioavailability of iron by researchers. Research is being carried out regularly so as to improve and measure bioavailability (Miller *et al.*, 1981; Narasinga *et al.*, 1994; Luten *et al.*, 1996; Chiplonkar *et al.*, 1999; Benkhedda *et al.*, 2010).

It is well known that iron from animal sources is better absorbed than that from vegetable or plant sources. Iron will be released from the protein complex in which it normally exists by the action of HCl and proteolytic enzyme in the stomach cum small intestine. Iron is maximally absorbed from duodenum than from jejunum. This is because of increased alkaline pH that forms insoluble ferric hydroxide-acid complexes, which cannot be absorbed. Divalent metal transporter 1 (DMT1) is a transporter protein that facilitates iron transport through epithelial cells of the intestine (Figure 13.1). In addition, DMT1 will also facilitate the uptake of trace minerals such as copper, zinc, lead, manganese, cadmium, and cobalt. The enterocytes release the iron present inside these cells into the bloodstream as ferroportin, and iron will then bind to a glycoprotein transporter, transferrin. Here, both ferroprotin and DMT1 will be present almost in all the cells involved in the transport of iron, macrophages being one among them (Fuqua *et al.*, 2012). Iron present in the form of ferrous ions is well-absorbed when it is taken on an empty stomach rather than when given along with the meal, because iron absorption will be reduced as a consequence of the same ligand binding processes, which influences the dietary non-haem iron that gets released from the food as iron (III), getting reduced in duodenum by cytochrome b1.

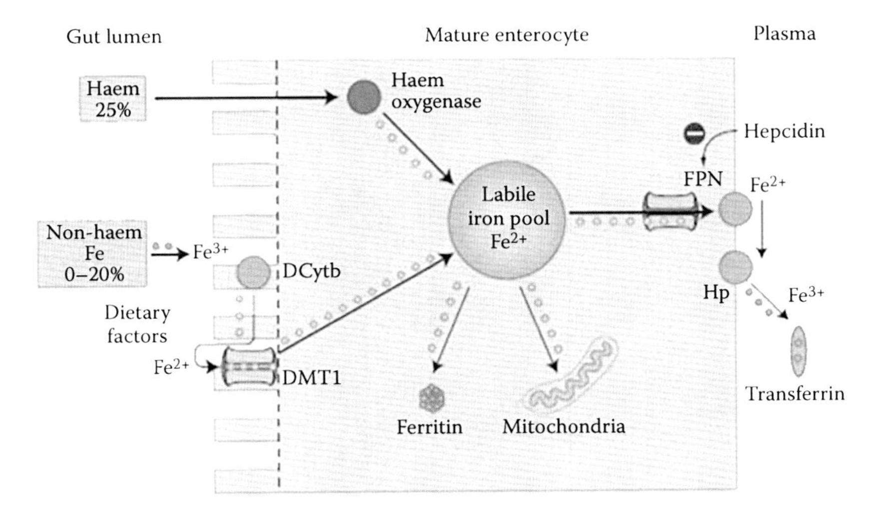

FIGURE 13.1 Molecular pathways of iron absorption. The figure refers to iron absorption by the epithelial cell. DMT1: divalent metal transporter 1; FPN: ferroportin; Hp: hephaestin; TF: transferrin; TFR: transferrin receptor.

DMT1 is upregulated during iron deficiency, which transports reduced iron through intestinal brush border membrane. When non haem iron enters the labile pool, a portion of it gets incorporated to ferritin, and some will get lost because of cells exfoliation. Amount of iron intended for retention will get transported across the serosal membrane with the help of ferroportin much before it is taken up by transport protein transferrin in of Fe^{3+} form (Figure 13.2). Copper containing ferroxidase, "hephaestin" predominantly expressed in small intestinal villi converts ferrous iron (Fe^{2+}) to ferric (Fe^{3+}) during the basolateral transfer step of iron absorption. The haem iron will bind to their receptors at the brush border membrane in the intestine released inside the cells by the enzyme haem oxygenase just before entering labile iron pool (Figure 13.2). After which, it follows a similar pathway as that of non-haem iron (Chiplonkar *et al.*, 1999; Benkhedda *et al.*, 2010).

Thus, the proteins important during iron metabolism are haemoglobin, DMT1, ferroportin (SLC40A1), hepcidin (i.e., a hormone produced by liver cells), growth differentiation factor (GDF), matriptase-2 (TMPRSS6), transferrin and transferrin receptors, twisted gastrulation protein, ferritin and haemosiderin (Miller *et al.*, 1981; Narasinga *et al.*, 1994).

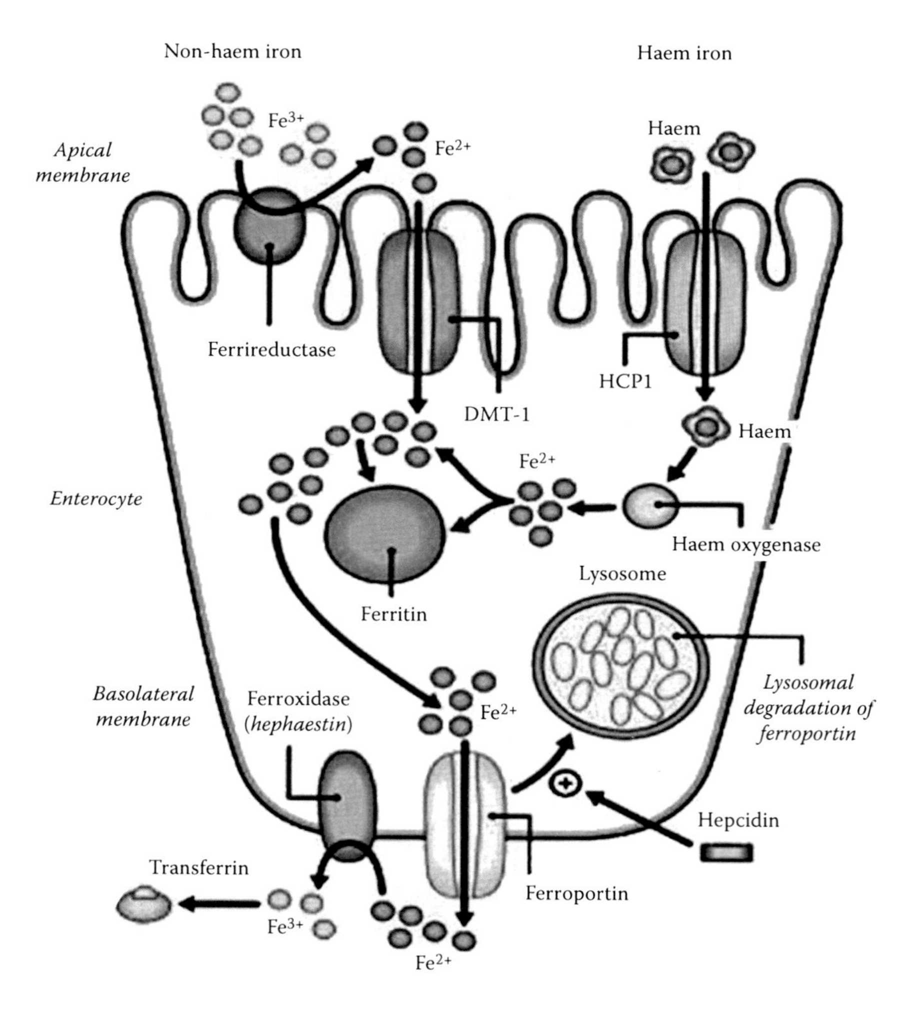

FIGURE 13.2 An overview picture of the mechanism involved in iron absorption and assimilation. Proteins important in iron metabolism are haemoglobin DMT1, ferroportin, hepcidin, matriptase-2, Transferrin and transferrin receptors, ferritin, and haemosidern.

13.1.5.1 Divalent Metal Transporter 1, or DMT1

DMT1, an electrogenic pump, needs proton as a co-transporter for transporting iron across the membranes. Expression of DMT1 protein is upregulated during iron deficiency and it also helps absorption of other divalent metal ions such as Zn^{2+}, Co^{2+}, Mn^{2+}, Cu^{2+}, and Pb^{2+}.

13.1.5.2 Ferroportin, or SLC40A1

SLC40A1 is a transmembrane domain protein and basolateral transporters of Fe. This protein is essential in release of iron from the macrophages and also present in intracellular organelles. It requires ceruloplasmin mainly for cell surface localization for ferroportin; its expression is regulated by another protein, hepcidin.

13.1.5.3 Hepcidin

Hepcidin is important for control of iron absorption and metabolism. Its expression is controlled by the HAMP gene and is one of the small peptides with several isoforms. It is known to be predominantly expressed in the liver. As stated earlier, it is involved in maintaining iron homeostasis. It binds to cell surface ferroportin, resulting in phosphorylation of tyrosine residue further internalization, ubiquitination, and degradation. All these processes occur in lysosomes. Hepcidin has a significant role in inhibition of iron absorption, release of iron from macrophages, and is transported across the placenta. It is found as bound to a protein, α-2–macroglobulin in plasma, with its route of clearance through the kidney. Quantification of Hepcidin in serum and urine is carried out by ELISA or mass spectrophotometer techniques. Its concentrations become low and sometimes undetectable in conditions such as haemochromatosis and anemia, and become high in some cases such as inflammation.

13.1.5.4 Growth Differentiation Factor and Twisted Gastrulation Protein

GDF 15, or growth differentiation factor, is one of the members of transforming growth factors, the TGF β super-family of proteins. GDF 15 expresses different properties as per the cell environment. Sometimes it activates macrophage, inhibits synthesis of hepcidin, and helps in proliferation of immature haemopoietic progenitors. It is unregulated during hypoxia and iron depletion but is independent of HIFs, or hypoxia-inducible transcription factors. TWSG1, twisted gastrulation protein, a second erythroid regulator of hepcidin expression, has also been identified.

13.1.5.5 TMPRSS6, or Matriptase-2

TMPRSS6 is a type 2 member of trans-membrane serine protease family, with its main expression being in the liver. This protein matriptase-2 bound to membrane regulates expression of hepcidin by cleaving membrane bound protein HJV and also by release of soluble fragments of HJV. Mutations in mice and humans have resulted in hepcidin up-regulation, causing blockade of iron transport from intestinal and macrophages into plasma and conditions such as hypochromic microcytic anemia.

13.1.5.6 Transferrin and Its Receptors

Transferrin is a single polypeptide chain found in plasma and extravascular fluid. This protein is synthesized predominantly in the liver, and its synthesis is inversely related to iron storage. Here, two ferric ions bind to each of the molecules. Nearly 4 mg of the body's iron gets carried by transferrin at a time; it plays a vital role in iron transport, since more than 30 mg of iron is being transported each day. Iron uptake by transferrin will require a protein that is close to specific receptors present on surface of the cell (Figure 13.3). Transferrin receptor gene code (TFRC) for TFR1 (transmembrane protein) is identified as CD71, and likewise, TFR2, the second receptor, binds to transferrin too. Transferrin binding with HFE, TFR1, and TFR2 are involved in hepcidin synthesis regulation.

Usually, 20 to 45% transferrin binding sites will be filled, that is, 0.1% of total iron in the body will be circulating as bound to transferrin. Almost all the absorbed iron will be used for

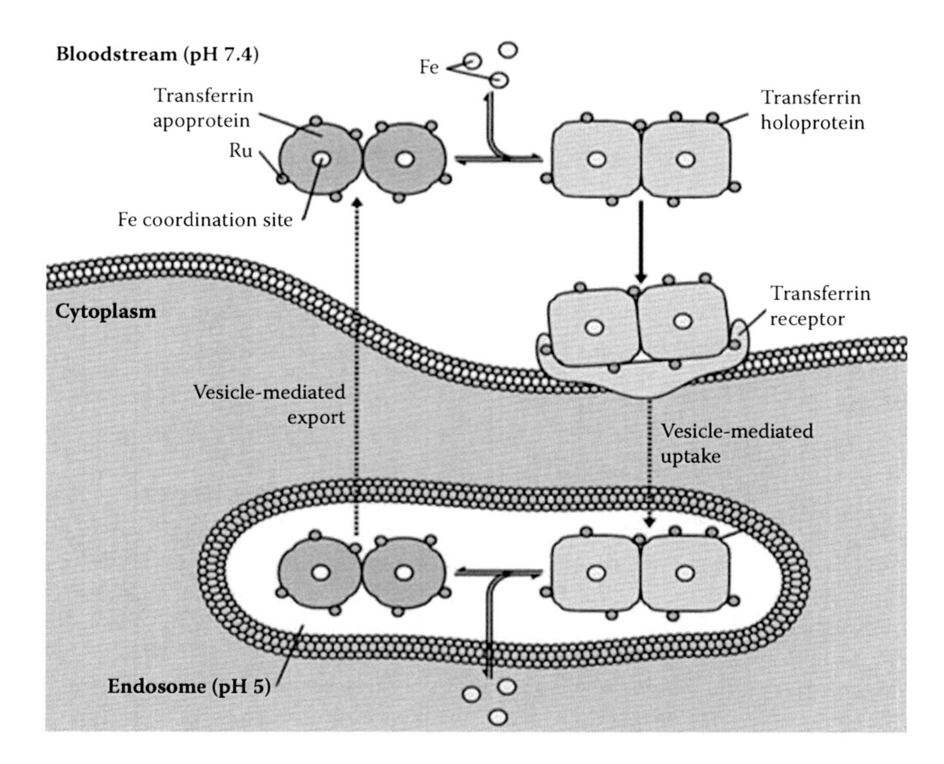

FIGURE 13.3 Transfer of iron from plasma into RBC. Uptaking of transferring iron is by receptor mediated endocytosis.

erythropoiesis by the bone marrow, wherein the membrane receptors on erythroid precursors avidly bind transferrin. As a result, 10–20% of absorbed iron goes to a storage pool that gets recycled by macrophages' engulfment and erythropoiesis, thus maintaining a balance between its storage and utilization. Other trace elements such as manganese, cobalt, etc. are also being transported in the same manner as that with iron (Nemeth, 2008).

13.1.5.7 Lactoferrin

Lactoferrin is a glycoprotein, which is structurally similar to that of transferrin. It is normally found in neutrophils and milk. It possesses a unique property of bacteriostasis at its secreting surfaces with the mode of action being deprivation of iron to the microorganisms that depend on it for their growth.

13.1.6 Regulation in Iron Absorption

Regulation of iron will take place in two plausible pathways:

1. Using diet: Excessive intake of dietary iron will not increase the absorption rate, since mucosal cells will get saturated with enough amount of iron, thus inhibiting the additional uptake of iron.
2. Stores regulator: If iron accumulation/storage in liver is sufficient, it triggers release of hepatic hepcidin that will decrease ferroportin release in intestinal mucosal cells, thus resulting in enterocytes for retaining the iron absorbed and then get sloughed off in a few days. When the body iron stores decrease, hepcidin gets diminished and the absorbed iron is put into circulation by intestinal mucosa.

In addition, the composition of diet will also influence iron absorption. For instance, citrate and ascorbate present in citrus fruits will form complexes with iron, resulting in its increased absorption, but tannins of tea will cause reduced absorption. Iron is readily absorbed in the form of haem from meat origin rather than as inorganic iron from vegetarian sources. Most of the non-haem dietary iron is found as Fe^{3+} that is reduced to Fe^{2+} before it gets absorbed. Duodenal microvilli contain ferric reductase that promotes absorption in ferrous iron (Hurrell, 2004).

Only a small amount of total iron in the body is being gained or lost every day, since most of the iron in the body is being recycled and used. When RBCs die, the iron from hemoglobin will get scavenged by the microphages and returned to the storage pool for reuse by the spleen. Intestinal absorption is also regulated in maintaining the iron homeostasis in the body. This is achieved by increased absorption signalled via decreased hepcidin, resulting due to hypoxia, inflammation, erythropoietic activity, and even by decreasing iron stores. Hepcidin synthesis is being influenced by the bone morphogenetic protein (BMP) pathway (Miller *et al.*, 1981; Narasinga *et al.*, 1994).

13.1.7 Iron Storage; Ferritin or Hemosiderin

Ferritin is a protein iron complex and is one primary iron storage form. It is a water soluble complex consisting of a spherical shaped apo-protein shell that encloses a core of ferric hydroxyphosphate (Theil, 2013). Ferritin in humans consists of 24 subunits with immunologically distinct two types, H and L subunits. Variation in these subunits will explain ferritin's heterogeneity that comes from different tissues. Using a special kind of electropheritic technique, IEF or isoelectric focusing, the two forms can be easily separated as ferritins, isolated from different origins such as spleen, liver, heart, and RBCs (https://bitesizebio.com). A small amount of ferritin (L subunit) is present in serum that normally contains little iron. M-ferritin is present in mitochondria. Expression of this type of ferritin can be correlated with tissues that have high mitochondrial number, rather than those involved in iron storage.

13.1.7.1 Hemosiderin

Hemosiderin is completely different from ferritin, as it is water insoluble. It is crystalline in nature that consists of protein-iron complex, when seen in light microscopy by staining with Prussian blue (Perl's) reaction. Sometimes, hemosiderin is found in amorphous form, with higher iron and protein ratio. This form gets originated by partial digestion of ferritin aggregates by lysosomal enzymes. In case of iron overload, hemosiderin levels become higher in both cell types (Saito *et al.*, 2012).

It is said that adult females and males have 2 and 6 g of total iron in the body, respectively. Out of this, 1.5–2 g gets found in RBCs and 0.5–1 g gets stored in spleen, bone marrow, and liver, and the remaining is found in iron-dependent enzymes and myoglobin.

13.1.8 Laboratory Testing Methods

i. Serum ferritin (indirect)
ii. Serum iron (direct)
iii. Serum iron binding capacity
iv. Complete blood count (CBC)
v. Bone marrow biopsy
vi. Liver biopsy

Simple tests that will give an idea on iron requirement and its storage in the body are direct method (serum iron) and indirect method (total iron binding capacity), accompanied by calculating saturation percentage of transferring. Serum ferritin is well-correlated with that of iron storage, but it gets elevated during the time of liver disease, inflammation, and conditions such as malignant neoplasms. Complete blood count will also give an indirect measure of the iron storage, for example, mean corpuscular volume (MCV) will decrease in case of iron deficiency. Iron storage can be obtained by study of bone marrow, liver, or both the biopsies. This is achieved by conducting simple iron staining of the above biopsies (Fleming and Bacon, 2005; Franchini and Veneri, 2005).

13.1.9 Iron and Immune Function

The immune system will regularly undergo permanent regeneration, and as a consequence, millions of immune cells are produced. The immune cell regeneration, enhanced by infection and recovery from infection, will depend on the rate of multiplication of the invading microbes with that of immune cells. The immune system will use both micronutrients and macronutrients required for synthesis of DNA, RNA, and protein. In all the age groups, especially in growing and old age groups, malnourishment will have a strong role on the immune system, as the reserved nutrition in the body gets limited. Clinical studies related to this, conducted earlier, disclose that iron plays an important role in regulation and function of T lymphocytes. Iron deficiency will result in cell-mediated immunity impairment associated with decreased activities of myeloperoxidase, antibacterial, and NK activities impairment (Theil, 2013).

Overall, iron is vital in efficient functioning of the immune system. Both physical and mental health depend on iron, especially during childhood and pregnancy, during which the developing child solely depends on mother's iron supplies.

13.1.10 Deficiency of Iron (Anemia)

Iron stored or present in the body is lost regularly during normal routine activities such as urination, defecation, in the form of sweat, and also during reconditioning of old cells of the skin. One should not forget about bleeding, during which iron is lost; hence, women require more iron when compared to men. In cases of declined iron storage, normal hemoglobin synthesis slows down, resulting in reduction in the transport of oxygen. This in turn leads to unwanted conditions like dizziness, fatigue, and lowered immunity, during which it is difficult to keep up with regular mundane activities (Yoshimura *et al.*, 1980). As there is no synthesis of iron that happens in our body to maintain the lost iron, it is quite necessary to consume enough dietary iron, that is, maintain an iron-rich diet as our daily menu to cope up with the requirement of loss due to the aforesaid reasons. Iron deficiency anemia is one of the most prevalent nutritional deficiency disorders in India and across the world (Figure 13.4). Anemia will affect people in almost all the age groups,

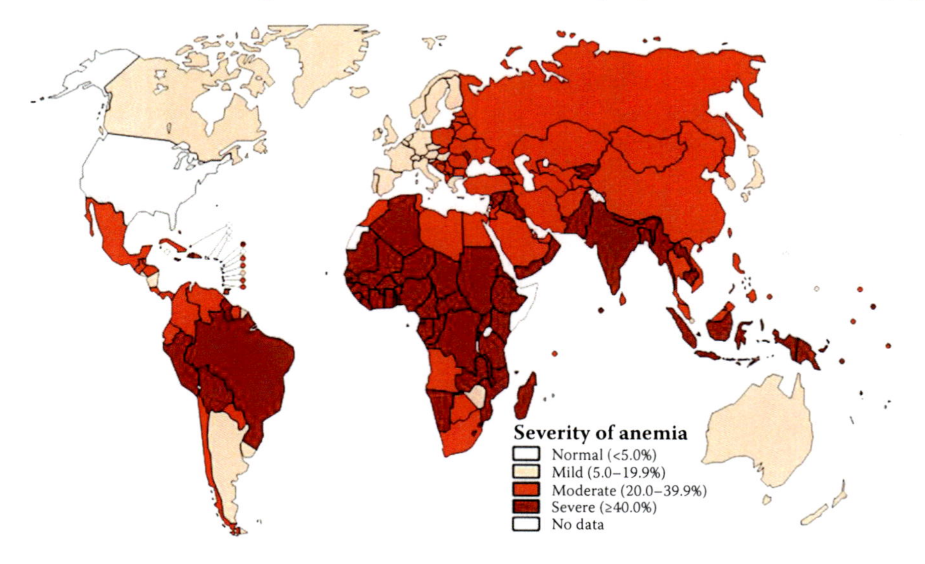

FIGURE 13.4 Worldwide prevalence of anemia.

but especially preschool female children, pregnant women, and girls of child-bearing age. It is said that around 25% of the total global population is suffering from anemia. Anemia is found to be more prevalent in the developing countries, owing to several factors. Studies by National Family Health Survey say that almost 50% of Indian women are diagnosed as anemic, which implies that every second Indian woman is anemic, and one of every five maternal deaths occurring in India is because of anemia.

Reproductive age group women during the menstrual cycle will require double the actual amount of iron. This can be achieved normally by increasing the absorption efficiency of iron from the gastrointestinal (GI) tract to accommodate the excess demand. The developing fetus will also draw Fe from its mother, as a result of which she requires extra iron. Infants also require 4–12 mg/L of iron in their formula, which is present in the breast milk and gets readily absorbed. Hence, the lactating mothers also require more iron in their diet (Andrews, 1999).

Trauma is one of the most important pathologic causes for blood loss, besides several GI-related pathological disorders, such as GI carcinomas, ulcers, inflammatory diseases, and hemorrhoids. Aspirin intake further aggravates such loss of blood by GI tract. Indeed, demand for iron will increase in case of hemorrhages due to the loss of RBCs or blood. Celiac disease (sprue) is one of the diseases that could impair iron absorption. Therefore, adults identified with iron deficiency go for endoscopic procedures for identification of the source of bleeding. This type of anemia will be characterized by decreased amount of hemoglobin resulting in less magnitude of mean corpuscular hemoglobin (MCH) owing to reduced size of RBCs. This condition is termed as *hypochromic microcytic anemia*. Total serum iron also gets diminished, whereas serum iron binding capacity will rather increase, with low transferrin saturation percent (5–10%). In individuals from high altitude regions, serum soluble transferrin receptors will be more expressed (Andrews, 1999; Clark, 2009).

Recent studies show that mean fractional absorption in subjects with iron deficiency was 17.5% vis-à-vis the normal value of 7.3%. This number is much higher than the absorption value, that is, 5% used earlier for calculating iron RDA. Consider that the absorption of iron will always be inversely proportional to that of its storage in the body. It is well known that Indians have less iron stored in the body when compared to people in Western countries; iron absorption in males could be estimated to be 5% and 8% in females (Rammohan *et al.*, 2012).

The values are in agreement with the recommendations of FAO/WHO, in which three bioavailability levels, that is, 5, 10, and 15%, are listed. Always our international recommendations will be 2 to 3 times lower when compared to Recommended Dietary Allowances (RDA) for Indians. This difference in RDA of iron for different age groups, genders, and categories of people across the world is mainly because of bioavailability of iron. Iron's bioavailability is dependent on its source and presence of ascorbic acid in excess in the diet. Haem iron is readily absorbed more than non-haem iron. Likewise, iron is absorbed more efficiently in the presence of vitamin C. Hence, it is important to increase the bioavailability of iron by changing its source and way of intake rather than only by increasing the iron content in the diet (Beck *et al.*, 2014). This will decently address the iron requirement for Indians. RDA for iron recommended in Indians by the committee is presented in Table 13.1.

The main cause for anemia will be either nutrition or infection. Among the nutritional causes that contribute to anemia are adolescent girls in the developing nations, who are always at higher rate of risk for morbidity and mortality. This is one of the shaping periods in their life, where there will be maximum level of changes that occur physically, psychologically, and even behaviorally. During this period of their life, there are more chances of developing nutritional anemia. Hence, on the whole, there is an urgent requirement of improving the nutritional status for adolescent girls by varied activities such as supplementation programs, community awareness, nutrition education through nutritional counseling, etc. (Patel *et al.*, 2017).

TABLE 13.1
Revised RDA Values for Iron Intake

		Iron (mg/day)	
Group	**Category**	**Revised**	**Old**
Men	Sedentary	17	28
	Moderate		
	Heavy		
Women	Sedentary	21	30
	Moderate		
	Heavy		
	Pregnant	35	38
	Lact. < 6 mths	21	30
	Lact. 6–12 mths		
Infants	0-6 mths	46	-
	6-12 mths	5	-
Children	1-3 yrs	9	12
	4-6 yrs	13	18
	7-9 yrs	16	26
Boys	10-12 yrs	21	34
	13-15 yrs	32	41
	16-17 yrs	28	50
Girls	10-12 yrs	27**	19
	13-15 yrs	27	28
	16-17 yrs	26	30

Source: ICMR, Nutrient Requirements and Recommended Dietary Allowance for Indians. A Report of the Expert Group of the Indian Council of Medical Research, New Delhi, 2010.
** Only RDA value increased.

13.1.11 Applications of Nanotechnology

Nanotechnology is said to be modulation of individual atom, molecule, or molecular cluster into a structure for creating a material or a device with a new or vastly different property.

Nanoencapsulation is the latest technology that has gained entry into food science. A vast number of applications have been found in food science including food processing, packaging, and delivery of nutrients to specific sites. We can design nanoencapsulated materials for sustained and targeted release by employing different mechanisms. Materials for encapsulation or coatings should be selected in an organized manner, so as to have the capacity to dissolve slowly or sometimes quickly in the stomach in the presence of acid, or, if needed, we can modify in such a way that it gets released only when a certain pH is reached, for example, the pH of the small intestine (i.e., alkaline pH). Coating or encapsulating materials can be made in such a manner that it dissolves quickly by absorbing the moisture in a food. Mechanical compression can release sensitive ingredients such as protected flavors. Nanoencapsulation enables food scientists/industrialists to accomplish the above mentioned points such as sustained release of the bioactive component at the targeted site for improved bioavailability of the encapsulated bioactive.

With the help of nanoencapsulation, we can regulate the following:

- Release of encapsulated material at a specific temperature or time in a targeted region, up and down the GI tract in simple manner, such as triggering-selected factors such as pH and enzyme activity.
- Release of active nutrients or biologics at precise absorption sites.

In addition, we can protect the nutrient(s) during processing and its targeting activities from different hazards like:

- Degradation by microbes
- pH and temperature changes
- Undesirable interactions
- Enzymatic degradation
- Interference of dust particles, since they are nano structured encapsulates
- Interaction of air, light, and metal to maintain its potency

If all the above aspects are achieved, we can deliver a product with following properties:

- Food or nutrient with longer stability, shelf life, and free-flowing
- Phytonutrients that can be included along with list of dietary supplements preparations that involve higher temperatures
- Nutrients to reload the antioxidant status associated with stress
- Nutrients with improved bioavailability
- Flavor/taste masking of nutrients with undesirable flavors and odors, during delivering desirable attributes
- Cost-effective nutrients
- Nutrients with consistent quality

A number of nanocarrier systems have been designed and investigated to improve the delivery and bioavailability of iron, including liposomes (Xia and Xu, 2005; Yuan *et al.*, 2013; Naveen and Khanum, 2014; Xu *et al.*, 2014; Zi *et al.*, 2015; Martinez-Gonzales *et al.*, 2016; Yuan *et al.*, 2017) and microcapsules (Kim *et al.*, 2003; Abbasi and Azari, 2011; Gupta *et al.*, 2015; Gupta *et al.*, 2016).

13.1.11.1 Emulsions

Macroemulsions are two immiscible fluids dispersed in each other. They are unstable thermodynamically and will get stabilized kinetically. Therefore, the emulsion's stability will be dependent on the emulsion's composition and its size. Normally, a conventional emulsion will have a droplet size larger than 1 mm, thus making it susceptible to gravitational forces. Its methods of synthesis can be varied to obtain different droplet size distributions.

Depending on the size of the emulsions, different terminologies are used, the droplet size of the emulsions ranges from 20 to 500 nm, that is, between conventional emulsions or microemulsions to nanoemulsions. Depending on the nature and size of the droplets in the emulsion, the terminology is fixed as ultrafine emulsions or mini emulsions or translucent emulsions or submicron emulsions and nanoemulsions. Because of the small size of the nanoemulsions, they appear to be transparent. Moreover, the Brownian motion of the particles prevents its sedimentation or creaming offering good stability to the emulsion (Patel and Joshi, 2012).

Nanoemulsions are known to be highly stable and can be diluted with water without changing the size or zeta potential of the particles (Figure 13.5). Nanoemulsions can be prepared by different methods, most common being the mechanical method. In this method, the emulsions are subjected to different mechanical disruption using high shear stirring, high pressure homogenizers, or ultrasound generators. All these processes will result in breaking up of large droplets of microemulsions into smaller ones provided the Laplace pressure is overcome. It is said that introduction of surfactant content at the interface will reduce the Laplace pressure resulting in stable small droplets of emulsion, called nanoemulsion. If, in the above process, more mechanical energy and surfactants are used, it becomes unfavorable for industrial applications (Jaiswal *et al.*, 2015).

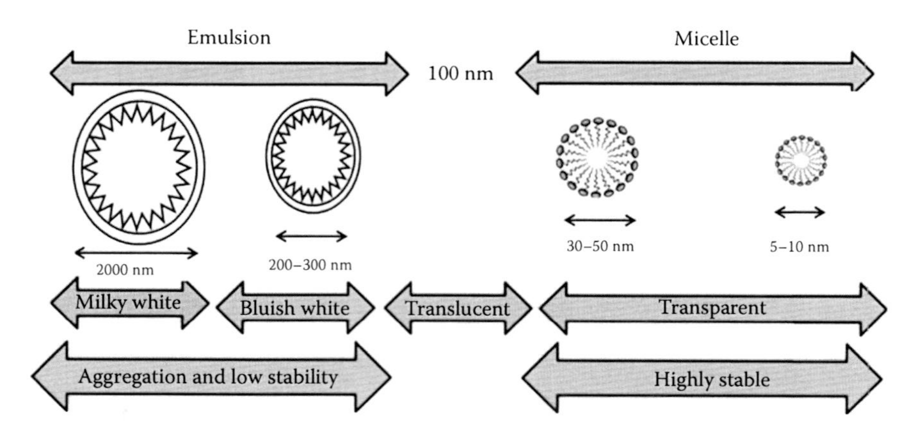

FIGURE 13.5 Various types of emulsions and their size ranges.

Hence, it is desired to prepare submicron emulsions. This can be achieved using low energy emulsification methods, where the surfactant's spontaneous curvature is changed. For instance, in case of nonionic surfactants, nanoemulsions can be prepared from changing the temperature of the emulsion. Just by increasing the temperature, transitional phase inversion can be achieved from oil-in-water (O/W) emulsion to water-in-oil (W/O) emulsion.

13.1.11.2 Transportation Using Liposomes

Liposomes will transport materials to different cells by the four methods listed below (Thomas and Joseph, 2001).

1. Direct transfer: Without the association of the contents of liposomes, the lipid moiety of the liposomes will get integrated with the cellular and subcellular membranes.
2. Endocytosis: Here, the whole of liposomes are eaten by the phagocytic cells present in the reticuloendothelial system, that is, macrophages and neutrophils.
3. Fusion: This will happen when the lipid bilayer of liposomes get fused with the plasma membrane of the cells during which the constituents of liposomes will get into the cytoplasm.
4. Adsorption: Here, the transfer of the contents will happen by simple electrostatic forces or nonspecific weak hydrophobic or by specific interactions with the components present on the cell surfaces.

13.1.11.3 Liposome Preparation

Liposomes can be prepared by various methods (Crommelin *et al.*, 2003):

1. Hand shaking method: In this method, the liposomes are being formed by introducing lipid molecules into an aqueous environment. This can be done when a thin dry lipid film gets hydrated. Here, the lamellae will swell and form finger-like projections, that is, myelin fingers. These myelin fingers should be broken by means of mechanical agitation like shaking or pipetting, or even by vortexing. During this process, the broken finger-like projections get resealed at the hydrophobic edges that result in formation of liposomes. Preparation of huge liposomes is being done by this method only.
2. Sonication methods: This method is usually used to prepare small or tiny liposomes that are unilamellar. The method uses two types of instrument for the preparation of small liposomes.
   a. Probe sonication method: In this method, the tip of the probe is put inside the emulsion or the mixture, for liposome preparation.
   b. Bath sonication method: Herein, the emulsion or the mixture itself is placed inside the sonicator bath in a glass beaker for liposome preparation.

3. Reverse phase evaporation method: It is a method where inverted micelles formation takes place. It is possible to prepare these by sonication of two different solvents as a mixture of an aqueous solution containing the water soluble molecule and organic solvent containing amphiphilic molecules. Afterward, the organic solvent is removed by flash evaporation resulting in a gel like or viscous solution forming inverted micelles. At one critical point in this technique, the gel will get collapsed and some of the inverted micelles may disintegrate. Multilayered liposomes are formed by excess of phospholipids present in the environment that will contribute in the formation of a complete bilayer around remaining micelles in the emulsion.
4. Freeze drying and rehydration method: Liposomes that are freeze-dried have good efficiencies to encapsulate. It is said that even macromolecule encapsulation is possible using this method. Herein, while dehydration process happens, both the material that has to be encapsulated and liposomes are brought close to each other. This will have better chances of getting encapsulated, as they are adhering to the surface of liposomes during rehydration. To achieve the best results, water should be added to the freeze-dried sample in a slow manner using a micropipette. After adding an aliquot, the mixture should be vortexed properly to ensure proper mixing and micelles formation. The volume of the water used to dehydrate should be always less when compared to the starting volume of the emulsion before drying. This will ensure maximum encapsulation efficiency.
5. Detergent depletion method: This is one of the methods commonly followed for proteoliposome formulations and sometimes for different varieties of liposomes preparation. Herein, case removal of detergents is achieved by well-known techniques such as dialysis, gel filtration, absorption, diffusion, etc. This will lead to the formation of homogenous liposomes. Mixed micelles can also be made using concentrated detergent solution achieved by the addition of detergent solution to multilamellar liposomes. Most popular detergents that can be used are alkyl (thio) glucoside, sodium cholate, and alkyloxypolyethylene.

13.1.11.4 Preparation of Micronutrient Incorporated Nanoemulsions

Iron with phosphaticylcholine, cholesterol, cholesterol acetate, or soyabean oil with the surfactant mixture, prepared by adding Chremophor EL and diethyleleglycolmonoethylether, mixed with distilled water and can be homogenized using high speed homogenizer.

13.1.11.5 Solubility of the Mineral

The solubility of active ingredients in different surfactants, co-surfactants, and oils can be determined from dissolving excess amounts of selected surfactants, co-surfactants, and oils in a stopped vial and mixed using vortex mixer. The equilibrated samples need to be centrifuged and the supernatant can be taken and filtered. The concentration of the active ingredient has to be noted for finding the encapsulation efficiency of the product.

13.1.11.6 Nanoemulsion's Characterization

Nanoemulsions can be characterized with techniques that are familiar such as:

1. TEM or transmission electron microscopy
2. Nanoemulsion droplet size analysis
3. Stability studies

13.1.12 Food Fortification with the Prepared Nanoemulsions

For fortification of nanoemulsions, a simple technique can be used. The nanoemulsions samples can be prepared using fresh pasteurized milk instead of using aqueous solutions. Following that, these can be homogenized, sonicated, and centrifuged for its final usage (Abbasi and Azari, 2011; Naveen and Khanum, 2014).

13.1.13 Bioavailability Test

In vivo bioavailability test can be carried out in rats weighing about 200 g to 350 g. The rats are given different samples of fortified nanoemulsions incorporated with the mineral depending on their body weight by force feeding method. After 3 hours, blood can be collected and iron can be estimated in blood to quantify the bioavailability. In addition the toxicity of the emulsions also must be studied (Naveen and Khanum, 2014).

13.2 CONCLUSIONS

Nanotechnology can be an excellent tool for increasing the absorption of iron. Soon iron-deficiency anemia will become a part of history in all nations, developing as well as developed. However, there are several questions that need to be answered before one can embark on such a mission. First, at present there are only a few studies reported in literature where nanoencapsulation of iron is used for food purposes. The available literature shows more than 200% increase in iron absorption when encapsulated into liposomes as compared to the unencapsulated iron. Tissue distribution studies reported a significant uptake and retention by brain, as brain tissue is quite rich in lipids. Third, only a few reports are available on vital organ function tests like kidney and liver. Fourth, what level of fortification should be permitted? Fifth, what will be the new RDA for encapsulated iron? There are several such questions that need to be addressed before advocating for nanoengineering of iron as the vital nutrient.

REFERENCES

Abbasi, S., and Azari, S. Efficiency of novel iron microencapsulation techniques: Fortification of milk. *Int J Food Sci Techno.* 2011; 46(9): 1927–1933.

Aggett, P. J. (2012). Iron. *In*: J. W. Erdman, I. A. Macdonald, S. H. Zeisel (eds.), *Present knowledge in nutrition.* 10th ed. Wiley-Blackwell Washington, DC, pp. 506–520.

Altamura, S., and Muckenthaler, M. U. J. Iron toxicity in diseases of aging: Alzheimer's disease, Parkinson's disease and atherosclerosis. *Alzheimers Dis.* 2009; 16(4): 879–895.

Andrews, N. C. Disorders of iron metabolism. *N Engl J Med.* 1999; 341: 1986–1995.

Beck, K. L., Conlon, C. A., Kruger, R., and Coad, J. Dietary determinants of and possible solutions to iron deficiency for young women living in industrialized countries: A review. *Nutrients.* 2014; 6(9): 3747–3776.

Benkhedda, K., L'Abbe, M. R., and Cockell, K. A. Effect of calcium on iron absorption in women with marginal iron status. *Br J Nutr.* 2010; 103: 742–748.

Chiplonkar, S. A., Agte, V. V., Tarwadi, K. V., and Kavadia, R. *In vitro* dialyzability using meal approach as na index for zinc and iron absorption in humans. *Biol Trace Elem Res.* 1999; 67: 249–256.

Clark, S. F. Iron deficiency anemia: Diagnosis and management. *Curr Opin Gastroenterol.* 2009; 25: 122–128.

Crommelin, D. J., Van, R. A. J. M. L., Wauben, M. H. M., and Storm, G. Liposomes in autoimmune diseases: Selected applications in immunotherapy and inflammation detection. *J Control Release.* 1999; 62: 245–251.

Crommelin, D. J. A., Bos, G. W., and Storm, G. Liposomes: Successful carrier systems for targeted delivery of drugs. *Business Briefing: Pharmatech.* 2003; 209–213.

Eaton, J. W., and Qian, M. Molecular bases of cellular iron toxicity. *Free Radic Biol Med.* 2002; 1; 32(9): 833–840.

Fleming, R. E., and Bacon, B. R. Orchestration of iron hemostasis. *N Engl J Med.* 2005; 352: 1741–1744.

Franchini, M., and Veneri, D. Recent advances in hereditary hemochromatosis. *Ann Hematol.* 2005; 84(6): 347–352.

Fuqua, B. K., Vulpe, C. D., and Anderson, G. J. Intestinal iron absorption. *J Trace Elem Med Biol.* 2012; 26 (2–3): 115–119.

Gupta, C., Chawla, P., and Arora, S. Development and evaluation of iron microencapsules for milk fortification. *CyTA-Journal of Food.* 2015; 13(1): 116–123.

Gupta, P. M., Perrine, C. G., Zuguo, M., and Kelley, S. S. Iron, anemia, and iron deficiency anemia among young children in the United States. *Nutrients.* 2016; 8(6): 330.

Hunt, J. R., Zito, C. A., and Johnson, L. K. Body iron excretion by healthy men and women. *Am J Clin Nutr.* 2009; 89: 1792–1798.

Hurrell, R., and Egli, I. Iron bioavailability and dietary reference values. *Am J Clin Nutr.* 2010; 91: 1461S–1467S.

Hurrell, R. F. Phytic acid degradation as a means of improving iron absorption. *Int J Vitam Nutr Res.* 2004; 74: 445–452.

Jaiswal, M., Dudhe, R., and Sharma, P. K. Nanoemulsion: An advanced mode of drug delivery system. *3 Biotech.* 2015; 5(2): 123–127.

Jones, P. Isoelectric focusing for separation of proteins and peptides. Available online https://bitesizebio.com/26811/isoelectric-focusing-separation-proteins-peptides/ (Accessed January 2018).

Kim, S. J., Ahn, J., Seok, J. S., and Kwak, H. S. Microencapsulated iron for drink yogurt fortification. *Asian-Australas J Anim Sci.* 2003; 16(4): 581–587.

Luten, J., Crews, H., Flynn, A., Van Dael, P., Kastenmayer, P., Hurrel, R. *et al.* Interlaboratory trial on the determination of the *in vitro* iron dialysability from food. *J Sci Food Agric.* 1996; 72: 415–424.

Martínez, G. R., Joan, E., and Maria, A. B. Liposomes loaded with hydrophobic iron oxide nanoparticles: Suitable T2 contrast agents for MRI. *Int J Mol Sci.* 2016; 17: 1209.

Miller, D. D., Schricker, B. R., Rasmussen, R. R., and Van Campen, D. An *in vitro* method for estimation of iron availability from meals. *Am J Clin Nutr.* 1981; 34: 2248–2256.

Murray, K. L. E., and Beard, J. (2010). Iron. *In*: P. M. Coates, J. M. Betz, M. R. Blackman *et al.* (Eds.), *Encyclopedia of dietary supplements* (2nd ed.), Informa Healthcare London and New York, pp. 432–438.

Narasinga Rao, B. S. Methods for the determination of biovailability of trace metals: A critical evaluation. *J Food Sci Technol.* 1994; 31: 353–361.

Naveen, S., and Khanum, F. Characterization and evaluation of iron nano-emulsion prepared by high speed homogenization. *IJBPAS.* 2014; 3(1): 45–55.

Nemeth, E. Iron regulation and erythropoiesis. *Curr Opin Hematol.* 2008; 15: 169–175.

Nemeth, E., and Ganz, T. Anemia of inflammation. *Hematol Oncol Clin North Am.* 2014; 28(4): 671–681.

Patel, R. P., and Joshi, J. R. An overview on nanoemulsion: A novel approach. *Int J Pharm Sci Res.* 2012; 3(12): 4640–4650.

Patel, S., Dhuppar, P., and Bhattar, A. Nutritional anemia status in adolescent girls in rural schools of Raipur, India. *Med Chem.* 2017; 7: 853–856.

Rammohan, A., Awofeso, N., and Robitaille, M. C. Addressing female iron-deficiency anaemia in India: Is vegetarianism the major obstacle? *ISRN Public Health.* 2012; 2012: 1–8.

Saito, H., Tomita, A., Ohashi, H., Maeda, H., Hayashi, H., and Naoe, T. Determination of ferritin and hemosiderin iron in patients with normal iron stores and iron overload by serum ferritin kinetics. *Nag J of Medl Sci.* 2012; 74(1–2): 39–49.

Serra, M. L., Pfrimer, K., Doreste, A. J., Ribas, B. L., Sanchez, Villegas, A. *et al.* Dietary assessment methods for intakes of iron, calcium, selenium, zinc and iodine. *Br J Nutr.* 2009; 102 Suppl 1: S38–S55.

Severance, S., and Hamza, I. Trafficking of heme and porphyrins in metazoa. *Chem Rev.* 2009; 109: 4596–4616.

Theil, E. C. Ferritin: The protein nanocage and iron biomineral in health and in disease. *Inorganic Chemistry.* 2013; 52: 1–23.

Thomas, W. L., and Joseph, R. R. (2001). *Remington the science and practice of pharmacy.* 20th ed., Mack Publishing Company, Easton, Pennsylvania. pp. 18042.

Xia, S., and Xu, S. Ferrous sulfate liposomes: Preparation, stability and application in fluid milk. *Food Res Int.* 2005; 38: 289–296.

Xu, Z., Liu, S., Wang, H., Gao, G., Yu, P., and Chang, Y. Encapsulation of iron in liposomes significantly improved the efficiency of iron supplementation in strenuously exercised rats. *Biol Trace Elem Res.* 2014; 162(1–3): 181–188.

Yoshimura, H., Inoue, T., Yamada, T., and Shiraki, K. Anemia during hard physical training (sports anemia) and its causal mechanism with special reference to protein nutrition. *World Rev Nutr Diet.* 1980; 35: 1–864.

Yuan, L., Ji, X., Chen, J., Xie, M., Geng, L., and Gao, R. Enhanced oral bioavailability and tissue distribution of ferric citrate through liposomal encapsulation. *CyTA-Journal of Food.* 2017; 15(1): pp. 136–142.

Yuan, L., Lina, G., Lan, G., Peng, Y., Xianglin, D., Jun, C. *et al.* Effect of iron liposomes on anemia of inflammation. *Int J Pharm.* 2013; 54: 82–89.

Zi, X., Shangyuan, L., Huijie, W., Guofen, G., Peng, Y., and Yanzhong, C. Encapsulation of iron in liposomes significantly improved the efficiency of iron supplementation in strenuously exercised rats. *Biol Trace Elem Res.* 2014; 162: 181–188.

Zimmermann, M. B., and Hurrell, R. F. Nutritional iron deficiency. *The Lancet.* 2007; 17; 370(9586): 511–520.

14 Exploring Nanoformulations of Pomegranate as Promising Nutraceuticals

Surbhi Dhawan, Sanju Nanda, Supriya Verma, Pradip Nirbhavane, and Bhupinder Singh**

CONTENTS

14.1 INTRODUCTION

Natural products have been popularly used owing to their effective prophylactic and curative effects on various ailments, coupled with lower adverse effects, better cost-effectiveness, and safety. One such highly explored natural source is pomegranate, *Punica granatum* L. It belongs to the kingdom Plantae, order Mrytales, and family Lythraceae, with the genus of the plant as *Punica* and species as *P. granatum* (Kumari *et al.*, 2012). The genus name, *Punica*, is a Roman name for Carthage, where the best pomegranates used to be grown (Bhandari, 2012). Pomegranate is known by different names in different languages, like *Anar* in Hindi and Persian, *Hinar* in Kurdish, *Nur* in Armenia, *Shí liú* in China, *Dadimah* in Sanskrit, and *Dalimba* in Marathi (Bhowmik *et al.*, 2013).

It is one of the ancient and highly distinctive fruits, considered as quite sacred by several religions, and even referred to in the Old Testament of the Bible, the Jewish Torah, and the Babylonian Talmud (Jurenka, 2008). Renowned as a symbol of fertility, abundance, and good luck, it was included in the

* Corresponding author.

ceremonies, art, and mythology of the Egyptians and Greeks, and in the personal emblem of the Holy Roman Emperor Maximilian (Shah *et al.*, 2011). Pomegranate is the oldest historic domesticated tree, with the highest rate of cultivation in the Mediterranean regions, including India, Iran, and Pakistan. Pomegranate is native of the areas ranging from the Himalayas in Northern India to Iran. It is also found in arid regions of the Southeast Asia, the East Indies, and the Tropical Africa. The tree is also cultivated for its fruit in the drier regions of California and Arizona (Bhandari, 2012).

The pomegranate tree with many spiny branches, mostly grows up to 12–16 feet and can be extremely long lived. The tree blossoms during the summer when the rainfall is minimal. The leaves are dark green in color and are 3–7 cm long and 2 cm broad. These have a glossy touch and are lance-shaped. The flowers are large, red, white, or variegated, and have a tubular calyx that eventually becomes the fruit. Flowers are generally 3 cm in diameter with 5–8 crumpled petals that remain on the fruit. The ripe pomegranate fruit can be up to five inches wide with a deep red leathery skin. It is grenade-shaped and crowned by the pointed calyx. The leathery skin is yellow overlaid with light, deep pink or rich red. The fruit contains many seeds (i.e., arils), separated by white, membranous pericarp, and each being surrounded by small amounts of tart, red juice. As the tree grows, old bark turns gray (Jurenka, 2008; Jarfeshany *et al.*, 2014; Aher and Rahane, 2016).

Best output is obtained in well-drained ordinary soil, but it can also flourish on calcareous or acidic loamy soil as well as rock-strewn gravel. These are mostly found in semi-arid, mild-temperate to subtropical climates. These are adapted naturally with the regions of cool winters and hot summers. However, fruit formation is adversely affected in the humid climate, and the tree can be severely injured by temperatures below 12°F. Pomegranates can bear considerable drought, but these must be irrigated to get good fruit production. The plants are tolerant to moderately saline water and soil conditions. Mostly, the trees are given 2 to 4 ounce applications of ammonium sulphate or other nitrogen fertilizer during the first two springs. After that, little fertilizer is needed, although the plants respond to an annual mulch of rotted manure or other compost. The fruits get ripened when a distinctive color appears, along with the production of metallic sound, when tapped. The fruits must be picked before over-maturity, when these tend to crack open, particularly during rain (Kumari *et al.*, 2012).

This exotic fruit has been popularized as a functional food and a source of nutraceuticals since the ancient times, and the trend is being continued today. It is recognized as a goldmine of nutrients and antioxidants. All the parts of the tree, including roots, leaves, stem, bark, seeds, fruit, flowers, and rind, as shown in Figure 14.1, are being used for various remedial purposes. Ayurveda treats pomegranate as a "pharmacy unto itself" since it can be used to cure innumerable ailments due to its anti-inflammatory, antioxidant, antitumor, and antiobesity properties (Jurenka, 2008). Presence of nutrients titled the pomegranate as "the energy power house." Around 153 valuable phytoconstituents, mostly nutrients, have been reported to be present in various parts of the tree. Hence, pomegranate has also been termed as a super fruit (Sharma and Maity, 2010).

The pomegranate fruit contains valuable compounds from different parts of the fruit, as described in Table 14.1. These can be further divided into several anatomical origins such as peel, seeds, and arils. Another important product obtained from pomegranate fruit is the juice that can be obtained from arils or from whole fruit.

The chemical composition of the fruits differs, depending on the cultivator, place, climate, maturity, cultivation practice, soil type, and storage conditions. Noteworthy variations in organic acids, phenolic compounds, sugars, water-soluble vitamins, and minerals of pomegranates have been reported over the years by researchers (Viuda-Martos *et al.*, 2010). About 50% of the total fruit weight corresponds to the peel, which is an important source of bioactive compounds such as phenolics, flavonoids, ellagitannins, proanthocyanidin compounds, minerals, mainly potassium, nitrogen, calcium, phosphorous, magnesium, sodium, and complex polysaccharides (Sharma and Maity, 2010; Viuda-Martos *et al.*, 2010). The edible part of the pomegranate fruit, around 50%, consists of 40% arils and 10% seeds. Arils contain 85% water, 10% total sugars, mainly fructose and glucose, and 1.5% of pectin, organic acids such as ascorbic acid, citric acid, and malic acid, and bioactive compounds such as phenolics and flavonoids, principally anthocyanins. The seeds are a

FIGURE 14.1 Pomegranate plant and its important parts and extractives (a) pomegranate arils, (b) pomegranate leaves, (c) pomegranate seed oil, (d) pomegranate peel, (e) pomegranate flower, (f) pomegranate juice.

TABLE 14.1
Various Constituents Found in Different Parts of Pomegranate and Their Indications

Plant Component	Constituents	Indications
Pomegranate juice	Anthocyanins, glucose, organic acid, ascorbic acid, ellagic acid, ellagitannins, gallic acid, caffeic acid, catechin, quercetin, rutin, minerals	Highly antioxidant property
Pomegranate seed oil	Conjugated linolenic acid, linoleic acid, oleic acid, stearic acid, punicic acid, eleostearic acid, catalpic acid	Antioxidant, anti-inflammatory, anticancer, skin rejuvenator
Pomegranate pericarp (peel, rind)	Luteolin, quercetin, kaempferol, gallic acid, ellagic acid glycosides, punicalagin, punicalin, pedunculagin	Antioxidant, anti-inflammatory, carbonic anhydrase inhibitor, treat dermatitis and hepatitis
Pomegranate leaves	Tannins (punicalin and punicafolin); and flavone glycosides, including luteolin and apigenin	Antimicrobial, astringent, antioxidant and anti-inflammatory
Pomegranate flower	Gallic acid, ursolic acid, triterpenoids, including maslinic and asiatic acid, other unidentified constituents	Antioxidant, anticancer, skin care
Pomegranate roots and bark	Ellagitannins including punicalin and punicalagin, numerous piperidine alkaloids	Antioxidant, anticancer, anti-inflammatory

Sources: Jurenka, J., *Altern Med Rev.*, 13(2): 128–144, 2008; Zahin, M. *et al.*, *Biomed Res Int.*, 467465, 2014; Gumienna, M. et al., *Eur Food Res Technol.*, 242(5): 631–640, 2016.

rich source of total lipids, while pomegranate seed oil (PSO) is comprised of 12% to 20% of total seed weight. The oil is characterized by a high content of polyunsaturated (n-3) fatty acids (PUFA) such as linolenic acid, linoleic acid, and other fatty acids such as punicic acid, oleic acid, stearic acid, and palmitic acid. The seeds also contain protein, crude fibers, vitamins, minerals, pectin, sugars, polyphenols, isoflavones (mainly genistein), the phytoestrogen, coumestrol, and the sex steroid, estrone (Viuda-Martos *et al.*, 2010).

Important constituents of pomegranate leaves are tannins such as punicalin and punicafolin and flavones glycosides such as luteolin and apigenin. Flowers are composed of ursolic acid and triterpenoids like maslinic acid and asiatic acid. Pomegranate roots and bark are predominated with piperidine alkaloids and ellagitannins. Research studies suggest that important therapeutically beneficial phytochemicals include ellagitannins (involving punicalagin), punicic acid, flavonoids, anthocyanidins, anthocyanins, and estrogenic flavones (Bhandari, 2012).

It is widely accepted that the beneficial health effects of fruits and vegetables in the prevention of diseases are due to the bionutrients these contain. The presence of significant amounts of bioactive compounds, such as phenolic acids, flavonoids, and tannins, assures considerable nutritional and therapeutic value of pomegranate fruit.

14.1.1 Traditional Uses

Since ancient times, all the parts of the tree have been used in folk medicine of several cultures. Recognized as a vital member of the traditional medicine, some of the traditional uses are listed below (Morariu *et al.*, 2006; Sharma and Maity, 2010; Chakraborty *et al.*, 2012; Kumari *et al.*, 2012):

- Doctors in Greece prescribed pomegranate juice as a remedy for inflammation, intestinal worms, persistent coughs, diarrhea, and dysentery.
- The Persians believed that the seeds conferred strength and invincibility on the battlefield, and in ancient China, the seeds were revered for their powers to promote longevity and immortality.
- People of the Georgian Republic in Russia used pomegranate for arresting chronic mucous discharges, passive hemorrhages, night sweats, and diarrhea.
- The juicy pomegranate fruit with its multitudinous seeds has been popular symbol of fertility and fecundity.
- It has also been prescribed to strengthen the human capillary system and prevent atherosclerosis, asthma, tonsillitis, and bronchitis.
- The medicinal properties of pomegranate roots are explicitly stated in *Charaka Samhita* (the oldest medical work) as a tenifuge.
- The Babylonians regarded pomegranate seeds as an agent of resurrection.
- Pomegranate fruit rind has been used as a tanning material by the Greeks.
- The Indians used pomegranate roots to treat tapeworm infestations.
- In the South Anatolia, Turkey, ashes of the fruit peel are used against skin infections.
- In the Unani medicine system, pomegranate flowers have been used as astringent, hemostatic, antibacterial, antifungal, antiviral, wound remedy, antidiarrheal, man sex power reconstituent, and antidiabetic.

14.1.2 Clinical Applications

Pomegranate is a complete fruit in itself owing to the tremendous health benefits of phytoconstituents present in all the parts of the plant. One of the issues of "Harvard Men's Health Watch" shares new scientific findings, suggesting that pomegranate may one day find an irresistible place in a healthful diet (Pantuck *et al.*, 2009). Some of the clinical and pharmaceutical benefits (Aviram *et al.*, 2008;

Bhandari, 2012; Zarfeshany *et al.*, 2014; Barathikannan *et al.*, 2016; Ramalingam *et al.*, 2016; Al-Megrin *et al.*, 2017; Bedel *et al.*, 2017) have been enumerated as listed here:

- **Anticancer activity:** The bioactive chemicals present in the pomegranates have the potential to prevent as well as treat cancer. Pomegranate juice extracts, pomegranate fruit extracts, PSO, and pomegranate juice have been explored to treat various types of cancers, including hepatocellular carcinoma, colon cancer, breast cancer, leukemia, and prostate cancer. Drinking 8 ounces of pomegranate juice a day has been documented to slow the progression of prostate cancer.
- **Cardio-protective potential:** Phytoconstituents of the pomegranate juice exert beneficial effects on the evolution of clinical vascular complications, coronary heart disease, and atherogenesis in humans. Studies on checking the efficacy of pomegranate flower to diminish cardiac fibrosis are also being reported.
- **Antidiabetic activity:** Pomegranate extract, rich in polyphenols, showed significant antagonizing effects on oxidative stress and lipid peroxidation in patients with Type 2 diabetes mellitus, thus preventing cardiovascular complications. Studies suggest that the anti-diabetic activity of the pomegranate extract may be due to the improved sensitivity of the insulin receptor.
- **Lipid-lowering potential:** Concentrated pomegranate juice and PSO consumption lower the lipid content and thus could mitigate the risk factors of heart diseases in hyperlipidemic patients.
- **Dermatological applications:** Several studies conducted in the past have revealed that pomegranate flower extract acts against UVA- and UVB-induced cell damage and, hence, pomegranate polyphenolics are widely indicated in topical applications. Ellagic acid-rich pomegranate extract, when ingested orally, showed inhibitory effect on pigmentation caused by UV irradiation. PSO is now known as antiaging oil, while pomegranate peel extracts have significant wound healing, antifungal, and antibacterial activities; hence, these can cure many skin infections.
- **Dental applications:** The hydroalcoholic extract from pomegranate fruits is found effective on dental plaque microorganisms. Local delivery of combination(s) of extracts from *Centella asiatica* and pomegranate considerably improved the clinical signs of chronic periodontitis and interleukin-1 beta level in maintenance patients. Moreover, gel containing pomegranate extract may be used as a topical antifungal agent for the treatment of candidosis associated with denture stomatitis.
- **Musculo-skeletal applications:** Dietary intake of pomegranate has been known to ease the symptoms of rheumatoid arthritis patients, attributable mainly to its antioxidant characteristics.
- **Gastro-protective potential:** Studies showed remarkable anti-*H. pylori* activity, useful to treat gastric ulcers. Pomegranate tannins are highly effective in treating gastric ulcer. The fruit enhances the secretion of adherent mucus and free mucus from the stomach wall, thus helping in curing gastric ulcers. This ultimately inhibits generation of reactive oxygen species (ROS), decreases the consumption of glutathione peroxidase and superoxide dismutase, and maintains the level of nitric oxide at a normal level.
- **Hepato-protective activity:** Due to the immense antioxidant property of pomegranate flower extract, it has been successfully explored for hepato-protection.
- **Fertility enhancement potential:** Pomegranate fruit has been known as a symbol of fertility and eternal life. *In vivo* animal studies have revealed that pomegranate juice intake increases the epididymal sperm concentration, sperm motility, spermatogenic cell density, diameter of seminiferous tubules, and germinal cell layer thickness.

14.1.3 Detailed Overview of Pomegranate Skin and Seeds

While the entire pomegranate is a repository of therapeutically beneficial important bionutrients and phytoconstituents, this chapter endeavors to highlight mainly pomegranate seeds and skin.

14.1.3.1 Pomegranate Seeds (Mekni *et al.*, 2014; Doostan *et al.*, 2017)

Considered to be a by-product of the food industry, pomegranate seeds account for about 20% (w/w) of the whole fruit, depending on the cultivator. Recent studies have proved that pomegranate seeds could potentially be a good source of nutrients and antioxidants, along with the presence of bioactive compounds, particularly polyphenols. There is a significant level of phenolic content in pomegranate seeds, in addition to sugars, vitamins, polysaccharides, polyphenols, and minerals. Though the overall lipid content is low, the lipid composition of pomegranate seeds is attracting increased attention, being rich in PUFA. Generally, consumers are conscious of saturated/unsaturated fatty acids ratio in the diet due to their beneficial biological effects and are notably interested in essential fatty acids, with emphasis on the health potential of unsaturated fatty acids. Conjugated linolenic acid is a collective term for the positional and geometric isomers of octadecatrienoic fatty acid. Natural seed oils of certain plants include conjugated trienoic fatty acids that are isomers of α-linolenic acid (9c12c15c-18:3), while PSO primarily contains punicic acid (9c11t13c-18:3). The molecular structures of punicic acid and α-linolenic acid are shown in Figure 14.2.

14.1.3.2 PSO (Ahangari *et al.*, 2012; de Melo *et al.*, 2014; Baccarin *et al.*, 2015)

PSO represents between 12% and 20% of the total seed weight and consists chiefly of conjugated octadecatrienoic fatty acid. Punicic acid, an isomer of this fatty acid, is found in high amounts in the oil and is synthesized *in situ* from linolenic acid, a non-conjugated octadecadienoic fatty acid present in lower amounts (about 7%) in PSO. PSO accounts for over 95% of the total fatty acids, practically all in the form of triglycerides (99%). The triglyceride composition of PSO is varying with the most important patterns being CLnA-CLnA-CLnA and CLnA-CLnAP. Certain minor compounds, such as sterols, steroids, and cerebroside, have also been found in PSO. Phytosterols are present in high concentrations (4.1–6.2 mg/kg) in PSO, mainly as β-sitosterol, campesterol, and stigmasterol, while tocopherols also occur, mainly as α-tocopherol (161–170 mg/100 g) and γ-tocopherol (80–93 mg/100 g).

Seeds were considered waste products, but the seed oil has obtained a lot of recent attention due to the presence of a rich amount of hydrophilic and lipophilic compounds. The seed oil has been used in ethnomedicine due to its pharmaceutical and nutraceutical components. Hence, it is established as a functional ingredient. PSO benefits both the exterior and the interior of the body, as determined by several research studies that it can boost health in many ways. Studies have been reported that those who drank pomegranate juice regularly had a significant increase in their prostate-specific antigen doubling time, leading to its use in the reinforced fight against prostate cancer (Viuda-Martos *et al.*, 2010; Peller *et al.*,2012). It has also shown to help in reducing the intensity and

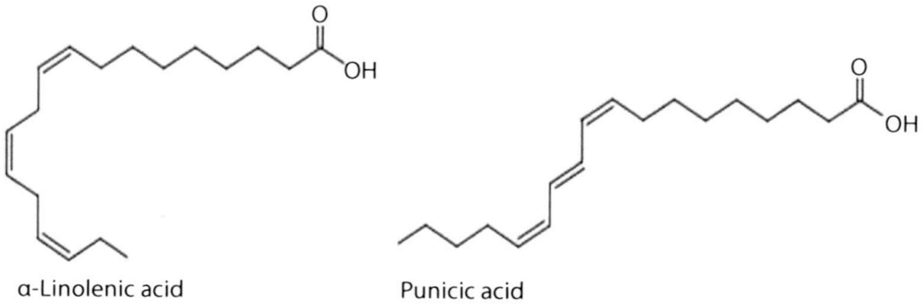

FIGURE 14.2 Molecular structures of α-Linolenic acid and Punicic acid.

incidences of Type 2 diabetes and several heart diseases. It may also help to prevent osteoarthritis by slowing the deterioration of the joints. There are evidences proving its beneficial effect on colon and breast cancer too. It possesses a wide array of biological properties, including antidiabetic, anti-obesity, antiproliferative, and anticarcinogenic against various forms of cancer. In particular, their natural characteristics of skin and hair care are worth taking note of. Studies have demonstrated that the oil deeply penetrates and hydrates the skin by supplying long-lasting moisture. It can revitalize dull and dry skin and hair to restore a more youthful look. The seed's oils are also a rich source of antioxidants, which fight against natural oxidation of skin tissue, combating free radicals with potential to damage our healthy cells. PSO also stimulates the creation of keratinocytes, the large cells seen in our skin's outer layers, producing younger and healthier skin, when these cells are regenerated.

14.1.3.3 Pomegranate Seed Extracts

The ethanolic seed extract shows the presence of triterpenoids and steroids, glycosides, saponins, alkaloids, tannins, carbohydrate, and vitamin C, whereas the aqueous seed extract only contains saponins and carbohydrates. Chloroform seed extract gives positive test for vitamin C only. A list of various bionutrients and phytoconstituents in pomegranate seed is compiled as Table 14.2.

TABLE 14.2
Various Bionutrients and Phytoconstituents in Pomegranate Seed

Class	Bionutrients/Phytoconstituents	Health Effects
Fatty acids	Punicic acid, α-linolenic acid, γ-linolenic acid, stearic acid, palmitic acid, oleic acid, palmitoleic acid, α-eleosteric acid, catalpic acid, caproic acid, caprylic acid, capric acid, lauric acid, myristic acid, gadoleic acid, behenic acid, arachidic acid	Have skin rejuvenating action, antioxidant and anti-inflammatory property, and could prevent and treat cancer and inflammatory diseases
Minerals	Iron, copper, sodium, magnesium, potassium, calcium, zinc, manganese, phosphorus	Boost bone health, relieve arthritis, dental ailments, treat insomnia and various types of cancers, improves immunity and maintain the balance of many hormones and enzymes
Anthocyanins	Cyanidin 3,5-diglucoside, delphinidin 3-glucoside, cyanidin 3-glucoside, Pelargonidin 3-glucoside	Antioxidant, anticarcinogenic, anti-inflammatory, neuroprotective, collagen protectant
Tannins	Punicalin, gallotannic acid, tannic acid	Antioxidant, antitumor, and antimicrobial
Alkaloids	Isopelletierine, pelletierine, methypelletierine, pseudopelletierine	Antitumor and antimicrobial
Polyphenols	Ellagic acid, gallic acid, punicalagin, luteolin	Protect RBCs against oxidative damage and possible disturbance of the lipid bilayer of biomembranes
Phytosterols	Campesterol, stigmasterol, sitosterol, Δ5-avenasterol, citrostadienol	Prevent cardiovascular diseases and cancer, and treat benign prostatic hyperplasia
Squalene and triterpenes	Squalene, cycloartenol, 24-methylene-cycloartenol, betulinol	Anticancer agents protect body from effects of aging, and maintain cholesterol level
Phospholipids	Phosphatidyl ethanolamine	Mitigate coronary heart disease, inflammation, and cancer

Sources: Viladomiu, M. *et al.*, *Evid Based Complement Alternat Med.*, 2013: 1–18, 2013; Verardo, V. *et al.*, *Food Res Int.*, 65(C): 445–452, 2014; Baccarin, T. *et al.*, *J Photochem Photobiol B.*, 153: 127–136, 2015; Barathikannan, K. *et al.*, *BMC Complement Altern Med.*, 16: 264, 2016.

14.1.3.4 Pomegranate Peel (Soni *et al.*, 2010; Bhandary *et al.*, 2012; Al-Rawahi *et al.*, 2014; Verardo *et al.*, 2014; Barathikannan *et al.*, 2016)

Just like pomegranate seeds, pomegranate peel is also considered waste by-product of the food industry. But recent studies on the biologically important phytochemicals present in the peel have indicated the importance of peel. About 50% of the total fruit weight corresponds to the peel, which is an important source of bioactive compounds such as phenolics, flavonoids, ellagic acid and proanthocyanidin compounds, minerals, mainly potassium, nitrogen, calcium, phosphorus, magnesium and sodium, and complex polysaccharides. These compounds are intense in pomegranate peel and juice, accounting to about 92% of the antioxidant activity allied with the fruit. The major class of pomegranate phytochemicals is the polyphenols that are predominant in the fruit and includes flavonoids (flavonols, flavanols, and anthocyanins), condensed tannins (pro-anthocyanidins), and hydrolysable tannins (ellagitannins and gallotannins). These tannins, being highly susceptible to both enzymatic and non-enzymatic hydrolysis, result in hydrolytic products such as glucose and ellagic acid or gallic acid. Pomegranate peel also includes organic and phenolic acids, sterols and triterpenoids, and alkaloids. Nevertheless, the health benefits of pomegranate peel are accredited with the pharmacological activities exhibited by bioactive phytochemicals like polyphenols. These functional compounds in pomegranate peels could be utilized by the food industry and pharma/nutraceutical industry. A list of diverse phytochemicals present in different pomegranate peels, that is, pericarp extracts, is reproduced in Table 14.3.

The peel of fruit is well-regarded for its astringent properties, as well as for its impressive antioxidant and antimicrobial potential. The aqueous peel extract of pomegranate is a potential source of antibacterial, antifungal, and antioxidant agents and could be used as a natural antioxidant and preservative in food and nonfood systems. It also exhibits good reducing power and iron-chelation capacity. Pomegranate peel extract possesses a strong potential for development as an anti-carcinogenic, antioxidant, and anti-inflammatory agent. Pomegranate peel extract could be used in preventing the incidence of long-term complication of diabetes (Ahmed *et al.*, 2014; Barathikannan *et al.*, 2016; Saffarzadeh-Matin *et al.*, 2017). Pomegranate peel extract, and to a lesser extent its fermented juice and seed cake extracts, stimulated type I procollagen synthesis and inhibited matrix metalloproteinase-1 (interstitial collagenase) production by dermal fibroblasts, but had no growth-supporting effect on keratinocytes. These results suggest heuristic potential of pomegranate fractions for facilitating

TABLE 14.3
List of Different Phytoconstituents in Pomegranate Peel Extracts

S. No.	Extraction Menstrum	Binutrients/Phyto-Constituents	Reference(s)
1.	Hexane	Coumaric acid	Eikani *et al.*, 2012; Khan *et al.*, 2017
2.	Ethanol	Phenolic constituents like punicalin, granatin A and B, lagerstanin, catechin, penigluconin, ellagic acid and ellagic acid derivatives	Wafa *et al.*, 2017
3.	Methanol	Phenolic constituents like gallic acid, catechin, epicatechin, chlorogenic acid andcaffeic acid	Abbasi *et al.*, 2008; Young *et al.*, 2017
4.	Ethyl acetate	Phenolics, flavonoids, γ-tocopherol, linoleic acid and esters	Barathikannan *et al.*, 2016
5.	Hydroalcoholic	Phenolic constituents like gallic acid, punicalagin and flavonoids like quercetin	Morzelle *et al.*, 2016
6.	Aqueous	Tannins, saponins, quinones, terpenoids, steroids, flavonoids, phenols, alkaloids, cardiac glycosides, coumarins and betacyanin	Sangeetha and Jayprakash, 2015

skin repair in a polar manner, with the aqueous extracts (especially of pomegranate peel) promoting regeneration of dermis.

The food rind powder has appreciable immune stimulatory activity and hepato-protective activity. It has also been traditionally used as anthelmintic, AntiTracheo-bronchitis, and for healing wounds, ulcers, bruises, stomatitis, vaginitis, and excessive bleeding. The fruit rind of pomegranate shows significant reduction in the intensity of diarrhea. The juice can be drunk and used as a mouth wash, douche, or enema.

In recent years, more medicinal values of pomegranate peel have been investigated such as abortifacient, analgesic, antiameobic, anticonvulsant, antifungal, antimalarial, antimutagenic, antiviral, antispasmodic, diuretic, and hypothermic.

14.2 NANOFORMULATIONS FOR IMPROVED EFFICIENCY

14.2.1 Need for Novel Drug Delivery Systems for Pomegranate Extracts and Oil

Pomegranate extracts and oil pose formulation challenges due to the presence of some active components that are highly prone to oxidation and instability. Other problems that are being encountered by the formulators involve the following:

- Efficient chloroform, petrol, acetone and methanolic extracts are available, which may not be appropriate for delivery per se.
- These are generally bulk drugs, thus calling for dose reduction.
- Currently, marketed formulations lack targeted drug delivery characteristics against various chronic diseases.
- Several side effects are associated with currently marketed formulations.
- Patient noncompliance is high due to large doses and less effectiveness with the conventionally available formulations.
- Most of the herbal origin drugs possess insoluble components, leading eventually to the lower bioavailability and increased systemic clearance, and requiring repeated administrations or higher dose(s).

To exploit the therapeutic benefits of the pomegranate constituents, a scientific approach needs to be applied. It has become necessary to deliver the components in a sustained manner to improve patient compliance and avoid repeated administration. This goal can be attained by designing novel drug delivery systems for herbal constituents. These nanoformulations would help to enhance patient compliance by reducing the number of doses, improve therapeutic value by reducing possible toxicity, and increase the bioavailability. In this context, nanostructured systems of herbal nutrients have shown immense potential, hitherto not much explored, for enhancing the activity and overcoming the problems associated with plant medicines. Hence, including novel nanocarriers in the traditional medicine system can be more efficacious to treat more chronic diseases such as asthma, diabetes, and cancer.

The formulators, while developing nanocoutured dosage forms like polymeric nanoparticles, nanospheres and nanocapsules, liposomes, proliposomes, solid lipid nanoparticles (SLNs) and nanoemulsions, were found to demonstrate stellar advantages for pomegranate constituents, as enumerated in Box 14.1.

Considering the immense importance of the phytoconstituents present in the different parts of the pomegranate, many of their formulations have been prepared such as emulsions, microencapsulation, and microemulsions, as depicted in Figure 14.3. To further increase the efficiency of these formulations, nanoformulations of PSO, pomegranate peel extract, pomegranate fruit extract, pomegranate juice have been reported such as nanoemulsions, nanoencapsulated particles, and nano lipidic carriers. A broad overview of the nanoformulations is presented below.

BOX 14.1 SALIENT MERITORIOUS FEATURES OF NANOSTRUCTURED FORMULATIONS OF POMEGRANATE EXTRACTS AND OIL

- These help to achieve targeted delivery of the bioactive(s).
- These tend to ameliorate the therapeutic activity of the bioactive.
- Their minuscule particle size and increased surface area ultimately improve dissolution of the drug in the gastrointestinal milieu and blood circulation.
- The concentration of bionutrients seems to persist at the sites for longer periods.
- The enhanced permeation and retention effect, that is, increased permeation through the barriers, result because of small size, while extended retention due to poor lymphatic drainage such as in tumor.
- These exhibit passive targeting to the disease site of action without the addition of any particular ligand moiety.
- These protect against plausible toxicity.
- These provide an enhanced stability and improved pharmacological activity.
- Their high levels provide protection from potential toxicity(ies).
- These provide protection against physical and chemical degradation.
- These protect against physical and chemical degradation of constituents.

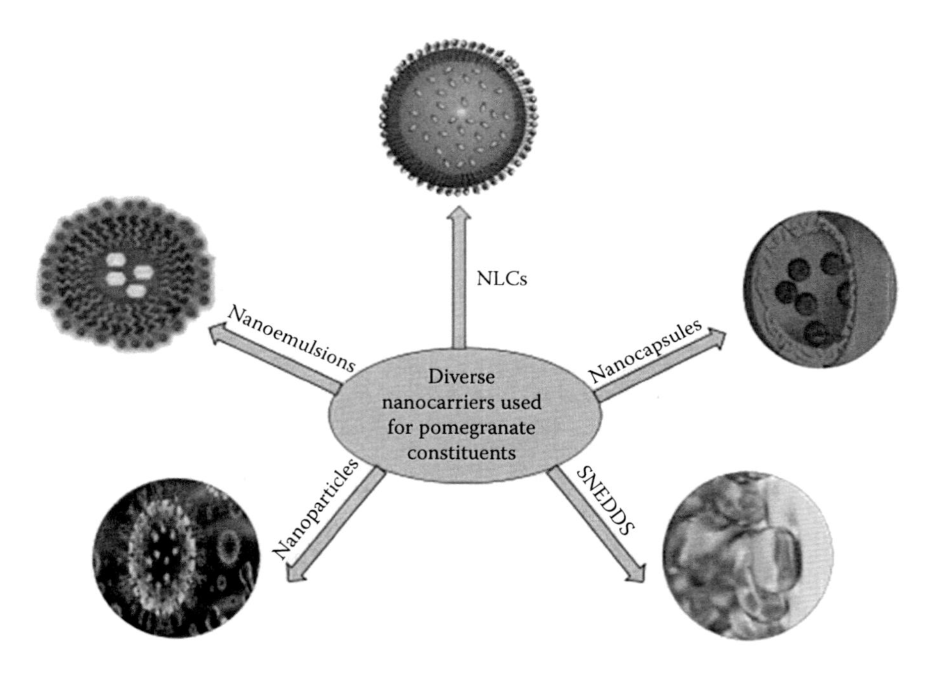

FIGURE 14.3 Different nanocarrier systems for pomegranate constituents.

14.2.2 Lipid-Based Nanostructured Drug Carriers

Important lipid-based nanocarriers systems include nanoemulsions, SNEDDS, and NLCs, as depicted in Figure 14.3. A further elaboration on these systems is described below.

14.2.2.1 Nanoemulsions

Nanoemulsion has been considered one of the most appropriate formulations for PSO. Many formulators have used pomegranate peel extract, in combination with PSO, to formulate nanoemulsions

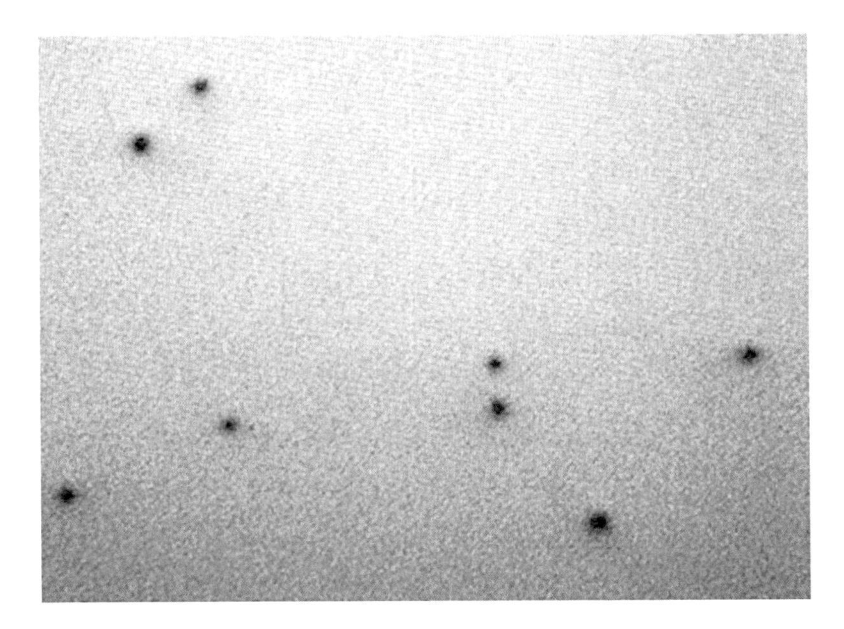

FIGURE 14.4 TEM of pomegranate seed oil nanoemulsions at 3000× resolution (Patent application no. 201711030008A).

to achieve the synergistic effect(s). The nanoemulsions are thermodynamically stable, transparent or translucent dispersions of oil and water, and stabilized using surfactants and co-surfactants. Oil phase can be modified depending on the properties required by the formulator, for example, using oily phase that improves penetration into the skin tissue for better effects. Nanoemulsions are also considered drug carriers to enhance the delivery of therapeutic agents. These are the recently explored nanostructured systems for systemic delivery of the active pharmaceuticals for immediate or controlled drug delivery, and/or targeting. These are known to show promising results in the field of cosmetics, diagnostics, drug therapies, and biotechnologies (Savardekar and Bajaj, 2016) (Figure 14.4). A nanoemulsion system has been developed in our laboratory of pomegranate seed oil using ultrasonication for topical delivery. Photoprotective potential of the developed formulations was established by determining anti-inflammatory, antioxidant, anti-elastase, antihyaluronidase, total phenolic content, and anti-collagenase activities. Low IC_{50} values for anti-elastase, anti-collagenase, and anti-hyaluronidase as 309 mg/ml, 4 mg/ml, and 95 mg/mL, respectively, corroborate nanoemulsions as a promising carrier for topical delivery of the oil (Dhawan and Nanda, 2018).

14.2.2.2 NLCs

Considering the challenges faced during herbal and nutraceutical drug formulations, the formulators have opted for NLCs too to surmount various difficulties. This delivery system uses physiological, biodegradable, and biocompatible lipid materials and surfactants. It has received acceptance from regulatory authorities to make use in different drug delivery systems (Iqbal *et al.*, 2012). Hence, the studies connoted that not only PSO can be widely used to formulate NLCs, with advantages like improved drug loading capacity, prevention of drug expulsion, and flexibility in modulating drug release, but also signified the versatility of NLCs, which be exploited for various routes of administration.

14.2.2.3 SNEDDS

SNEDDS are anhydrous pre-concentrate mixtures of lipids, emulsifiers, and cosolvents or co-emulsifiers. By virtue of their composition and nanometric size range, the SNEDDS are known to enhance the oral biopharmaceutical potential of various drugs and bioactives by facilitating their dissolution in gastrointestinal media, inhibiting metabolism by CYP450 enzymes in gut enterocytes, bypassing the efflux of P-gp glycoproteins, and above all, circumnavigating hepatic first-pass metabolism by potentiating drug absorption through transportation via intestinal lymphatic pathway. Lately, the enormous potential of diverse lipid-based nanoformulations has been intensively exploited to improve oral bioavailability of numerous phytoactives and nutraceuticals (Singh *et al.*, 2014). Table 14.4 presents a holistic and updated account of diverse lipidic nanoformulations of pomegranate seed oil and pomegranate extract.

14.2.3 Nanoparticles

Gold, silver, and copper have been widely used for the formulation of stable dispersions of nanoparticles of pomegranate peel extract, leaf extract, and PSO, useful in the areas of medical application such as cancer diagnosis and therapy (Victor and Roberto, 2015). These metallic nanoparticles have distinctive add-on advantage, as they improve the physicochemical properties of the nanoparticles. The noble metal nanoparticles have unique optical and physical properties that make them useful for various purposes like diagnosis, drug delivery, photo-thermal ablation, medical imaging, as biomarkers and antimicrobial coatings. Apart from these, the nanoparticles tend to improve the efficiency of the active components too. Overall, these metals are accepted because of the following properties:

- No adverse immunological response by the body
- No recorded toxicity
- Inertness and stability

The main objective is to formulate nanoparticles with controlled particle size, surface properties, and release of bioactive agents, ultimately helping to achieve their site-specific action at therapeutically optimal rate and dose regimen (Mohenraj and Chen, 2006).

14.2.4 Nanocapsules

Nanoscale shells, prepared from biocompatible polymeric material(s), are fundamentally the nanocapsules (Kothamasu *et al.*, 2012). These are nano-vesicular systems that exhibit typical core-shell structure of a polymeric membrane that encapsulates an inner liquid core at the nanoscale. The core can contain the active material in the form of a liquid, solid, or molecular dispersion. In addition, depending on the preparation method and materials used in the formulation, the core can be adjusted as lipophilic or hydrophobic (Mora-Huertas *et al.*, 2010). Table 14.5 outlines an updated account of literature of various polymeric nanocarriers of pomegranate constituents.

14.3 PATENT REVIEW

Besides the enormous published literature on nanonutraceutical of pomegranate, there are a few patent reports available too; a brief account is enumerated here.

Mousa *et al.* (2011) invented a formulation and method for treating osteoporosis and bone fracture in animals. A composition was proposed for formulating a nanoformulation using a combination of natural products and claimed to be useful to treat osteoporosis and other bone-related disorders. The active ingredients included *Lepidium* extracts, calcium, and vitamin D, along with antioxidants like pomegranate extract, flavonoids, or their combinations. Different methods were employed to prepare nanoformulations, including encapsulation of natural products using ionic gelation method and a blend of alginate-chitosan polymers. In the second case, hybrid cross-linked polyvinylpyrrolidone hydrogel nanoparticles were prepared by polymerization reactions conducted in a reverse micellar environment, employing sodium bisethylhexylsulpho succinate, as the surfactant for micelle formation, under the atmosphere of nitrogen gas. In the third case, chitosan-grafted

TABLE 14.4
An Updated Bird's Eye View of Various Lipid-Based Nanostructured Carriers of Pomegranate Seed Oil and Pomegranate Extract

S. No.	Formulation Components	Purpose	Outcomes	References
A)		Nanoemulsions		
1.	Nanoemulsions of ethyl acetate fraction of pomegranate peel extract using PSO or medium chain triglyceride oil	To study *in vitro* skin permeation and retention of the major pomegranate peel polyphenols	After 8 h of skin permeation, all polyphenolic compounds were mostly retained in the skin. Retention of gallic acid in the stratum corneum showed promising results.	Baccarin and Lemos-Senna, 2017
2.	The combination of ketoprofen and PSO (1.5% and 3%) with Span 80 and Tween 80 as surfactants	To improve the anti-nociceptive effect	Physicochemical characteristics of the nanoemulsions and photo-stability of ketoprofen was improved. Ketoprofen pharmacological potential was enhanced.	Ferreira *et al.*, 2016a
3.	Nanoemulsions with PSO concentration as 1.5 and 3%	To evaluate the cytotoxicity of PSO NEs on human blood cells and antiglioma effects against C6 cells	No genotoxic and oxidative damage was observed in lipid peroxidation, protein carbonylation and catalase activity evaluations for nanoemulsions. Besides, *in vitro* antitumor activity indicated that nanoemulsions were promising for glioma treatment.	Ferreira *et al.*, 2016b
4.	PSO, long-chain triglyceride and medium-chain triglyceride	To compare the use of long-chain triglyceride and medium-chain triglyceride in the nanoemulsions of PSO dosage forms	Physical stability of PSO nanoemulsions was better with long-chain triglyceride as oil phase than with medium-chain triglyceride as oil phase.	Yuliani and Hartini, 2016
5.	O/W nanoemulsions containing chitosan was prepared. A blend of surfactants, like polyoxyethylene sorbitan monooleate, and organic sunscreens like benzophenone-3, diethylamino hydroxybenzoyl were selected	To analyze the photoprotective effect of chitosan	The developed photoprotective and antioxidant nanoemulsion containing chitosan was stable for at least 6 months. Additionally, chitosan acted by promoting retention of the formulation in epidermis.	Cerqueira-Coutinho *et al.*, 2015
6.	Polyphenol-rich ethyl acetate fraction with soy lecithin and Tween 80	To analyze the photoprotection provided by PSO nanoemulsion entrapping the polyphenol-rich ethyl acetate fraction against UVB-induced DNA damage in the keratinocyte HaCaT cell line	Nanoemulsions protected the cell DNA against UVB-induced damage with no phototoxic effect. The sun protection factor determined for ethyl acetate fraction-loaded nanoemulsion was relatively high.	Baccarin *et al.*, 2015

(*Continued*)

TABLE 14.4 (CONTINUED)
An Updated Bird's Eye View of Various Lipid-Based Nanostructured Carriers of Pomegranate Seed Oil and Pomegranate Extract

S. No.	Formulation components	Purpose	Outcomes	References
7.	Nanoemulsions - PSO along with Tween 80 and glycerol monooleate were used as surfactants, ultrasonication technique was used	To evaluate the potential of PSO nanoemulsions in multiple sclerosis	PSO nanoemulsion showed reduction in dose burden at much lower dose as compared to natural PSO alone. Thus, novel formulations of natural antioxidants such as nano-PSO may be considered for the treatment of patients of demyelinating diseases.	Binyamin *et al.*, 2015
8.	Nanoemulsions formed with PSO and ketoprofen using pullulan as a polymeric stabilizer	To evaluate *in vitro* antitumor activity of PSO nanoemulsions against glioma cell	Nanoemulsions were able to delay the photodegradation profile of ketoprofen under UVC radiation, regardless of the concentration of pullulan. Nanoemulsion showed significant antitumor activity.	Ferreira *et al.*, 2015
9.	PSO with Tween 80, glycerol monooleate and glycerol as surfactants, using ultrasonication technique	To evaluate nanoemulsions for the prevention and treatment of neurodegenerative diseases	Nano-PSO significantly delayed disease presentation when administered to asymptomatic TgMHu2ME199K mice and postponed disease aggravation in already sick mice.	Mizrahi *et al.*, 2014
B)		**NLCs**		
1.	Prepared NLCs using oils (PSO, wheat germ oil, etc., individually, and in combination	To formulate NLCs for sun protection along with a photoprotective agent that absorbs the UVA radiations, i.e., diethylamino hydroxybenzoyl hexyl benzoate	The NLCs especially NLCs of PSO showed significant sun protection activity, which were incorporated into a cream base. Out of all the formulations, PSO-based allopathy cream with the combination of wheat germ oil showed best results.	Badea *et al.*, 2014

(Continued)

TABLE 14.4 (CONTINUED)
An Updated Bird's Eye View of Various Lipid-Based Nanostructured Carriers of Pomegranate Seed Oil and Pomegranate Extract

S. No.	Formulation components	Purpose	Outcomes	References
2.	NLCs for topical purposes	To formulate NLCs to enhance the penetration of ellagic acid-rich pomegranate peel extract and to determine the anti-tyrosinase activity	The results showed strong anti-tyrosinase activity of ellagic acid-rich pomegranate peel extract, along with prolonged release of ellagic acid from NLCs up to 12 h. Ellagic acid-rich pomegranate peel extract is known to improve bioavailability of active ingredients.	Tokton *et al.*, 2014
C)		**SNEDDS**		
1.	Prepared SNEDDS of ellagic acid with polyethylene glycol, polysorbate, caprylic/capric triacylglycerol at the ratio of 45/45/10 wt.%	To formulate food-grade self-nanoemulsifying system to improve the dissolution and absorption of ellagic acid	Ellagic acid SNEDDS formulation reported 6-fold improvement in bioavailability with almost 10-fold increase in C_{max}.	Wang *et al.*, 2017
2.	Prepared SEDDS using PSO and resveratrol	To formulate SEDDS for the synergistic effect of PSO with resveratrol, PSO was employed as oil phase for the first time to develop SEDDS and to yield better therapeutic outcomes	Water solubility of RES was enhanced 20 times and stability of resveratrol in intestinal fluid was significantly improved. The antiinflammatory activity and cytotoxicity against MCF-7 cancer cells indicated much higher potency and safety of resveratrol SNEDDS-PSO than those of resveratrol SNEDDS-isopropyl palmitate.	Lu *et al.*, 2015

TABLE 14.5
An Updated Overview of Polymeric Nanoparticulate Systems of Various Pomegranate Constituents

S. No.	Formulation Components	Purpose	Outcome	References
A)		Nanoparticles		
1.	Synthesized silver nanoparticles of pomegranate peel	To develop eco-friendly method of synthesis of silver nanoparticles	The synthesized silver nanoparticles showed enhanced inhibitory effects on aflatoxin B1 production in *A. flavus*	Al-Othman *et al.*, 2017
2.	Synthesized copper nanoparticles of pomegranate peel extract, where copper is served as a reducing as well as capping agent	To find an eco-friendly alternative to chemical and physical methods for biomedical applications and research	The *in vitro* studies proved superior antibacterial activities of Cu nanoparticles against opportunistic pathogens, like *Micrococcus luteus* and *Pseudomonas aeruginosa*, and also physicochemical stability of copper nanoparticles	Kaur *et al.*, 2016
3.	Synthesized silver nanoparticles of an aqueous extract of pomegranate leaves, which served as a reducing and capping agent	To evaluate the antibacterial activity against *Pseudomonas spp.*, *Bacillus cereus*, *Staphylococcus albus*, and *Proteus pathogens*	The particles showed significant antibacterial effects. Hence, these could be used for various medical and industrial applications	Nisha *et al.*, 2015
4.	Fabrication of silver nanoparticles in PSO using laser ablation technique	To enhance the thermal properties of PSO nanoparticles	The photo-acoustic technique indicated that the thermal effusivity of the silver nanoparticles of PSO got distinctively enhanced vis-à-vis PSO alone	Sadrolhosseini *et al.*, 2015
5.	Gold nanoparticles in PSO using laser synthesized ablation	To improve the dermatological activity of PSO using gold nanoparticles	Laser irradiation did not show any destructive effects on PSO, indicating that laser ablation can be considered as a green method for synthesis of Au-NPs in PSO	Sadrolhosseini *et al.*, 2014

(*Continued*)

TABLE 14.5 (CONTINUED)
An Updated Overview of Polymeric Nanoparticulate Systems of Various Pomegranate Constituents

S. No.	Formulation Components	Purpose	Outcome	References
6.	Silver nanoparticles of pomegranate peel extract was used as reducing and capping agent	To design simple and eco-friendly method to prepare nanoparticles exhibiting antibacterial properties	These silver nanoparticles showed maximum antibacterial activity against *Staphylococcus aureus*: using this economically viable and eco-friendly method	Shanmugavadivu *et al.*, 2014
7.	Prepared green nanoparticles using pomegranate juice and peel extract	To find an eco-friendly alternative to chemical and physical methods for biomedical applications and research	Characterization using TEM and UV spectroscopy indicated the effectiveness of the method	Chauhan *et al.*, 2011
B)		**Nanocapsules**		
1.	Preparation and evaluation of hydrogel containing silibilin-loaded pomegranate oil based nanocapsules suspension	As an alternative for the treatment of cutaneous UVB radiation-induced damages	It proved satisfactory characteristics for cutaneous application, demonstrating prolonged antiedematogenic effects with hydrogel prepared with free isolated compounds	Marchiori *et al.*, 2017a
2.	Develop nanocapsule suspensions with pomegranate oil as oil core for silibinin	To encapsulate and assess their toxicity *in vitro* and radical scavenging activity	Nanocapsules-controlled silibinin release at least 10 times vis-à-vis free silibilin in methanolic solution. *In vitro* scavenging capacity was statistically higher than from free silibilin and the silibilin nanocapsules	Marchiori *et al.*, 2017b

poly (lactic-co-glycolic acid) nanoparticles were prepared by double emulsion-diffusion evaporation technique employing sonication.

Morariu (2006) invented a composition for improving the appearance of aged skin, characterized by wrinkles, loss of elasticity, and hyperpigmentation caused by chronoaging and/or photoaging of skin. The main mechanisms involved to resolve this issue include inhibiting particularly skin damage resulting from reactive carbonyl species, glycation of skin proteins, formation of advanced glycation end products, and formation of advanced lipoxidation end products. Formulation was prepared that included benfotiamine, pyridoxamine, and a dermatologically acceptable carrier (e.g., as a lotion, cream, ointment, soap, or the like) so as to facilitate topical application, with added advantages of antioxidants like pomegranate extract, tocopherols, tocotrienols, astaxanthin, carnosine, macqui berry anthocyanins, billberry anthocyanins, blueberry anthocyanins, ellagic acid, silymarin, quercetin, naringenin, rosmarinic acid, and clove essential oil. The composition was formulated into liposomes or nanosomes that were found to be useful in targeting DNA repair associated with photoaging. Several techniques, including coacervation, fat-coating, and interface polymerization or inclusion complexes with cyclodextrins, have been employed. Nanoemulsions were also prepared that provided the benefit of reduced surfactant requirement, ultimately preventing intolerance and stickiness, when applied to skin.

Orza (2017) invented antiaging nanoformulations, composed of mesoporous silica nanoparticles, plant extracts like pomegranate oil, fennel oil, rosemary oil, chamomile oil, and rosehip oil, and biologically active agents like aspartic acid, vitamins, and others. The preparation of nanocarriers involved a new method of encapsulating the active ingredients within a multilayer nanocarrier. The invention proposed the multi-encapsulation of the ingredients in a core of silica biocompatible shell, followed by the co-encapsulation of other active ingredients using a polysaccharide, protein, or a biodegradable polymer. These nanocarriers were incorporated into various topical systems such as creams, serums, eye cream, and mask called as "nano-layer-by-layer" delivery system. This novel delivery system could cross the skin barrier and show cell growth, remove wrinkles and rejuvenate the skin, and achieve controlled release at the desired site.

Recently, Dhawan *et al.* (2017) reported novel nanoemulsion-based formulation of PSO for topical delivery employing QbD-driven approach. The developed nanoemulsion was subsequently incorporated into a gel base and evaluated for its photo-protective potential. A patent for the same invention has been filed with Indian Patent Office (Application no. 201711030008 A).

14.4 CONCLUSION

Considering the wonder of the pomegranate fruit, it has been widely explored for diverse activities like antioxidant, antiaging, neuroprotective, and anticancer. However, several challenges regarding their phytoconstituents still need to be addressed, including their stability, solubility, and bioavailability. Various nanostructured systems have proved to be much more efficient carrier systems of PSO and pomegranate seed extract to successfully overcome such unfavorable challenges. Pomegranate extracts and oil are extensively being formulated into nanoformulations for their corresponding health and therapeutic benefits. The authors, in this chapter, have discussed several formulations such as nanoemulsions, nanoparticles, SNEDDS, nanocapsules, and NLCs of pomegranate peel extract and PSO. Other parts of the pomegranate are also being explored for their therapeutic advantages, which can also be further formulated into nano-based compositions to improve their therapeutic worth.

REFERENCES

Abbasi, H., Rezaei, K., and Rashidi, L. Extraction of essential oils from the seeds of pomegranate using organic solvents and supercritical CO_2. *J Am Oil Chem Soc.* 2008; 85: 83–89.

Ahangari, B., and Sargolzaei, J. Extraction of pomegranate seed oil using subcritical propane and supercritical carbon dioxide. *Theor Found Chem En.* 2012; 46(3): 258–265.

Aher, Y. L., and Rahane, S. A review of pomegranate cultivation. *Int Res J Eng Tech.* 2016; 3(4): 2083–2088.

Ahmed, A. T. J., Belal, S. K. M., and Salem, A. G. E. Protective effect of pomegranate peel extract against diabetic-induced renal histo-pathological changes in albino rats. *IOSR J Dental Med Sci.* 2014; 13(10): 94–105.

Al-Megrin, W. A. *In vivo* study of pomegranate (*Punica granatum*) peel extract efficacy against *Giardia lamblia* in infected experimental mice. *Asian Pac J Trop Biomed.* 2017; 7(1): 59–63.

AL-Othman, M. R., EL-Aziz, A. R. M., Mahmoud, M. A., and Hatamleh, A. A. Green biosynthesis of silver nanoparticles using pomegranate peel and inhibitory effects of the nanoparticles on aflatoxin production. *Pak J Bot.* 2017; 49(2): 751–756.

Al-Rawahi, A. S., Edwards, G., Al-Sibani, M., Al-Thani, G., Al-Harrasi, A. S., and Rahman, M. S. Phenolic constituents of pomegranate peels (*Punica granatum* L.) cultivated in oman. *European J Med Plants.* 2014; 4(3): 315–331.

Aviram, M., Volkova, N., Coleman, R., Dreher, M., Reddy, M. K., Ferreira, D. *et al.* Pomegranate phenolics from the peels, arils, and flowers are antiatherogenic: Studies *in vivo* in atherosclerotic apolipoprotein E-deficient mice and *in vitro* cultured macrophages and lipoproteins. *J Agric Food Chem.* 2008; 56(3): 1148–1157.

Baccarin, T., and Lemos-Senna, E. Potential application of nanoemulsions for skin delivery of pomegranate peel polyphenols. *AAPS PharmSciTech.* 2017; 18(8): 3307–3314.

Baccarin, T., Mitjans, M., Ramos, D., Lemos-Senna, E., and Vinardell, M. P. Photoprotection by *Punica granatum* seed oil nanoemulsion entrapping polyphenol-rich ethyl acetate fraction against UVB-induced DNA damage in human keratinocyte (HaCaT) cell line. *J Photochem Photobiol B.* 2015; 153: 127–136.

Barathikannan, K., Venkatadri, B., Khusro, A., Al-Dhabi, N. A., Agastian, P., Arasu, M. V. *et al.* Chemical analysis of *Punica granatum* fruit peel and it's *in vitro* and *in vivo* biological properties. *BMC Complement Altern Med.* 2016; 16: 264.

Bedel, H. A., Turgut, N. T., Kurtoğlu, A. U., and Usta, C. Effects of the nutraceutical, punicic acid. *Indian J Pharm Sci.* 2017; 79(3): 328–334.

Bhandari, P. R. Pomegranate (*Punica granatum* L). Ancient seeds for modern cure? Review of potential therapeutic applications. *Int J Nutr Pharmacol Neurol Dis.* 2012; 2(3): 171–184.

Bhandary, S. K., Kumari, S. N., Bhat, V. S., Sharmila, K. P., and Bekal, M. P. Preliminary phytochemical screening of various extracts of *punica granatum* peel, whole fruit and seeds. *Nitte Uni J Health Sci.* 2012; 2(4): 34–38.

Bhowmik, D., Gopinath, H., Kumar, B. P., Duraivel, S., Aravind, G., and Kumar, K. P. S. Medicinal uses of *Punica granatum* and its health benefits. *J Pharmacogn Phytochem.* 2013; 1(5): 28–35.

Binyamin, O., Larush, L., Frid, K., Keller, G., Friedman-Levi, Y., Ovadia, H. *et al.* Treatment of a multiple sclerosis animal model by a novel nanodrop formulation of a natural antioxidant. *Int J Nanomedicine.* 2015; 20(10): 7165–7174.

Cerqueira-Coutinho, C., Santos-Oliveira, R., dos Santos, E., and Mansur, C. R. Development of a photoprotective and antioxidant nanoemulsion containing chitosan as an agent for improving skin retention. *Eng Life Sci.* 2015; 15(6): 593–604.

Chakraborty, M., Dipak, G., Axay, P., and Jagdish, K. Phytochemical and pharmacological profile of *Punica Granatum*: An overview. *Int Res J Pharm.* 2012; 3(2): 65–68.

Chauhan, S., Upadhyay, M. K., Rishi, N., and Rishi, S. Phytofabrication of silver nanoparticles using pomegranate fruit seeds. *Int J Nanomater Biostruct.* 2011; 1(2): 17–21.

de Melo, I. L. P., de Carvalho, E. B. T., and Mancini-Filho, J. Pomegranate seed oil (*Punica Granatum* L.): A source of punicic acid (Conjugated α-Linolenic Acid). *J Hum Nutr Food Sci.* 2014; 2(1): 1024.

Dhawan, S., and Nanda, S. A nanoformulation of pomegranate seed oil and uses thereof. Application no. 201711030008 A. The Patent Office Journal No. 36/2017 dated 08/09/2017.

Dhawan, S., and Nanda, S. *In vitro* estimation of photo-protective potential of pomegranate seed oil and development of a nanoformulation. *Curr Nutr Food Sci.* 2018; 14: 1–16.

Eikani, M. H., Golmohammad, F., and Homami, S. S. Extraction of pomegranate (*Punica granatum* L.) seed oil using superheated hexane. *Food Bioprod Process.* 2012; 90(1): 32–36.

Ferreira, L. M., Cervi, V. F., Gehrcke, M., Elita, F., Azambuja, J. H., Braganhol, E. *et al.* Ketoprofen-loaded pomegranate seed oil nanoemulsion stabilized by pullulan: Selective antiglioma formulation for intravenous administration. *Colloids Surf B Biointerfaces.* 2015; 130: 272–277.

Ferreira, L. M., Sari, M. H. M., Cervi, V. F., Gehrcke, M., Barbieri, A. V., and Zborowski, V. A. Pomegranate seed oil nanoemulsions improve the photostability and *in vivo* antinociceptive effect of a non-steroidal anti-inflammatory drug. *Colloids Surf B Biointerfaces.* 2016a; 144: 214–221.

Ferreira, L. M., Gehrcke, M., Cervi, V. F., Eliete, P., Bitencourt, R., and Ferreira, E. Pomegranate seed oil nanoemulsions with selective antiglioma activity: Optimization and evaluation of cytotoxicity, genotoxicity and oxidative effects on mononuclear cells. *Pharm Biol.* 2016b; 54(12): 2968–2977.

Gumienna, M., Szwengiel, A., and Górna, B. Bioactive components of pomegranate fruit and their transformation by fermentation processes. *Eur Food Res Technol.* 2016; 242(5): 631–640.

Iqbal, M. A., Md, S., Sahni, J. K., Baboota, S., Dang, S., and Ali, J. Nanostructured lipid carriers system: Recent advances in drug delivery. *J Drug Target.* 2012; 20(10): 813–830.

Jurenka, J. Therapeutic applications of pomegranate (*Punica granatum* L.): A review. *Altern Med Rev.* 2008: 13(2): 128–144.

Kaur, P., Thakur, R., and Chaudhury, A. Biogenesis of copper nanoparticles using peel extract of Punica granatum and their antimicrobial activity against opportunistic pathogens. *Green Chem Lett Rev.* 2016; 9(1): 33–38.

Khan, I., Rahman, H., Abd El-Salam, N. M., Tawab, A., Hussain, A., Khan, T. A. *et al. Punica granatum* peel extracts: HPLC fractionation and LC MS analysis to quest compounds having activity against multidrug resistant bacteria. *BMC Complement Altern Med.* 2017; 17(1): 247.

Kothamasu, P., Kanumur, H., Ravur, N., Maddu, C., Parasuramrajam, R., and Thangavel, S. Nanocapsules: The weapons for novel drug delivery systems. *Bioimpacts.* 2012; 2(2): 71–81.

Kumari, K., Dora, J., Kumar, A., and Kumar, A. Pomegranate (*Punica granatum)*—Overview. *Int J Pharm Chem Sci.* 2012; 1(4): 1218–1222.

Lu, L. Y., Liu, Y., Zhang, Z. F., Gou, X. J., Jiang, J. H., Zhang, J. Z. *et al.* Pomegranate seed oil exerts synergistic effects with trans-Resveratrol in a self-nanoemulsifying drug delivery system. *Biol Pharm Bull.* 2015; 38(10): 1658–1662.

Marchiori, M. C. L., Rigon, C., Camponogara, C., Oliveira, S. M., and Cruz, L. Hydrogel containing silibinin-loaded pomegranate oil based nanocapsules exhibits anti-inflammatory effects on skin damage UVB radiation-induced in mice. *J Photochem Photobiol B.* 2017a; 170: 25–32.

Marchiori, M. C. L., Rigon, C., Copetti, P. M., Sagrillo, M. R., and Cruz, L. Nanoencapsulation improves scavenging capacity and decreases cytotoxicity of silibinin and pomegranate oil association. *AAPS PharmSciTech.* 2017b; 18(8): 3236–3246.

Mekni, M., Dhibi, M., Kharroubi, W., Hmida, R. B., Cheraif, I., and Hammami, M. Natural conjugated and trans fatty acids in seed oils and phytochemicals in seed extracts issued from three Tunisian pomegranate (*Punica granatum.* L) cultivars. *Int J Curr Microbiol App Sci.* 2014; 3(8): 778–792.

Mizrahi, M., Friedman-Levi, Y., Larush, L., Frid, K., Binyamin, O., Dori, D. *et al.* Pomegranate seed oil nanoemulsions for the prevention and treatment of neurodegenerative diseases: The case of genetic CJD. *Nanomedicine.* 2014; 10(6): 1353–1363.

Mohenraj, V. J., and Chen, Y. Nanoparticles—A Review. *Trop J Pharm Res.* 2006; 5(1): 561–573.

Mora-Huertas, C. E., Fessi, H., and Elaissari, A. Polymer-based nanocapsules for drug delivery. *Int J Pharm.* 2010; 385: 113–142.

Morariu, M. Topical compositions comprising benfotiamine and pyridoxamine. US 2006/0045896 A1; 2006.

Morzelle, M. C., Salgado, J. M., Telles, M., Mourelle, D., Bachiega, P., Buck, H. S. *et al.* Neuroprotective effects of pomegranate peel extract after chronic infusion with amyloid-β peptide in mice. *PLOS One.* 2016; 11(11): 1–20.

Mousa, S. A., Qari, M. H., and Ardawi, M. S. Compositions and methods of natural products in nanoformulations for the prevention and treatment of osteoporosis. US 2011/0104283 A1; 2011.

Nisha, M. H., Tamileswari, R., Jesurani, S., Kanagesan, S., Hashim, M., and Alexander, S. C. P. Green synthesis of silver nanoparticles from pomegranate (*Punica granatum*) leaves and analysis of anti-bacterial activity. *Int J Adv Technol Eng Sci.* 2015; 3(6): 8.

Orza, A. L. Novel anti-wrinkle and anti-aging nano formulations and method of preparation using novel nano co-delivery system. US 20170157005 A1; 2017.

Pantuck, A. J., Zomorodian, N., Rettig, M., Aronson, W. J., Heber, D., and Belldegrun, A. S. Long-term follow-up of phase 2 study of pomegranate juice for men with prostate cancer shows durable prolongation of prostate-specific antigen doubling time. *J Urol.* 2009; 181(4 Suppl): 295.

Peller, C. J., Ye, X., Wozniak, P. J., Gillespie, B. K., Sieber, P. R., Greengold, R. H. *et al.* A randomized phase II study of pomegranate extract for men with rising PSA following initial therapy for localized prostate cancer. *Prostate Cancer Prostatic Dis.* 2013; 16(1): 50–55.

Sadrolhosseini, A. R., Noor, A. S. M., Husin, M. S., and Sairi, A. Green synthesis of gold nanoparticles in pomegranate seed oil stabilized using laser ablation. *J Inorg Organomet Polym Mater.* 2014; 24(6): 1009–1013.

Sadrolhosseini, A. R., Rashid, S. A., Noor, A. S. M., and Mehdipour, L. A. Fabrication of silver nanoparticles in pomegranate seed oil with thermal properties by laser ablation technique. *Dig J Nanomater Biostruct.* 2015; 10(3): 1009–1018.

Saffarzadeh-Matin, S., and Khosrowshahi, F. M. Phenolic compounds extraction from Iranian pomegranate (*Punica granatum*) industrial waste applicable to pilot plant scale. *Ind Crop Prod.* 2017; 108: 583–597.

Sangeetha, R., and Jayaprakash, A. Phytochemical screening of *Punica granatum* Linn. peel extracts. *J Acad Ind Res.* 2015; 4(5): 160–162.

Savardekar, P., and Bajaj, A. Nanoemulsions—A review. *Int J Res Pharm Chem.* 2016; 6(2): 312–322.

Shah, M., Shah, S., and Patel, M. The aspects of *Punica granatum*. *J Pharm Sci Bio-Sci Res.* 2011; 1(3): 154–159.

Shanmugavadivu, M., Kuppusamy, S., and Ranjithkumar, R. Synthesis of pomegranate peel extract mediated silver nanoparticles and its antibacterial activity. *Am J Adv Drug Deliv.* 2014; 2(2): 174–182.

Sharma, J., and Maity, A. Pomegranate phytochemicals: Nutraceutical and therapeutic values. *Fruit Veg Cereal Sci Biotech.* 2010; 4(2): 56–76.

Singh, B., Bandopadhyay, S., Kapil, R., Singh, R., and Katare, O. Self-emulsifying drug delivery systems (SEDDS): Formulation development, characterization, and applications. *Crit Rev Ther Drug Carr Syst.* 2009; 26(5): 427–451.

Singh, B., Jain, A., and Beg, S. Lipid-based nanostructured DDS for oral intake. *Chronicle Pharmabiz.* 2014; 34–36.

Soni, H., Nayak, G., Mishra, K., Singhai, A. K., and Pathak, A. K. Pharmacognostic and phytochemical evalution of peel of *Punica granatum*. *Int J Pharmacognosy Phytochem Res.* 2010; 2(2): 56–58.

Tokton, N., Ounaroon, A., Panichayupakaranant, P., and Tiyaboonchai, W. Development of ellagic acid rich pomegranate peel extract loaded nanostructured lipid carriers (NLCs). *Int J Pharm Pharm Sci.* 2014; 6(4): 259–265.

Verardo, V., Garcia-Salas, P., Baldi, E., Segura-Carretero, A., Fernandez-Gutierrez, A., and Caboni, M. F. Pomegranate seeds as a source of nutraceutical oil naturally rich in bioactive lipids. *Food Res Int.* 2014; 65(C): 445–452.

Victor, S. U., and Roberto, V. J. Gold and silver nanotechology on medicine. *J Chem Biochem.* 2015; 3(1): 21–33.

Viladomiu, M., Hontecillas, R., Lu, P., and Bassaganya-Riera, J. Preventive and prophylactic mechanisms of action of pomegranate bioactive constituents. *Evid Based Complement Alternat Med.* 2013; 2013: 1–18.

Viuda-Martos, M., Fernández-López, J., and Pérez-Álvarez, J. A. Pomegranate and its many functional components as related to human health: A review. *Compr Rev Food Sci Food Saf.* 2010; 9: 635–654.

Wafa, B. A., Makni, M., Ammar, S., Khannous, L., Hassana, A. B., Bouaziz, M. *et al.* Antimicrobial effect of the Tunisian Nana variety *Punica granatum* L. extracts against Salmonella Enterica (serovars Kentucky and Enteritidis) isolated from chicken meat and phenolic composition of its peel extract. *Int J Food Microbiol.* 2017; 241: 123–131.

Wang, S., Chou, C., and Su, N. A food-grade self-nanoemulsifying delivery system for enhancing oral bioavailability of ellagic acid. *J Func Foods.* 2017; 34: 207–215.

Young, J. E., Pan, Z., Teh, H. E., Menon, V., Modereger, B., Pesek, J. J. *et al.* Phenolic composition of pomegranate peel extracts using a liquid chromatography-mass spectrometry approach with silica hydride columns. *J Sep Sci.* 2017; 40: 1449–1456.

Yuliani, S. H., and Hartini, M. Comparison of physical stability properties of pomegranate seed oil nanoemulsion dosage forms with long-chain triglyceride and medium-chain triglyceride as the oil phase. *Tradit Med J.* 2016; 21(2): 93–98.

Zahin, M., Ahmad, I., Gupta, R. C., and Aqil, F. Punicalagin and ellagic acid demonstrate antimutagenic activity and inhibition of benzo (a)pyrene induced DNA adducts. *Biomed Res Int.* 2014: 467465.

Zarfeshany, A., Asgary, S., and Javanmard, S. H. Potent health effects of pomegranate. *Adv Biomed Res.* 2014; 3:100.

Index

Page numbers followed by f and t indicate figures and tables, respectively.

O

P

T

U